HVAC
Level Three

Trainee Guide
Fourth Edition

PEARSON

Boston   Columbus   Indianapolis   New York   San Francisco   Upper Saddle River
Amsterdam   Cape Town   Dubai   London   Madrid   Milan   Munich   Paris   Montreal   Toronto
Delhi   Mexico City   São Paulo   Sydney   Hong Kong   Seoul   Singapore   Taipei   Tokyo

NCCER

President: Don Whyte
Director of Product Development: Daniele Dixon
HVAC Project Manager: Rob Richardson
Senior Manager: Tim Davis
Quality Assurance Coordinator: Debie Ness

Desktop Publishing Coordinator: James McKay
Permissions Specialist: Amanda Werts
Production Specialist: Megan Casey
Editor: Chris Wilson

Writing and development services provided by Topaz Publications, Liverpool, NY

Lead Writer/Project Manager: Troy Staton
Desktop Publisher: Joanne Hart
Art Director: Alison Richmond

Permissions Editors: Toni Burke, Andrea LaBarge
Writers: Troy Staton, Thomas Burke, John Tianen

Pearson Education, Inc.

Editorial Director: Vernon R. Anthony
Executive Editor: Alli Gentile
Editorial Assistant: Douglas Greive
Program Manager: Alexandrina B. Wolf
Operations Supervisor: Deidra M. Skahill
Art Director: Jayne Conte
Director of Marketing: David Gesell
Executive Marketing Manager: Derril Trakalo
Marketing Manager: Brian Hoehl
Marketing Coordinator: Crystal Gonzalez

Composition: NCCER
Printer/Binder: Courier/Kendallville, Inc.
Cover Printer: Lehigh-Phoenix Color/Hagerstown
Text Fonts: Palatino and Univers

Credits and acknowledgments for content borrowed from other sources and reproduced, with permission, in this textbook appear at the end of each module.

10 9 8 7 6 5 4 3 2 1

PEARSON

ISBN-13: 978-0-13-375083-6
ISBN-10: 0-13-375083-3

Preface

To the Trainee

Heating, ventilating, and air-conditioning (HVAC) systems regulate the temperature, humidity, and the total air quality in residential, commercial, industrial, and other buildings. This also extends to refrigeration systems used to transport food, medicine, and other perishable items. Other systems may include hydronics (water-based heating systems), solar panels, or commercial refrigeration. HVAC technicians maintain and repair such systems. As a technician, you must be able to maintain, diagnose, and correct problems throughout the entire system. You must know how to follow blueprints or other specifications to maintain any system. You may also need working knowledge of welding, basic pipefitting, electrical, and sheet metal practices for the installation of ducts.

Nearly all buildings and homes in the United States alone use forms of heating, cooling, and/or ventilation. The increasing development of HVAC technology causes employers to recognize the importance of continuous education and keeping up to speed with the latest equipment and skills. Hence, technical school training or apprenticeship programs often provide an advantage and a higher qualification for employment.

NCCER's program has been designed by highly qualified subject matter experts with this in mind. Our four levels present an apprentice approach to the HVAC field, including theoretical and practical skills essential to your success as an HVAC technician. As the population and the number of buildings grow in the near future, so will the demand for HVAC technicians.

We wish you the best as you begin an exciting and promising career. This newly revised HVAC curriculum will help you enter the workforce with the knowledge and skills needed to perform productively in either the residential or commercial market.

New with *HVAC Level Three*

NCCER is proud to release the newest edition of *HVAC Level Three* in full color with updates to the curriculum that will engage you and give you the best training possible. In this edition, you will find that the layout has changed to better align with the learning objectives. There are also new end-of-section review questions to complement the module review. The text, graphics, and special features were enhanced to reflect advancements in HVAC technology and techniques. This edition contains six troubleshooting modules, as well as new modules *Zoning, Ductless, and Variable Refrigerant Flow Systems* and *Customer Relations*.

We invite you to visit the NCCER website at **www.nccer.org** for information on the latest product releases and training, as well as online versions of the *Cornerstone* magazine and Pearson's NCCER product catalog.

Your feedback is welcome. You may email your comments to **curriculum@nccer.org** or send general comments and inquiries to **info@nccer.org**.

NCCER Standardized Curricula

NCCER is a not-for-profit 501(c)(3) education foundation established in 1996 by the world's largest and most progressive construction companies and national construction associations. It was founded to address the severe workforce shortage facing the industry and to develop a standardized training process and curricula. Today, NCCER is supported by hundreds of leading construction and maintenance companies, manufacturers, and national associations. The NCCER Standardized Curricula was developed by NCCER in partnership with Pearson, the world's largest educational publisher.

Some features of the NCCER Standardized Curricula are as follows:

- An industry-proven record of success
- Curricula developed by the industry for the industry
- National standardization providing portability of learned job skills and educational credits
- Compliance with the Office of Apprenticeship requirements for related classroom training (*CFR 29:29*)
- Well-illustrated, up-to-date, and practical information

NCCER also maintains a National Registry that provides transcripts, certificates, and wallet cards to individuals who have successfully completed a level of training within a craft in NCCER's Curricula. *Training programs must be delivered by an NCCER Accredited Training Sponsor in order to receive these credentials.*

Special Features

In an effort to provide a comprehensive, user-friendly training resource, we have incorporated many different features for your use. Whether you are a visual or hands-on learner, this book will provide you with the proper tools to get started in the HVAC trade.

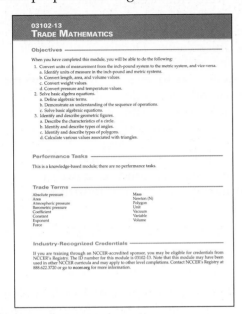

Color Illustrations and Photographs

Full-color illustrations and photographs are used throughout each module to provide vivid detail. These figures highlight important concepts from the text and provide clarity for complex instructions. Each figure reference is denoted in the text in *italic type* for easy reference.

Figure 31 Scroll compressor operation.

Introduction

This page is found at the beginning of each module and lists the Objectives, Performance Tasks, Trade Terms, and Required Trainee Materials for that module. The Objectives list the skills and knowledge you will need in order to complete the module successfully. The Performance Tasks give you an opportunity to apply your knowledge to the real-world duties that heavy equipment operators perform. The list of Trade Terms identifies important terms you will need to know by the end of the module. Required Trainee Materials list the materials and supplies needed for the module.

Special Features

Features provide a head start for those entering HVAC field by presenting technical tips and professional practices from operators in a variety of disciplines. These features often include real-life scenarios similar to those you might encounter on the job site.

The Meter

Originally, the meter was defined as ¹⁄₁₀,₀₀₀,₀₀₀ of Earth's meridional quadrant (the distance from the North Pole to the equator). This was how the meter got its name. This distance was etched onto a metal bar that is kept in France. In 1866, the United States legalized the use of the metric system and placed an exact copy of the metal bar into the US Bureau of Standards. In October 1983

Notes, Cautions, and Warnings

Safety features are set off from the main text in highlighted boxes and are organized into three categories based on the potential danger of the issue being addressed. Notes simply provide additional information on the topic area. Cautions alert you of a danger that does not present potential injury but may cause damage to equipment. Warnings stress a potentially dangerous situation that may cause injury to you or a co-worker.

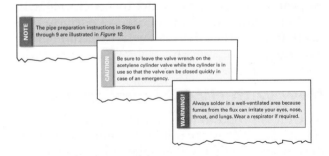

Going Green

Going Green looks at ways to preserve the environment, save energy, and make good choices regarding the health of the planet. Through the introduction of new construction practices and products, you will see how the "greening of America" has already taken root.

Did You Know?

The *Did You Know?* features offer hints, tips, and other helpful bits of information from the trade.

Did You Know?

Copper is one of the most plentiful metals. It has been in use for thousands of years. In fact, a piece of copper pipe used by the Egyptians more than 5,000 years ago is still in good condition. When the first Europeans arrived in the new world, they found the Native Americans using copper for jewelry and decoration. Much of this copper was from the region around Lake Superior, which separates northern Michigan and Canada.

One of the first copper mines in North America was established in 1664 in Massachusetts. During

Step-by-Step Instructions

Step-by-step instructions are used throughout to guide you through technical procedures and tasks from start to finish. These steps show you not only how to perform a task but also how to do it safely and efficiently.

Step 1 Cut the pipe to the required length (*Figure 53A*). Then ream the inside edge of the pipe with a reamer or sharp knife; **chamfer** the outside edge of the pipe with a suitable reamer, single-cut file, or sand cloth (*Figure 53B*). Examine the pipe end; if cracks are noted, the pipe should be cut off at least 2" away from the end of the crack.

Step 2 Test fit the pipe and fitting (*Figure 53C*). The pipe should fit more tightly as it reaches about two-thirds of the socket depth. Once the primer and cement are applied, the surfaces will be mildly dissolved and slippery, and the pipe and fit-

Trade Terms

Each module presents a list of Trade Terms that are discussed within the text and defined in the Glossary at the end of the module. These terms are denoted in the text with **bold, blue type** upon their first occurrence. To make searches for key information easier, a comprehensive Glossary of Trade Terms from all modules is located at the back of this book.

In the trades, there are two primary categories of copper tubing. The two categories are based on the temper, or hardness of the material. **Annealed** copper tubing is often referred to as soft copper. It is generally sold in manageable rolls of various lengths. Both 25-foot and 50-foot rolls are common. Annealed tubing is very easy to bend and form. As long as it is done with the proper tools and technique, it can be formed without suffering significant damage. **Hard-drawn** tubing, commonly known in the trade as hard copper, is fabricated by drawing, or pulling, the copper through a series of dies to decrease its size to the desired diameter. This process tends to harden the tubing, making it more rigid. Both annealed and hard-drawn tubing are

Review Questions

Review Questions are provided to reinforce the knowledge you have gained. This makes them a useful tool for measuring what you have learned.

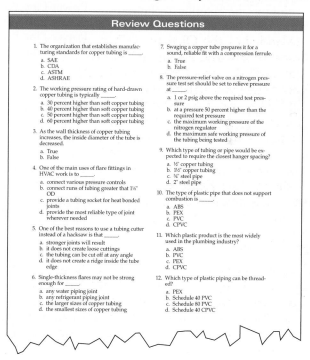

V

NCCER Standardized Curricula

NCCER's training programs comprise more than 80 construction, maintenance, pipeline, and utility areas and include skills assessments, safety training, and management education.

Boilermaking
Cabinetmaking
Carpentry
Concrete Finishing
Construction Craft Laborer
Construction Technology
Core Curriculum:
    Introductory Craft Skills
Drywall
Electrical
Electronic Systems Technician
Heating, Ventilating, and
    Air Conditioning
Heavy Equipment Operations
Highway/Heavy Construction
Hydroblasting
Industrial Coating and Lining
    Application Specialist
Industrial Maintenance Electrical
    and Instrumentation Technician
Industrial Maintenance
    Mechanic
Instrumentation
Insulating
Ironworking
Masonry
Millwright
Mobile Crane Operations
Painting
Painting, Industrial
Pipefitting
Pipelayer
Plumbing
Reinforcing Ironwork
Rigging
Scaffolding
Sheet Metal
Signal Person
Site Layout
Sprinkler Fitting
Tower Crane Operator
Welding

Maritime

Maritime Industry Fundamentals
Maritime Pipefitting

Green/Sustainable Construction

Building Auditor
Fundamentals of Weatherization
Introduction to Weatherization
Sustainable Construction
    Supervisor
Weatherization Crew Chief
Weatherization Technician
Your Role in the Green
    Environment

Energy

Alternative Energy
Introduction to the Power Industry
Introduction to Solar Photovoltaics
Introduction to Wind Energy
Power Industry Fundamentals
Power Generation Maintenance
    Electrician
Power Generation I&C
    Maintenance Technician
Power Generation Maintenance
    Mechanic
Power Line Worker
Power Line Worker: Distribution
Power Line Worker: Substation
Power Line Worker: Transmission
Solar Photovoltaic Systems
    Installer
Wind Turbine Maintenance
    Technician

Pipeline

Control Center Operations, Liquid
Corrosion Control
Electrical and Instrumentation
Field Operations, Liquid
Field Operations, Gas
Maintenance
Mechanical

Safety

Field Safety
Safety Orientation
Safety Technology

Supplemental Titles

Applied Construction Math
Careers in Construction
Tools for Success

Management

Fundamentals of Crew Leadership
Project Management
Project Supervision

Currículos en Español

Acabado de concreto: nivel uno,
    nivel dos
Aislamiento: nivel uno, nivel dos
Albañilería: nivel uno
Andamios
Aparejamiento básico
Aparajamiento intermedio
Aparajamiento avanzado
Carpintería:
    Introducción a la carpintería,
    nivel uno; Formas para
    carpintería, nivel tres
Currículo básico: habilidades
    introductorias del oficio
Electricidad: nivel uno, nivel dos,
    nivel tres, nivel cuatro
Encargado de señales
Especialista en aplicación de
    revestimientos industriales:
    nivel uno, nivel dos
Herrería: nivel uno, nivel dos,
    nivel tres
Herrería de refuerzo: nivel uno
Instalación de rociadores: nivel
    uno
Instalación de tuberías: nivel uno,
    nivel dos, nivel tres, nivel cuatro
Instrumentación: nivel uno, nivel
    dos, nivel tres, nivel cuatro
Orientación de seguridad
Mecánico industrial: nivel uno,
    nivel dos, nivel tres, nivel
    cuatro, nivel cinco
Paneles de yeso: nivel uno
Seguridad de campo
Soldadura: nivel uno, nivel dos,
    nivel tres

Currículos en Português

Currículo essencial: Habilidades
    básicas para o trabalho
Instalação de encanamento
    industrial: nível um, nível dois,
    nível três, nível quatro

Acknowledgments

This curriculum was revised as a result of the farsightedness and leadership of the following sponsors:

ABC of Wisconsin

Builders Association of North Central Florida

Industrial Management and Training Institute, Inc.

Lee Company

Madison Area Technical College

Progress Energy

Santa Fe College

This curriculum would not exist were it not for the dedication and unselfish energy of those volunteers who served on the Authoring Team. A sincere thanks is extended to the following:

Corky Dotson

Art Grant

Daniel Kerkman

Joseph Pietrzak

Bill Snyder

Norman Sparks

John Stronkowski

NCCER Partners

American Fire Sprinkler Association

Associated Builders and Contractors, Inc.

Associated General Contractors of America

Association for Career and Technical Education

Association for Skilled and Technical Sciences

Carolinas AGC, Inc.

Carolinas Electrical Contractors Association

Center for the Improvement of Construction Management and Processes

Construction Industry Institute

Construction Users Roundtable

Construction Workforce Development Center

Design Build Institute of America

GSSC – Gulf States Shipbuilders Consortium

Manufacturing Institute

Mason Contractors Association of America

Merit Contractors Association of Canada

NACE International

National Association of Minority Contractors

National Association of Women in Construction

National Insulation Association

National Ready Mixed Concrete Association

National Technical Honor Society

National Utility Contractors Association

NAWIC Education Foundation

North American Technician Excellence

Painting & Decorating Contractors of America

Portland Cement Association

SkillsUSA®

Steel Erectors Association of America

U.S. Army Corps of Engineers

University of Florida, M. E. Rinker School of Building Construction

Women Construction Owners & Executives, USA

Contents

Module Ten

Steam Systems

Focuses on the use of steam for storing and moving energy in HVAC systems. It reviews the fundamentals of water that relate to steam and describes the basic steam system cycle. It discusses a steam system's operational components—steam boilers and their accessories and controls; steam system loads, including heat exchangers/converters and terminal devices. Steam system valves and piping are covered in detail, including common types of steam system piping arrangements, the components of a condensate return/feedwater system, steam and condensate pipe sizing; and pressure-reducing valves and thermostatic valves. Various types of steam traps and their maintenance requirements are also presented. (Module ID 03306-13; 10 Hours)

Module Eleven

Retail Refrigeration Systems

Provides trainees with guidance related to retail refrigeration systems. Developing the necessary skills to understand the applications, principles, and troubleshooting of retail refrigeration systems are vital to the future success of trainees in the HVAC/R trade. (Module ID 03304-13; 15 Hours)

Module Twelve

Customer Relations

Presents the importance of establishing good relations with customers and provides guidance on how to achieve that goal. The module focuses on the ways that a technician can make a good first impression and describes ways to communicate in a positive way with customers. It further covers the elements of a service call and provides guidance for dealing with different types of problem customers. (Module ID 03316-13; 5 Hours)

Glossary

Index

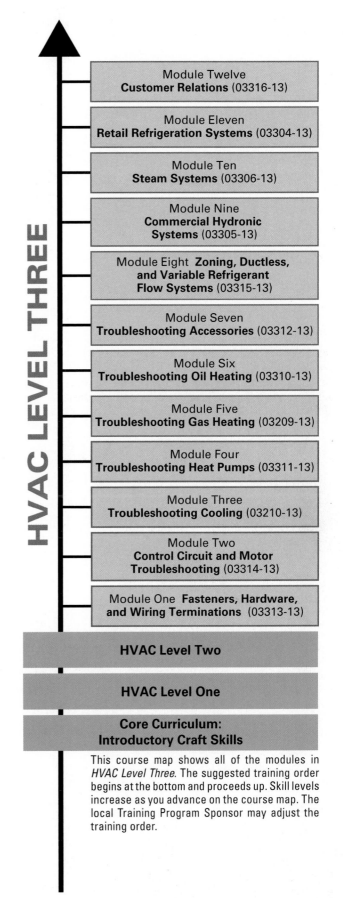

This course map shows all of the modules in *HVAC Level Three*. The suggested training order begins at the bottom and proceeds up. Skill levels increase as you advance on the course map. The local Training Program Sponsor may adjust the training order.

03313-13

Fasteners, Hardware, and Wiring Terminations

OVERVIEW

During the installation and maintenance of HVAC equipment, technicians work with a variety of fasteners, hardware, and wiring terminations. These components must be carefully selected for the specific application and installed according to accepted practices. Using the correct components ensures that the assembled system will perform properly, and will not fail because of the wrong choice of component or its incorrect installation.

Module One

Trainees with successful module completions may be eligible for credentialing through NCCER's National Registry. To learn more, go to **www.nccer.org** or contact us at **1.888.622.3720**. Our website has information on the latest product releases and training, as well as online versions of our *Cornerstone* magazine and Pearson's product catalog.

Your feedback is welcome. You may email your comments to **curriculum@nccer.org**, send general comments and inquiries to **info@nccer.org**, or fill in the User Update form at the back of this module.

This information is general in nature and intended for training purposes only. Actual performance of activities described in this manual requires compliance with all applicable operating, service, maintenance, and safety procedures under the direction of qualified personnel. References in this manual to patented or proprietary devices do not constitute a recommendation of their use.

Objectives

When you have completed this module, you will be able to do the following:

1. Identify the types, uses, and installation methods of fasteners.
   a. Identify various types and the uses of threaded fasteners.
   b. Identify and explain how to use taps, dies, and screw extractors.
   c. Explain how to install and torque fasteners to a specific value.
   d. Identify and explain how to install various types of toggle and anchor bolts.
   e. Identify various types of non-threaded fasteners.
2. Identify and describe the installation of various types of vibration isolators.
   a. Identify and describe vibration isolators used to support and suspend equipment and piping.
   b. Identify and describe methods used to restrain equipment during seismic events.
   c. Explain how to select and install various vibration isolators.
3. Identify common low- and line-voltage electrical termination hardware and explain how to properly terminate wiring connections.
   a. Identify common low- and line-voltage electrical terminating hardware.
   b. Explain how to properly terminate low- and line-voltage wiring connections.

Performance Tasks

Under the supervision of your instructor, you should be able to do the following:

1. Torque threaded hardware to a specific torque value.
2. Select the appropriate drill bit and install an anchor in brick or concrete block.
3. Terminate line- and low-voltage wiring on a compressor contactor.

Trade Terms

AL-CU
Break-away torque
Connector
Lug
Nominal size
Run-down resistance
Seismic activity
Seismic restraint

Set or seizure
Tensile strength
Termination
Thread classes
Thread series
Thread standards
Torque
Vibration isolator

Industry-Recognized Credentials

If you are training through an NCCER-accredited sponsor, you may be eligible for credentials from NCCER's Registry. The ID number for this module is 03313-13. Note that this module may have been used in other NCCER curricula and may apply to other level completions. Contact NCCER's Registry at 888.622.3720 or go to **www.nccer.org** for more information.

Contents

Topics to be presented in this module include:

Figures and Tables

Figures and Tables

SECTION ONE

1.0.0 FASTENERS

Objectives

Identify the types, uses, and installation methods of fasteners.

   a. Identify various types and uses of threaded fasteners.
   b. Identify and explain how to use taps, dies, and screw extractors.
   c. Explain how to install and torque fasteners to a specific value.
   d. Identify and explain how to install various types of anchor bolts.
   e. Identify various types of non-threaded fasteners.

Performance Tasks 1 and 2

Torque threaded hardware to a specific torque value.

Select the appropriate drill and install an anchor in brick or concrete block.

Trade Terms

Break-away torque: The torque required to loosen a fastener. This is usually lower than the torque to which the fastener has been tightened.

Nominal size: A means of expressing the size of a bolt or screw. It is the approximate diameter of a bolt or screw.

Run-down resistance: The torque required to overcome the resistance of associated hardware, such as locknuts and lock washers, when tightening a fastener.

Set or seizure: In the last stages of rotation in reaching a final torque, the fastener may lock up; this is known as seizing or set. This is usually accompanied by a noticeable popping effect.

Tensile strength: The maximum stress a material can endure before being pulled apart.

Thread classes: Threads are distinguished by three classifications according to the amount of tolerance the threads provide between the bolt and nut. The different classes are designated by a number and letter combination that indicates the degree of fit for a thread. Higher numbers indicate a tighter fit.

Thread series: The combinations of diameter and pitch in threads which are applied to specific thread diameters to identify the coarseness or fineness of a thread.

Thread standards: An established set of standards for machining threads.

Torque: Applying a specific amount of twisting force to a threaded fastener.

Fasteners are used to install many types of parts and equipment. Common fasteners include bolts, screws, pins, and retainers. It is important that the correct fastener be selected. For example, bolts and screws are graded based on strength. Even if the size of a bolt is correct, it may fail if an inferior grade (strength) is installed.

1.1.0 Threaded Fasteners

Threaded fasteners are one of the most common types of fasteners. Many are assembled with nuts and washers while others are installed in threaded holes.

Types of threaded fasteners that will be covered in this section include the following:

- Machine bolts, machine screws, stud bolts, and cap screws
- Nuts
- Toggle bolts
- Inserts
- Flat and lock washers
- Set screws
- Screws

Before reviewing the different types and shapes of threaded fasteners, it is important to understand the various types of threads used. In addition, technicians must understand the various grades of threaded fasteners. The thread type and grade used will depend upon the application for the fastener.

1.1.1 Thread Designations

Fastener threads are made to established thread standards. The most common standard is the Unified or American National Standard. There are three series defined by the Unified standard. These series are based on the number of threads per inch for a fastener of a certain diameter. The three series are the following:

- *Unified National Coarse (UNC) Thread* – UNC thread is used for bolts, screws, nuts, and other general applications. Fasteners with UNC threads are used for rapid assembly and disassembly of parts where corrosion or slight damage may occur.

- *Unified National Fine (UNF) Thread* – UNF thread is used for bolts, screws, nuts, and other uses where a finer thread than UNC is required.
- *Unified National Extra Fine (UNEF) Thread* – UNEF thread is used for thin-walled tubes, nuts, ferrules, and couplings.

Bolt and screw threads are designated by a standard method (*Figure 1*). The standard designations include the following:

- Nominal size (diameter).
- Number of threads per inch.
- Thread series symbol.
- Thread class. The thread class determines how tightly the male (external) and female (internal) threads fit each other. There are three classes, each of which is applied to external and internal threads. Classes 1A, 2A, and 3A apply to external threads, such as a bolt. Classes 1B, 2B, and 3B apply to internal threads, such as a nut. Higher numbers indicate a tighter fit. Therefore, Class 3A and 3B threads provide the tightest fit. Class 1 is the rarest, and is only used when quick disassembly is important. The loose fit provides limited strength and resistance to separation. Class 2 is the most common. Class 3 threads are needed when close tolerances and maximum strength are priorities.
- Left-hand thread symbol. Unless shown, the threads are right hand. This symbol is used only for left-hand fasteners.

Metric screw threads based on the American National Standard are also in common use. Metric M-profile threaded screws are a coarse-thread series of fasteners used for general fastening purposes. Metric MJ-profile threaded screws have slightly different thread characteristics, and are used with aircraft parts and other high-stress applications requiring extra strength.

NOMINAL SIZE (DIAMETER)
NO. OF THREADS PER INCH
THREAD SERIES SYMBOL
THREAD CLASS SYMBOL
LEFT-HAND THREAD
(NO DESIGNATION IF R.H. THREAD)

¾ – 10 – UNC – 2A – LH

03313-13_F01.EPS

Figure 1 Thread designations.

Coarse pitch threads (fewer threads per inch) are used in routine applications and offer greater stripping strength when used with larger diameter fasteners. Coarse threads are easier to form and nuts are easier to start and run down on the threaded section. Fine pitch threads have a higher tensile strength, and resist loosening when subjected to shock or vibration. Fine threads are useful if a thin nut is needed or where very fine adjustments must be made.

1.1.2 Threaded Fastener Grade Designations

The strength and quality of a fastener can be determined by special grade markings on the head of the fastener. These markings are standardized by the Society of Automotive Engineers (SAE) and ASTM International. Grade markings are sometimes called line markings.

Figure 2 shows the SAE and ASTM markings for steel bolts and screws.

While there are several grades of bolts, the three most commonly used are SAE Grade 2, Grade 5, and Grade 8. Grade 2 is made from low-carbon steel and is used in low-stress applications. For example, a Grade 2 bolt might be adequate for a bracket supporting a sheet metal duct. Grades 5 and 8 bolts use higher-quality steel that has been quenched and annealed. They are used in applications subjected to higher stress or pressure. These higher-grade bolts would be used to secure the cylinder heads of a semi-hermetic compressor, as one example. Higher stress or pressure dictates whether a grade 8 is used rather than a grade 5. Never substitute a lower grade bolt for a higher grade. Always choose bolts that have grade markings when the application has any importance. Bolts without grade markings are likely of poor quality and may fail.

1.1.3 Machine Bolts, Machine Screws, Stud Bolts, and Cap Screws

Machine bolts (*Figure 3*) are used to assemble parts that do not require close tolerances. The tolerance is the amount of variation allowed from a standard. Machine bolts are made with diameters ranging from ¼" to 3" and with lengths from ½" to 30". A machine bolt is tightened and released by turning its mating nut that is usually furnished along with the bolt.

Machine screws are used for general assembly. They have slotted or recessed heads. Machine screws are available in diameters from #6 through #12 (0.060" to 0.2160"). Fractional diameters range from ¼" to ½", and vary in length from 1/8" to 3". They are also made in metric sizes.

ASTM AND SAE GRADE MARKINGS FOR STEEL BOLTS & SCREWS

Grade Marking	Specification	Material
(plain hex head)	SAE-GRADE 0	STEEL
	SAE-GRADE 1 ASTM-A 307	LOW CARBON STEEL
	SAE-GRADE 2	LOW CARBON STEEL
(one line marking)	SAE-GRADE 3	MEDIUM CARBON STEEL, COLD WORKED
(A 449 marking)	SAE-GRADE 5	MEDIUM CARBON STEEL, QUENCHED AND TEMPERED
	ASTM-A 449	
(A 325 marking)	ASTM-A 325	MEDIUM CARBON STEEL, QUENCHED AND TEMPERED
(BB marking)	ASTM-A 354 GRADE BB	LOW ALLOY STEEL, QUENCHED AND TEMPERED
(BC marking)	ASTM-A 354 GRADE BC	LOW ALLOY STEEL, QUENCHED AND TEMPERED
(five radial lines)	SAE-GRADE 7	MEDIUM CARBON ALLOY STEEL, QUENCHED AND TEMPERED ROLL THREADED AFTER HEAT TREATMENT
(six radial lines)	SAE-GRADE 8	MEDIUM CARBON ALLOY STEEL, QUENCHED AND TEMPERED
	ASTM-A 354 GRADE BD	ALLOY STEEL, QUENCHED AND TEMPERED
(A 490 marking)	ASTM-A 490	ALLOY STEEL, QUENCHED AND TEMPERED

ASTM SPECIFICATIONS
A 307 – LOW CARBON STEEL EXTERNALLY AND INTERNALLY THREADED STANDARD FASTENERS.
A 325 – HIGH STRENGTH STEEL BOLTS FOR STRUCTURAL STEEL JOINTS, INCLUDING SUITABLE NUTS AND PLAIN HARDENED WASHERS.
A 449 – QUENCHED AND TEMPERED STEEL BOLTS AND STUDS.
A 354 – QUENCHED AND TEMPERED ALLOY STEEL BOLTS AND STUDS WITH SUITABLE NUTS.
A 490 – HIGH STRENGTH ALLOY STEEL BOLTS FOR STRUCTURAL STEEL JOINTS, INCLUDING SUITABLE NUTS AND PLAIN HARDENED WASHERS.

SAE SPECIFICATION
J 429 – MECHANICAL AND QUALITY REQUIREMENTS FOR THREADED FASTENERS.

03313-13_F02.EPS

Figure 2 Grade markings for steel bolts and screws.

HEX HEAD BOLT

HEX HEAD CAP SCREW

SQUARE HEAD BOLT

HEX SOCKET HEAD CAP SCREW

CONTINUOUS THREAD
(ALL THREAD) STUD

DOUBLE-END
STUD

MACHINE BOLTS, CAP SCREWS, AND STUD BOLTS

FLAT

OVAL

PAN

ROUND

FILLISTER

TRUSS

HEX
SOCKET

TORX®
SOCKET

SLOTTED

CROSS RECESS
TYPE 1

CROSS RECESS
TYPE 2

CLUTCH

HEX
SOCKET

TORX®
SOCKET

MACHINE SCREWS

03313-13_F03.EPS

Figure 3 Machine bolts, machine screws, cap screws, and stud bolts.

A machine screw normally mates with an internally threaded hole into which it is tightened or released. They can also be used with matching nuts.

Cap screws are generally used on assemblies that need a finished appearance. They pass through a clearance hole in one part of the assembly and are screwed into a threaded hole in the other part. This clamps the parts together when the cap screw is tightened. They are made to close tolerances with machined or semi-finished bearing surfaces under the head. Cap screws come in coarse and fine threads, and in diameters from ¼" to 2". Lengths from ⅜" to 10" are available. They are also made in metric sizes.

Stud bolts are headless bolts, threaded either along the entire length or on both ends. One end can be screwed into a threaded hole. The part to be clamped is fitted over the other end of the stud, and a nut and washer are screwed on to fasten the two parts together.

Threaded rod is nothing more than a metal rod that has threads on both ends, or is threaded over its entire length. Standard lengths are available, such as 6 foot or 10 foot lengths, but longer rods can be cut to any length. The rod is usually made of low-carbon steel since the rods are often used in low-load applications. Typical HVAC applications include use as a pipe hanger or as part of a trapeze assembly for supporting a duct (*Figure 4*).

1.1.4 Nuts

Nuts used with most threaded fasteners have either hex (hexagonal) or square shapes and are used with bolts having the same shaped head. *Figure 5* shows different types of nuts.

Figure 4 Threaded rods supporting ducts.

Some special-purpose nuts include the following:

- *Acorn nut* – These are used when appearance is important or when there are exposed sharp threads on the fastener.
- *Castellated (or castle) and slotted nuts* – After the nut is tightened, a cotter pin is fitted into one set of slots and through a hole in the bolt. The cotter pin keeps the nut from loosening.
- *Self-locking nut* – This has a nylon insert or is slightly deformed so it cannot work loose. Once a self-locking nut is used and removed, it should be thrown away and replaced with a new one.
- *Wing nut* – These are used where frequent adjustments and service are necessary. Wing nuts allow for loosening and tightening without the use of a wrench.
- *Jam nut* – This is a thin nut used to lock a standard nut in place. Note that jam nuts are installed on the fastener first, leaving it under the standard nut. The jam nut is often found on top, which is incorrect.
- *U-nuts and J-nuts* – These are used with soft, thin panel materials, or with thin sheet metal. The U-nut and J-nut have a threaded hole or stamped opening that accepts threads. They are slipped onto the edge of thin, relatively soft material to provide a quick method of attachment.
- *Cage nut* – These are square nuts enclosed in a stamped sheet metal box. The cage has a spring clip at the bottom which can be clipped into a square hole in a rack. The nut is loose enough inside the cage to allow for some small misalignment when mounting equipment.

1.1.5 Thread Repair Inserts

Thread repair inserts, or heli-coils (*Figure 6*), are a special kind of fastener used to provide high-strength threads in soft metals and plastics. They are also used to replace damaged or stripped threads in a tapped hole. Inserts are made in standard sizes and forms including metric sizes. The insert, which is larger in diameter than the tapped hole, is compressed during installation and allowed to spring back, permanently anchoring the insert in the tapped hole.

1.1.6 Flat and Lock Washers

Flat washers (*Figure 7*) provide an enlarged surface used to distribute the load from bolt heads and nuts. Flat washers are made in light, medium, heavy-duty, and extra heavy-duty series.

Fender washers are similar to flat washers with the difference being they have a larger outside diameter to bridge oversized holes or other wide-clearance requirements.

Lock washers are used to keep bolts or nuts from working loose. They are placed between the flat washers and the bolts or nuts. Some common types of lock washers include the following:

- *Split ring* – Commonly used with bolts and cap screws
- *External* – Provides the greatest resistance
- *Internal* – Used on small screws
- *Internal-external* – Used for oversized mounting holes
- *Countersunk* – Used with flat or oval-head screws

1.1.7 Set Screws

Set screws (*Figure 8*) are usually made of heat-treated steel and are used to secure objects within or against another object. For example, set screws are commonly used to fasten a pulley or gear to a shaft. Set screws are classified by head styles and point styles.

Set screws are usually headless (also known as blind), meaning the screw is fully threaded yet has no head. This type of screw is usually driven using fluted, hex socket, slotted or square heads. Once the set screw is screwed into the threaded hole, it uses a clamping force through the bottom tip that protrudes through the hole using friction as a fastener.

The cup point is the most common type of set screw. It is well suited for use in high vibration applications because it is very durable and shows a high resistance to wear while having strong grip capabilities. The knurled cup point is also designed to lock to prevent the screws from working loose during vibration.

Figure 5 Nuts. (1 of 2)

1.1.8 Screws

Screws are used to fasten materials in a secure manner or used if the fastened materials have to be taken apart. The two most common screw head types are round and flat. Slots types and head types are varied to fit both hand and powered drivers (*Figure 9*). Screw sizes are based on the diameter or body of the screw and are given in gauge numbers from No. 2 to No. 24. Screws larger than ¼" diameter are sized in fractions of an inch. Lengths range from ¼" to 6". Screws can be made of different metals and are often coated or plated to reduce corrosion.

Lag screws are used for heavy-duty fastening of objects to wood structures. When used with a lead lag shield, lag screws can be used to fasten objects to concrete or masonry (*Figure 10*).

Thread-forming screws (*Figure 11*) are used mainly to fasten light-gauge metal parts. They form a thread as they are driven. This eliminates tapping the hole. Some also drill their own hole, using a tip that is shaped like a common high-speed drill bit.

CAGE NUTS

J-NUT   U-NUT

03313-13_F05B.EPS

Figure 5 Nuts. (2 of 2)

HELI-COIL
(THREADED
INSERT)

03313-13_F06.EPS

Figure 6 Thread repair insert.

FLAT   SPLIT RING   EXTERNAL

INTERNAL   INTERNAL-   COUNTERSUNK
             EXTERNAL

03313-13_F07.EPS

Figure 7 Flat and lock washers.

Thread-cutting screws cut threads into the metal as they are driven into a pilot hole. They are made of hardened steel and are used to join heavy-gauge sheet metal and nonferrous metal parts.

Self-drilling sheet metal screws are commonly used in the installation of sheet metal duct systems. These screws increase productivity by decreasing the number of steps required for the job. However, standard thread-forming screws often have a very sharp point for a tip. When they are used on thinner gauges of sheet metal, pre-drilling is not necessary. The sharp point helps penetrate the metal to get the screw started.

1.2.0 Taps, Dies, and Screw Extractors

Taps and dies are used to cut threads in materials such as metal, plastic, and hard rubber. Screw extractors are used to remove fasteners that have broken off flush to a surface or below it.

1.2.1 Taps

Taps (*Figure 12*) are used to cut internal threads in materials; dies are used to cut external threads on bolts and rods. Taps are used to thread holes that go all the way through a piece of material, or holes that end inside the material (blind holes). Common types of taps include the following:

- *Taper taps* – Taper taps are used when starting a tapping operation or when tapping through holes. They have a chamfer length of 8 to 10 threads.
- *Plug taps* – Plug taps are used following the taper tap. They have a chamfer length of 3 to 5 threads.
- *Bottoming taps* – Bottoming taps are always used after the taper and plug taps have already been used. They are used for threading the bottom of a blind hole and have a chamfer length of only 1 to 1½ threads.

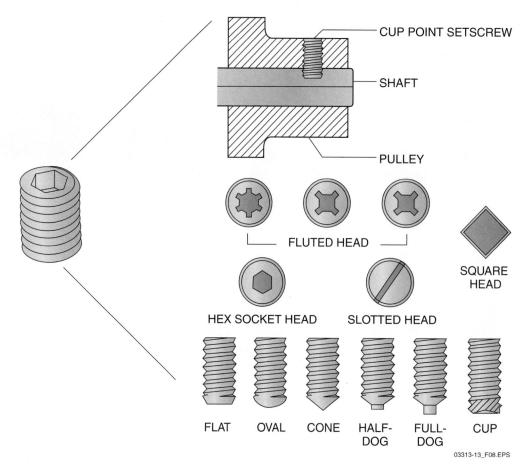

CUP POINT SETSCREW

SHAFT

PULLEY

FLUTED HEAD

SQUARE HEAD

HEX SOCKET HEAD

SLOTTED HEAD

FLAT   OVAL   CONE   HALF-DOG   FULL-DOG   CUP

03313-13_F08.EPS

Figure 8 Set screws.

- *Pipe taps* – Pipe taps are used where extremely tight fits are necessary, as with pipe fittings. The diameter of the tap from end to end of the threaded portion is tapered at the rate of ¾ inch per foot.

1.2.2 Dies

Dies are made in a variety of shapes and sizes and are solid or adjustable. Common types of dies include the following:

- Pipe dies will cut American Standard Pipe Threads or other pipe (such as metric) threads.
- Rethreading dies (die nuts) are used for repairing damaged or rusty threads on bolts or screws. They are usually hexagon-shaped and can be turned using any type of wrench that will fit.
- Screw adjusting dies are adjusted by a screw in the die.
- Open adjusting dies are adjusted by three screws in the die holder.
- Two-piece rectangular dies are adjusted by setscrews and held in ordinary or ratchet diestocks.

Taps and dies can be obtained in complete sets that include various taps and dies, as well as diestocks, tap wrenches, guides, and the screwdrivers and wrenches necessary to loosen and tighten adjusting screws. *Figure 13* shows a common tap and die set.

1.2.3 Fastener Extraction

A bolt may become seized if it has been in place for a long time, was forced into the bottom of a hole, or became jammed in place. In these cases, the head may break off or the nut may become stripped. This can leave the bolt or screw locked in place. In order to remove the fastener, a special tool called a screw extractor is used. Screw extractors are commonly referred to as easy-outs (*Figure 14*). To use a screw extractor, a hardened drill of the correct size is used to drill a hole centered and down the axis of the fastener. The screw extractor is a tapered tool with a spiral that is the reverse of the thread of the fastener. The taper is inserted into the fastener and turned in the direction that would be used to unscrew the fastener. The spiral grips the inside of the hole, and breaks loose the fastener, allowing the piece to be removed. Before attempting to break the fastener loose, satu-

Figure 9 Screw head types and screw drives.

Figure 11 Thread-forming and thread-cutting screws.

Figure 10 Lag screw.

rate the threads with penetrating oil and let the oil soak in. A light tap with a hammer may help loosen the fastener. Applying heat with a torch and/or applying dry ice to the joint may also help loosen the fastener. If there is enough of the broken fastener available, a nut may be welded to the stud and a wrench used to back out the broken fastener.

1.3.0 Torquing Threaded Fasteners

To properly torque threaded fasteners, technicians must know the type of fastener in use and the correct torque to be applied. *Figure 15* shows typical torque specifications for fasteners. The

Fastener Extraction—Art or Science?

Extracting a seized or broken fastener can be a challenge. Special tools such as screw extractors are helpful, but there are many other ways that stubborn fasteners can be removed using readily available tools and materials. Technicians often have to get creative. For example, if the hex head of a bolt is rounded off, a file can be used to flatten two of the sides enough so they can be gripped with a wrench. A hacksaw can be used to cut a slot in the top of a broken bolt so that a screwdriver can be used to remove the stud. Penetrating oil can help break the bond of a rusted connection. The use of heat or cold can cause metal to expand or contract, helping to break the bond holding the bolt. The vibrations caused by light hammer taps can also help break the bond.

(A) TAP

(B) PIPE

(C) PLUG

(D) TAPER

(E) BOTTOMING

Figure 12 Taps.

03313-13_F13.EPS

Figure 13 Tap and die set.

03313-13_F14.EPS

Figure 14 Screw extractor.

TORQUE IN FOOT POUNDS

FASTENER DIAMETER	THREADS PER INCH	MILD STEEL	STAINLESS STEEL 18-8	ALLOY STEEL
¼	20	4	6	8
5/16	18	8	11	16
⅜	16	12	18	24
7/16	14	20	32	40
½	13	30	43	60
⅝	11	60	92	120
¾	10	100	128	200
⅞	9	160	180	320
1	8	245	285	490

SUGGESTED TORQUE VALUES FOR GRADED STEEL BOLTS

GRADE		SAE 1 OR 2	SAE 5	SAE 6	SAE 8
TENSILE STRENGTH		64,000 PSI	105,000 PSI	130,000 PSI	150,000 PSI
GRADE MARK					
BOLT DIAMETER	THREADS PER INCH	FOOT POUNDS TORQUE			
¼	20	5	7	10	10
5/16	18	9	14	19	22
⅜	16	15	25	34	37
7/16	14	24	40	55	60
½	13	37	60	85	92
9/16	12	53	88	120	132
⅝	11	74	120	169	180
¾	10	120	200	280	296
⅞	9	190	302	440	473
1	8	282	466	660	714

03313-13_F15.EPS

Figure 15 Torque specifications.

specifications are usually stated in inch-pounds for small fasteners or foot-pounds for larger fasteners. However, the proper torque to be applied may be specified in manufacturer' assembly instructions or similar sources of information. Tightening sequences are also important to ensure proper fit and alignment. This section describes the guidelines you should follow when torquing threaded fasteners.

1.3.1 Torque Wrenches

Torque wrenches (*Figure 16*) are used to tighten fasteners to the specified value. The click-type torque wrench is likely the most common type used in the HVAC trade. The correct torque is set by twisting the wrench handle in or out, aligning the index mark with the desired torque value. When tightening a fastener and the torque setting has been reached, a slight break or give can be felt in the torque wrench handle. The beam-type torque wrench does not require an initial setting. It is simply pulled until the needle deflects enough to align with the mark for the desired torque value.

When assembling equipment using a threaded fastener, each fastener should be inspected for damaged threads. If damaged, replace the fastener with a new one. If the threads are dirty or rusty, clean them with a wire brush. In most cases, no type of lubricant, including oil, should be used. This can change the torque value. A fastener to be torqued should only be lubricated if the specification for the toque value instructs the installer to do so. Always follow the manufacturer's instructions and recommended torque values when installing fasteners on equipment.

(A) BEAM-TYPE TORQUE WRENCH

(B) CLICK-TYPE TORQUE WRENCH

(C) DIGITAL TORQUE WRENCH

03313-13_F16.EPS

Figure 16 Torque wrenches.

If the fastener is being installed in a threaded hole, inspect the threads in the hole for damage. If the threads are damaged, the hole should be re-tapped. Be sure to use the correct size and type of tap for the fastener being used. If the hole is blind, a bottom tap will be required.

The correct size torque wrench for a job is one that will read between 25 and 75 percent of its scale when the required torque is applied. This allows for adequate capacity and provides satisfactory accuracy. Avoid using an oversized torque wrench because the scale divisions are too coarse,

making it difficult to get an accurate reading. Using a wrench that is too small will not allow for extra capacity in the event of **set or seizure** or **run-down resistance**. All threaded fasteners must be clean and undamaged in order to get accurate readings. The calibration of torque wrenches must be checked periodically to guarantee accuracy.

The following terms, sometimes used in manufacturers' service literature, must be understood when using a torque wrench:

- **Break-away torque** – The torque required to loosen a fastener. This is generally lower than the torque to which it has been tightened. For a given size fastener, there is a direct relationship between tightening torque and breakaway torque. This relationship is determined by actual test. Once known, the tightening torque can be checked by loosening and checking breakaway torque.
- *Set or seizure* – In the last stages of rotation in reaching a final torque, seizing or set of the fastener may occur. When this happens, there is usually a noticeable popping sound and vibration. To break the set, back off and then again apply the tightening torque. Accurate torque settings cannot be made if the fastener is seized.
- *Run-down resistance* – The torque required to overcome the resistance of associated hardware, such as locknuts and lock washers, when tightening a fastener. To obtain the proper torque value where tight threads on locknuts produce a run-down resistance, add the resistance to the required torque value. Run-down resistance must be measured on the last rotation or as close to the makeup point as possible.

After determining the correct torque value, set the controls or indicator on the wrench to that value. This is not required if using a beam-type torque wrench. Place the torque wrench on the object to be tightened (nut or bolt) and hold the head of the wrench with one hand to support and align the wrench and object being tightened (*Figure 17*). Secure the other end of the fastener with a second wrench if required. As the object is tightened, watch or listen for the indication that the proper torque has been achieved. Once the torquing is complete, always return the torque wrench setting to zero for storage.

1.3.2 Tightening Sequences

When tightening bolts on flanges and similar surfaces, the bolts must be tightened to the proper torque and in the proper sequence. Tighten each

SQUARE FOUR-BOLT

CIRCULAR FOUR-BOLT

NONCIRCULAR MULTI-BOLT

CIRCULAR MULTI-BOLT

03313-13_F18.EPS

Figure 18 Common fastener-tightening sequences.

DIRECTION OF ROTATION DEPENDS ON DIRECTION
OF THREADS ON BOLT

03313-13_F17.EPS

Figure 17 Proper use of a torque wrench.

fastener only a small amount at a time until snug, following the proper tightening sequence for the bolt pattern and fastener type. After all the fasteners are snug, tighten to the final torque. This prevents warping or damaging the flange or machine part and helps prevent leaks. The numbers in *Figure 18* show the proper tightening sequences for some common bolt patterns.

When flanged ends (*Figure 19*) are bolted together, they must be properly aligned to reduce stress on the flange faces and on the fasteners as they are being torqued.

> **CAUTION**
>
> Tighten fasteners only small amounts at a time, following the proper tightening sequence for the bolt pattern and fastener type before reaching the final torque. Failure to follow the correct tightening sequence or over-tightening can result in damage to the fasteners, any sealing gaskets, or the object being fastened.

1.4.0 Toggle Bolts and Anchor Bolts

Toggle bolts and anchor bolts are specialized fasteners used to secure objects in place. Toggle bolts are used to fasten a part to a hollow wall or panel. Anchor bolts are used to fasten parts and equipment to concrete and/or masonry.

Reusing Nuts and Bolts

When new nuts and bolts are torqued, a certain amount of stretching and/or deforming takes place. If the nut/bolt combination is taken apart and reused, additional stretching and deforming takes place which further weakens the combination. In low-stress applications this may not be a problem. However, in high-stress environments such as the head bolts on a semi-hermetic compressor, it may be wise to replace the bolts with new. Compressor manufacturers sell replacement cylinder head bolts in kit form.

Note that jam nuts should be torqued differently. When a jam nut is used, the thin jam nut is placed on the fastener first, not last. It is then torqued to 25 to 50 percent of the final torque required. The second, standard nut is then placed on top and torqued to the final torque value. While the upper nut is torqued, the jam nut must be held with a spanner wrench or other wrench thin enough to prevent interference with the rotation of the upper nut.

MATING
FLANGES
PROPERLY
ALIGNED

03313-13_F19.EPS

Figure 19 Flanged coupling alignment.

1.4.1 Toggle Bolts

As a general rule, toggle bolts (*Figure 20*) should only be used to support light loads. In many cases, the surface will be sheet rock or a similar wall or siding material. Regardless of the strength of the bolt, the wall material may not be sound enough to reliably support a great deal of weight.

To install a toggle bolt, first drill the correct size hole in the wall or panel. The hole must accommodate the wings of the toggle. Remove the screw from the toggle bolt, insert it in the object to be fastened, and then re-insert the screw into the toggle assembly. Compress the wings of the toggle toward the head of the bolt, and insert it into the hole. Once this is done, the spring-loaded wings will deploy and engage the back side of the wall. Use a screwdriver to draw the wings against the backside of the wall and secure the object (*Figure 21*).

1.4.2 Anchor Bolts

Types of concrete and masonry anchors include the following:

- Concrete inserts
- Non-expanding concrete fasteners
- Expanding concrete fasteners
- Adhesive anchors
- Tapcon® screws

Concrete inserts (*Figure 22*) are imbedded in wet concrete and the hanger extension rod is screwed into the anchor after the concrete has set. The nut is welded to the insert. Other attachments can be added, depending on the requirements of the system.

REGULAR TOGGLE BOLT

ROUND HEAD   FLAT HEAD   MUSHROOM HEAD   BUTTON HEAD   SLOTTED HEAD

HEAD TYPES

⟵ 11" ⟶

FIXTURE HANGER TOGGLE BOLT

TIE-WIRE TOGGLE BOLT

03313-13_F20.EPS

Figure 20 Toggle bolts.

03313-13_F21.EPS

Figure 21 Toggle bolt installation.

(A)

(B)

CONCRETE INSERT

NUT

HANGER EXTENSION ROD

(C)

03313-13_F22.EPS

Figure 22 Concrete inserts.

Non-expanding concrete anchors (*Figure 23*) are placed in a hole drilled in hardened concrete and are held in place with grout or epoxy. Expanding concrete anchors (*Figure 24*) however, are placed in snug-fitting holes drilled in hardened concrete. As the nut is tightened, it causes the wedge at the base of the anchor to be drawn up into the sleeve. The sleeve expands as a result, wedging it tightly against the walls of the hole.

Bolts can be secured in holes drilled in concrete using an adhesive, like a non-expanding anchor. It is available as pre-packaged capsules of epoxy and hardener. Both components are packed in a glass or plastic capsule that is inserted into the hole. The bolt is inserted to break the capsule and twisted around to mix the two components. The bolt is then left in place until the epoxy hardens (*Figure 25*). Use care when working with the epoxy as it may contain glass fragments that can cause cuts.

Tapcon® screws (*Figure 26*) can be used for light and medium-duty applications where something must be fastened to concrete or masonry. Introduced in 1976, these screws are driven directly into concrete. Various diameters and lengths are available. After a pilot hole of the correct diameter is drilled in the masonry, the screw is driven in, threading its way into the material.

1.4.3 Installing Non-expanding Concrete Anchors

When installing anchors in hardened concrete, make sure the area where the equipment is to be fastened is smooth so that the equipment will have solid footing. Uneven footing might cause the equipment to twist, warp, not tighten properly, or vibrate during operation.

03313-13_F23.EPS

Figure 23 Non-expanding concrete anchor.

03313-13_F24.EPS

Figure 24 Expanding concrete anchor.

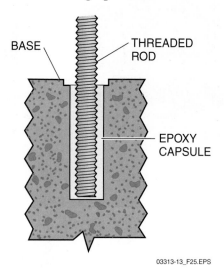

03313-13_F25.EPS

Figure 25 Adhesive anchor.

03313-13_F26.EPS

Figure 26 Tapcon® screws.

Carefully inspect the drill and bit to make sure they are in good operating condition. If they are not, you could be injured by parts that chip and fly off during operation.

When installing a non-expansion anchor bolt in hardened concrete, it is installed in a drilled hole filled with a filler material (grout). When installing this type of anchor, the drill bit must be slightly larger in diameter than the head of the fastener and the flat washer used with the fastener. The flat washer must be able to fit down inside the hole, just below surface level.

Also, the anchor bolt should extend out of the hole far enough for the threads and a little of the unthreaded bolt to be above the surface level. Before selecting a bolt length, confirm how much of the bolt needs to protrude from the concrete surface. For example, if a vibration-isolating pad will be placed on the floor under the equipment foot, this will require a longer bolt length. During the drilling process, the drill can be lubricated with water. *Table 1* summarizes the diameter (drill bit size) and depth of the opening to be drilled to install a grouted-in-place non-expanding anchor.

After the hole is drilled, the bolt installed, and the hole filled with grout, make sure the bolt is left straight after working the anchor around in the hole and installing the washer. The washer centers the bolt and holds it until the grout hardens. If the grout hardens and the bolt is not straight, it will be unusable and the job will have to be repeated after the bolt is removed. Allow the grout to fully dry before mounting anything to the anchor bolt.

1.4.4 Installing Expanding Concrete Anchors

When drilling the hole for an expanding-type concrete anchor, the diameter of the hole should be only slightly larger than the diameter of the anchor. The anchor manufacturer will specify the correct diameter drill bit needed for each size anchor. This is the size that should be used. If the hole is too large or out-of-round, the anchor may not be able to wedge itself tightly in the hole. Using the correct size hole, as specified by the bolt manufacturer, ensures that the specifications of strength and hold will remain valid.

The depth of the hole must be such that the anchor does not bottom out prematurely or protrude above the hole. *Figure 27* shows an expanding-type concrete anchor set in a hole of the proper diameter and depth. When drilling the hole, concrete dust may accumulate in the bottom of the hole, preventing complete insertion of the anchor. Use air to blow out any accumulated dust.

1.4.5 Installing Tapcon® screws

Tapcon® screws are driven into masonry or concrete in much the same way wood screws are driven into wood (*Figure 28*). Using this type of fastener requires that a pilot hole of the correct diameter and depth be drilled in the masonry. Failure to use the correct diameter pilot drill can result in two problems. If the pilot hole is oversized, the threads of the screw cannot properly engage the masonry or concrete. Under stress, the screw may pull out. If the pilot hole is undersized, it is dif-

Table 1 Hole Sizes for Non-Expanding Concrete Anchors

DIAMETER OF BOLT TO BE FASTENED	DIAMETER OF OPENING	MINIMUM DEPTH OF OPENING
1/4"	3/4"	3"
3/8"	1 1/4"	4"
1/2"	1 1/2"	4"
5/8"	1 3/4"	6"
3/4"	1 3/4"	6"
1"	2"	8"
1 3/4"	2 1/2"	12"

*Concrete must have minimum compression strength of 3000 pounds per square inch.

03313-13_T01.EPS

Figure 27 Installed expanding-type concrete anchor.

ficult to drive the screw. Over-torquing the screw often results in the head of the fastener breaking off. The manufacturer will specify the correct diameter masonry drill to use. In many cases, the correct drill is packaged with the screws.

1.5.0 Non-Threaded Fasteners

Non-threaded fasteners have many uses. This section describes the following types of non-threaded fasteners:

- Retainer rings
- Pins
- Keys
- Rivets

Figure 28 Hanger installed with Tapcon® screws.

1.5.1 Retainer and Lock Rings

Retainer and lock rings (*Figure 29*) are used for both internal and external fastening. Some retainer rings are seated in grooves. Others are self-locking and do not require a groove. Special pliers are used to remove internal and external rings. Self-locking rings snap in place.

Hammer Drills

Hammer drills are commonly used to drill holes in masonry and concrete. The combination of the twisting of the drill bit accompanied by hammer-like blows makes drilling holes in masonry and concrete much easier than is possible with a conventional electric drill.

03313-13_SA01.EPS

BASIC INTERNAL RING

BASIC EXTERNAL RING

INVERTED INTERNAL RING

INVERTED EXTERNAL RING

INTERNAL SELF-LOCKING RING

EXTERNAL SELF-LOCKING RING

TRIANGLE SELF-LOCKING RING

(A)

LOCK RING PLIERS

(B)

03313-13_F29.EPS

Figure 29 Retainer and lock rings and lock ring pliers.

1.5.2 Pins

Pin fasteners (*Figure 30*) are used to align mating parts, to hold gears and pulleys on shafts, and to secure slotted nuts.

Some common pins and their uses are as follows:

- *Dowel pins* – These are fitted into reamed holes to position mating parts. They also support a portion of the load placed on the parts.
- *Taper pins* – These are used to fasten gears, pulleys, and collars to a shaft.
- *Cotter pins* – These are fitted into a hole drilled through a shaft. Cotter pins are used to prevent parts from slipping on or off the shaft, and also to keep slotted nuts from working loose.

1.5.3 Keys

Keys (*Figure 31*) are metal parts used to prevent a gear or pulley from rotating on a shaft. One half of the key fits into a seat or groove on the shaft. The other half fits into a keyway in the hub of a gear or pulley.

1.5.4 Rivets

Rivets are used to permanently join two pieces of material. Two common types of rivets are the tinner's rivet and the blind, or pop, rivet (*Figure 32*). To install a tinner's rivet, the rivet is inserted into a hole that has been drilled or punched. The small end of a tinner's rivet is hammered or set in the form of a head. Hollow tinner's rivets are clinched at the small end with a special tool.

HARDENED AND GROUND DOWEL PIN

FULL-LENGTH TAPER

TAPER PIN

HALF-LENGTH TAPER

CLEVIS PIN

FULL-LENGTH WITH PILOT

COTTER PIN

REVERSE TAPER

03313-13_F30.EPS

Figure 30 Pin fasteners.

Tinner's rivets are sized in ounces or pounds per 1,000 rivets. For example, a 6-ounce rivet means that 1,000 of these rivets will weigh 6 ounces. As the weight increases, so does the diameter and length of the rivet.

A manual pop rivet tool (*Figure 33*) can be used to install blind rivets in a shop or at the job site. Blind rivets are used to fasten sheet metal, fiberglass, and plastics. They are available in various lengths and diameters. Blind rivets are used when the joint can only be reached from one side. To install a blind rivet, the rivet is inserted into a

COUNTERSUNK   TINNER'S   FLATHEAD   ROUNDHEAD

TINNER'S RIVETS

RIVET TOOL

FULL MANDREL

BREAK MANDREL

BLIND RIVET

BLIND RIVET INSTALLATION

03313-13_F32.EPS

Figure 32 Rivets.

previously drilled or punched hole and then the rivet is set, or popped, using the tool.

GIB HEAD KEY

PRATT & WHITNEY KEY

WOODRUFF KEY

KEYWAY

SQUARE KEY

KEY SEAT

03313-13_F31.EPS

Figure 31 Keys.

03313-13_F33.EPS

Figure 33 Pop rivet tool.

Removing Rivets

Chisel or grind off the heads of tinner's rivets or steel blind rivets; remove the shaft by striking through with a punch. Drill off blind or countersunk rivet heads using a high-speed electric drill, then punch out the shaft.

Additional Resources

Air Conditioning Systems, Principles, Equipment, and Service, Latest Edition. Upper Saddle River, NJ: Prentice Hall.

1.0.0 Section Review

1. Fasteners over 1" in diameter almost always use _____.

    a. coarse threads
    b. fine threads
    c. metric threads
    d. metric fine threads

2. Which type of die is often shaped like a hexagon?

    a. Open adjusting
    b. Screw adjusting
    c. Rethreading
    d. Two-piece

3. When using a correct sized torque wrench, the specified torque reading will show on the indicator in a range between _____.

    a. 40 to 60 percent of full scale
    b. 75 to 100 percent of full scale
    c. 25 to 75 percent of full scale
    d. 50 to 80 percent of full scale

4. Non-expanding concrete anchors are held in place with _____.

    a. folding wings
    b. wedges
    c. adhesive
    d. grout

5. To keep a slotted nut from working loose, use a _____.

    a. key pin
    b. cotter pin
    c. spring pin
    d. dowel pin

SECTION TWO

2.0.0 VIBRATION ISOLATORS

Objectives

Identify and describe the installation of various types of vibration isolators.

   a. Identify and describe vibration isolators used to support and suspend equipment and piping.

   b. Identify and describe methods used to restrain equipment during seismic events.

   c. Explain how to select and install various vibration isolators.

Trade Terms

Seismic activity: Movement of the earth such as during an earthquake.

Seismic restraint: A device or method used to prevent or restrict the movement of equipment during a seismic event.

Vibration isolator: A device that prevents the transmission of vibrations from operating machinery to surrounding objects.

HVAC equipment contains components such as fans, blowers, and compressors, all of which contain electric motors. The electric motors and the devices they drive generate normal noises and vibrations that can be objectionable. If equipment that generates vibrations is in direct contact with other objects, those vibrations can be transferred to the structure, and to pipes and ducts connected to the equipment. For example, an attic-mounted air handler that is installed directly on the attic floor can transmit vibrations to the structural members of the floor, allowing the vibrations to be heard throughout the structure. To reduce the transmission of unwanted noise and vibration, various kinds of **vibration isolators** are installed on the equipment and on the ducts and pipes connected to the equipment.

2.1.0 Types of Vibration Isolators

Vibration isolators can be grouped based on the type of vibration they are designed to reduce or eliminate. Vibration isolators can be broken into four distinct groups of devices:

- Devices that reduce vibration caused by compressors, fans, and blower motors

- Devices that reduce vibration transmission between objects
- Devices that reduce vibration transmission through ducts
- Devices that reduce vibration transmission through pipes

2.1.1 Compressor, Fan, and Blower Motor Vibration Isolators

Welded hermetic compressors installed in packaged units and split-systems are mounted on grommet-like vibration isolators made of rubber (*Figure 34*). A bolt with an integral washer passes through the center of the grommet and holds the compressor to the unit's base pan. The rubber prevents metal-to-metal contact between the base pan and compressor, keeping any vibrations the compressor generates from being transferred to the rest of the unit.

It is very important to note that isolators used in this way are of little or no value if the hold-down bolt is tightened. The hold-down bolts of compressors are tightened at the factory for shipment of the unit only. However, during the installation or initial startup, the bolts should be loosened so that they are not applying any pressure on the rubber grommets. Crushing the rubber grommet by tightening the bolt eliminates most or all of the isolation characteristics.

A similar arrangement is used to isolate a direct-drive blower motor from the blower housing

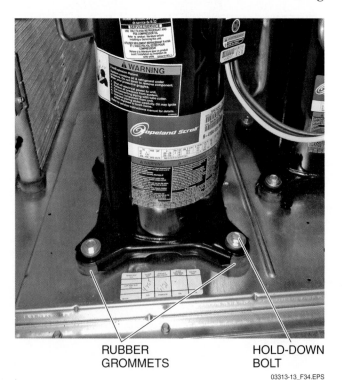

RUBBER GROMMETS   HOLD-DOWN BOLT

03313-13_F34.EPS

Figure 34 Hermetic compressor vibration isolators.

(*Figure 35*). Three grommet-like isolators are attached to the three motor mounts that are located 120 degrees apart around the motor. These bolts must be tight to hold the motor in place, but not tight enough to damage the isolator.

Heavier semi-hermetic compressors are mounted on stronger isolators that use springs instead of rubber (*Figure 36*). These compressors installed in equipment are often equipped with some means to prevent the compressor from moving during shipment. This may take the form of a removable shipping bracket and/or mounting hardware that keeps the mounting springs compressed. Before starting the compressor, the bolts compressing the mounting springs are backed out to allow the mounting springs to decompress and provide isolation. Even though the compressor may be spring-mounted within a unit, equipment manufacturer's sometimes recommend that the entire unit be mounted on spring-type isolators to further reduce the transmission of vibrations.

2.1.2 Vibration Isolators that Reduce Vibration Between Objects

Equipment vibration can be transmitted between the equipment and the object to which it is mounted. For example, a roof-mounted packaged unit can transmit vibrations to the structural members of the roof. Relatively lightweight residential air handlers installed directly on ceiling joists in an attic can be isolated by placing vibration-isolating pads (*Figure 37*) under key locations of the unit. This particular product is constructed of two layers of grooved rubber sheet, separated by a cork center.

When working with all vibration isolators, it is essential to ensure that the isolator is working within its designed load range. The pad shown in *Figure 37* is rated at 50 pounds per square inch. That means that only 4 square inches of pad area is needed for a 200-pound unit. If too much weight is applied, the pad will likely crush down too tightly against its cork center and reduce its effectiveness. If it is under-loaded, it will also fail to provide good isolation. This concept is common to all vibration isolation products. To perform the desired function, they must be properly loaded.

Residential air handlers installed in attics can also be suspended from roof rafters using multiple threaded rods that are coupled through vibration-isolating hangers (*Figure 38*). Each threaded rod contains its own vibration isolator (*Figure 39*). This prevents any vibration from being transmitted through the threaded rods to the roof rafters.

In *Figure 39*, note the spring deflection scale. This provides a quick visual means of determining how much the spring is deflected under the load. Loading a spring isolator to somewhere near mid-range is usually ideal. For example, if a 50-pound load on each corner of a suspended load is expected, vibration isolators with a maxi-

VIBRATION
ISOLATOR

DIRECT-DRIVE

03313-13_F35.EPS

Figure 35 Direct-drive blower vibration isolators.

CAPSCREW
LOCKWASHER
PLATEWASHER
SPRING CUP
COMPRESSOR
FOOT

03313-13_F36.EPS

Figure 36 Semi-hermetic compressor vibration isolator.

03313-13_F37.EPS

Figure 37 Vibration-isolating pad.

Figure 38 Air handler suspended with vibration isolators.

mum loading of roughly 100 pounds would likely be chosen.

Heavy HVAC equipment, such as a chiller, is mounted on heavy-duty spring-mounted vibration isolators (*Figure 40*). Although it may appear that the upper pad of the isolator is touching the top of the base, it is actually riding on top of the spring. Since the corner and/or side-to-side loads of a large piece of equipment are usually different, it is not unusual to see two, or even three, different sizes of vibration isolators supporting a unit.

2.1.3 Vibration Isolators that Reduce Vibration Through Ducts

In metal duct systems, any noise or vibration generated by the equipment can be transmitted throughout the structure. Because of the materials used, flexible ducts and ducts made with rigid fiberglass board do not easily transmit vibrations and do not require vibration isolation. The simplest way to prevent vibrations from the air handler into a metal duct system is to install flexible (non-metallic) connectors in the supply and return air ducts as they enter/leave the air handler (*Figure 41*). Flexible connectors can be installed in vertical and horizontal sections of duct. In commercial applications, metal ducts can be suspended with threaded rod using spring-mounted vibration isolators. Lining sheet metal ducts with fiberglass duct liner can also reduce the transmission of noise and vibrations.

Figure 39 Vibration-isolating hanger.

Figure 40 Heavy-duty vibration isolator.

RETURN DUCT　　SUPPLY DUCT

FLEX (CANVAS) CONNECTOR

AIR HANDLER OR FURNACE

**VIBRATION/NOISE CONTROL
AT AIR HANDLER**

**VIBRATION/NOISE/MOVEMENT
CONTROL IN DUCT RUNS**

03313-13_F41.EPS

Figure 41 Ductwork vibration and noise control devices.

2.1.4 Vibration Isolators that Reduce Vibration Through Pipes

Pipes, like metal duct systems, can transmit any noise or vibration generated by the equipment throughout the structure. Vibrations can be reduced at the source (a pump or compressor) or in the piping system itself. Compressors, especially larger semi-hermetic compressors, can generate noise and vibration at start-up and during operation. The starting torque of the motor can cause the compressor to move as it starts. If rigid pipes are connected to the compressor, they can be stressed by the starting torque and transmit noise and vibrations through the connected piping. To reduce this problem, a flexible metal vibration eliminator (*Figure 42*) can be installed in the suction and/or discharge lines of the compressor. The vibration eliminator is typically made of corrugated bronze with a braided bronze jacket. Stainless steel versions are also available. A wide range of lengths and diameters are available so they can be installed in almost any refrigeration

03313-13_F42.EPS

Figure 42 Flexible metal vibration eliminator.

system. Copper fittings on either end allow the fitting to be brazed into copper refrigerant lines. The corrugations in the body of the device make it flexible and able to absorb equipment vibrations. Note that the isolator in *Figure 42* is designed specifically for refrigerant line use.

Piping suspended by threaded rods (*Figure 43*) can also include vibration isolators. Again, sizing each vibration isolator for the expected load is essential to get the desired performance.

03313-13_F43.EPS

Figure 43 Piping suspended with vibration isolators.

2.2.0 Seismic Restraints

In areas with high seismic activity, local codes often require that HVAC equipment be fitted with some form of seismic restraint to prevent movement if an earthquake occurs. The reason for this is quite simple. If gas-fired equipment moves too much, it can break gas and electrical connections, resulting in a fire. If heavier equipment breaks loose, it can cause additional structural damage and possibly injure bystanders. Equipment can be restrained using angle iron, straps, bumpers, vibration-isolating devices, cables, and other methods.

The variety of seismic restraint requirements for all the different types of HVAC equipment is beyond the scope of this module. For more detailed information on this subject, refer to *FEMA 412, Installing Seismic Restraints for Mechanical Equipment* published by the Federal Emergency Management Agency.

2.2.1 Seismic Restraints for Residential Equipment

There are several important requirements for equipment installed in a residential environment. For example, residential gas furnaces and water heaters with input ratings less than 100,000 Btuh must use flexible connectors between any rigid gas pipe and the appliance. Furnaces and water heaters can be attached to the structure with metal straps bolted to the structure. Furnaces can also be restrained with bumpers. Residential split-system outdoor units should be bolted to a concrete pad using a concrete anchor.

Bumpers can be used to prevent horizontal movement on equipment that is not top-heavy and prone to tipping over. The bumpers shown in *Figure 44* are bolted to concrete.

2.2.2 Seismic Restraints for Light Commercial Equipment

Packaged rooftop units that are curb-mounted can be converted in the field to be resistant to seismic events or the unit can be installed on a factory-built seismic curb like the one shown in *Figure 45*. Equipment such as ceiling-mounted air handlers can be suspended and constrained with a combination of threaded rod hangers with vibration isolators and steel cables (*Figure 46*).

2.2.3 Seismic Restraints for Hydronic and Commercial Refrigeration Equipment

Pumps used in hydronic systems are restrained using a number of devices. In *Figure 47*, vibration isolators and snubbers retain the pump on its pad.

Flexible connectors in the pump's supply and return water lines along with vibration-isolated pipe hangers help prevent breakage of the pipes in case of seismic movement. *Figure 48* shows a combination coil spring isolator and seismic restraint that can be used for various kinds of heavy floor-mounted mechanical equipment.

2.3.0 Selecting Vibration Isolators and Materials

The selection of the correct vibration isolator or seismic restraint depends on the type of equipment, its weight, and how and where it is to be installed. Local codes often dictate what devices or materials are required for a particular application. Job specifications also dictate the type of vibration isolation that an application requires.

Vibration isolators and seismic restraints that can be purchased as components include the following:

- Suspension-type vibration isolators
- Spring-type vibration isolators
- Grommet-type vibration isolators
- Vibration-isolating pads
- Flexible metal refrigerant pipe
- Factory-built seismic roof curbs

Manufacturers of these products and devices usually provide printed selection information so that the correct device can be found for any application. In addition to common information such as dimensions, they can provide weight limitations or range of movement limitations. *Figure 49* shows one particular type of vibration isolator, along with its selection data. Note that a single basic assembly can accommodate a number of different springs. The spring design and construction characteristics determine its capacity. The springs are often color-coded, making it easy to visually confirm which spring has been provided in an installation. Note that the isolators at the lower end of the chart, with the highest weight rating, actually are equipped with two springs. A core spring is provided inside the main, outer spring.

> ### Did You Know?
> # The Great San Francisco Earthquake
>
> In 1906, a devastating earthquake struck the City of San Francisco. Many of the city's water mains were broken, resulting in a loss of water pressure. With little or no water pressure available to fight the many fires that started after the earthquake, fires raged out of control, destroying the city.

TWO ANCHORS PER ANGLE
OR AS REQUIRED BY
CONSTRUCTION DRAWINGS

ELASTOMERIC PAD
ADHERES TO
STEEL ANGLE

ANCHORS

EQUIPMENT

PLAN VIEW

03313-13_F44.EPS

Figure 44 Bumpers.

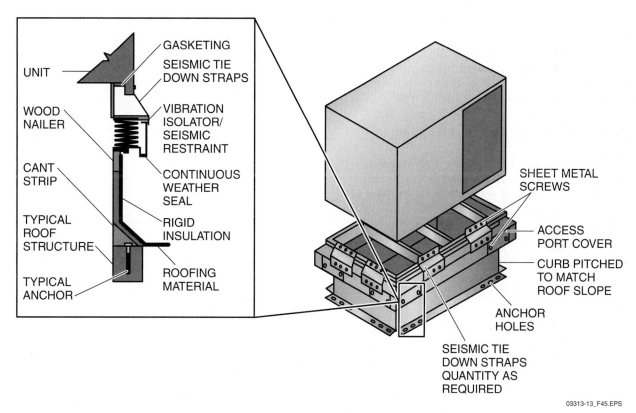

GASKETING

SEISMIC TIE
DOWN STRAPS

UNIT

VIBRATION
ISOLATOR/
SEISMIC
RESTRAINT

WOOD
NAILER

CONTINUOUS
WEATHER
SEAL

CANT
STRIP

RIGID
INSULATION

TYPICAL
ROOF
STRUCTURE

ROOFING
MATERIAL

TYPICAL
ANCHOR

SHEET METAL
SCREWS

ACCESS
PORT COVER

CURB PITCHED
TO MATCH
ROOF SLOPE

ANCHOR
HOLES

SEISMIC TIE
DOWN STRAPS
QUANTITY AS
REQUIRED

03313-13_F45.EPS

Figure 45 Factory-built seismic curb.

Seismic Gas Shut-Off Valve

The danger of fire when gas lines rupture during an earthquake is obvious. A new gas shut-off valve is now available to help reduce this danger. Installed in the main gas line after the gas meter, this valve shuts off gas flow to the entire structure if an earthquake of magnitude 5.4 or above is detected. If shutoff occurs, the valve must be manually reset.

Figure 46 Suspended equipment seismic restraints.

Figure 47 Water pump seismic restraints.

Figure 48 Combination seismic restraint.

Other vibration isolators or seismic restraints must be field-fabricated using common hardware items such as the following:

- Angle iron
- Metal straps
- Steel cable
- Threaded rod
- Nuts, bolts, washers, etc.

The size and quality of these materials is often spelled out in local codes or job specifications. For example, the SAE grade of bolts required for a restraint or the diameter of threaded rod required to suspend a fan coil may be specified.

Vibration isolators and seismic restraints use similar devices and methods to achieve similar results. An excellent resource for learning what to use for a particular application can be found in *FEMA 412, Installing Seismic Restraints for Mechanical Equipment* published by the Federal Emergency Management Agency. Tables based on equipment type offer suggestions on how to properly install a piece of equipment.

Did You Know?
Earthquake-Resistant Buildings

Many new skyscrapers being built in earthquake zones around the world are being built on giant suspension springs. This allows the building to sway instead of collapse when an earthquake occurs.

(A)

TYPE 1" DEFLECTION C, CI, CIP & CIW RATINGS

Size	Rated Capacity (lb)	Rated Defl (in)	Mount Constant (lb/in)	Spring Color/Stripe
A-45	45	1.60	28	Blue
A-75	75	1.50	50	Orange
A-125	125	1.33	94	Brown
A-200	200	1.15	174	Black
A-310	310	1.00	310	Yellow
A-400	400	1.00	400	Green
A-510	510	1.00	510	Red
A-625	625	1.00	625	White
B-65	65	2.10	31	Brown
B-85	85	2.10	40	White/Black
B-115	115	2.00	57	Silver
B-150	150	2.00	75	Orange
B-280	280	1.60	174	Green
B-450	450	1.31	344	Red
B-750	750	1.12	670	White
B-1000	1000	1.00	1000	Blue
B-1250	1250	1.00	1250	Gray
B-1650	1650	1.00	1650	Black
C-1000	1000	1.00	1000	Black
C-1350	1350	1.00	1350	Yellow
C-1750	1750	1.00	1750	Black
C-2100	2100	1.00	2100	Yellow*
C-2385	2385	1.00	2385	Yellow**
C-2650	2650	1.00	2650	Red*
C-2935	2935	1.00	2935	Red**
D-3500	3500	1.00	3500	Black*
D-4200	4200	1.00	4200	Yellow*
D-4770	4770	1.00	4770	Yellow**
D-5300	5300	1.00	5300	Red*
D-5870	5870	1.00	5870	Red**
E-7000	7000	1.00	7000	Black*
E-8400	8400	1.00	8400	Yellow*
E-9540	9540	1.00	9540	Yellow**
E-10600	10600	1.00	10600	Red*
E-11740	11740	1.00	11740	Red**
F-12600	12600	1.00	12600	Yellow*
F-14310	14310	1.00	14310	Yellow**
F-15900	15900	1.00	15900	Red*
F-17610	17610	1.00	17610	Red**
G-18900	18900	1.00	18900	Yellow*
G-21465	21465	1.00	21465	Yellow**
G-23850	23850	1.00	23850	Red*
G-26415	26415	1.00	26415	Red**

*with RED core spring   **with GREEN core spring

(B)

03313-13_F49.EPS

Figure 49 Vibration isolator selection information.

For example, assume an air handler is to be mounted on a concrete floor. For simple vibration isolation, open- or housed-spring devices can be used. If the installation must be resistant to seismic activity, housed-spring isolators cannot be used because they cannot resist uplift. If open-spring devices are used and the installation must be seismic-compliant, snubbers or bumpers must be used to restrict back-and-forth and side-to-side motion. Equipment manufacturers may provide specific instructions and/or kits for installing their equipment to be seismic compliant.

Seismic Building Codes

In many parts of the United States, concerns about earthquakes are not reflected in local building codes. However, in states such as California, all aspects of building construction must consider the effect an earthquake can have on a structure and what must be done during the construction of the building to reduce or eliminate those effects.

Additional Resources

FEMA 412, Installing Seismic Restraints for Mechanical Equipment, Latest Edition. Washington, DC, Federal Emergency Management Agency.

2.0.0 Section Review

1. Flexible vibration isolators installed in ducts are mounted in _____.

   a. vertical sections of duct
   b. horizontal sections of duct
   c. supply air ducts only
   d. vertical or horizontal ducts

2. Seismic restraints are placed on HVAC equipment to prevent _____.

   a. movement
   b. vibrations
   c. fires
   d. noises

3. The product catalog for an open-spring vibration isolator is likely to contain information about how much it can safely move during normal operation.

   a. True
   b. False

SECTION THREE

3.0.0 WIRING TERMINATIONS

Objectives

Identify common low- and line-voltage electrical termination hardware and explain how to terminate wiring connections.

  a. Identify common low- and line-voltage electrical terminating hardware.
  b. Explain how to properly terminate low- and line-voltage wiring connections.

Performance Task 3

Terminate low- and line-voltage wiring on a compressor contactor.

Trade Terms

AL-CU: An abbreviation for aluminum and copper, commonly marked on electrical connectors to indicate the device is suitable for use with conductors made with either metal.

Connector: A device used to physically and electrically connect two or more conductors.

Lug: A wiring terminal used on wire sizes No. 8 and larger. Also a term used to describe a screw-type terminal that simply applies pressure to a stripped wire, without the need for a crimp-on terminal.

Terminal: A device used for connecting wires to an object.

Termination: The connection of a wire to its destination.

A properly made termination should last as long as the insulation on the wire itself. A poorly made termination will always be a source of trouble because it will overheat under load and eventually fail.

The basic requirements for a good electrical termination include the following:

- It must be mechanically and electrically secure.
- It must be insulated as well as or better than the existing insulation on the conductors.
- These characteristics should last as long as the conductor is in service.

There are many different types of wiring terminations. The selection of the proper type for a given application often depends on how and where the termination is used. Terminations are normally made with a solder-less pressure connector to save time.

3.1.0 Connectors and Terminations

A variety of wire terminations for stranded wire are shown in *Figure 50*. These connectors are available in various sizes to accommodate wire sizes No. 22 AWG and larger. They can be installed with manual crimping tool. The range is normally stamped on the tongue of each terminal.

Crimp-type terminations for wires smaller than No. 8 AWG are normally made to accept at least two wire sizes and are often color-coded. For example, one manufacturer's color code is red for No. 18 or No. 20 wire, blue for No. 16 or No. 14 wire, and yellow for No. 12 or No. 10 wire. Crimp-type terminations for wire sizes No. 8 and larger, commonly called lugs, are made to accept one specific conductor size. Mechanical compression-type connectors and lugs are made to accommodate a range of different wire sizes.

Compression-type connectors for connecting conductors to screw terminals for low- and line-voltage circuits include those in which hand tools indent or crimp tube-like sleeves on the end of a conductor. Proper crimping action changes the size and shape of the connector and deforms the conductor strands enough to provide good electrical conductivity and mechanical strength.

Figure 51 shows the basic structure of a crimp connector. The crimp barrel receives the wire and is crimped to secure it in place. The Vs or dimples inside the barrel improve the wire-to-terminal conductivity and also increase the termination tensile strength. Most crimp connectors have nylon or vinyl insulation covering the barrel to reduce the possibility of shorting to adjacent terminals. The insulation is color coded according to the connector's wire range to reduce the problem of wire-to-connector mismatch. An inspection hole is provided at the end of the barrel to allow visual inspection of the wire position. For the smaller wire sizes, a sleeve is crimped over the conductor insulation in the process of crimping the barrel, providing strain relief for the conductor.

Conductor Terminations and Splices

Poor electrical connections are responsible for a large percentage of equipment burnouts and fires. Many of these failures are a direct result of improper terminations, poor workmanship, and the use of improper devices.

RING TONGUE

RING TONGUE (SLOTTED)

HOOK SLOT

OFFSET RING TONGUE

RECTANGULAR

FLANGED FORK

LOCKING FORK

FORK

FEMALE SPADE TERMINAL

BENT TONGUE

FLAG

03313-13_F50.EPS

Figure 50 Crimp-on terminations.

CRIMP BARREL (THIS PORTION CRIMPED TO THE WIRE)

BRAZED OR BUTTED SEAM

INSPECTION HOLE

STUD SIZE (SIZE VARIES)

STUD SIZE

CONDUCTOR RANGE

TONGUE (STYLE VARIES) CONNECTS TO THE OTHER MEMBER

COLOR-CODED TO INDICATE WIRE SIZE

HIGH CONDUCTIVITY TIN-PLATED COPPER

SERRATIONS PROVIDE MAXIMUM CONTACT AND TENSILE STRENGTH AFTER CRIMPING

INSULATION

FUNNEL ENTRY ASSURES RAPID WIRE INSERTION

03313-13_F51.EPS

Figure 51 Basic crimp connector structure.

Aluminum has certain properties that are different from copper and special precautions must be taken to ensure reliable terminations and connections. Standard copper connectors cannot be safely used on aluminum wire. Aluminum connectors are designed to bite through the aluminum oxide film that coats aluminum conductors as the connector is applied to a conductor. The conductor should also be wire brushed and coated with an oxide-inhibiting compound to ensure a reliable joint. Never use connectors designed strictly for use on copper conductors on aluminum conductors. Connectors listed for use on both metals will be marked **AL-CU**.

3.2.0 Installing Terminations

Conductors must be stripped to the proper length and the correct termination selected before the termination is applied to the conductor.

3.2.1 Preparing Conductors for Termination

Before any connection can be made, the ends of the conductors must be properly cleaned and stripped. This ensures a low-resistance connection and avoids contaminating the termination. Stripping is the removal of insulation from the end of the conductor. Conductors should only be stripped using the appropriate stripping tool. This will help to prevent cuts and nicks in the wire, which can reduce the conductor area and weaken the conductor.

Poorly stripped conductors can result in damage that can lead to a stress concentration in the damaged area. Heat, rapid temperature changes, mechanical vibration, and motion can aggravate the damage, causing faults in the circuitry or even total failure. Lost strands are a problem in crimp-type terminals, while exposed strands present a safety hazard.

A variety of factors determine how precisely a conductor can be stripped, including the wire size and type of insulation. Wires should only be stripped using the appropriate stripping tool. The specific tool used depends on the size and type of wire being stripped. *Figure 52* shows a common type of wire stripper for small conductors. It can be used to cut, strip, and crimp wires from No. 22 through No. 10 AWG.

To use this tool, insert the conductor into the proper size knife groove, then squeeze the tool handles. The tool cuts the insulation, allowing the conductor to be easily removed without crushing its stripped end. If the knife groove used is for a smaller wire size than the wire being stripped,

Figure 52 Wire stripper/crimper.

the conductors beneath the insulation will be damaged. With conductors with fine strands, several of the strands will likely be cut along with the insulation. The length of the strip is regulated by the amount of wire extending beyond the blades when it is inserted in the knife groove.

Figure 53 shows two types of production-grade stripping tools. The upper tool (A) can be used to strip conductors from No. 20 to No. 10 AWG. The lower tool (B) strips wires from No. 18 to No. 6 AWG. Note that the lower tool has front entry jaws for use in tight spaces.

(A)

(B)

Figure 53 Production-grade wire strippers.

Figure 54 shows the positioning of the wire in the crimp barrel when stripped to the proper length. The conductor insulation must be in the belled mouth of the terminal. This relieves stress on the strands or wire and increases the strength of the connection. Allowing conductor strands to protrude out of the inspection hole more than $1/32$" will interfere with the terminal screw. Cutting the strands too short will reduce the contact surface area.

03313-13_F55.EPS

Figure 55 Manual crimping tool.

3.2.2 Installing a Crimp Connector

Most connectors used in low- and line-voltage circuits in HVAC equipment can be applied with a simple plier-type crimping tool (*Figure 55*). These hand tools are similar in construction to ordinary mechanics' pliers except the jaws are shaped for crimping.

To install a crimp connector, first insert the stripped end of the wire completely into the connector. Position the crimping tool in place over the connector, and then squeeze the tool to fully crimp the connector. Make sure that the crimping tool jaws are fully closed, indicating that a full compression crimp has been made.

STRIPPED TOO LONG

TERMINAL INSULATION

NO INSULATION SUPPORT

STRIPPING THAT IS TOO LONG WILL INTERFERE WITH THE TERMINAL SCREW

STRIPPED TOO SHORT

STRIPPING THAT IS TOO SHORT DOES NOT PROVIDE ENOUGH CONTACT SURFACE

$1/32$" MAXIMUM

INSULATION

CONDUCTOR

PROPER STRIPPING LENGTH WITH INSULATION INSIDE THE TERMINAL

03313-13_F54.EPS

Figure 54 Proper stripping length.

Additional Resources

Air Conditioning Systems, Principles, Equipment, and Service, Latest Edition. Upper Saddle River, NJ: Prentice Hall.

3.0.0 Section Review

1. Crimp-type terminations for wire sizes smaller than No. 8 AWG are normally made to accept at least _____.

    a. a single wire size
    b. stranded wire only
    c. two wire sizes
    d. copper wires only

2. When is it acceptable to use a very sharp knife blade to strip insulation from a conductor?

    a. If the stripper blade is dull
    b. If the conductor is solid
    c. It is never acceptable
    d. If the conductor is copper

SUMMARY

HVAC technicians are constantly assembling mechanical equipment and taking it apart. Knowing what fastener to use and how to correctly install the fastener helps ensure that the equipment will operate without breakdowns.

Vibrations caused by equipment operation must be prevented from becoming objectionable or from causing damage. Vibration isolators and vibration dampers are used to minimize this problem. In areas where seismic activity occurs, mechanical equipment must be prevented from uncontrolled movement that could damage the equipment and/or cause personal injury when a seismic event occurs.

Being able to correctly apply terminations on conductors ensures that electrical equipment will experience trouble-free operation.

1. In a situation where stress is extremely high, which grade of bolt should be used?

   a. Grade 1
   b. Grade 2
   c. Grade 5
   d. Grade 8

2. Bolts and screws used on parts that must have a finished appearance are called _____.

   a. machine bolts
   b. machine screws
   c. cap screws
   d. stud bolts

3. A jam nut is used _____.

   a. when frequent removal is required
   b. to lock a standard nut
   c. when appearance is important
   d. to lock a split washer

4. Which type of lock washer is used with bolts and cap screws?

   a. Split ring
   b. External
   c. Countersunk
   d. Internal

5. When threading the bottom of a blind hole, use a _____.

   a. pipe tap
   b. plug tap
   c. bottoming tap
   d. taper tap

6. When installing a ½" Grade 8 steel bolt, it should be tightened to _____.

   a. 37 foot-pounds
   b. 60 foot-pounds
   c. 85 foot-pounds
   d. 92 foot-pounds

7. The force required to loosen a fastener is called _____.

   a. run-down resistance
   b. break-free resistance
   c. break-away torque
   d. loosen-up torque

8. How are concrete inserts installed?

   a. Set in place with epoxy
   b. Set in place with grout
   c. Screwed directly into concrete
   d. Imbedded in wet concrete

9. Blind rivets are used when a joint can only be accessed from one side.

   a. True
   b. False

10. Welded hermetic compressors used in residential systems minimize compressor vibration by being mounted on _____.

    a. rubber grommets
    b. open springs
    c. rubber pads
    d. vibration isolators

11. Bumpers are used as seismic restraints on equipment to prevent _____.

    a. tip-over
    b. up and down movement
    c. side-to-side movement
    d. swaying

12. Local building codes may specify the SAE grade of bolts that must be used in a seismic restraint.

    a. True
    b. False

13. The wire size range for a crimp-on terminal can be found _____.

    a. on the terminal packaging
    b. stamped on the tongue
    c. printed on the insulation
    d. printed on the conductor

14. Lugs are used with _____.
   a. stranded wires
   b. No. 8 and larger wires
   c. solid wires
   d. No. 10 and smaller wires

15. When crimping a conductor on a terminal, conductor strands should protrude out the inspection hole no more than _____.
   a. ¼"
   b. ⅛"
   c. ³⁄₃₂"
   d. ¹⁄₃₂"

Trade Terms Introduced in This Module

AL-CU: An abbreviation for aluminum and copper that is commonly marked on electrical connectors to indicate the device is suitable for use with conductors made with either metal.

Break-away torque: The torque required to loosen a fastener. This is usually lower than the torque to which the fastener has been tightened.

Connector: A device used to physically and electrically connect two or more conductors.

Lug: A wiring terminal used on wire sizes No. 8 and larger. Also a term used to describe a screw-type terminal that simply applies pressure to a stripped wire, without the need for a crimp-on terminal.

Nominal size: A means of expressing the size of a bolt or screw. It is the approximate diameter of a bolt or screw.

Run-down resistance: The torque required to overcome the resistance of associated hardware such as locknuts and lock washers, when tightening a fastener.

Seismic activity: Movement of the earth such as during an earthquake.

Seismic restraint: A device or method used to prevent or restrict the movement of equipment during a seismic event.

Set or seizure: In the last stages of rotation in reaching a final torque, the fastener may lock up; this is known as setting or seizing. This is usually accompanied by a noticeable popping effect.

Tensile strength: The maximum stress a material can endure before being pulled apart.

Terminal: A device used for connecting wires to an object.

Termination: The connection of a wire to its destination.

Thread classes: Threads are distinguished by three classifications according to the amount of tolerance the threads provide between the bolt and nut, and are designated by a number and letter combination that indicates the degree of fit for a thread. Higher numbers indicate a tighter fit.

Thread series: The combinations of diameter and pitch in threads that are applied to specific thread diameters to identify the coarseness or fitness of a thread.

Thread standards: An established set of standards for machining threads.

Torque: Applying a specific amount of twisting force to a threaded fastener.

Vibration isolator: A device that prevents the transmission of vibrations from operating machinery to surrounding objects.

Additional Resources

This module presents thorough resources for task training. The following resource material is suggested for further study.

Air Conditioning Systems, Principles, Equipment, and Service, Latest Edition. Upper Saddle River, NJ: Prentice Hall.

FEMA 412, Installing Seismic Restraints for Mechanical Equipment, Latest Edition. Washington, DC, Federal Emergency Management Agency.

Figure Credits

Section Review Answer Key

Answer	Section Reference	Objective
Section One		
1. a	1.1.1	1a
2. c	1.2.2	1b
3. c	1.3.1	1c
4. d	1.4.2	1d
5. b	1.5.2	1e
Section Two		
1. d	2.1.3	2a
2. a	2.2.0	2b
3. a	2.3.0	2c
Section Three		
1. c	3.1.0	3a
2. c	3.2.1	3b

NCCER CURRICULA — USER UPDATE

NCCER makes every effort to keep its textbooks up-to-date and free of technical errors. We appreciate your help in this process. If you find an error, a typographical mistake, or an inaccuracy in NCCER's curricula, please fill out this form (or a photocopy), or complete the online form at **www.nccer.org/olf**. Be sure to include the exact module ID number, page number, a detailed description, and your recommended correction. Your input will be brought to the attention of the Authoring Team. Thank you for your assistance.

Instructors – If you have an idea for improving this textbook, or have found that additional materials were necessary to teach this module effectively, please let us know so that we may present your suggestions to the Authoring Team.

NCCER Product Development and Revision

13614 Progress Blvd., Alachua, FL 32615

Email: curriculum@nccer.org
Online: www.nccer.org/olf

❏ Trainee Guide ❏ Lesson Plans ❏ Exam ❏ PowerPoints Other _____

Craft / Level: _____ Copyright Date: _____

Module ID Number / Title: _____

Section Number(s): _____

Description: _____

_____

_____

_____

_____

Recommended Correction: _____

_____

_____

_____

_____

Your Name: _____

Address: _____

_____

Email: _____ Phone: _____

03314-13

Control Circuit and Motor Troubleshooting

OVERVIEW

Most malfunctions that occur in HVAC equipment are caused by failure in the power distribution or control system. Technicians must have an understanding of how the various components function and interpret the schematics, wiring diagrams, and other service literature. In addition, new technologies such as variable frequency motor drives and electronically commutated motors are adding more complexity. The successful troubleshooter is able to read schematics and wiring diagrams to understand the underlying operating sequences.

Module Two

Trainees with successful module completions may be eligible for credentialing through NCCER's National Registry. To learn more, go to **www.nccer.org** or contact us at **1.888.622.3720**. Our website has information on the latest product releases and training, as well as online versions of our *Cornerstone* magazine and Pearson's product catalog.

Your feedback is welcome. You may email your comments to **curriculum@nccer.org**, send general comments and inquiries to **info@nccer.org**, or fill in the User Update form at the back of this module.

This information is general in nature and intended for training purposes only. Actual performance of activities described in this manual requires compliance with all applicable operating, service, maintenance, and safety procedures under the direction of qualified personnel. References in this manual to patented or proprietary devices do not constitute a recommendation of their use.

Objectives

When you have completed this module, you will be able to do the following:

1. Identify and describe the operation of common HVAC control circuit devices.
   a. Identify and describe the operation of relays, contactors, and motor starters.
   b. Identify and describe the operation of other common safety and control circuit devices.
2. Describe the operation, installation, and testing of various thermostats and temperature controls.
   a. Describe the operation of various thermostats and temperature controls.
   b. Identify and describe how to troubleshoot thermistors.
   c. Explain how to install and wire thermostats.
   d. Explain how to troubleshoot the functions of a thermostat.
3. Describe the sequence of operation for basic HVAC systems.
   a. Describe the sequence of operation of a basic cooling-only system.
   b. Describe the sequence of operation for a common heating and cooling system.
   c. Describe the operation of basic pneumatic control systems.
4. Explain how to troubleshoot common control circuits and load components.
   a. Identify basic safety practices related to troubleshooting HVAC power and control circuits.
   b. Explain how to approach HVAC-related problems and prepare for troubleshooting.
   c. Explain how to test high-voltage power sources.
   d. Explain how to troubleshoot control circuits and low-voltage power sources.
   e. Explain how to troubleshoot both resistive and inductive loads, including motors and their related devices.
   f. Explain how to troubleshoot various hydronic control system components.
5. Describe the operation of variable frequency drives (VFD) and their selection considerations.
   a. Describe the operation of a VFD.
   b. Identify VFD parameters that can be programmed.
   c. Describe the important considerations for the selection of a VFD.
   d. Explain dynamic motor braking processes.
6. Identify and describe how to service electronically commutated motors (ECMs).
   a. Identify and describe the operation of ECMs.
   b. Describe how to install and set-up an ECM.
   c. Describe how to troubleshoot an ECM.

Performance Tasks

Under the supervision of your instructor, you should be able to do the following:

1. Wire, check the operation of, and adjust the heat anticipator/cycle rate of a thermostat.
2. Interpret control circuit diagrams.
3. Perform electrical tests and/or troubleshooting procedures on the following:
   - Single- and three-phase power sources
   - Fuses and circuit breakers
   - Resistive loads
   - Relays and/or contactors
   - Motor windings
   - Start and run capacitors
   - Start relays and thermistors

Trade Terms

Actuator
Base speed
Bleed control
Cooling compensator
Deadband
Differential
Droop
Electric-pneumatic (E-P) relay

Fault isolation diagram
Insulated gate bipolar transistor (IGBT)
Invar®
Label diagram
Ladder diagram
Pneumatic-electric (P-E) relay
Wiring diagram

Industry-Recognized Credentials

If you are training through an NCCER-accredited sponsor, you may be eligible for credentials from NCCER's Registry. The ID number for this module is 03314-13. Note that this module may have been used in other NCCER curricula and may apply to other level completions. Contact NCCER's Registry at 888.622.3720 or go to **www.nccer.org** for more information.

Contents

Topics to be presented in this module include:

Figures and Tables

Figures and Tables

1.0.0 HVAC CONTROL DEVICE OPERATION

Objectives

Identify and describe the operation of common HVAC control circuit devices.

a. Identify and describe the operation of relays, contactors, and motor starters.
b. Identify and describe the operation of other common safety and control circuit devices.

Performance Task 3

Perform electrical tests and/or troubleshooting procedures on the following:

Relays and/or contactors

Trade Terms

Ladder diagram: A troubleshooting aid that depicts a simplified schematic diagram of the equipment. The load lines are arranged like the rungs of a ladder between vertical lines representing the voltage source. Normally, all the wire color and physical connection information is eliminated from the diagram to make it easier to use by focusing on the functional, not the physical, aspects of the equipment.

Most HVAC control systems are designed to automatically maintain the desired heating, cooling, and ventilation conditions. The controls for a small system such as a window air conditioner are very simple—a couple of control switches, a thermostat, and a couple of relays. As the system gets larger and provides more features, the controls become more complicated. For example, add gas heating to a packaged cooling unit, and the size and complexity of the control circuits more than double. Make it a heat pump instead, and the control complexity continues to increase.

The use of electronic controls is now widespread across all HVAC product lines. It is rare to find a modern unit that does not contain some kind of electronic control. Manufacturers put these controls in their equipment to increase reliability and energy efficiency. Electronic controls also allow engineers to design more versatility into the products than was possible with electromechanical controls.

Large commercial systems may use pneumatic and electronic controls in conjunction with conventional electrical controls. These systems may have thirty or forty control devices, whereas a window air conditioner has just a handful.

The good news is that there are only a few different kinds of control devices. Once different controls are recognized and the role each plays is understood, it will not matter how many are used to control a particular system.

All automatic control systems have the following basic characteristics in common:

- One or more sensing elements (thermistor, thermostat, pressure switches, or humidistat) measure changes in temperature, pressure, and humidity.
- One or more control mechanisms translate these changes into energy that can be used by devices such as motors and valves.
- The connecting wiring, pneumatic piping, and mechanical linkages transmit the energy to the motor, valve, or other controlled devices that need to respond.
- The controlled devices then use the energy to achieve some change. For example, motors operate compressors, fans, or dampers. Valves control the flow of gas to burners or cooling coils and permit the flow of air in pneumatic systems. Valves also control the flow of liquids in chilled-water systems.
- The sensing elements in the control detect the change in conditions and again signal the control system.
- The controls then stop the motor, close the valves, or terminate the action of the component being used. As a result, the call for change is ended.

Although this may sound like a complex arrangement, it is actually quite simple. A need to change the environmental conditions is transmitted through the control system to devices that can make those changes happen.

1.1.0 Relays, Contactors, and Motor Starters

HVAC control system circuits contain relays, contactors, and motor starters. These devices are physically different in size and configuration, but their principles of operation are basically the same. Because of their extensive use in control and power distribution circuits, it is extremely important to understand how these devices operate. Without this understanding, it will be difficult to read schematics and troubleshooting circuits. Refer to the *Appendix* at the back of this module for common schematic symbols.

1.1.1 Basic Relays

A relay operates to stop or permit the flow of electricity to a load. Sometimes, a relay is used to reroute the flow of electricity in a different direction. Since thermostats and other control devices are unable to carry a significant electrical load, the load of some devices must be transferred to the somewhat-heavier electrical contacts of a relay. Relays are used to make and break other electrical circuits when they are themselves energized.

Relays can be hard-wired into a circuit, or plug-in relays (*Figure 1*) can be used. Plug-in relays have a mating socket or base into which they fit, making troubleshooting and replacement easier. All wiring terminations are then made at the socket or base. Instead of having to connect and disconnect wiring, a new relay can simply be snapped into place.

In today's equipment, relays that were once stand-alone components are now often soldered to a PC board. If a relay soldered on an electronic control is suspected of being defective, the entire electronic control will have to be replaced.

In its basic form, the relay consists of two parts: an electromagnetic coil and a set of contacts. When the coil is energized with the proper voltage, it causes the position of the relay's contacts to change. Contacts are identified as being either normally closed or normally open. This refers to their position when the relay coil is de-energized. All relay contacts shown on schematic diagrams are shown with the relay in the de-energized position.

Figure 2 shows the open and closed contacts of a typical relay. It also shows the schematic symbol for normally closed and normally open relay contacts. Remember this is the position of the contacts when the relay coil is de-energized. A normally closed set of contacts will allow electric current to flow through the contacts when the relay is de-energized. A normally open set of contacts close only when the relay is energized.

Figure 3A shows the schematic symbol for a simple relay consisting of the relay coil and a set of contacts. As shown, this relay is a normally closed relay. It is also classified as a single-pole, single-throw (SPST) relay because it only has one set of contacts and one current path. When the relay coil is de-energized as shown, current present at terminal 3 can travel through the closed contacts to terminal 4 for subsequent application to the remainder of control circuit. Terminal 3 is the pole described by the words single-pole. Since there is only one potential current path, it is further described as single-throw.

When voltage is applied across the coil of the relay via terminals 1 and 2, the coil energizes and the relay contacts open, preventing any current applied at terminal 3 from flowing through the contacts to terminal 4. Remember, when the coil is energized, the normally closed contacts open. The last thing that needs to be determined about a relay is its coil voltage. Most relay coils used in HVAC control circuits operate on 24VAC, but that should never be assumed.

It is important to understand that no electrical connection exists between the coil of a relay and its contacts within the relay housing. Only in rare cases is there an internal connection. Opening and closing of the relay contacts happens because of electromechanical linkage between the coil and contacts. *Figure 3B* shows a mechanical representation of the relay just described. As shown, terminals 1 and 2 are still the coil connections, and terminals 3 and 4 are still the contact connections. When power is applied to the relay coil, an electromagnetic field is created by the coil. This electromagnetic field pulls the plunger up, causing the normally closed contacts to open. There are other ways to accomplish this in relays, but this diagram shows what is going on inside the relay mechanically.

Typically, the power applied to the coil might be 24VAC, while 240VAC may be applied through the contacts. This is one common situation but there

03314-13_F01.EPS

Figure 1 Examples of plug-in relays.

Relays

The invention of the relay is credited to Samuel Morse, one of the inventors of the telegraph. Morse developed the relay to boost signal strength. An incoming signal activated an electromagnet, which closed a battery circuit, thereby transmitting the signal to the next relay.

Figure 2 Normally open and normally closed relay contacts.

(A)
SCHEMATIC
SYMBOLS

(B)
ELECTROMECHANICAL
REPRESENTATION

03314-13_F03.EPS

Figure 3 Single-pole, single-throw (SPST) relay.

are a number of other coil/contact voltage combinations. There are some applications in which the same voltage is applied to both the coil and the contacts of a relay. In this case, the connection between the two is always made using terminals outside of the relay. Again, it is important to remember that there is no electrical connection between the coil and contacts of a relay within the housing of most relays.

There are other types of relays. The counterpart of the first relay is the SPST normally open relay. This relay is similar to the SPST normally closed relay, except that when the coil is de-energized, the contacts are normally open. When the relay coil is energized, the contacts close.

The next relay is the single-pole, double-throw (SPDT) relay (*Figure 4*). In the SPST relay described earlier, the power applied at terminal 3 was applied to terminal 4, or not applied to terminal 4,

SINGLE-POLE, DOUBLE-THROW
(SPDT) RELAY

03314-13_F04.EPS

Figure 4 Single-pole, double-throw relay.

depending on whether the contacts where closed or open, respectively.

With a double-throw relay, power can be directed between two different terminals. As shown, with the coil de-energized, power coming in on the common terminal 3 is applied through the normally closed set of contacts to terminal 5. If the coil is energized, the positions of the relay contact sets change, causing the normally open contacts to close and the normally closed contacts to open. This causes the power applied from terminal 3 to be redirected from terminal 5 to terminal 4 for subsequent application to a different branch of the unit's control circuits. Terminal 3 is still the single pole found in this relay.

The next relay is the double-pole, double-throw (DPDT) relay. As shown in *Figure 5*, another set of double-throw contacts is added to a basic SPDT relay to form this type of relay. When the coil is energized, it causes both sets of contacts to change position simultaneously. The two poles are terminals 3 and 6. It is important to remember that there is no electrical connection between the first set of contacts and the second set of contacts. In *Figure 5*, the coil is designated as R1 and would be found on a wiring diagram labeled as such. The four sets of contacts would also be identified on a wiring diagram with the labels shown here.

With an understanding of the concept of adding contact sets, it is easy to see how poles can be added to a relay. Applications may be found that require the use of a 3PDT (three-pole, double-throw) relay or even a 4PDT (four-pole, double-throw) relay. Regardless of the relay coil/contact configuration, each pole is electrically isolated from the coil and from all other poles. When the coil is energized, the position of all the poles changes simultaneously.

Since each pole in a relay is electrically isolated from each other and from the coil, the poles from the same relay can be wired into different branches of the equipment's control circuits. Even though the coil and poles are physically housed in the same relay assembly, they are commonly shown in different areas of schematic and ladder diagrams used for troubleshooting. Schematic and ladder diagrams are covered later in this module.

Because the relay coil and contacts are often shown in different areas of a schematic or ladder diagram, we need a method of designating which relay is associated with which contact set. An example is shown in *Figure 5*, where the coil is designated R1 (Relay 1) and all related contact sets are also identified with numbers that begin with R1. As shown, the first pole has contacts R1-1 and R1-2 (relay 1/contact set 1, and relay 1/contact set 2). The second pole has contacts R1-3 and R1-4. This method of relay designation is also used for three- and four-pole relays. For example, if a schematic or ladder diagram contains a normally open contact set designated R1-3, that means that the coil of relay R1 controls the contacts. If the R1 coil is energized, the R1-3 contacts will close. Likewise, all the other sets of R1 contacts are also going to change position.

There are many different designations for coils. For example, CR (control relay), CC (compressor contactor), or FR (fan relay) may be seen on diagrams. There are dozens of possible designations and they vary widely among manufacturers. Always consult the legend provided with the unit diagram.

1.1.2 Time-Delay Relays

The purpose of a time-delay relay is to delay the normal operation of a compressor or motor for a predetermined length of time after the control system has been energized. The length of the delay depends on the time built into the relay mechanism, and may vary from a fraction of a second to several minutes. Many time-delay relays have an adjustable delay period. A common use for a time-delay relay is to delay the startup and shutdown of a heating blower to avoid cold air movement and extract all the remaining heat from the heat exchanger.

Electrical systems containing several motors may also use time delays to start the motors one at a time in order to limit the inrush current. For example, the schematic drawing in *Figure 6* shows four compressor control circuits with time-delay relays for sequencing the starting of each motor. In this type of motor design, if all three stages of the thermostat are closed and electrical power is supplied to the units, compressor contactor coil 1 and time-delay relay 1 become energized. Compressor 1 will start immediately. After the time delay, the contacts of time-delay relay 1 close to energize compressor contactor coil 2, which starts

03314-13_F05.EPS

Figure 5 Double-pole double-throw (DPDT) relay.

compressor 2. Time-delay relay 2 also becomes energized, which eventually starts compressor 3. All four compressors are eventually started in a staged manner.

1.1.3 Compressor Short-Cycle Timers

Attempting to start a compressor against high head pressure can damage the compressor. When a refrigerant system shuts down, it should not be restarted until the pressures in the system have had time to equalize. Short-cycling can be caused by a momentary power interruption or by an occupant changing the thermostat setting.

A compressor short-cycle protection circuit contains a timing function that prevents the compressor contactor from re-energizing for a specified period of time after the compressor shuts off. This type of relay is referred to as a delay-on-break model, since the delay period begins when the contactor control circuit opens. A delay-on-make model begins its timing period when the circuit is first energized. If a delay-on-make model is used, the timing period will not start until the system calls for the compressor to start. This results in an undesirable delay in most cases; the timing period should begin when the compressor shuts down.

Anti-short-cycle periods are typically 2 to 5 minutes. Short-cycle timers are available as self-contained modules that can be direct-wired into a unit. They are often sold as optional and after-market accessories. They are also a common built-in feature of electronic programmable room thermostats. The feature can usually be turned on and off through the programming or set-up menu.

1.1.4 Lockout Control Relays and Circuits

The purpose of the lockout relay in a control circuit is to prevent the automatic restart of the HVAC equipment. If the lockout relay has been activated, the system may be reset only by interrupting the power supply to the control circuit; for example, turning the room thermostat function switch OFF and then ON, or turning the main power switch off and then on again.

In the circuit in *Figure 7*, the lockout relay coil, due to its high resistance, is not energized during normal operation. This high resistance, or impedance, is by design. However, when any one of the safety controls opens the circuit to the compressor contactor coil, current flows through the lockout relay coil, causing it to become energized and to open its contacts. These contacts remain open, keeping the compressor contactor circuit open until the power is interrupted after the safety control has reset. Performance depends on the resistance of the lockout relay coil being much greater than the resistance of the compressor contactor coil.

Figure 7 Lockout relay used in an HVAC control circuit.

Figure 6 Four compressor control circuits with time-delay relays.

If the lockout relay becomes defective, it should be replaced with an exact duplicate to maintain the proper coil circuit resistance. This type of relay is sometimes called an impedance relay.

It is permissible to add a control relay coil in parallel with the contactor coil when a system demands another control. The resistance of the contactor coil and the relay coil in parallel decreases the total resistance and does not affect the operation of the lockout relay.

> **CAUTION**
>
> Never put additional lockout relays, lights, or other load devices in parallel with the lockout relay coil. Doing so might defeat the lockout and cause equipment damage.

1.1.5 Solid-State Relays

Unlike the electromechanical relays just discussed, solid-state relays (*Figure 8*) have no moving parts. They use electronic devices such as triacs (a special type of current gateway) to control switching action and light-emitting diodes (LEDs) to report their status. Solid-state relays are used in general control applications, as well as in motor starting circuits.

1.1.6 Contactors and Motor Starters

Contactors (*Figure 9*) are a type of heavy-duty relay. They are used to start and stop high-current loads. They are used to start motors and compressors where overload protection is provided is not required or is provided separately. A contactor does not provide overload protection.

A contactor is a normally open, single-throw device with one or more poles. It operates in the same way as a relay and is often represented on a

03314-13_F09.EPS

Figure 9 Typical contactor.

schematic in a similar way. When the coil is energized, the movable contacts are closed against the stationary contacts, thus completing the circuit. When power is removed from the coil, the contacts open, stopping the flow of electric current to the related load.

Many contactors have auxiliary contacts or can be fitted with accessory contact sets. This allows the contactor to also perform like a relay. For example, when the contactor is energized and the main contacts close, an auxiliary set of contacts could open to remove power from a crankcase heater circuit.

A motor starter (*Figure 10*) is used to stop and start motors and provide overload protection. They are essentially contactors with an overload relay attached. Thermal overloads or current sensors in the starter sense excessive current flowing to a motor and protect the motor from overload. If more current is flowing than the motor is designed to handle, the overload device breaks the contactor control circuit and de-energizes the motor. Once the overload relay has been tripped, a manual reset is required to restore normal operation.

The starter shown in *Figure 10* requires the installation of a thermal element for each phase of

03314-13_F08.EPS

Figure 8 Solid-state relay.

Electronic Solid-State Relays

Some advantages of solid-state relays are longer life, higher reliability, high-speed switching, and high resistance to shock and vibration. The absence of mechanical contacts eliminates contact bounce, arcing when contacts open, and hazards from explosives and flammable gases.

THERMAL OVERLOADS ARE INSERTED HERE

RESET BUTTON

03314-13_F10.EPS

Figure 10 Motor starter with attached overload relay.

power. These elements must be very carefully sized to accommodate the motor being protected. Manufacturers provide charts for this purpose. The selection is primarily based on two things: the full load current rating of the motor, and the temperature of the area where the starter is located. Since the overload device is thermal, the temperature of the installation area must be considered. If the area is considerably warmer than the area where the motor is, the starter may trip before the motor has reached its maximum load rating. Electronic overloads offer a ranged dial to set the desired overload current value.

Like contactors, motor starters often have auxiliary contact sets. Some may have one or more sets of auxiliary contacts built in to the main body. Combination starters combine motor overload protection with circuit protection and a means of disconnect. These typically include a standard starter and a fused or non-fused switch or circuit breaker in the same enclosure.

1.1.7 Switch and Contactor/Relay Contact Checks

Once a switch or contactor/relay contact has been identified as the possible cause of an electrical problem, the contacts should be tested to confirm their position. Switches and contactor/relay contacts can easily be tested by making a continuity measurement (with power off) to determine whether the contacts are open or closed (*Figure 11A*). If the switch contacts are open, the ohmmeter indicates an infinite resistance reading. If the switch contacts are closed and well-mated, the ohmmeter indicates continuity (near-zero or zero ohms).

Note that the power must be removed from the contact structure to make a continuity check. As such, the check can only confirm the status of normally open and normally closed switch contacts. It does not reflect the status of the contacts when the system is powered up.

Switches and contactor/relay contacts can also be tested to determine whether the contacts are open or closed by placing voltmeter leads across energized contacts. This is shown in *Figure 11B*. If the switch contacts are open, the voltmeter indicates line voltage (240V in this example). If the switch contacts are closed, the voltmeter indicates zero volts.

WARNING!

A live circuit can cause an electrical shock. Always use caution when measuring voltage in a live circuit.

1.2.0 Control Circuit Switches and Safety Devices

Refrigerant system compressor control circuits normally include several different types of safety switches. These include various types of pressure and temperature switches.

Auxiliary Contact Sets

Auxiliary contacts generally mount on one or both sides of a contactor. Rather than being electrically energized, like a relay, a tab on the movable part of the contactor physically engages the contact linkage and pushes the contacts closed. Many different configurations are available, and more than one set can often be added. The set shown here is double-pole, single-throw.

03314-13_SA01.EPS

ZERO OHMS =
CLOSED CONTACTS

INFINITE RESISTANCE =
OPEN CONTACTS

(A) CONTINUITY TESTING

03314-13_F11A.EPS

Figure 11 Relay contact continuity and voltage checks. (1 of 2)

ZERO VOLTS =
CLOSED CONTACTS

240V =
OPEN CONTACTS

(B) VOLTAGE TESTING

03314-13_F11B.EPS

Figure 11 Relay contact continuity and voltage checks. (2 of 2)

Many systems use one or more pressure switches in the control circuit. Some are safety devices designed to protect the compressor. Others are used to activate condenser fans, prove the operation of an induced draft fan, and similar functions. Smaller commercial refrigeration systems may even use a pressure switch to control the compressor instead of a thermostat.

In all HVAC systems, temperatures must be kept within a specified range. Electromechanical devices such as thermostats have been used for years to sense temperature. In modern equipment, thermistors have taken over the role of sensing temperature because of their versatility. Electromechanical thermostats only open or close a set of contacts at a preset temperature, limiting their versatility. Thermistors, on the other hand, provide actual and current temperature values to electronic devices, allowing the system to respond in a variety of ways. For example, an outdoor tem-

perature thermistor can simultaneously provide outdoor temperature information to both a room thermostat and a defrost control board. The two devices use the temperature information in different ways. The room thermostat uses the information to display the outdoor temperature to the user, while the heat pump control board uses the same information to decide if supplemental heat is required.

1.2.1 Pressure Switches

Most pressure switches use a bellows mechanism that presses against switch contacts to open or close them. While many pressure switches have a fixed pressure setting, others are adjustable (*Figure 12*). Differential pressure switches, like those used as oil pressure safety switches, require two pressure inputs. They are designed to monitor the difference between two pressures and react accordingly. In an oil pressure safety switch, the difference between the pressure in the compressor crankcase and the oil pump outlet are monitored.

High-pressure safety switches are designed to open if the compressor head pressure is too high. Low-pressure safety switches are designed to open if the suction pressure is too low. Common causes of high head pressure are dirty condenser coils or a failed condenser fan motor. Common causes of low suction pressure include a loss of charge and low evaporator airflow. A pressure safety switch is normally closed, and is wired in series with the compressor contactor control circuit. While low-pressure safety switches often reset automatically, high-pressure safety switches typically require a manual reset.

Some manufacturers also use a type of low-pressure switch called a loss-of-charge switch that removes power to the equipment if the refrigerant charge is low. Loss-of-charge switches are often used in heat pumps. The setting of these switches is usually far below any normal operating pressures.

Relays, Contactors, and Motor Starters

Like any switching device, relays, contactors, and motor starters have a limited life. Normal failure modes include contact burning and sticking. Be careful when selecting replacement contactors because they are rated differently for different uses. Some are rated for inductive loads, others are rated for resistive loads, and still others are rated for both types of loads.

(A) FIXED PRESSURE
SWITCHES

(B) ADJUSTABLE LOW-
PRESSURE SWITCH

(C) DIFFERENTIAL
PRESSURE SWITCH

24V

HPCO   LPCO

COMPRESSOR
CONTACTOR
COIL

C

THERMOSTAT   COMPRESSOR
OVERLOADS

PRESSURE SWITCHES

HPCO = HIGH PRESSURE CUTOUT
LPCO = LOW PRESSURE CUTOUT

(D) PRESSURE SWITCH
SYMBOLS

03314-13_F12.EPS

Figure 12 Pressure switches.

Newer gas furnaces that use an induced draft fan often have a pressure switch in the heating control circuit to monitor and prove the operation of the inducer motor. Some furnaces do use a centrifugal switch in the back of the motor to prove operation. However, technically speaking, such a switch only proves the motor is turning; it does not prove that any amount of air is actually moving through the heat exchanger. A differential pressure switch can prove a sufficient amount of air is moving by comparing the pressure in the furnace enclosure to the pressure in the latter part of the heat exchanger.

The pressure switch (*Figure 13*) is the more common type of inducer-proving switch. It has tubing connected to the housing, and in some cases, to the burner enclosure. When the inducer motor is operating, a pressure is created that causes the pressure

switch to operate, allowing the ignition sequence to continue. To check the operation of an inducer pressure switch, disconnect the pressure tube from the switch and set the furnace thermostat to call for heat. The furnace burner should not ignite.

1.2.2 Freezestats

A freezestat is an electromechanical safety switch designed to provide protection against freezing. For example, in a chiller, water temperature must not be allowed to get to the point that it could freeze and severely damage the chiller. The freezestat switch is usually a normally closed bimetal switch. On a chiller, it is typically installed on the chilled water line leaving the chiller (*Figure 14*). They are typically set to open at a temperature no lower than 35°F.

INDUCER
PRESSURE
SWITCH

PRESSURE
TUBE

03314-13_F13.EPS

Figure 13 Inducer pressure switch.

Freezestats are also commonly found on water coils that are subjected to outside air. For example, a hot water coil may be found in an outside air duct. The coil is used to partially condition the outside air before it mixes with return air and passes over the primary heating and cooling coils. A freezestat would be used to close a set of outside air dampers if the boiler fails and the hot water coil is in danger of freezing.

Freezestats are used for hydronic purposes, such as a chiller safety control. They are simple devices with a sensing element inserted into a piping well. For coils, a freezestat typically has a very long capillary tube that is laced across the face of the coil surface. This type of capillary is called an averaging element, as it does not read the temperature at a single spot.

1.2.3 Outdoor Temperature Thermostat

Outdoor temperature thermostats are widely used with HVAC equipment. Like freezestats, they open/close at a predetermined temperature. Some common uses of outdoor temperature thermostats include the following:

- Allowing the operation of heat pump supplemental electric heaters below a pre-determined outdoor temperature. Multiple outdoor temperature thermostats can stage on additional electric heaters as the temperature continues to drop.
- Allow the operation of a compressor crankcase heater only when the outdoor temperature drops below a pre-determined temperature, typically about 65°F.
- On a dual-fuel heat pump, it switches heating between heat pump and gas or oil furnace operation as the outdoor temperature drops below or climbs above the heat pump's balance point.
- Provide both capacity and energy control for an air conditioner by staging additional compressor capacity on or off as outdoor temperature rises and falls.

IN   OUT

REFRIGERANT

CHILLER
BARREL

CHILLED
WATER
RETURN

CHILLED WATER SUPPLY

FREEZESTAT

03314-13_F14.EPS

Figure 14 Chilled water freezestat.

- Provide capacity control for a furnace or boiler by modulating burner operation or staging additional burners on or off as outdoor temperature rises and falls.
- Prevent equipment operation when outdoor temperatures become excessively hot or cold. For example, air-to-air heat pumps should not be operated in the heating mode when outdoor temperature exceeds 65°F. Some air-to-air heat pumps use an outdoor thermostat to stop operation at below-zero temperatures when the heat pump is relatively ineffective.

1.2.4 Furnace Fan Control

A furnace fan control (*Figure 15*) is a temperature-actuated switch that, when heated, will close a set of contacts to start the furnace blower motor. The sensing element of the fan switch is positioned inside one of the heat exchangers where the temperature is the highest.

The fan switch is normally actuated by a bimetal element that opens or closes the contacts in response to a temperature change. The fan control may be set to bring the fan on at about 120°F and to stop the fan at about 100°F.

In operation, the burner provides heat to the heat exchangers for as long as a minute or two to warm the air before the fan is started. This avoids blowing cold air into the conditioned space. When the thermostat is satisfied, the main burner stops providing heat, but the fan continues to operate until the air temperature has been reduced. This removes excess heat from the heat exchangers and improves efficiency by using residual heat.

In modern equipment, most manufacturers now use a timed on and off delay of the furnace blower motor instead of a temperature-actuated control.

Air handlers that use electric resistance heaters do not use temperature-activated switches to activate the fan. A fan relay is used to power the fan because air must immediately flow across the energized electric heaters to prevent them from overheating.

1.2.5 Furnace Limit Controls

A limit control is also a heat-actuated switch with a bimetal sensing element positioned near the heat exchangers. This is a heating safety control that is normally wired in series with the low-voltage control circuit of the furnace. If the temperature inside the heating chamber or plenum reaches approximately 200°F, power in the control circuit will be interrupted to prevent furnace overheating.

On oil burners, the limit switch is commonly wired in series with the power supply serving the furnace primary control. In an oil burner, the power will be shut off to the primary control, which stops burner operation. The blower motor will continue to function, since it does not rely on control power from the primary control to operate.

Electric heat limit switches (*Figure 16*) consist of a bimetal sensing element that is enclosed in a disc-shaped housing that resembles a large coin. Radiant heat from the electric heating elements will cause the sensing element to open if there is inadequate airflow across the heating elements. In contrast to a gas or oil furnace that contains a sin-

Figure 15 Complete 24V furnace control circuit.

(A) AUTO-RESET LIMIT SWITCH

(B) MANUAL-RESET LIMIT SWITCH

03314-13_F16.EPS

Figure 16 Limit switches.

gle limit switch, a fan coil with multiple banks of electric heaters will contain several limit switches, all wired in series.

Limit switches protect from transient events. For example, a dirty air filter may cause a furnace to cycle on and off on the limit switch. When the dirty filter is replaced, the problem goes away because most furnace limit switches are automatic reset. There are other events that can occur to a furnace that require it to be shut down until the problem is corrected. For example, a soot-blocked heat exchanger may cause flames to roll out the front of the burner. This condition cannot be allowed to repeat for obvious reasons. To prevent such events from reoccurring, manual-reset limit switches are used. A flame-rollout switch is an example of a manual-reset limit switch. The switch must be reset manually before operation will resume. A service technician will usually have to reset the switch to troubleshoot the furnace, observe the problem, and then correct it. Manual reset switches of any type are easily identified by the presence of a reset button, usually colored red. Fusible links, also called thermal fuses or one-shot devices, contain a link that melts at a low temperature. When the link melts, it opens the control circuit and shuts down the unit. Fusible links are rarely used on fossil fuel equipment. They are most often used as additional protection with limit switches on electric heat assemblies.

Fusible links must be replaced, with one of the correct melting temperature, to restore operation. On air handlers with electric heating elements, the same types of devices provide similar protection.

1.2.6 Combination Fan/Limit Switch

A combination fan/limit switch uses a common bimetal element that is spiral-wound to activate a fan switch and a limit switch. Both switches are contained in the same housing. The element is inserted so that it is between sections of a furnace heat exchanger or installed in the warm air plenum directly above the furnace. Heat acts on the spiral-wound bimetal element, causing a rotating motion that activates the fan switch when the selected fan-on temperature is reached. If the fan fails to come on or airflow is restricted, the temperature will continue to rise until the bimetal element opens the limit switch, typically set around 200°F.

This type of combination switch often comes factory-equipped with a jumper that enables the device to operate on line voltage only (*Figure 17*). With the jumper removed, the fan switch operates on line voltage, but the limit switch is in the low-voltage circuit. Since the two sets of contacts can be isolated from each other, they can function in different circuits at different voltages.

NO JUMPER; LIMIT
SWITCH BREAKS LOW-VOLTAGE
CIRUCUIT

WITH JUMPER; LIMIT
SWITCH BREAKS LINE VOLTAGE TO
CONTROL TRANSFORMER

03314-13_F17.EPS

Figure 17 Combination fan/limit switch wiring options.

Additional Resources

Refrigeration and Air Conditioning: An Introduction to HVAC/R. Latest Edition. Larry Jeffus. Pearson/Prentice Hall. Upper Saddle River, NJ.

Fundamentals of HVAC/R. Latest Edition. Carter Stanfield and David Skaves. Prentice Hall. Upper Saddle River, NJ.

1.0.0 Section Review

1. A technician measures 230V across a set of normally open relay contacts, and the relay coil is energized. What does this indicate?

   a. Normal operation
   b. The relay contacts are shorted
   c. There is continuity through the contacts
   d. The contacts are still open; the relay is defective

2. A furnace limit control is wired into a gas furnace's control circuit in _____.

   a. parallel with the fan switch
   b. series with the fan switch
   c. series in the control circuit
   d. parallel with the transformer primary

2.0.0 THERMOSTATS AND TEMPERATURE CONTROLS

Objectives

Describe the operation, installation, and testing of various thermostats and temperature controls.

a. Describe the operation of various thermostats and temperature controls.
b. Identify and describe how to troubleshoot thermistors.
c. Explain how to install and wire thermostats.
d. Explain how to troubleshoot the functions of a thermostat.

Performance Task 1

Wire, check the operation of, and adjust the heat anticipator/cycle rate of a thermostat.

Trade Terms

Cooling compensator: A fixed resistor installed in a thermostat to act as a cooling anticipator.

Deadband: A temperature band, usually 3°F, that separates heating and cooling in an automatic changeover thermostat.

Differential: The difference between the cut-in and cut-out points of a thermostat.

Droop: A mechanical condition caused by heat that affects the accuracy of a bimetal thermostat.

Invar®: An alloy of steel containing 36 percent nickel. It is one of the two metals in a bimetal device.

The room thermostat is the temperature-control device commonly used in residential and light commercial HVAC systems to maintain indoor comfort. It does this by cycling the cooling equipment on if the temperature rises in the conditioned space, or by cycling on the heating equipment if there is a temperature drop. The type of room thermostat chosen, the location where it is installed, and how it is adjusted are factors that can affect its ability to maintain space temperature at or near its setpoint.

2.1.0 Thermostat and Temperature Control Operation

Programmable electronic thermostats using electronic sensing elements are becoming increasingly popular; however, thermostats with bimetal sensing elements are still being manufactured and used in line-voltage thermostats. Although not quite as precise as the electronic thermostat, bimetal devices are considerably less expensive and have proven to be effective and reliable for simple applications. Most bimetal thermostats will maintain the space temperature within ±2° of the setpoint. A good electronic thermostat will maintain the temperature within ±1°.

A bimetal element (*Figure 18*) is composed of two different metals bonded together; one is usually copper or brass. The other, a special metal called **Invar®**, contains 36 percent nickel. When heated, the copper or brass has a more rapid expansion rate than the Invar®, and changes the shape of the element. The movement that occurs when the bimetal changes shape is used to open or close switch contacts in the thermostat.

While bimetal elements are constructed in various shapes, the spiral-wound element is the most compact in construction and the most widely used. In *Figure 18*, for example, a glass bulb containing mercury (a conductor) is attached to a coiled bimetal strip. When the bulb is tipped in one direction, the mercury makes an electrical connection between the contacts and the switch

03314-13_F18.EPS

Figure 18 Bimetal sensing elements.

---

is closed. When the bulb is tipped in the opposite direction, the mercury moves to the other end of the bulb and the switch is opened. Mercury-bulb thermostats are still in use but are being phased out. Digital non-programmable thermostats are replacing mercury-bulb types.

Thermostat switching actions should take place rapidly to prevent arcing, which could cause damage to the switch contacts. A magnet is used to provide rapid action to help eliminate the arcing potential in some bimetal thermostats. These are referred to as snap-action thermostats. Mercury is not used to complete the circuit.

WARNING!	Mercury is toxic. Even short-term exposure may result in damage to the lungs and central nervous system. Mercury is also an environmental hazard. Do not dispose of thermostats containing mercury bulbs in regular trash. Contact your local waste management or environmental authority for disposal/recycling instructions.

Most residential and small commercial thermostats are of the low-voltage type. With low-voltage control circuits, there is less risk of electrical shock and less chance of fire from short circuits. Low-voltage components are also less expensive and less likely to produce arcing, coil burnout, and contact failure. However, line voltage thermostats are also used, usually to control simple electric heating equipment.

A very few self-generating systems still exist that generate about 750mV from the pilot flame. This is just enough power to operate the gas valve; these systems have no fan or other significant electrical loads. The thermostats in millivolt systems are similar in construction and design to 24V thermostats; however, they are not interchangeable.

2.1.1 Heating-Only Thermostats

A wall-mounted heating-only thermostat typically contains a temperature-sensitive switch as shown in *Figure 19*. The thermostat shown here is the bimetal, snap-action type.

When the temperature in the conditioned space reaches the thermostat setpoint, the thermostatic switch will open. Because of residual heat in the heat exchangers, the continued operation of the fan, and the time it takes for the bimetal to respond to room conditions, the temperature can overshoot the setpoint. To avoid the discomfort that this condition might cause, the thermostat contains an adjustable heat anticipator that causes the thermostat to open before the temperature in

Figure 19 Heating-only thermostat.

Did You Know?
Millivolt Heating Systems

Millivolt heating systems were developed decades ago when natural gas was first brought to major cities. At that time, many homes were heated with coal using gravity-type duct systems. Those gravity furnaces did not require electrical power to operate since they did not use a blower to circulate the heated air. The millivolt system allowed utilities to retrofit existing coal furnace to burn natural gas. This kept costs to the customer low because the existing furnace could be used and electrical power did not have to be brought to the furnace. The millivolt system generated just enough power to open and close the gas valve. As late as the 1970s, these systems could still be encountered. No doubt a few of these relics still exist, but they are becoming extremely rare.

the space reaches the setpoint. The heat anticipator is a small resistor in series with the switch contacts that gets warm when the heating circuit is energized. The anticipator warms the bimetal strip, causing the contacts to open earlier than they would otherwise. Heat anticipators are typically adjustable, and are set based on the current draw of the heating control circuit flowing through the thermostat.

2.1.2 Cooling-Only Thermostats

The cooling thermostat (*Figure 20*) is the opposite of a heating thermostat. When the bimetal coil heats up and unwinds, the switch closes its contacts and starts the compressor. The cooling-only thermostat contacts close on a rise of temperature, while the heating-only thermostat contacts open on a rise.

Cooling thermostats contain a device called a **cooling compensator** to help improve indoor comfort. The cooling compensator (*Figure 21*) is a fixed resistance in parallel with the thermostatic switch. No current flows through the compensator when cooling is on because it has a much higher resistance than the switch contacts. In this case, the contacts are essentially a short circuit.

Figure 20 Cooling-only thermostat.

Figure 20 Cooling-only thermostat.

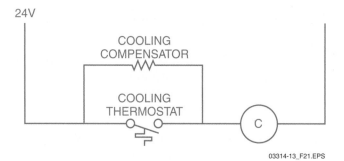

Figure 21 Cooling compensator.

When the thermostat circuit is open, however, a small current can flow through the compensator and the contactor coil.

Because of the size of the compensator, the current is not enough to energize the contactor. The heat created by the current flowing through the compensator makes the thermostatic switch contacts close sooner than they would without the compensator. In this way, the cooling compensator accounts for the lag between the call for cooling and the time when the system actually begins to cool the space.

2.1.3 Heating-Cooling Thermostats

When heating and cooling are combined for year-round comfort, it is impractical to use a separate thermostat for each mode. Therefore, the two are combined into one heating-cooling thermostat (*Figure 22*).

A switch provides a means to switch between the heating and cooling modes. The same switch (the Mode button in *Figure 22*) can also be set to OFF. The button in the center controls the function of the fan. In the Auto position, the indoor fan cycles on and off with the heating or cooling equipment. In the On position, the indoor fan runs continuously, regardless of whether the unit is in heating or cooling.

Figure 22 Heating-cooling thermostat.

This thermostat is also equipped with a Filter Reset button. A timer in the thermostat can be set for 30 to days. At the end of the selected time period, the thermostat will remind the user to change the filter. After it is changed, the user resets the reminder and the time interval starts over.

2.1.4 Automatic Changeover Thermostats

The disadvantage of a manual-changeover heating-cooling thermostat is that the building occupant must determine whether heating or cooling is needed at a given time and set the thermostat switch accordingly. In some climates, that is very impractical; the need can change several times a day.

The automatic-changeover thermostat automatically selects the mode, depending on the heating and cooling setpoints. Automatic changeover thermostats look very similar to other heat/cool thermostat. The one exception is the addition of another position on the function switch. In addition to Heat, Cool, and Off, an Auto position is available. This should not be confused with the Auto position on the fan function switch. The occupant can still select either heating or cooling. However, when the thermostat function switch is in the Auto position, the thermostat selects the operating mode. In electromechanical thermostats, all that is necessary is for one of the thermostatic switches to close, indicating that the conditioned space is too warm or too cold. With electronic thermostats, the temperatures at which heating or cooling is selected is determined by the occupant and entered into the control. The logic in the microprocessor then determines which operating mode is needed.

Older electromechanical thermostats contain a built-in mechanical **differential**, which is the difference between the cut-in and cut-out points of a thermostat. The differential is normally 2°F. For example, if the heating setpoint is 70°F, the furnace may turn on at 70°F and run until the temperature is 72°F. Automatic changeover thermostats also have a minimum interlock setting, commonly known as the **deadband**. The deadband is a built-in feature that prevents the heating and cooling setpoints from being any closer together than 3°F and fighting each other to control the space temperature. The deadband can be adjusted on many electronic models to suit unique applications. However, the 3°F deadband is a common default value and serves most applications well.

2.1.5 Two-Stage Thermostats

A two-stage indoor thermostat (*Figure 23*) is normally used to control equipment that has more than one stage of heating or cooling capacity. Examples of such equipment include heat pumps, furnaces with staged burners, and cooling equipment using multiple compressors or compressors with capacity control. For example, when a heat pump calls for heat, the first stage of the room thermostat controls the compressor, providing compression heat. If compression heat cannot satisfy the room thermostat, the second stage of the room thermostat, usually preset 1°F to 2°F off of the first stage, energizes supplementary electric heat to satisfy the room thermostat.

2.1.6 Programmable Thermostats

Programmable thermostats are self-contained controls with all features and functions located in the unit mounted on the wall. Early programmable thermostats looked like conventional thermostats and contained a motor-driven time clock that raised and lowered temperatures at desired intervals. This technology is now obsolete.

Modern electronic programmable thermostats (*Figure 24*) use microprocessors and integrated circuits to provide a wide variety of control and energy-saving features. Their control panels use touch-screen technology and their indicators are digital readouts rather than analog needle positioning. Different thermostats offer different features; the more sophisticated and expensive the thermostat, the more features it generally offers.

03314-13_F23.EPS

Figure 23 Heat pump thermostat.

Figure 24 Electronic programmable thermostat.

The following are some of the features available on electronic thermostats:

- *Better temperature control* – Temperatures are sensed electronically, allowing temperatures to be held within a much tighter range than possible with bimetal thermostats. Some versions have built-in humidity control.
- *Installation versatility* – Wireless versions of electronic thermostats are available for applications where it is difficult or impossible to pull thermostat cable. A built-in transmitter sends information to a remote receiver connected to the equipment being controlled. The model shown in *Figure 24* is capable of wireless communication.
- *Override control* – This feature allows the occupant to override the program when desired. For example, the night setback on Monday night could be overridden so the thermostat is not lowered before the football game is over.
- *Multiple programs* – This feature allows the occupant to design and select different schedules for different conditions. For example, a special schedule for a vacation away from home could be programmed.
- *Battery backup* – This feature prevents program loss in the event of a power failure.
- *Staggered startup for multi-unit systems* – This feature avoids excessive current drain. This is an important feature in office buildings, shopping malls, and hotels.
- *Maintenance tracking* – This feature indicates when maintenance is to be performed (for example, when to replace filters).

The potential savings available from programmable thermostats are significant. For example, it is estimated that a setback of 10°F for both day-

time and nighttime can result in a 20 percent energy savings. A 5°F setback will yield a 10 percent energy savings.

2.1.7 Line Voltage Thermostats

Most modern thermostats operate at low voltages. There are also thermostats that operate at line voltages (115 volts or higher). They are commonly used for controlling electric baseboard heaters.

Line voltage thermostats may be controlled by a bimetal sensing element or a hydraulic sensing bulb. The latter controls the thermostat by means of pressure. *Figure 25* shows a thermostat that is actuated by pressure on a bellows. The sensing bulb contains refrigerant which increases in pressure as the temperature increases. The increasing pressure acts to expand the bellows, thus pushing the switch toward the closed position. Note that bellows-type thermostats were also used in low-voltage applications as well. Bellows-type thermostats are now obsolete.

Line voltage thermostats often suffer from a mechanical condition known as thermal offset, commonly called **droop**. Droop causes the thermostat control point to drift away from the selected setpoint. Because it is caused by heat, droop most often occurs in line voltage thermostats, which are subject to heat from high current. It may also occur in 24V thermostats that use anticipators.

2.2.0 Thermistors

Thermistors are temperature-sensitive semiconductor devices. Their resistance varies in a predictable way with variations in temperature. This allows them to be used in a variety of HVAC control applications. Thermistors have either a positive or a negative coefficient of resistance. If the resistance increases as temperature rises, it has a

Figure 25 Remote bulb thermostat.

positive coefficient of resistance. If resistance increases as the temperature drops, it has a negative coefficient of resistance. Thermistors (*Figure 26*) come in different sizes and shapes to suit a variety of applications.

Thermistors are used to sense temperature changes. They are also used as motor protective devices. Some typical applications of thermistors are: electronic thermometers, room thermostats, duct sensors, electronic expansion valve sensors, heat pump temperature sensing, and selected control circuits.

Thermistors are used in temperature differential controls, such as the defrost control in some heat pumps. Two thermistors are used; one senses coil temperature and the other senses air temperature. Manufacturers often provide resistance charts based on temperature for checking thermistors used in their products (*Figure 27*).

Another use of thermistors is as a start-assist device for single-phase compressors (*Figure 28*). A ceramic thermistor with a steep-slope positive temperature coefficient (PTC) is wired in parallel with the compressor's run capacitor. This can increase the starting torque by 200 to 300 percent. Procedures for checking a compressor start-assist thermistor are covered later in this module. Of course, this type of thermistor is much larger than a thermistor used to simply sense temperature.

03314-13_F26.EPS

Figure 26 Thermistors.

03314-13_F28.EPS

Figure 28 Thermistor compressor start-assist.

THERMISTOR CURVE

03314-13_F27.EPS

Figure 27 Resistance versus temperature chart.

When a thermistor is used as a sensing device for the operation of a thermal-electric (TE) expansion valve, both the expansion valve and the thermistor can be checked as follows: With the leads of the voltmeter placed across the terminals of the electric valve (*Figure 29*) and the system operating at or near peak load, the voltmeter reading should be within the range of 15V to 20V. With the system operating at lower loads, the reading should remain between 8V and 14V. The size of the expansion valve in relation to the capacity of the system will be directly related to the time that the valve registers the voltage limits. Therefore, the manufacturer's recommendations should be consulted. The readings should be observed for two or three minutes.

Like all thermistors, those used on the internal windings for overload protection of three-phase motors can be checked by taking a resistance reading through them. An ohmmeter should be used and connected as illustrated in *Figure 30*.

A short or ground in the thermistor will be indicated by a reading of zero or a value approaching zero. An open would be indicated by a reading of infinity. The procedure is as follows:

Step 1 Shut off the disconnect switch and allow the equipment to cool.

Step 2 Connect the ohmmeter leads as indicated in *Figure 30* and record the resistance in ohms.

The values will change with a change in temperature, but at room temperature (75°F) the reading should be about 75 ohms. Again, the resistance of the winding temperature sensors will vary by

03314-13_F30.EPS

Figure 30 Overload thermistor test.

manufacturer and application, so unit specifications should be consulted before condemning or accepting the devices.

Thermistors will sometimes fail or provide incorrect information. The built-in diagnostic feature of the system in which thermistors are connected will often evaluate information coming from the thermistors and report a fault when it senses an invalid input. A thermistor is not likely to fail on its own. If a thermistor fault is reported, it could mean that the thermistor has been damaged, that the thermistor lead is damaged or loose, or that the thermistor is not making good thermal contact preventing it from providing an accurate indication. The resistance of a thermistor itself can be tested with an ohmmeter. The manu-

03314-13_F29.EPS

Figure 29 TE expansion valve testing.

facturer's literature will usually contain a graph or table that shows expected resistance readings for a range of temperatures. *Table 1* is an example of such a troubleshooting aid. All thermistors do not have the same characteristics, so it is important to acquire the right information for the system being serviced.

2.3.0 Thermostat Installation

Even the best, most sensitive thermostat cannot perform correctly if it is poorly installed. Selecting the proper location for the thermostat is the first step in any installation procedure.

2.3.1 Installation Guidelines

The thermostat should be installed in the conditioned space. It should be installed about five feet above the finished floor on a solid inside wall that is free from vibration caused by events like slamming doors. Vibrations can make thermostat contacts chatter. For proper operation, electromechanical thermostats must be mounted level on a wall that is plumb. Electronic thermometers can function properly if they are not level or plumb, but they should be mounted in that manner for appearance sake.

The following practices should be observed when installing a thermostat:

• The installer must be a trained technician.

• The manufacturer's instructions must be followed. Failure to comply could lead to damage or a hazardous condition.
• Make sure the thermostat is suitable for the application.
• Check for proper operation in all modes when the installation is complete.

CAUTION

Thermostats containing solid-state devices are sensitive to static electricity. Before handling the instrument, discharge any static electricity by touching a metal doorknob or any object that is grounded. Touch only the front cover when holding the device.

Think About It
Thermistors

Thermistors are of two types: positive temperature coefficient (PTC) thermistors and negative temperature coefficient (NTC) thermistors. The resistance of a PTC thermistor increases with an increase in temperature; while the resistance of a NTC thermistor decreases with an increase in temperature. Based on the definitions given for PTC and NTC thermistors, what type of thermistor is being represented by the data shown in *Table 1*?

Table 1 Thermistor Values

TEMPERATURE (°F)	RESISTANCE (Ohms)	TEMPERATURE (°F)	RESISTANCE (Ohms)	TEMPERATURE (°F)	RESISTANCE (Ohms)
-60	362,640	45	11,396	150	1,020
-55	297,140	50	9,950	155	929
-50	245,245	55	8,709	160	844
-45	202,841	60	7,642	165	768
-40	168,250	65	6,749	170	699
-35	139,960	70	5,944	175	640
-30	116,820	75	5,249	180	585
-25	98,420	80	4,644	185	535
-20	82,665	85	4,134	190	490
-15	69,685	90	3,671	195	449
-10	58,915	95	3,265	200	414
-5	50,284	100	2,913	205	380
0	42,765	105	2,600	210	350
5	36,475	110	2,336	215	323
10	31,216	115	2,092	220	299
15	26,786	120	1,879	225	276
20	23,164	125	1,689	230	255
25	19,978	130	1,527	235	236
30	17,276	135	1,377	240	219
35	14,980	140	1,244		
40	13,085	145	1,126		

03314-13_T01.EPS

Use care when unpacking the new thermostat since rough handling can damage the thermostat. Save all literature and hardware for later reference and use.

Locate the thermostat about five feet above the floor in an area with good air circulation at room temperature. Avoid locations that create the following conditions:

- Drafts or dead air spots
- Hot or cold air from ducts or diffusers
- Radiant heat from direct sunlight or hidden heat from appliances
- Concealed supply ducts and chimneys (to avoid radiant heat)
- Unheated areas behind the thermostat, such as an outside wall or garage

To replace an existing heating thermostat, first turn off furnace power to de-energize the control circuit. Before removing the existing thermostat, record the thermostat wire color and corresponding thermostat terminal designation. This helps prevent miswiring of the new thermostat. Install the new wall plate or subbase (*Figure 31*). The subbase provides terminals for the connection of the wiring. If the wall plate or subbase must be

mounted on an outlet box, use the proper adapter ring or cover plate. Ensure that the thermostat is level (*Figure 32*).

2.3.2 Thermostat Wiring

The thermostat is connected through multi-conductor thermostat wire to a terminal strip or junction box in the air conditioning unit. A standard coding method is used in most of the HVAC industry to designate popular wiring terminals and wire colors. *Figure 33* shows the color code and corresponding terminals of a heat/cool thermostat. This color code/terminal designation is fairly standard among thermostat and equipment manufacturers. However, it is not safe to assume that any existing system is wired according to that color code. There are additional wire colors and terminal designations used with more complex thermostats. For example, when there are multiple stages of heating or cooling, those terminals are designated with the appropriate letter plus a number, such as W_1 and W_2. Wiring terminals are often provided for the connection of external temperature sensors. In addition to the traditional terminals found on most room thermostats, many modern electronic thermostats also contain a C (Common) terminal. These modern thermostats require power for the display screen and other functions. By connecting the thermostat to R and C, 24VAC power is always available to the device.

Thermostat wiring should be done in accordance with national and local electrical codes. All wiring connections should be tight. To avoid damaging the conductor, always use a stripping tool designed to strip small-gauge wire. Color-coded wiring should always be used for easy reference should system troubleshooting ever be required. Thermostat wires with multiple conductors are readily available from

VERTICAL OUTLET BOX

HORIZONTAL OUTLET BOX

03314-13_F31.EPS

Figure 31 Thermostat mounting.

03314-13_F32.EPS

Figure 32 Thermostat leveling.

THERMOSTAT WIRING CODES

TERMINAL DESIGNATION	WIRE COLOR	FUNCTION
R	RED	POWER
G	GREEN	FAN CONTROL
Y	YELLOW	COOLING CONTROL Y1 = STAGE 1 Y2 = STAGE 2
W	WHITE	HEATING CONTROL W1 = STAGE 1 W2 = STAGE 2
O (B)	ORANGE (BROWN)	HEAT PUMP REVERSING VALVE CONTROL

03314-13_F33.EPS

Figure 33 Thermostat wiring.

any HVAC parts distributor. Normally, 18-gauge thermostat wire is adequate for most installations. However, long runs of thermostat wire can produce a voltage drop that might prevent equipment operation. If a run of thermostat wire is over 100 feet, use 16-gauge or heavier wire to reduce voltage drop. Splicing should be avoided whenever possible. When splicing is necessary, soldered splices are recommended. When wires are stapled to the structure, make sure that the staple does not go through the wire insulation. Seal the wire opening in the wall space behind the thermostat so it is not affected by drafts within the stud cavity.

2.3.3 Checking Current Draw

Knowing the current draw of the control circuit of a furnace is required to properly adjust the heat anticipator. Heat anticipators are common on electromechanical thermostats; electronic models use a different approach to eliminating swing and droop.

The current draw of devices in the heating control circuit may be printed on the device nameplate or directly on the body of the control (*Figure 34*). However, such devices do not always operate at the rated current, and other factors can, and do, influence the current draw of the complete heating control circuit. As a result, it is always best to measure the current draw in the system to ensure the initial setting of the heat anticipator is correct.

> **WARNING!**
> To prevent electrical shock or equipment damage, make sure the power is off before connecting the wiring for the current draw adjustment.

To test the current draw of the heating control circuit, shut off the power and connect a clamp-on ammeter of the appropriate range (0 to 2 amperes) around a wire connected between the R and W terminals of the thermostat (*Figure 35*). Since the current is so low, wrap 10 passes of wire around the jaws of the clamp-on ammeter to provide a more accurate reading. The actual reading is then divided by 10 to find the actual current draw of the circuit. For example, if the meter reading is 4.0 amps, dividing by 10 results in a circuit current of 0.40 amps. When making the current reading, operate the system for a minute or so first, take the reading, and then shut off the furnace.

2.3.4 Adjusting Heat Anticipators and Cycle Rates

The heat anticipator is adjustable and should be set at the actual amperage value measured in the control circuit, or at a value specified by the equipment manufacturer. Small variations from the required setting can be made to improve system performance. The pointer on an adjustable heat anticipator is set to match the current draw in the furnace's control circuit. For example, if the amperage draw of a control circuit is 0.4A, the anticipator should be set at 0.4A. The heat anticipator helps lengthen or shorten the burner on-time by

24V, 60Hz, 0.4A

03314-13_F34.EPS

Figure 34 Gas valve current draw.

24V

TO GAS,
OIL, AND
ELECTRICAL
UNITS

03314-13_F35.EPS

Figure 35 Amperage check.

artificially adding heat to the bimetal coil. More heat shortens the cycle while less heat lengthens the cycle.

Electronic programmable thermostats do not use a traditional heat anticipator to control the length of time the burner operates. Instead, they control cycle rate, which is defined as the number of times per hour that a burner operates. Different electronic thermostat manufacturers have different methods of setting the cycle rate. A common method involves manipulating the various function buttons while observing the display screen. The following is an example of how one thermostat manufacturer sets the cycle rate.

- Consult the furnace manufacturer's literature for the correct cycle rate.
- Press the Up and Down buttons simultaneously for three seconds until the function number and option number are displayed.
- Press the Up and Down buttons simultaneously for one second until the desired function

number is displayed. For this thermostat, function number 3 is the heating cycle rate.

- Once function 3 is displayed, press the Up or Down button to change the option number. Each option number represents a different cycle rate. This thermostat offers cycle rates from two to six per hour. Select the option recommended by the furnace manufacturer.
- Press the Up and Down buttons for three seconds to save the changes and exit the menu.

Like the traditional heat anticipator, adjustments can be made if the cycle rate does not satisfy the user. However, starting with the manufacturer's suggested setting, or even with the default setting of the thermostat manufacturer, often results in satisfactory operation.

2.3.5 Final Check

The final step is to confirm that the thermostat functions properly in all operating modes. With the furnace power on, place the thermostat in the Heat mode. Place the fan switch in the Auto position. Adjust the setpoint temperature to at least 5°F above the room temperature. The burner should come on followed by the blower after a short delay. Then adjust the setpoint temperature to 5°F below the room temperature. The main burner should shut off, but the blower will likely continue to run for one to two minutes. Next, set the thermostat to the cooling mode.

> **CAUTION**
>
> To avoid compressor damage, do not check cooling operation unless the outside temperature is at least 50°F and the crankcase heater has been on for at least 24 hours.

Set the thermostat to the Cool position, and the fan switch to Auto. Adjust the temperature setpoint to at least 5°F below the room temperature. The cooling equipment should come on either immediately, or after any start delay, if the unit is so equipped. The indoor fan should come on immediately. After the cooling system has started, adjust the setpoint to at least 5°F above the room temperature. The cooling equipment should turn off immediately or very soon. Depending on the equipment in use, the indoor fan may cycle off with the compressor, or it may run for an additional period of time for increased energy efficiency.

With the system switch now in the Off position, adjust the thermostat setpoint up and then down. The system should not respond for heating or cooling. Leave the mode selection switch set to Off. Then set the fan function switch to

On. The indoor fan should come on and continue to run until the switch is placed back in the Auto position.

2.3.6 Mercury Thermostat Disposal

Mercury is a toxic metal that was and still is used in thermostats. It is estimated that of all the known mercury contamination in the United States, about 8 percent of that can be traced to mercury thermostats that have been improperly disposed. Today, manufacturers are free to continue to make mercury thermostats as long as they have a program in place to collect and recycle old thermostats. As of 2013, several states have laws in place to deal with the disposal and recycling of mercury thermostats. These state laws have several things in common including:

- Mandatory recycling of all mercury thermostats
- Specific procedures to follow during disposal/recycling
- Designated entities to handle collection
- Designated methods of handling /shipping to prevent accidental spillage

Since state laws vary, always consult a local hazardous waste disposal authority for guidance. Across the nation, many HVAC wholesalers have voluntarily agreed to act as collection points for mercury thermostats. Contractors and the general public can drop off mercury thermostats at these locations, free of charge. Some states offer a cash bounty for each mercury thermostat turned in.

2.4.0 Thermostat Troubleshooting

Troubleshooting thermostats involves visual checks, electrical checks, and electronic troubleshooting for electronic thermostats. When troubleshooting and servicing electronic thermostats, be sure to avoid damaging their microprocessors and integrated circuits. As a first step, visually check the thermostat for problems that could affect its operation.

- Location is correct and is the control is free from vibration.
- Check that function switches operate freely.
- Mercury-bulb thermostats must be level.
- Make sure the mercury bulb or bimetal element moves freely.

If control circuit voltage (usually 24V) is inadequate, it can prevent normal system operation and lead to the erroneous conclusion that the thermostat is defective. Measure the control circuit voltage between the R and C terminals on the furnace or air handler to check that adequate control power is available. If no voltage is measured, troubleshoot the control transformer circuit. If the voltage is lower than about 21V, it will affect the operation of control relays and other control components. Check the control transformer circuit for the cause of low or absent voltage.

Various functions of the thermostat can be checked by removing the thermostat from the wall and using jumper wires to simulate the different thermostat functions. Use an insulated jumper wire(s) with alligator clips connected across the thermostat's R (24V) terminal and fan (G), heat (W), or cool (Y) terminal, as applicable. The jumpers simulate the switches found within a thermostat. If a thermostat is suspected to be the cause of a system not operating properly, the thermostat should be removed from its subbase and the jumpers placed across the subbase terminals. The jumper wire testing can also be conducted at the indoor unit low-voltage terminal board as well. However, the thermostat should first be removed from the subbase to avoid any possibility of damaging it. If the use of jumpers confirms that the thermostat is defective, replace the thermostat. Do not attempt to field-repair the device.

2.4.1 Fan Switch Operation Checks

If the thermostat cannot control indoor fan operation, connect the jumper wire across subbase terminals R and G (*Figure 36*). If the indoor fan runs with the jumper in place, this proves that the rest of the circuit is intact and functional. Therefore, the thermostat Fan switch or thermostat internal wiring is defective. If the fan does not run, troubleshooting should move to the indoor unit.

2.4.2 Cooling Operation Checks

If the thermostat cannot control cooling operation, connect jumper wires across subbase terminals R and Y and R and G (*Figure 37*). If the cooling unit and indoor fan run, the thermostat's cooling and/or fan circuits or related wiring is defective. If not, then troubleshooting should move to the indoor unit.

2.4.3 Heating Operation Checks

If the thermostat cannot control heating operation, connect a jumper wire across subbase terminals R and W (*Figure 38*). If the furnace runs, the thermostat's heating circuits or related wiring is defective. If the furnace or other heat source does not run, the problem is likely elsewhere.

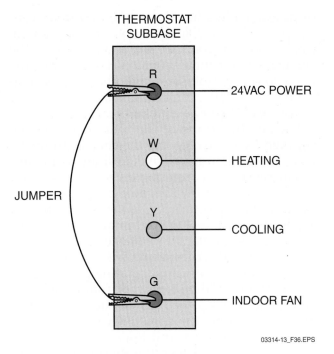

Figure 36 Thermostat fan switch function check.

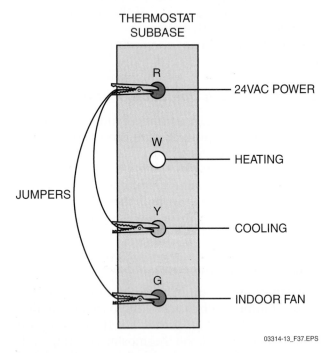

Figure 37 Thermostat cooling function check.

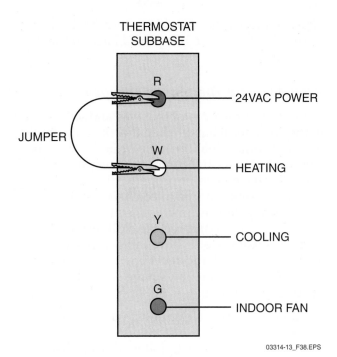

Figure 38 Thermostat heating function check.

2.4.4 *Electronic Thermostats*

Due to their low cost, easy availability, and high reliability, electronic programmable thermostats are now the thermostat of choice in both residential and light commercial applications. These thermostats use microprocessors and integrated circuits to provide a wide variety of control and energy-saving features. Their control panels often use touch-screen technology and their indicators are digital readouts. Different thermostats offer different features; the more sophisticated the thermostat, the more features it generally offers.

Troubleshooting electronic thermostats is similar to troubleshooting electromechanical thermostats. If the thermostat is suspected as causing problems, it can be removed and jumper wires used to simulate compressor operation. Since electronic thermostats require power to operate, a loss of power can affect their operation. Power checks should include the following:

- High-voltage power is available at the heating/ cooling equipment.

- Low-voltage (24V) power is available.
- Batteries (if used) are full strength.

Many electronic thermostats have a variety of programming options. This makes them highly flexible and allows them to be applied in a number of system types. When troubleshooting suspected thermostat problems, it is important to first check the thermostat programming. A minor error in the setup can result in erratic operation and mislead technicians to suspect problems elsewhere in the system. Acquire the manufacturer's literature for the thermostat in service and review the setup before troubleshooting begins.

Electronic thermostats are often self-diagnosing. For example a digital display screen that is dead or blank may indicate a loss of power. Often, error signals or other indications of a problem will be displayed on the screen. For example, the image of a battery on the screen often indicates a weak or dead battery. Other error codes identify faulty components such as an open or shorted temperature sensor. Because of their diversity and complexity, manufacturers of electronic programmable thermostats often provide aids in their service literature to use when troubleshooting their products. *Figure 39* is an example of a manufacturer's troubleshooting guide for an electronic programmable heat pump thermostat.

TROUBLESHOOTING

Reset Operation

If a voltage spike or static discharge blanks out the display or causes erratic thermostat operation, reset the thermostat by pressing the ⌃, ⌄ and PRGM buttons simultaneously. If the thermostat has power, has been reset and still does not function correctly contact your heating/cooling service person or place of purchase.

Symptom	Possible Cause	Corrective Action
No Heat/No Cool/No Fan (common problems)	1. Blown fuse or tripped circuit breaker. 2. Furnace power switch to Off. 3. Furnace blower compartment door or panel loose or not properly installed.	Replace fuse or reset breaker. Turn switch to On. Replace door panel in proper position to engage safety interlock or door switch.
No Heat	1. System Switch not set to Heat. 2. Loose connection to thermostat or system 3. Heating System requires service or thermostat requires replacement.	Set System Switch to Heat and raise setpoint above room temperature. Verify thermostat and system wires are securely attached. Diagnostic: Set System Switch to Heat and raise the setpoint above room temperature. Within five minutes the thermostat should make a soft click sound. This sound usually indicates the thermostat is operating properly. If the thermostat does not click, try the reset operation listed above. If the thermostat does not click after being reset, contact your heating and cooling service person or place of purchase for a replacement. If the thermostat clicks, contact the furnace manufacturer or a service person to verify the heating system is operating correctly.
No Cool	1. System Switch not set to Cool. 2. Loose connection to thermostat or system. 3. Cooling System requires service or thermostat requires replacement	Set System Switch to Cool and lower setpoint below room temperature. Verify thermostat and system wires are securely attached. Same procedures as diagnostic for No Heat condition except set the thermostat to Cool and lower the setpoint below the room temperature. There may be up to a five minute delay before the thermostat clicks in Cooling if the compressor lock-out option is selected in the configuration menu (Item 6).
Heat, Cool or Fan Runs Constantly	1. Possible short in wiring. 2. Possible short in thermostat. 3. Possible short in Heat/Cool/Fan system. 4. Fan Switch set to Fan On.	Check each wire connection to verify they are not shorted or touching together. No bare wire should stick out from under terminal screws. Try resetting the thermostat as described below. If the condition persists, the manufacturer of your system or service person can instruct you on how to test the Heat/Cool/ system for correct operation. If the system operates correctly, replace the thermostat.
Furnace Cycles Too Fast or Too Slow (narrow or wide temperature swing)	1. The location of the thermostat and/or the size of the Heating System may be influencing the cycle rate.	Item 4 in the Configuration Menu is the adjustment that controls the cycle rate. If an acceptable cycle rate is not achieved using the FA (Fast) or SL (Slow) adjustment, contact a local service person for additional suggestions.
Cooling Cycles Too Fast or Too Slow (narrow or wide temperature swing)	1. The location of the thermostat and/or size of the Cooling System may be influencing the cycle rate.	The cycle rate for cooling is fixed and cannot be adjusted. Contact a local service person for suggestions.
Thermostat Setting and Thermometer Disagree	1. Thermostat thermometer setting requires adjustment.	The thermometer can be adjusted +/-4 degrees as listed in item 8 of the Configuration Menu. No other adjustment is possible.
Clock Loses or Gains Time	1. Loss of power to thermostat.	The thermostat will maintain its program in memory even with no power but the clock time will be incorrect when power is restored. See No Heat/No Cool/No Fan (common problems) above for items to check in the system.
Heat or Cool Starts Early	1. EMR activated.	See Configuration Menu (Item 3).
Thermostat Does Not Follow Program	1. AM or PM set incorrectly in program. 2. AM or PM set incorrectly on the clock. 3. Voltage spike or static discharge. 4. See "Heat or Cool Starts Early" above.	Check current clock and program settings including the AM or PM designations for each time period. If a voltage spike or static discharge occurs use the Reset Operation listed above.
Blank Display and/or Keypad Not Responding	1. Voltage Spike or Static Discharge.	If a voltage spike or static discharge occurs use the Reset Operation listed above.
Thermostat has HP/SS switch and Configuration Menu has selection for HP or SS	1. Earlier version of thermostat model.	If switch is present, it must be in HP position for proper fan operation. If selection appears in Configuration Menu, it must be set for HP.

03314-13_F39.EPS

Figure 39 Electronic-programmable thermostat troubleshooting guide.

Additional Resources

Refrigeration and Air Conditioning: An Introduction to HVAC/R. Latest Edition. Larry Jeffus. Pearson/Prentice Hall. Upper Saddle River, NJ.

Fundamentals of HVAC/R. Latest Edition. Carter Stanfield and David Skaves. Prentice Hall. Upper Saddle River, NJ.

2.0.0 Section Review

1. The common deadband that prevents heating and cooling setpoints from being too close together is _____.
   a. 1°F
   b. 2°F
   c. 3°F
   d. 5°F

2. A thermistor with a negative coefficient of resistance will experience increased resistance as temperature drops.
   a. True
   b. False

3. Before handling an electronic programmable thermostat, always _____.
   a. read the instructions
   b. ground the thermostat
   c. install the batteries
   d. ground yourself

4. To ensure correct system and thermostat operation, the control circuit voltage should not be lower than _____.
   a. 21VAC
   b. 24VAC
   c. 200VAC
   d. 230VAC

SECTION THREE

3.0.0 HVAC SYSTEM OPERATING SEQUENCES

Objectives

Describe the sequence of operation for basic HVAC systems.

a. Describe the sequence of operation of a basic cooling-only system.
b. Describe the sequence of operation for a common heating and cooling system.
c. Describe the operation of basic pneumatic control systems.

Performance Task 2

Interpret control circuit diagrams.

Trade Terms

Actuator: The portion of a regulating valve that converts one type of energy, such as pneumatic pressure, into mechanical energy (for example, opening or closing a valve).

Bleed control: A valve with a tiny opening that permits a small amount of fluid to pass.

Electric-pneumatic (E-P) relay: A three-way pneumatic solenoid valve.

Pneumatic-electric (P-E) relay: A pressure switch in which a pneumatic signal causes an electrical change.

Control circuits can be relatively simple, yet they are sometimes difficult for the new technician to understand. The function of most control circuits is simply to control the start and stop of various loads. To understand a basic control circuit, knowing when the load should be energized and de-energized is important. This is called the sequence of operation.

The sequence of operation determines the types of devices used in the control circuits of each HVAC system. When sitting down to design a control circuit, the first thing an engineer must know is the required sequence of operation—for example, what load is supposed to be energized and when it should occur. Then a control circuit sequence can be designed.

3.1.0 Cooling-Only System Sequence

An automatic air conditioning circuit is simple in nature, and is a good place to begin an analysis of control circuits. *Figure 40* shows a very basic cooling system control circuit.

Power from the power plug is applied to the On-Off switch. When the switch is in the ON position, power is passed to the cooling thermostat (TC). When the thermostat calls for cooling and closes, power is passed to both the compressor and fan motor, causing both motors to and run. When the thermostat is satisfied and opens, the motors lose power and shut off. This describes the sequence of operation for this circuit.

Figure 41 shows a control circuit that is more typical of a basic comfort cooling system. Some additional features have been added to this circuit. A transformer has been added because 24VAC is required for the control circuit. This makes the control circuit components and wiring less expensive and easier to install. An indoor fan motor (IFM) has also been added to the circuit. Because the control circuit is now operating at 24VAC, the control relays (C and IFR) are added to provide control for the motors.

The following is the sequence of operation of the circuit shown in *Figure 41*.

- Line voltage is applied at terminals L1 and L2, making it present at the contactors (C), the indoor fan relay (IFR), and the primary of the system transformer at all times. With the transformer energized, the 24VAC output from its secondary is applied through one control cir-

03314-13_F40.EPS

Figure 40 Basic cooling system control circuit.

03314-13_F41.EPS

Figure 41 Typical cooling system control circuit.

cuit path to the On terminal of the Fan switch and by a second control circuit path to the thermostat On-Off switch.

- When the Fan switch is in the On position, power is applied through the first control circuit path to the coil of the indoor fan relay (IFR), energizing the relay. This causes the related IFR contacts to close, applying line voltage to the indoor fan motor (IFM) and causing it to run. This mode of operation allows the occupant to use the indoor fan for ventilation without operating the compressor for cooling. Some occupants will allow the fan to operate continuously even as the compressor cycles on and off to cool the structure.
- With the On-Off switch closed, the thermostat calling for cooling, and the Fan switch placed in the Auto position, both the contactor coil (C) and the indoor fan motor coil (IFR) are energized. This causes the compressor, outdoor fan, and indoor fan motors to be turned on simultaneously as required for the cooling mode of operation. Most users with a cooling-only system choose to allow the indoor fan to cycle on and off with the compressor.

Regardless of the complexity, the basic control arrangement shown in *Figure 41* is at the heart of the control circuits used with cooling equipment. Any additional circuits will represent special features used to improve equipment safety or operating efficiency.

3.2.0 Heating and Cooling System Sequence

While straight cooling systems do exist, most HVAC systems provide both heating and cooling. *Figure 42* is the diagram of a combined heating/cooling unit. The circuit is more complex, but at its heart is the simple control circuit previously described. The additional features are not that complicated. First of all, heating controls have been added near the bottom of the diagram, along with a heating-cooling thermostat. The cooling control has more extras, such as a compressor short-cycle protection circuit, a crankcase heater, and a current-sensitive overload device.

The heating/cooling unit shown in *Figure 42* is typical of a product found in both residential and light commercial applications. Some of the features of the heating/cooling unit include:

- A common room thermostat provides temperature control in both operating modes.
- A two-speed indoor fan that runs on high speed for cooling and low speed for heating. In heating, the indoor fan is controlled by the time-delay relay.
- Operation of the inducer fan must be proven before the gas valve is turned on. The inducer pressure switch (PS) closes when the induced-draft motor is running at the required speed. If the induced-draft fan stops, the pressure switch will open, disabling the gas supply. The induced-draft motor (IDM) is energized by the induced-draft relay (IDR) as soon as the thermostat closes.
- The heating section has two additional safety devices in series with the gas valve. A flame rollout switch (RS) will open if burner flames escape from the burner. This represents an unsafe condition. The limit switch (LS) is a heat-sensitive switch that disables burner operation if excessive heat builds up around the heat exchanger.

Use the circuit diagram in *Figure 42* for the following operating sequences.

3.2.1 Cooling Sequence of Operation

Assume that line-voltage and control circuit voltage is adequate and available to the unit. To start cooling operation in a heat/cool unit, set the room thermostat mode selector switch to COOL. Placing the mode selector switch in this position locks out all heating system functions. Set the fan function switch to AUTO so that the indoor fan cycles on and off with the compressor. As room temperature rises, thermostat contacts TC close, supplying 24VAC

Figure 42 Circuit diagram of a cooling/gas-heating system.

LEGEND

C	Compressor contactor	OL	Overload
COMP	Compressor motor	PRI	Primary
CH	Crankcase heater	PS	Pressure switch
GV	Gas valve	RS	Flame rollout switch
IDM	Induced-draft motor	SEC	Secondary
IDR	Induced-draft relay	TC	Cooling thermostat
IFM	Indoor fan motor	TDR	Time delay relay
IFR	Indoor fan relay	TH	Heating thermostat
LS	Limit switch	TRAN	Transformer

03314-13_F42.EPS

power to the coil of the indoor fan relay (IFR) and to the logic circuit of the short-cycle timer. The indoor fan relay energizes instantly, closing IFR contacts across terminals 2 and 4. This supplies line-voltage power to the indoor fan motor (IFM) and it starts and runs on high speed. At the same time, IFR contacts across terminals 4 and 5 open. This isolates the low-speed windings of the indoor fan motor that are used for the heating mode, preventing any possibility of the motor having both windings energized. Once the logic in the short-cycle timer times out (2 to 5 minutes), the contacts across terminals T1 and T2 close to energize the contactor (C). In the line-voltage side of the circuit, contactor contacts L1 – T1 and L2 – T2 close to supply power to the compressor (COMP) and outdoor fan motor (OFM). The outdoor fan motor(s) are usually tied directly to compressor operation, except when one or more motors are staged on by pressure in large units. The unit will continue to run until the room thermostat is satisfied. When TC contacts open, the contactor and indoor fan relay both de-energize, causing all motors in the system to shut off.

3.2.2 Heating Sequence of Operation

Assume that line-voltage and control circuit voltage is adequate and available to the unit. To start heating operation in a heat/cool unit, set the room thermostat mode selector switch to Heat. Placing the mode selector switch in this position locks out all cooling system functions. Set the fan function switch to Auto so that the indoor fan cycles on and off with the burners.

As room temperature falls, thermostat contacts TH close, supplying 24VAC power to the coil of the induced-draft relay (IDR). In the line-voltage side of the circuit, the normally-open contacts of the induced-draft relay close, applying power to the induced-draft motor (IDM).

Assume that no unsafe conditions exist and the flame rollout switch (RS) and limit switch (LS)

are both closed. When the induced-draft motor comes up to speed and there are no problems in the system vent, the pressure switch (PS) closes to provide power to the gas valve and pilot igniter. The gas valve feeds gas to the pilot assembly where it is ignited by a high-voltage spark. Once the pilot igniter module verifies that the pilot is sufficient to light the main burners, the gas valve opens and the burners light. At the same time, the gas valve is energized, the time-delay relay (TDR) is energized to begin timing. In the line-voltage side, TDR contacts are in series with normally closed contacts of the indoor fan relay (IFR) and in series with the indoor fan motor (IFM). The purpose of the time delay is to allow the air surrounding the heat exchanger time to warm up so that cold air is not blown into the structure. Once the delay times out, the indoor fan motor starts and runs on low speed. The normally-open contacts of the indoor fan relay (IFR) isolate the high-speed windings of the indoor fan motor. The unit will continue to run until the room thermostat is satisfied. When TH contacts open, the gas valve and induced-draft relay are de-energized, shutting off the burners. The time-delay relay is also de-energized. A time-delay occurs after the relay is de-energized. This allows the indoor fan motor to continue to run after the burner shuts off to recover residual heat left in the heat exchanger.

3.3.0 Pneumatic Controls

Pneumatic control systems use compressed air to supply energy for the operation of valves, motors, relays, and other pneumatic control components. They are used primarily in larger commercial air conditioning systems. Pneumatic control circuits consist of air piping, valves, orifices, and similar mechanical devices. Thermostats control the pressure in an air line; the air in the line can, in turn, operate the pneumatic motors, which in turn operate dampers, valves, and switches.

Think About It
Analyzing a Circuit Diagram

Use the diagram in *Figure 42* to determine the answers to these questions.

1 Does the induced-draft motor turn off if the rollout switch opens?
2 If you connect a voltmeter between L2 and terminal 4 of the IFR while the compressor contactor is de-energized, what voltage, if any, would the meter read?
3 Is the fan relay (IFR) energized or de-energized in the heating mode?

The following advantages are available through the use of pneumatic control systems, especially in commercial and industrial applications:

- Pneumatic equipment is well suited to modulating operation, but simpler two-position operation can also be provided.
- A great variety of control sequences and combinations are available while using relatively simple pieces of equipment.
- Pneumatic equipment is relatively free of operating difficulties as long as the air supply is kept clean and dry.
- Pneumatic systems limit explosion and shock hazards.
- Pneumatic systems can work to a high degree of precision.

3.3.1 Basic Components

Pneumatic control systems (*Figure 43*) consist of the following components:

- A source of clean, dry, compressed air
- Air lines called mains that deliver air from the supply to the controlling devices
- Controlling devices, such as thermostats, humidistats, relays, and switches or controllers
- Branch circuits or branch lines that deliver air from the controlling devices to the controlled devices

A. TO ACTUATOR
B. THERMOSTAT
C. TO OTHER PARTS OF SYSTEM
D. HIGH-PRESSURE GAUGE
E. COMPRESSOR UNIT
F. LOW-PRESSURE GAUGE
G. PRESSURE SWITCH
H. RELIEF VALVE
I. PRESSURE REGULATOR AND FILTER
J. DRAIN

03314-13_F43.EPS

Figure 43 Pneumatic system.

- Operators or **actuators**, such as valves and motors, which are the controlled components of the system

The air source is an electrically driven compressor that is connected to the storage tank in which the pressure is maintained between fixed limits (usually 20 to 30 psi). Air leaving the tank is conditioned to remove oil, water, and dust. In many installations, a small refrigeration unit is included to condense out any entrained moisture. Oil or moisture in the control air lines can disable the control system. Pressure-reducing valves control the air pressure to the main that feeds the controller (thermostat, etc.).

The controller regulates the positioning of the controlled device. It does this by taking air from the supply main at a constant pressure and delivering it through the branch line to the controlled component at a pressure that is varied according to the change in the measured condition.

For example, a change in the temperature of the conditioned space causes the thermostat to change the air pressure in the branch line. This change in pressure then causes the controlled component to move toward the open or closed position, depending on whether the room temperature has increased or decreased. The valve may be a chilled or hot water valve. As the air pressure to it changes, the position of the valve changes to regulate the flow of water as needed.

There are four types of controllers or thermostats:

- A direct-acting thermostat increases the pressure of its branch line when the air temperature increases in the conditioned space.
- A reverse-acting thermostat decreases the pressure of its branch line when the air temperature increases in the conditioned space.
- A gradual thermostat gradually changes its branch line pressure when the air temperature changes in the space. This type of thermostat can maintain any pressure from 0 psi to 15 psi. The vast majority of pneumatic wall thermostats are the gradual type.
- A positive-acting thermostat abruptly changes its branch line pressure when the room temperature changes. In this instance, the branch line pressure is either 0 psi or 15 psi. The thermostat does not maintain pressures between these values. It is a two-position (binary) control, and therefore is either fully open or fully closed. Humidity and pressure controllers may operate in a similar manner.

A bleed thermostat control (*Figure 44*) reacts to temperature and positions the vane near or away from a bleed port. Thus, the pressure in the branch is relative to how much air is allowed to bleed off. **Bleed controls**, if used directly, do not provide a wide range of control. Therefore, they are frequently coupled to a relay that is separately furnished with air for activating purposes; the bleed thermostat simply controls the relay action. Bleed controls maintain a constant drain on the compressed air source. In *Figure 44*, note the restrictor. This slows down the flow of air through the main to the bleed thermostat and the load. When the bleed port is unblocked, air pressure cannot build up downstream. When the bleed port is blocked, pressure slowly builds downstream of it and a control or load reacts.

Non-bleed thermostat controllers (*Figure 45*) use air only when the branch line pressure is being increased. The air pressure is regulated by thermostat valves. The valves eliminate the constant bleeding of air that is present in the bleed thermostat. The exhaust and air main valves are controlled by the action of the bellows resulting from changes in space temperature. The exhaust or bleeding action is relatively small and occurs only on a pressure increase. Assuming the example is a direct-acting thermostat, an increase in temperature will allow the thermostat to increase the pressure applied to the actuator or relay. As the temperature then decreases, the air main port will be closed and air in the branch line will be able to trickle out of the exhaust.

Relays are installed between a controller and a controlled device. They are used to perform a function that cannot be accomplished by the controller. A diverting relay, for example, can supply branch air in one position and exhaust it in another position. It can also supply air to either of two branches without exhausting the other branch, or it can shut off the branch air without exhausting the branch.

The controlled devices are mostly pneumatic damper actuators and valve operators (*Figure 46*).

The device moves according to changes in the branch line pressure. A pneumatic actuator or valve operator contains one of three mechanisms: a bellows (*Figure 47*); a diaphragm; or a cylinder and piston. The damper linkage or valve stem is connected to the bellows and when the branch line pressure decreases, the bellows expand due to the internal spring pressure. As the branch line pressure increases, the bellows contract as a result of the increase in air pressure. The expansion and contraction of the bellows, piston, or diaphragm moves the linkage so that the damper opens and closes.

A normally open damper is installed so that it moves toward the open position as the air pressure in the damper motor decreases. A normally closed damper is a damper that is installed so that it moves toward the closed position when the air pressure in the damper motor decreases. Both arrangements are shown in *Figure 48*.

As the branch line pressure changes, the movement of the bellows actuates the lever arm or valve stem. The spring exerts an opposing force so that a balanced, controlled position can be stabilized. The actuator arm can be linked to a variety of functions, but movement of a damper is likely the most common one.

03314-13_F45.EPS

Figure 45 Non-bleed thermostat.

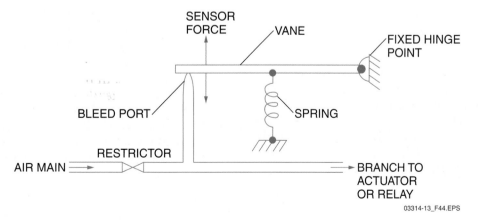

03314-13_F44.EPS

Figure 44 Bleed-type thermostat.

DAMPER ACTUATOR

VALVE OPERATOR

03314-13_F46.EPS

Figure 46 A pneumatic damper actuator and a valve operator.

03314-13_F47.EPS

Figure 47 Pneumatic actuator mechanism.

(A) NORMALLY OPEN

03314-13_F48A.EPS

Figure 48 Normally ~~closed~~ *open* damper.(1 of 2)

There will always be some crossover between the air devices and the electrical system. The devices most widely used for this process are pneumatic-electric (P-E) relays and electric-pneumatic (E-P) relays. P-E relays are simply pressure switches in which a pneumatic signal causes electrical contacts top open or close. P-Es are used to turn on electrical devices such as pumps and electric heaters using air pressure. All P-Es have switch differentials (the pressure change that causes the switch to make and break).

(B) NORMALLY CLOSED

03314-13_F48B.EPS

Figure 48 Normally closed damper.(2 of 2)

E-P relays are basically three-way solenoid valves. An air input is switched between one of two possible output ports. They are generally used as interlocking devices where a circuit is enabled under certain conditions that are signaled electrically. *Figure 49* shows a system using both and E-P and a P-E relay. When an E-P is de-energized, the common (input) and normally open ports are connected, and the normally closed port is blocked. When the E-P is energized, the common port is connected with the normally closed port, and the normally open port is blocked.

*MORE THAN ONE PRE-HEAT COIL IS REQUIRED WHEN DESIGN INLET TEMPERATURE IS BELOW –5°F

A. TWO POSITION DAMPER MOTOR
B. ELECTRIC-PNEUMATIC RELAY
C. MODULATING STEAM OR WATER VALVES
D. PNEUMATIC-ELECTRIC RELAY
E. MODULATING TEMPERATURE CONTROLLERS

03314-13_F49.EPS

Figure 49 Control system using E-P and P-E relays.

Additional Resources

Refrigeration and Air Conditioning, An Introduction to HVAC/R. Latest Edition. Larry Jeffus. Air Conditioning and Refrigeration Institute. Upper Saddle River, NJ: Prentice Hall.

3.0.0 Section Review

1. In most basic cooling-only systems, occupants often choose to allow the compressor and indoor fan to cycle on and off together.

   a. True
   b. False

2. Refer to *Figure 42*. What effect does a limit switch have on the operation of the induced-draft motor?

   a. It prevents its operation.
   b. It allows its operation after the switch resets.
   c. It has no effect.
   d. It allows its operation after the pressure switch resets.

3. A damper installed so that it moves toward the open position as air pressure in the damper motor decreases is called a _____.

   a. normally open damper
   b. pneumatic solenoid
   c. normally closed damper
   d. three-way solenoid

SECTION FOUR

4.0.0 TROUBLESHOOTING CIRCUITS AND LOAD COMPONENTS

Objective

Explain how to troubleshoot common control circuits and load components.

a. Identify basic safety practices related to troubleshooting HVAC electrical and electronic circuits.
b. Explain how to approach HVAC-related problems and prepare for troubleshooting.
c. Explain how to test high-voltage power sources.
d. Explain how to troubleshoot control circuits and low-voltage power sources.
e. Explain how to troubleshoot both resistive and inductive loads, including motors and their related devices.
f. Explain how to troubleshoot various hydronic control system components.

Performance Tasks 2 and 3

Interpret control circuit diagrams.

Perform electrical tests and/or troubleshooting procedures on the following:

- Single- and three-phase power sources
- Fuses and circuit breakers
- Resistive loads
- Relays and/or contactors
- Motor windings
- Start and run capacitors
- Start relays and thermistors

Trade Terms

Fault isolation diagram: A troubleshooting aid usually contained in the manufacturer's installation, startup, and service instructions for a particular product. Fault isolation diagrams are also called troubleshooting trees. Normally, fault isolation diagrams begin with a failure symptom then guide the technician through a logical decision-action process to isolate the cause of the failure.

Label diagram: A troubleshooting aid usually placed in a convenient location inside the equipment. It normally depicts a wiring diagram, a component arrangement diagram, a legend, and notes pertaining to the equipment.

Wiring diagram: A troubleshooting aid, sometimes called a schematic, that provides a picture of what the unit does electrically and shows the actual external and internal wiring of the unit.

Troubleshooting HVAC systems covers a wide range of electrical and mechanical problems. Obviously, not all problems fit easily into these two categories. For example, a loose or corroded wire may cause a compressor in a cooling system to cycle on and off intermittently. Although the problem is electrical, it creates a mechanical refrigeration problem.

Another example is a compressor that fails from shorted windings because of the acids formed during a previous burnout. This is an example of a mechanical refrigeration problem that looks like an electrical problem.

The methods used to troubleshoot electrical problems and components are very similar regardless of the type of HVAC equipment being serviced. This section describes those procedures that are common to troubleshooting electrical problems in most types of HVAC equipment.

4.1.0 Safety Practices

Practically all electric shocks are due to human error, rather than equipment failure. Nearly all deaths due to electric shock are due to the worker's failure to observe safety precautions, failure to repair equipment with electrical defects, or failure to remedy all defects found by tests and inspections. These same issues can also lead to damage to both electrical and electronic systems and components.

4.1.1 Electrical Safety Precautions

The following is a list of recommended safety practices:

- Always turn off power at the main disconnect before working on an electrical circuit or device. Lock and tag the power switch.
- Use a voltmeter to verify that the power to the unit is actually off. Remember that even though the power may be switched off, there is still potential at the input side of the shutoff switch.
- Never switch equipment on or off while standing in or touching a wet surface or area.
- Notify the local utility whenever a power line is touching the ground.

- If it is necessary to work on live electrical wiring, try to use one hand only. If a shock is received while using only one hand, current will probably flow through the hand and arm, then down the side and through the feet to the ground. If a shock is conducted through both hands and arms, the electrical path would be through the heart and lungs and could be fatal.
- Use protective equipment such as rubber gloves and insulated boots.
- Use tools with dielectric insulation.
- Remove metal jewelry such as rings and watches.

The *National Electrical Code®* (*NEC®*), when used together with the electrical code for your local area, provides the minimum requirements for the installation of electrical systems. Always use the latest edition of the code as your on-the-job reference. It specifies the minimum provisions necessary for protecting people and property from electrical hazards. In some areas, different editions of the code may be in use, so be sure to use the edition specified by your employer.

4.1.2 Precautions When Working with Solid State Controls

Solid-state equipment and/or components can be damaged if improperly handled or installed.

> **WARNING!**
>
> Solid-state equipment may contain devices that have potentially hazardous leakage current in the OFF state. Power should be turned off and the device disconnected from the power source before working on the circuit or load.

> **CAUTION**
>
> Some solid-state devices can be damaged by electrostatic charges. These devices should be handled in the manner specified by the manufacturer.

The following guidelines should be followed when handling, installing, or otherwise working with solid-state equipment or components:

- Microprocessors and other integrated circuit chips are very sensitive to, and can be damaged by, static electricity from sources such as lightning or people. When handling circuit boards, avoid touching the components, printed circuit, and connector pins. Always ground yourself before touching a board. Disposable grounding wrist straps are sometimes supplied with boards containing electrostatic-sensitive components. If a wrist strap is not supplied with the equipment, a wrist strap grounding system should be used. It allows handling of static-sensitive components without fear of electrostatic discharge. Store unused boards inside their special metallized shielded storage bags or a conductive tote box (*Figure 50*).
- Solid-state devices can be damaged by the application of reverse polarity or incorrect phase sequence. Input power and control signals must be applied with the polarity and phase sequence specified by the manufacturer.
- The main source of heat in many solid-state systems is the energy dissipated in the output power devices. Ensure that the device is operated within the maximum and minimum ambient operating temperatures specified by the manufacturer. Follow the manufacturer's recommendations pertaining to the selection of enclosures and ventilation requirements. For existing equipment, make sure that ventilation passages are kept clean and open.
- Moisture, corrosive gases, dust, and other contaminants can all have adverse effects on a system that is not adequately protected against atmospheric contaminants. If these contaminants are allowed to collect on printed circuit boards, bridging between the conductors may result in a malfunction of the circuit. This could lead to noisy, erratic control operation or a permanent malfunction. A thick coating of dust could also prevent adequate cooling of the board or heat sink, causing a malfunction.
- Excessive shock or vibration may cause damage to solid-state equipment. Follow all manufacturers' instructions for mounting.
- Electrical noise can affect the operation of solid-state controls. Sources of electrical noise include machines with large, fast-changing voltages or currents when they are energized or de-energized, such as motor starters, welding equip-

METALLIZED
SHIELDING BAG

CONDUCTIVE
TOTE BOX

03314-13_F50.EPS

Figure 50 ESD protection bag and tote box.

ment, SCR-type adjustable speed devices, and other inductive devices. The following are basic steps to help minimize electrical noise:

– Maintain sufficient physical separation between electrical noise sources and sensitive equipment to ensure that the noise will not cause malfunctions or unintended actuation of the control.
– Maintain a physical separation between sensitive signal wires and electrical power and control conductors. This separation can be accomplished by using conduit, wiring trays, or methods recommended by the manufacturer.
– Use twisted-pair wiring in critical signal circuits and noise producing circuits to minimize magnetic interference.
– Use shielded wire to reduce the magnitude of the noise coupled into the low-level signal circuit by electrostatic or magnetic coupling.
– Follow provisions of the *NEC*® with respect to grounding. Additional grounding precautions may be required to minimize electrical noise. These precautions generally deal with ground loop currents arising from multiple ground paths. The manufacturer's grounding recommendations must also be followed.

4.1.3 OSHA Lockout/Tagout Rule

In addition to many other safety rules, OSHA released *29 CFR 1926.417, Lockout and Tagging of Circuits*, many years ago. This rule covers the specific procedure to be followed for the "servicing and maintenance of machines and equipment in which the unexpected energization or startup of the machines or equipment, or releases of stored energy, could cause injury to employees. This standard establishes minimum performance requirements for the control of such hazardous energy."

The purpose of the OSHA procedure is to make sure that machinery is isolated from all potentially hazardous energy, and tagged and locked

out before employees perform any servicing or maintenance activities where the unexpected energization, startup, or release of stored energy could cause injury.

> **WARNING!**
>
> The OSHA procedure provides only the minimum requirements for the lockout/tagout procedure. Employees should consult the lockout/tagout procedures in their local area as well as their company's own policies and procedures. Remember, lives depend on this procedure. It is critical that the correct procedures for a job site be used.

4.1.4 Lockout/Tagout Procedure

To prepare for a lockout/tagout, make a survey to locate and identify all isolating devices (*Figure 51*). Be certain which switch(es), valve(s), or other energy-isolating devices apply to the equipment to be locked and tagged. Remember that more than one energy source may be involved.

The following procedure outlines the general steps for lockout/tagout (note that this procedure is provided only as an example). Each employer will designate who is qualified to conduct the procedure. Not every employee has the proper training to do so.

Step 1   Notify all affected employees of the lockout/tagout and why it is necessary. The authorized employee will know the type of energy that the machine or equipment uses and the associated hazards.

Step 2   If the machine or equipment is running, shut it down using the normal procedure.

> ## Ambient Temperature Surrounding Solid-State Devices
>
> Do not overlook other sources of heat in or around an enclosure containing solid-state devices that might raise the ambient temperature to undesirable levels. Look for and avoid heat sources such as power supplies, transformers, radiated heat, adjacent furnaces, and sunlight.

03314-13_F51.EPS

Figure 51 Condensing unit with wall-mounted disconnect switch.

For example, press the stop button or open the toggle switch.

Step 3 Operate the switch, valve, or other energy-isolating device(s) so that the equipment is isolated from its energy source(s). Stored energy must be dissipated or restrained by repositioning, blocking, or bleeding down. Examples of stored energy are springs, elevated machine members, rotating flywheels, hydraulic systems, and air, gas, steam, or water pressure.

Step 4 Lock and tag the energy isolating devices with assigned individual lock(s) and tag(s).

Step 5 After making sure that no personnel are exposed, operate the start button or other normal operating controls to make certain the equipment will not operate. This will confirm that the energy sources have been disconnected. The equipment is now locked and tagged.

All equipment must be locked and tagged to protect against accidental operation when such operation could cause injury to personnel. Never try to operate any switch, valve, or energy-isolating device when it is locked and tagged.

> **CAUTION**
> Return operating controls to their neutral or Off position after testing.

4.1.5 *Restoring Machines or Equipment*

After service and/or maintenance is complete, the equipment is ready for normal operation. Use the following procedure to restore the machines or equipment to their normal operating condition.

Step 1 Check the area around the machines or equipment to make sure that all personnel are at a safe distance.

Step 2 Remove all tools from the machine or equipment and reinstall any guards.

Step 3 Again making sure that all personnel are in the clear, remove all lockout/tagout devices.

Step 4 Operate the energy isolating devices to restore energy to the machine or equipment.

4.2.0 Using an Organized Approach to Electrical Troubleshooting

Troubleshooting can be defined as a procedure through which a technician locates the source of a problem. Once the problem is identified, the technician then makes the repairs and/or adjustments to correct the cause of a problem so that it will not recur.

Troubleshooting can be divided into the five basic elements listed here. The vast majority of problems can usually be found quickly if a systematic approach is used. This includes the following:

- Customer interviews
- Physical examination of the system
- Basic system analysis
- Use of manufacturer's troubleshooting aids
- Fault isolation in equipment problem area

4.2.1 *Customer Interviews*

The troubleshooting procedure should begin with the technician learning all that can be learned about the customer's complaint by talking to the customer and the service dispatcher. Talking with the customer prior to working on the equipment is always recommended because it can provide valuable information on equipment operation that can aid in the troubleshooting process. It can also identify the source of a problem that is not related to the HVAC equipment, thereby eliminating unnecessary equipment maintenance.

The first evidence of trouble with the HVAC system is often a complaint from an individual who is too cold, too hot, or may be bothered by drafts. In many cases, the problem behind such a complaint may not be an equipment or control system malfunction, but a personal comfort problem. If the system appears to be operating correctly but individuals are complaining about comfort problems, the technician should check for one or more of the following conditions before assuming the HVAC equipment is malfunctioning:

- *Air distribution and circulation problems* – Persons outside of the immediate area that is controlled by a thermostat may feel too hot or cold. The thermostat senses only the temperature at its location. Temperature levels in all the other areas controlled by the same thermostat are subject to variation that can be caused by poor air distribution and/or room air circulation.
- *False heat loads* – Direct sunlight on the thermostat or heat from lamps, appliances, and pipes can cause overcooling of a zone. Direct

sunlight or artificial sources of radiant heat can also cause heating discomfort.

- *Covered grilles and diffusers* – Occupants frequently cover part or all of a discharge grille, causing improper heating or cooling.
- *Occupant locations* – Occupants located adjacent to outside walls or windows may be subject to air infiltration or radiant cooling or heating from a wall.
- *Insufficient conditioned air supply* – May be caused by poor air distribution design or fan speed, dirty filters, or lack of the proper amount or size of return air outlets.
- *Overcrowding* – Overheating will result if more people or mechanical equipment occupy a conditioned area than the space was designed to hold.
- *System size* – Extreme weather conditions may exceed the capacity of the heating or cooling equipment.
- *Drafts* – Forced-air systems rely on the movement of air to deliver the desired conditioning. To many people, even a slight air motion may be uncomfortable. This problem may be alleviated by moving the occupant's work station or redirecting the airflow.
- *Stale air* – A stuffy or smoky atmosphere usually results from an insufficient fresh air supply, air that is too humid, or air that has inadequate exhaust. Stale air can also be caused by overcrowding or low air circulation.

4.2.2 Physical Examination of the System

Many problems can be identified by simply using senses such as sight, touch, and smell to check the system. This can be done with power off or power on.

- Look for evidence of leaks and physical damage. Oil stains on coils or tubing may indicate a refrigerant leak.
- Look for dirt accumulation on filters and coils that can affect system operation.
- Listen for unusual sounds or feel for vibrations that could indicate a malfunction.
- Check for odors, especially the smell of overheating or gas.

4.2.3 Basic System Analysis

The proper diagnosis of a problem requires knowledge of proper system operation. If a unit is unfamiliar, study the manufacturer's service literature to become familiar with the equipment's modes and sequence of operation.

The second part of the diagnosis is to find out what the unit is actually doing, and what symptoms are exhibited by the improperly operating unit. Finding out what the unit is doing is accomplished both by carefully listening to the customer's complaints and by analyzing the operation of the unit. As applicable, this means making electrical, temperature, pressure, and/or airflow measurements at key points in the system. The set of measured values can then be compared with a set of typical readings for a properly operating system as previously recorded on system operating logs or in the manufacturer's service literature. This process can often quickly pinpoint the system problem.

4.2.4 Use of Manufacturers' Troubleshooting Aids

To aid in the isolation of faults, many manufacturers provide troubleshooting information marked on the equipment or contained in the service instructions for a particular product. This information typically includes the following:

- **Label diagrams**
- Troubleshooting tables
- **Fault isolation diagrams**
- Diagnostic equipment and tests

Label diagrams (*Figure 52*) are usually placed in a convenient location inside the equipment, typically on the inside of a control circuit access panel. They normally show a component arrangement diagram, wiring diagram, legend, and notes pertaining to the equipment.

The component arrangement diagram shows where the components are physically located in the unit. It is useful because it helps locate and identify the components shown on the wiring diagram. The legend identifies the meanings of the symbols and abbreviations used on the label diagram.

The wiring diagram, sometimes called a schematic, provides a picture of what the unit does electrically and shows the actual external and internal wiring of the unit. Many label diagrams also contain a ladder diagram (*Figure 53*). Ladder diagrams do not usually show wire color and physical connection information. This makes it more useful by focusing on the functional, not the physical, aspects of the equipment. Wiring and ladder diagrams are the primary troubleshooting aids for isolating electrical problems.

Troubleshooting tables and/or fault isolation diagrams are usually contained in the manufacturer's installation, startup, and service instruc-

WIRING DIAGRAM

COMPONENT ARRANGEMENT DIAGRAM

NOTES

LEGEND

03314-13_F52.EPS

Figure 52 Typical label diagram.

03314-13_F53.EPS

Figure 53 Ladder diagram.

tions for a particular product. As shown in *Figure 54*, troubleshooting tables are intended to guide technicians to a corrective action based on observations of system operation. Fault isolation diagrams (*Figure 55*), also called troubleshooting trees, normally start with a failure symptom observation and take technicians through a logical decision-action process to isolate the failure.

Many manufacturers incorporate electronically controlled or semi-automatic testing features in their equipment to help isolate malfunctions. Depending on the equipment, these built-in diagnostic devices can be simple or complex. Some units contain microprocessor controllers that can run a complete check of all system functions, and then report back the results by means of a system numeric display or flashing LED display. Normally, these built-in test functions isolate a fault to a functional problem area. For example, if a test indicates a compressor failure, this means that the failure has been isolated to the compressor and its related control circuits and wiring. The technician must perform additional troubleshooting within the problem area to find out exactly where the fault is located.

In addition to built-in diagnostic equipment, many manufacturers have developed a series of stand-alone electronic module testers that are available to troubleshoot the different electronic control modules commonly used in the manufacturer's product line. The module tester is usually plugged into the control module in the equipment under test.

Troubleshooting involves the testing of each module control circuit using a sequential troubleshooting process that is performed per the manufacturer's instructions provided with the module tester.

Malfunction	Probable Cause	Corrective Action
Compressor motor and condenser motor will not start, but fan/coil unit (blower motor) operates normally	Check the thermostat system switch to ascertain that it is set to Cool.	Make necessary adjustments to settings.
	Check the thermostat to make sure that it is set below room temperature.	Make necessary adjustments.
	Check the thermostat to see if it is level. Most thermostats must be mounted level; any deviation will ruin their calibration.	Remove cover plate, place a spirit level on top of the thermostat base, loosen the mounting screws, and adjust the base until it is level; then tighten the mounting screws.
	Check all low-voltage connections for tightness.	Tighten.
	Make a low-voltage check with a voltmeter on the condensate float switch; the condensate may not be draining.	The float switch is normally found in the fan/coil unit. Repair or replace.
	Low air flow could be causing the trouble, so check the air filters.	Clean or replace.
	Make a low-voltage check of the antifrost control.	Replace if defective.
	Check all duct connections to the fan-coil unit.	Repair if necessary.
Compressor, condenser, and fan/coil unit motors will not start		Adjust as necessary.

03314-13_F54.EPS

Figure 54 Typical troubleshooting table.

Figure 55 Typical fault isolation diagram.

4.2.5 Fault Isolation in the Equipment Problem Area

Once troubleshooting aids have isolated a problem to a functional equipment area, it may be necessary to make additional measurements and use a step-by-step process of elimination to isolate the specific cause of the problem. This is usually the case for the more difficult problems encountered.

4.3.0 Testing High-Voltage Power Sources

Input power distribution circuits serve as the power source for the entire unit. They operate at either single-phase or three-phase line voltages, and act to distribute the input power to the various loads in the unit. Power circuits usually consist of the field-installed power wiring from

---

Think About It
Wiring Diagram Exercise

Using the schematic diagram of the packaged heat pump, answer the following questions:

1. Where would voltmeter leads be placed to check for the correct incoming voltage to the unit?

2. Placing voltmeter leads across terminals R and X of the low-voltage terminal strip yields a reading of 24V. What does this indicate?

3. During defrost, the defrost relay (DR) is energized. During a normal defrost, voltmeter leads are placed across terminal 2 of the defrost relay and terminal X on the low-voltage terminal strip. What should the voltmeter read?

4. Assume a room thermostat is connected to the low-voltage terminal strip. Across which terminals on the low-voltage terminal strip would voltmeter leads have to be placed to confirm that a call for second-stage heat was coming from the room thermostat? What voltage should be measured?

5. During defrost, the defrost relay (DR) is energized. Knowing this, what happens to the outdoor fan motor during defrost?

6. It must be known if the demand defrost control is receiving the correct input power. Across which terminals of the demand defrost control should power be checked? What is the expected voltage reading?

7. The compressor in this unit occasionally won't start and trips the power supply circuit breaker. Compressor current draw during startup must be checked. What test instrument is required and where in the circuit would the instrument be attached or applied?

03314-13_SA02.EPS

8. HC #1 and HC #2 relays control the operation of electric resistance heating elements. How do those relay coils receive power during a defrost cycle? Are both energized?

9. If the compressor fails in this heat pump, it is still possible to heat the home. How does the occupant make this happen? At what terminals on the low-voltage terminal strip would you expect to find 24VAC that proves the thermostat is calling for the electric heat to operate?

10. Since this unit energizes the defrost relay during heat pump defrost operation, what does this fact tell about the state of the reversing valve solenoid (RVS) during cooling operation? Is the RVS energized or de-energized during the cooling mode?

---

the main electrical service to a disconnect switch located near the unit, and from the disconnect switch to the unit. The input power and distribution circuits include protective devices such as fuses and/or circuit breakers.

All HVAC equipment is designed to operate within a specific range of system voltages including a safety factor, typically ±10 percent. This safety factor is added to compensate for temporary supply voltage fluctuations that might occur. Continuous operation of HVAC equipment outside the intended voltage range can damage the equipment. When the contactor closes in a cooling unit and applies power to loads such as the compressor and outdoor fan motor, the voltage level may drop about 3 percent from the measured open circuit voltage (contactor open). This is acceptable as long as the voltage does not drop below the manufacturer's stated minimum voltage.

4.3.1 Effects of High and Low Voltage

Too high or too low an operating voltage can cause overheating and possible failure of motors and other devices. The power supply voltage applied to the equipment should be measured and checked against the supply voltage indicated on the unit nameplate (*Figure 56*).

If low voltage exists at the equipment, the voltage should be measured at the electrical service entrance to make sure that a voltage drop does not exist in the branch circuit or feeder that supplies the HVAC equipment. If a voltage drop exists, it may be necessary to install larger wires between the service and the equipment.

Operating voltages applied to motors in the equipment must be maintained within limits from the voltage value given on the motor nameplate. The voltage tolerances used for most HVAC motors are as follows:

- *Single-voltage rated motors* – The input supply voltage should be within ±10 percent of the nameplate voltage. For example, a motor with a nameplate single voltage rating of 230V should have an input voltage that ranges between 207V (–10 percent of 230V) and 253V (+10 percent of 230V).
- *Dual-voltage rated motors* – The input supply voltage should be within ±10 percent of the nameplate voltage. For example, a motor with a nameplate dual voltage rating of 208V/230V should have an input voltage that ranges between 187V (–10 percent of 208V) and 253V (+10 percent of 230V).

4.3.2 Voltage Phase Imbalance

Voltage imbalance becomes very important when working with three-phase equipment. A small imbalance in phase-to-phase voltage can result in a much greater current imbalance. This current imbalance increases the heat generated in the motor windings. Both current and heat can cause nuisance overload trips and may cause motor failure. For this reason, the voltage imbalance between any two legs of the supply voltage applied to a three-phase system or motor should not exceed 2 percent. If a voltage imbalance of more than 2 percent exists at the input to the HVAC equipment, correct the problem in the building or utility power distribution system before op-

Figure 56 Single-phase input voltage checks.

erating the equipment. *Figure 57* shows how the amount of voltage imbalance is determined in a three-phase system using the following formula:

Percent imbalance =
maximum deviation from average
÷ average voltage × 100

The current imbalance in any one leg of a three-phase system should not exceed 10 percent. A current imbalance may occur without a voltage imbalance. This can occur when an electrical terminal, contact, etc. becomes loose or corroded, causing a high resistance in the leg. Since current follows the path of least resistance, the current in the other two legs will increase, causing more heat to be generated in the devices supplied by those legs. The current imbalance in a three-phase system is determined in the same way as voltage imbalance, but average current is substituted for average voltage.

4.3.3 Fuse/Circuit Breaker Checks

Fuses and/or circuit breakers are normally the first components checked when a unit is totally inoperative. The best way to test a fuse is by measuring continuity (*Figure 58*). To check fuses, always open the unit disconnect switch, then remove the fuses using an insulated fuse puller. Test the fuses for continuity using an analog or digital multimeter (VOM/DMM). If a short (zero ohms) exists across the fuse, it is usually good. If an open (infinite resistance) exists across the fuse, it is defective. A blown fuse is usually caused by some abnormal overload condition, such as a short circuit within the equipment or an overloaded motor. Replacing a blown fuse without

1)

PHASE	MEASURED READING
L1	215V
L2	221V
L3	224V

2) $\text{AVERAGE VOLTAGE} = \dfrac{215 + 221 + 224}{3} = 220V$

3) INDIVIDUAL PHASE IMBALANCE FROM AVERAGE
   L1 TO L2 = 220 − 215 = 5V
   L2 TO L3 = 221 − 220 = 1V
   L3 TO L1 = 224 − 220 = 4V

4) 5V = MAXIMUM IMBALANCE

5) % IMBALANCE = $\dfrac{\text{MAXIMUM IMBALANCE}}{\text{AVERAGE VOLTAGE}} \times 100$

   % IMBALANCE = $\dfrac{5V}{220V} \times 100 = 2.27\%$ (OUT OF BALANCE)

MAXIMUM VOLTAGE IMBALANCE BETWEEN ANY TWO LEGS MUST NOT EXCEED 2%.

(A) CALCULATING VOLTAGE IMBALANCE

1)

PHASE	MEASURED READING
L1	25A
L2	27A
L3	26A

2) $\text{AVERAGE CURRENT} = \dfrac{25 + 27 + 26}{3} = 26A$

3) INDIVIDUAL PHASE IMBALANCE FROM AVERAGE
   L1 TO L2 = 25 − 27 = 2A
   L2 TO L3 = 27 − 26 = 1A
   L3 TO L1 = 26 − 25 = 1A

4) 2A = MAXIMUM IMBALANCE

5) % IMBALANCE = $\dfrac{\text{MAXIMUM IMBALANCE}}{\text{AVERAGE CURRENT}} \times 100$

   % IMBALANCE = $\dfrac{2A}{26A} \times 100 = 7.7\%$ (IN BALANCE)

MAXIMUM CURRENT IMBALANCE BETWEEN ANY TWO LEGS MUST NOT EXCEED 10%.

(B) CALCULATING CURRENT IMBALANCE

03314-13_F57.EPS

Figure 57 Three-phase voltage and current checks.

USE INSULATED
FUSE PULLER TO
REMOVE FUSE

230
VOLTS

1-PHASE

L2

L1

UNIT

ZERO READING = GOOD FUSE

MEASURABLE OR INFINITE RESISTANCE
READING = OPEN FUSE

**(A) CHECKING BY CONTINUITY
(DE-ENERGIZED)**

230 VOLT
1-PHASE

**(B) CHECKING BY VOLTAGE
(ENERGIZED)**

03314-13_F58.EPS

Figure 58 Fuse checks.

locating and correcting the cause can result in damage to the equipment.

Fuses can also be tested with the circuit energized. This is accomplished by shutting off power to the unit, then disconnecting the wires from the load side of each fuse. This eliminates the possibility of current being fed back to the meter through a short circuit within the unit. Set the multimeter to measure AC voltage on a range that is higher than the highest voltage expected. Turn on the power and place one of the multimeter test leads on the line side of a fuse. Touch the other test lead to the load side of another fuse. If voltage is measured on the load side of a fuse, the fuse is good; if not, the fuse is defective. Repeat this procedure so that all fuses are measured with one test lead on the load side and the other test lead on the input side of a different fuse. This method tests one fuse at a time. If the measurement is performed with

both test leads on the load side of the fuses, and the multimeter shows no reading, it indicates that a fuse is blown, but not which one.

Before testing a circuit breaker (*Figure 59*), set the circuit breaker to OFF at the distribution panel. If required, remove the panel that covers the circuit breaker to expose the body of the breaker and the wires connected to its terminals. Set the multimeter to measure AC voltage on a range that is higher than the highest voltage expected. Measure the voltage applied to the circuit breaker input terminals:

- A to neutral or ground (single-pole breaker)
- A to B (two-pole breaker)
- A to B, B to C, and C to A (three-pole breaker)

2-POLE

3-POLE

A   B

A   B   C

A1   B1

A1   B1   C1

MEASURED INPUT AND OUTPUT VOLTAGES
SHOULD BE THE SAME

(A) VOLTAGE TESTING

A   B

B1

A1

(B) CURRENT TESTING

03314-13_F59.EPS

Figure 59 Circuit breaker checks.

Make sure that the breaker is closed by first setting it to the off position, then setting it to the on position. Measure the voltage at the circuit breaker output terminals:

- A1 to neutral or ground (single-pole breaker)
- A1 to B1 (two-pole breaker)
- A1 to B1, B1 to C1, and C1 to A1 (three-pole breaker)

The measured input and output voltages should be the same under a load. With no load applied, there will rarely be any difference at all. Defects, in the form of a voltage drop, will be revealed when a significant load is applied. If the voltage is significantly lower than that measured at the input to the circuit breaker, visually inspect the circuit breaker for loose wires and terminals or signs of overheating. If none are found, the circuit breaker should be replaced.

If the circuit breaker shows signs of overheating, or trips when voltage is applied to the equipment, reset it, then check the current flow through the breaker using an AC clamp-on ammeter.

Set up the AC clamp-on ammeter to measure AC current on a range that is higher than the highest current expected. Check the ampere rating marked on the breaker. It is usually stamped on the breaker lever or body. One wire at a time, measure the current flow in the wires connected to the circuit breaker output terminals:

- A1 (single-pole breaker)
- A1, B1 (two-pole breaker)
- A1, B1, and C1 (three-pole breaker)

If the circuit breaker trips at a current below its rating or is not tripping at a higher current, the circuit breaker should be replaced. Be sure that the breaker is not being tripped because of high ambient temperature.

> **NOTE**
>
> Some service entrance panels will only accept replacement circuit breakers of the same brand as the panel manufacturer. Other service entrance panels will accept circuit breakers made by other manufacturers.

4.4.0 Control Circuits and Low-Voltage Power Sources

Control circuits provide a link between loads and the input power. Control circuits start, stop, or otherwise control the operation of a load. They usually contain one or more control devices such as relays, switches, and thermostats that work to apply or remove power from the loads. The more complex the system, the more control devices it will have. When a load such as a compressor motor is not working, it must be determined whether the problem is in the load itself, or in the circuits controlling the load.

Control circuits generally operate at 24VAC. This low voltage is obtained by using a control transformer to step down the line voltage. Because most control circuits operate at 24VAC, they are often called the low-voltage circuits. In some larger systems using three-phase line voltages of 240VAC and higher, a control voltage of 120VAC or higher may be used for some of the control devices. For this reason, always measure the control circuit voltage. Never assume the control circuit voltage is a low voltage.

4.4.1 Isolating to a Faulty Circuit via the Process of Elimination

Isolating a faulty functional circuit (input power distribution circuit, load circuit, or control circuit) can be relatively easy based on an analysis of the equipment operation and a process of elimination. Also, talking to the customer prior to working on the equipment is always recommended because it can provide valuable information that can aid in the troubleshooting process.

For example, on a service call on the heat pump system shown in *Figure 60*, the customer complains that the unit is running but is blowing warm air instead of cool air. A preliminary check of system operation reveals that the compressor is not running when the thermostat is calling for cooling.

Begin fault isolation through the process of elimination. Because the indoor fan motor runs, it is possible to immediately eliminate the input power distribution circuits as the source of the problem. Next, find out if the compressor will run in the reverse cycle heating mode by setting the thermostat to call for heating. If it is too warm, thermostat TH-1 can be jumped to simulate a call for heating. For the purpose of this example, assume the compressor runs. The compressor load circuit and everything in the compressor contactor energizing path, including the control relay (CR) can now be eliminated.

This isolates the problem to the only devices left in the control circuit that are unique to the cooling mode. By studying the diagram in *Figure 60*, those components are identified as the cooling thermostat (TC) and the related COOL control switch. The thermostat can be eliminated as the cause of the problem if the unit works when the control switch is set to the Auto mode.

Figure 60 Heat pump wiring ladder diagram.

03314-13_F60.EPS

Circuit Breakers

Circuit breakers are available in a wide variety of sizes and types to suit various applications. Some types of circuit breakers are listed as heating, air conditioning, and refrigeration (HACR) circuit breakers. HACR breakers have a built-in time delay that allows a higher-than-rated current to momentarily flow in the circuit. This compensates for the large starting current drawn by such loads. The equipment being protected must also be marked by the manufacturer as suitable for protection by this type of breaker.

03314-13_SA03.EPS

4.4.2 Isolating to a Faulty Circuit Component

Once the source of an electrical problem has been isolated to the malfunctioning load or control circuit, the next step is to make a series of voltage measurements across the components in the malfunctioning circuit to find the faulty component. In *Figure 61*, the measurements can start from the line or control voltage side of the circuit and move toward the load device, such as a motor or a relay coil. Measurements are made until either no voltage is observed, or until the voltage has been measured across all the components in the circuit. Note that when there are many devices in the circuit under test, the measurements can be made by starting at the midpoint in the circuit (divide by two), then working towards either the source of line or control voltage or the load device, depending on whether voltage was or was not measured at the mid-point. As a result of taking these voltage measurements, one of the following situations should exist.

A. Open low-pressure switch contacts.

B. Open relay contacts – check related coil control circuit.

C. Load device (contactor coil) is defective.

At some point within the circuit under test, no voltage will be indicated on the meter. This pinpoints an open component or set of switch contacts between the last measurement point and the previous measurement point. *Figure 61A* is an example of this situation. In this case, the contacts of the low-pressure switch are open, preventing the compressor contactor coil (C) from energizing.

If the open is caused by a set of contactor or relay contacts, determine if the related contactor or relay coil is not being energized or is defective. *Figure 61B* is an example of this situation. In this case, the contacts (CR) in the control circuit are open, preventing the compressor contactor (C) from energizing. These contacts close when the control relay coil (CR) is energized. Before assuming that the problem is caused by the open contacts (CR), troubleshoot the control circuit containing the related relay coil (CR) to find out if the coil is energized or de-energized.

If the coil is de-energized, further troubleshoot its control circuit to find out why. For example, if the thermostat cooling switch contacts are open, the relay coil (CR) will not be energized.

If voltage is measured at the contactor coil, motor, or other load device, and the device is not working, the load device is most likely at fault. Turn off the power to the unit, then disconnect the device from the circuit and test it to confirm

(A) OPEN LOW PRESSURE SWITCH CONTACTS

(B) OPEN CR RELAY CONTACTS – CHECK RELATED COIL CONTROL CIRCUIT

(C) LOAD DEVICE (CONTACTOR COIL) IS OPEN

03314-13_F61.EPS

Figure 61 Isolating to a faulty circuit component.

that it is defective. *Figure 61C* shows an example of this situation in which 24V power is applied to the compressor contactor (C), but it is not energized. In this case, the contactor is probably defective.

4.4.3 Low-Voltage Power Source Testing

Control transformers are usually checked by measuring the voltages across the secondary and primary windings (*Figure 62*). Typically, the secondary winding is measured first. The multimeter should be set to measure AC voltage on a range that is higher than the control voltage expected. If the voltage measured across the secondary winding is within ±10 percent of the required voltage, (21.6V to 26.4V) the transformer is good. If voltage is lower than 21.6V or no voltage is measured at the secondary winding, the voltage across the primary winding must be measured.

Some low-voltage transformers are equipped with more than one primary connection to accommodate different primary voltages. For example, a transformer may have a connection for a 208V power supply and a 240V power supply. The factory connection will be for one of the two voltages. It is the start-up technician's responsibility to check that this connection is correct. For example, if the transformer is factory-connected for 240V and the unit's supply voltage is 208V, the secondary voltage will be incorrect.

If the voltage measured at the primary winding is within ±10 percent of the required voltage, and the secondary voltage is low or zero, check the following:

- Check the transformer for overheating or physical damage
- Check that all transformer electrical connections are tight.

- If the primary has more than one voltage selection lead, check that the lead matching the unit's supply voltage is connected.
- Check the fuse/circuit breaker in the transformer's secondary. If the fuse is external, replace it. Reset the circuit breaker if tripped. If the fuse is internal to the transformer, replace the transformer.

Continuity checks can confirm if either set of windings is open. To perform a continuity check, shut off power and disconnect one lead on each winding of the transformer (primary and secondary) to isolate the transformer from the rest of the circuit. Use an ohmmeter set on a low scale to measure continuity. Replace any transformer that has open windings. If the transformer appears to be fused internally, do not attempt to remove and replace the fuse. Any transformer that appears damaged or shows signs of overheating should be replaced.

A transformer protection circuit breaker may be panel-mounted within sight of the device it is protecting. It may have a small reset button that pops up to indicate the breaker is tripped. This type of protection is not common on smaller transformers however. Some transformers may have a built-in circuit breaker instead of a fuse.

4.5.0 Resistive and Inductive Loads

Loads are devices that consume power to do work. Compressor motors, fan motors, heater elements, and the primary winding of transformers are all loads normally found in cooling and heating units. Because the input power distribution circuits and the load circuits are both energized and operate at the input voltage level, they are often called the high-voltage circuits (*Figure 63*).

Electric crankcase heaters and electric heater elements are just two examples of resistive loads found in HVAC equipment. Inductive loads include contactor, relay, and motor starter coils. They also include control transformers, solenoid valves, and some gas valves. Compressor motors, fan motors, and other motors are also inductive loads. As a general rule, HVAC systems contain more inductive loads than resistive.

Once a resistive or inductive load has been identified as the probable cause of an electrical problem, it should be tested to confirm that it is good or defective. The best way to test a resistive or inductive load is by measuring the resistance across the terminals of the device. Before measuring resistance, make sure to electrically isolate the component being measured by disconnecting the wiring leads. This is important in order to

03314-13_F62.EPS

Figure 62 Control transformer checks.

THERMOSTAT IS OPEN TO ISOLATE RESISTIVE LOAD (CRANKCASE HEATER) FROM CIRCUIT

WIRE DISCONNECTED TO ISOLATE INDUCTIVE LOAD (CONTACTOR COIL) FROM CIRCUIT

METER READS

MEASURABLE RESISTANCE = GOOD LOAD

ZERO RESISTANCE = SHORTED LOAD

INFINITE RESISTANCE = OPEN LOAD

03314-13_F64.EPS

Figure 64 Resistive and inductive load resistance checks.

03314-13_F63.EPS

Figure 63 High-voltage and low-voltage circuits.

achieve an accurate resistance reading, and to ensure power is not accidently applied to the ohmmeter. Also, the meter will read the resistance of other components that are connected in parallel with the component to be measured. In *Figure 64*, a reading of zero ohms indicates a shorted load, while a reading of infinite resistance indicates an open load. In either case, the device should be replaced. When the multimeter indicates a measurable resistance (continuity), it usually indicates that the device is good. If a low resistance is measured on a contactor, relay, or starter coil, place one meter probe to ground or to the unit frame. Touch the other probe to each coil terminal. If a resistance is measured from either terminal to ground, it is grounded and must be replaced.

4.5.1 Resistive Load Testing

Resistive loads used in HVAC tend to be heaters. Crankcase heaters, electric heating elements, and even thermostat heat anticipators are all examples of common resistive heating loads. The thermistor is an example of a resistive device that is not

a load. With the exception of thermistors, the resistance of heating devices is fixed. For example, a crankcase heater has a certain resistance value in ohms that, when subjected to a given voltage, will produce a certain amount of heat. Resistive loads in HVAC equipment rarely fail by changing their resistance. Instead they tend to fail by becoming open circuits. For example, when a section of a heating element melts, the resistive circuit is no longer intact. Once open, they do not produce heat. The lack of heat is often the first symptom that a resistive load has failed. Assuming power is still available to a resistive load, a clamp-on ammeter placed around one of the power leads supplying the heater can confirm if it is drawing current. If a resistive load has voltage available, but is not drawing current, it is probably open.

That failure mode can be confirmed by shutting off power, disconnecting at least one side of the heater from the circuit, and checking continuity with an ohmmeter.

Manufacturers rarely provide the resistance value of resistive loads. For example, the voltage and power (watts) of a crankcase heater is often provided. On the rare occasion that the resistance of a crankcase heater is needed, measure the resistance of a similar heater that is known to be good.

4.5.2 Basic Motor Testing

In addition to the standard safety precautions that must be taken when working on electrical equipment, the following precautions must be adhered to when troubleshooting compressors and other motors.

The operation and uses of motors in HVAC equipment has been studied previously, but is reviewed briefly here. Six types of single-phase motors are commonly used in HVAC equipment: shaded-pole; split-phase; permanent split capacitor; capacitor start; capacitor start, capacitor run; and electronically commutated motors (ECM). Permanent split capacitor (PSC) and capacitor start, capacitor run (CSR) motors are most often

used in single-phase hermetic compressors because of their good running characteristics and high efficiency. Indoor and outdoor fan motors and blower motors are usually single- or multi-speed PSC motors. Shaded-pole motors are typically used in low-torque applications such as small direct-drive fan and blower motors. The use of electronically commutated motors (ECMs) is now very common in fan and blower motor applications. Complete coverage of ECMs is provided later in this module.

Both PSC and CSR motors (*Figure 65*) have at least three external terminals leading to two internal windings. The main or run winding (R) contains relatively few turns of heavy wire. The start winding (S) contains a greater number of turns of lighter wire. The point where the two

(A) PSC MOTOR

(B) CSR WITH A START RELAY

(C) CSR WITH A CENTRIFUGAL SWITCH

03314-13_F65.EPS

Figure 65 PSC and CSR motors.

Measuring Resistance

The actual resistance value measured for functional resistive and inductive loads can vary widely depending on the type of device. Ideally, the exact resistance value for the device can be found in the manufacturer's service literature. Another way to judge the resistance reading is by comparing the resistance of the device being tested with that of a similar device that is known to be good.

windings meet internally is called the common (C). The arrangement of the motor windings used in both the PSC and CSR motors are the same. The configuration of the motor as a PSC or CSR motor is determined by the run and/or start circuit components used with the motor. The PSC motor has a run capacitor permanently connected across the run and start windings. The CSR uses an extra capacitor called a start capacitor to aid in starting. As shown, the start capacitor and the contacts of a start relay (SR) are connected in parallel with the run capacitor. When the motor is turned on and reaches about 75 percent of full speed, these contacts open and remove the start capacitor from the circuit. The start relay method of removing the start capacitor from the circuit is commonly used with hermetic and semi-hermetic compressor motors. A start relay can be used with all CSR motors; however, in non-compressor applications, a centrifugal switch is frequently used to disconnect the start capacitor from the circuit when the motor comes up to speed. Start capacitor failures are often the cause of compressor and other motor problems.

Multi-speed PSC motors (*Figure 66*) used to drive fans and/or blowers in HVAC equipment are capable of operating at two or more speeds. The motor's speed can be changed by switching the motor leads, terminal taps, or by the use of speed control switches or relays. In many heating/cooling units, the motor speed is selected automatically by the control circuits, as determined by the mode of operation. Normally, slower fan speeds are used with heating modes of operation, and higher speeds are used with cooling modes of operation.

There are many types of multi-speed motors. As shown, the speed is changed by connecting the line voltage either to the low-speed tap (LO), medium-speed tap (MED), or high-speed tap (HI) of the motor. The specific taps used are selected when installing the unit. A control relay with contacts located in the motor load circuit is normally used to prevent more than one motor winding speed tap from being energized at the same time, a condition that would destroy the motor.

Because multi-speed motors use tapped windings, series-connected winding sections, and/or other wiring configurations that enable operation at different speeds, they may fail in such a way that the motor will not run at one or more speeds, but will run at other speeds. When troubleshooting multi-speed motors, it is important to eliminate the speed selection circuits external to the motor as the cause of the problem before condemning the motor itself.

Three-phase motors (*Figure 67*) are generally used when high starting torque is needed or when the motor requirements are greater than 1 horsepower. All have at least three internal windings, with each winding having an equal resistance and the same number of wire turns. Six- and nine-lead, three-phase motors are also found in large applications where part winding start is necessary to reduce the initial inrush current at motor startup. Three-phase motors have good starting and running characteristics and high efficiency. Three-phase motors require no external starting relays or capacitors.

4.5.3 Open, Shorted, or Grounded Winding Checks

Motors are tested for open, shorted, and/or grounded windings with the equipment power off, the capacitors discharged, and the motor leads disconnected from all the related components, including the run and start capacitors and the start relay.

Testing for shorted or open windings is done by measuring the resistance of the windings with the ohmmeter portion of a multimeter. Be sure to use a meter that has an accurate low range (R × 1Ω) resistance scale because the resistance of some undamaged windings can be as low as one-half ohm. Perform the test with the multimeter set to measure resistance on the R × 1Ω scale. Make sure that the meter is zeroed. One lead of

03314-13_F66.EPS

Figure 66 PSC multi-speed motor.

03314-13_F67.EPS

Figure 67 Three-lead, wye-connected, single-voltage, three-phase motor.

the multimeter is connected to one of the motor leads, as shown in *Figure 68*. Touch the other lead to the remaining motor leads, one lead at a time, and observe the meter indication. If the multimeter reads a measurable resistance, the windings are probably good. If the multimeter reads zero resistance at one or more leads, the motor has a shorted winding; if it reads an infinite resistance, it has an open winding. Some rules of thumb commonly used to judge the condition of windings are as follows:

- If testing a single-phase PSC/CSR compressor motor, the resistance of the start winding is typically three to five times that of the run winding. For non-compressor motors, the resistances can vary widely depending on the design of the motor.
- When testing a single-phase multi-speed motor run winding, the highest resistance is normally measured between the common lead and the low (LO) speed lead, and the lowest resistance between the common lead and the high (HI) speed lead. The resistance measured between the common lead and the medium speed lead (MED) should be somewhere between that measured for the LO and HI leads.
- If testing a three-phase motor, the motor windings are usually judged to be good if the resistance measured across each winding is nearly identical to the other two windings.

When checking a motor with an internal motor protection device, always make sure that the motor has had adequate time to cool off so that the protective device has time to reset. Note that this may take an hour or more.

Testing a motor for grounded windings is done by measuring the resistance of the windings with a multimeter. Perform the test with the multimeter set up to measure resistance on the R × 10,000 scale. Connect one lead of the multimeter to a good ground connection, such as the motor or compressor frame or compressor discharge/suction line. Poor electrical contact because of a coat of paint, layer of dirt, or corrosion can cause an inaccurate measurement and hide a grounded winding. The other meter lead is then touched to each of the motor leads, one lead at a time, while watching the meter indication (*Figure 69*). An infinite or high resistance should be measured from each lead to ground. If a high resistance reading is indicated, it should not be less than 1,000Ω per volt. For example, on a 230V motor, the resistance should not be less than 230,000Ω (230V × 1,000Ω/V = 230,000Ω). This indicates that the motor winding is not grounded.

If a low or zero resistance, or a measurable resistance that is less than 1,000V per volt is measured, this usually indicates that the motor winding is grounded. However, if testing a hermetic or semi-hermetic compressor motor, note that erroneous resistance readings to ground can be measured if liquid refrigerant is present in the compressor shell. In this instance, it is recommended that the refrigerant be recovered from the system. Then retest the compressor before condemning it.

4.5.4 Identifying Unmarked Terminals of a PSC/CSR Motor

Sometimes the terminals on a single-phase motor are not marked, or are hard to identify. The terminals can be identified by using a multimeter to measure the resistance of the motor windings. First, the multimeter is used to find the two

NORMAL READING = MEASURABLE RESISTANCE

MAKE SURE TO ALLOW TIME FOR PROTECTIVE DEVICE TO RESET

- INFINITE RESISTANCE = OPEN WINDING(S)
- ZERO OHMS = SHORTED WINDING(S)

03314-13_F68.EPS

Figure 68 Motor open or shorted winding check.

NORMAL READING = INFINITE OR HIGH RESISTANCE (1,000 OHMS PER VOLT)

03314-13_F69.EPS

Figure 69 Grounded winding check.

terminals across which the highest resistance is measured (*Figure 70*). These are the run (R) and start (S) terminals. The remaining terminal is the common (C) terminal. Next, put one lead of the multimeter on the common (C) terminal, and find which of the remaining terminals gives the highest resistance reading. This is the start winding (S) terminal. The remaining terminal is the run winding (R) terminal. The C to R resistance reading will be the lowest.

4.5.5 Start and Run Capacitor Checks

The start and/or run circuits on single-phase motors use capacitors. Capacitors affect the wattage, amperage draw, torque, speed, and efficiency of a motor. Run capacitors are connected in the motor

IDENTIFY THE COMMON (C) TERMINAL

IDENTIFY THE START (S) AND RUN (R) TERMINALS

03314-13_F70.EPS

Figure 70 Identifying unmarked terminals of a PSC/CSR motor.

circuit at all times; therefore, they are referred to as continuous-duty capacitors. They are usually larger in physical size, but have lower capacitance ratings than start capacitors. Because run capacitors are in the circuit at all times, they are typically filled with a dielectric fluid that acts to dissipate heat. A shorted capacitor may provide a visual indication of its failure. The pop-out hole at the top of a start capacitor may appear bulged or blown. A run capacitor may be bulged and/or leaking. If a capacitor is found to be defective, always replace it with one specified by the manufacturer.

If the exact capacitance value of a capacitor is required, use a capacitor tester (*Figure 71*). Follow the tester manufacturer's instructions to perform the test. Typically, the measured microfarad value for a start capacitor should be ±20 percent of the value shown on the capacitor label. If the measured value is outside the range of ±20 percent, replace the capacitor. For a run capacitor, the measured value should be ±10 percent of the value shown on the capacitor. If the measured value is outside the range of ±10 percent, replace the capacitor.

4.5.6 Start Relay Checks

The start relay is used to remove the start capacitor from the motor starting circuit when the motor reaches about 75 to 80 percent of its operating speed. Start relays are made that can be actuated by either current or voltage. The start relays used with HVAC equipment motors are normally voltage-actuated relays (potential relays); therefore, the remainder of this discussion will cover the testing of a voltage-actuated start relay.

Start relays tend to fail with their contacts closed. This results in the start capacitor remaining in the start circuit, causing the motor's start

MFD VALUE CHECK

40 MFD 400V

03314-13_F71.EPS

Figure 71 Capacitor checks.

winding to overheat and fail. It may also result in failure of the start capacitor. When it is necessary to replace a start relay, an identical replacement must be used. Substitution of a relay with a different pickup voltage can cause damage to the start capacitor or motor start winding. Also, the replacement relay must be positioned and wired exactly as the original.

Start relays can be tested by measuring the motor start winding current with a clamp-on ammeter. The use of an analog clamp-on ammeter is recommended because it is easier to observe the current reading. If using a digital clamp-on ammeter, one with a MIN/MAX current capability must be used. Testing begins by first finding the full load amps (FLA) rating for the motor as marked on the compressor or motor nameplate. The clamp-on ammeter is then set up to measure AC current on a range scale that is higher than the motor FLA.

With the power turned off, the clamp-on ammeter jaws are placed around the wire that connects the motor start capacitor to the start relay contacts, as shown in *Figure 72*. Power is turned on to the unit while watching the clamp-on ammeter indication to observe the current flow in the start capacitor circuit as the motor starts.

When the start relay is operating properly, the clamp-on ammeter current indication should momentarily indicate current flow, and then fall back to zero as the motor comes up to speed. This shows that the start relay is good because its contacts have opened.

If current continues to be read on the clamp-on ammeter after the motor is up to speed, the relay contacts are stuck closed. This means the relay is defective and should be replaced. The start capacitor and/or motor start winding can be damaged when the start relay contacts are stuck closed.

If no current is shown on the clamp-on ammeter, the relay contacts may be stuck open, the related start capacitor may have failed open, or the related wiring may be open. In this instance, the relay contacts can be checked for continuity with the unit power turned off. Contacts that are stuck open will measure infinite resistance.

4.5.7 Start Thermistor Checks

Start thermistors can be used to provide additional starting torque for PSC compressors. The start thermistor is a temperature-sensitive device that changes its electrical resistance as a result of a change in temperature. Positive temperature coefficient (PTC) thermistors increase their resistance with an increase in temperature. PTC thermistors are commonly used in the start circuits of PSC motors. *Figure 73* shows a PSC compressor motor with a PTC start thermistor. As shown, the PTC thermistor is placed across (parallel to) the run capacitor.

At room temperature, the PTC thermistor resistance is very low, about 25Ω to 50Ω. When the compressor is turned on, the application of voltage provides an initial surge of high current through the start winding, because the low resistance of the PTC thermistor is effectively bypassing (shorting) the run capacitor. This surge results in increased motor starting torque. The temperature increase created by the high current causes the PTC thermistor resistance to increase very rapidly to several thousand ohms, blocking current flow and effectively removing the thermistor from across the run capacitor. The motor then

Figure 73 Start thermistor check.

Figure 72 Start relay check.

runs as a normal PSC motor. A small leakage current through the thermistor keeps the thermistor heated and its resistance high while the motor is running. Circuit operation remains this way until the motor is turned off. After a cool-down period, the thermistor's resistance will once again be the low value needed to start the motor.

The thermistor is tested with the equipment power off, the capacitors discharged, and the thermistor under test isolated from the remainder of the circuit. Testing the thermistor is done by making a resistance measurement. Before attempting to measure the resistance, wait at least 10 minutes to allow the thermistor to cool to ambient temperature. The cold resistance of any PTC thermistor should be about 100 to 180 percent of the thermistor ohm rating. For example, a thermistor rated at 25Ω should have a cold resistance of 25Ω to 45Ω. If the PTC thermistor resistance is much lower or more than 200 percent higher than its rating, the thermistor should be replaced.

Note that, if the resistance of a start thermistor is acceptable and yet the compressor still does not start consistently, a start capacitor and start relay may have to be installed to ensure consistent compressor starting energy is provided. PTC start thermistors are not as powerful a starting aid as a capacitor and start relay.

4.6.0 Hydronic Controls

In many parts of the United States, hot water or hydronic heat is very popular. With this type of heat, hot water (or in some cases, steam) is circulated through pipes to radiators in different parts of the building. Boilers are pressurized vessels and have unique safety and installation requirements. The nature of hydronic heat dictates specialized components and controls not found in other areas of HVAC.

A simple hydronic system, such as the one shown in *Figure 74*, contains a hot water boiler where the water is heated, a circulating pump to move the heated water throughout the system, a tank to absorb water as it expands when heated, a device called an aquastat that controls the water temperature, and a relief valve to bleed excess pressure so that boiler pressures do not reach explosive levels. The water may be heated by an oil or gas burner. These burners and many of the controls associated with them are similar to the burner controls on gas or oil-fired warm air furnaces.

4.6.1 Aquastat

The aquastat performs several important functions. In its simplest form, it is nothing more than

Figure 74 Simple hydronic system.

a limit switch set to prevent water from exceeding a certain temperature. An aquastat, such as the one shown in *Figure 75*, is usually mounted on the boiler and contains a temperature-sensing element that is inserted in a well in the boiler jacket. With such an aquastat, the water in the boiler will stay cold until there is a call for heat. The burner will then ignite and start to warm the water. The burner will operate until the room thermostat is satisfied or the water in the boiler reaches the high limit temperature setting.

More complex aquastats provides additional functions. In addition to a high limit setting, they contain a low limit setting that prevents the water in the boiler from getting too cold. This feature is

Figure 75 Aquastat.

necessary if domestic hot water is provided by a heat exchanger, called a tankless coil, that is inserted in the boiler jacket. This type of aquastat also contains a circulator control that prevents circulator operation until the water is warm enough to provide heat from the radiators.

Other specialized aquastats, especially those used on oil-fired boilers, have the oil burner primary control built into the aquastat. *Figure 76* shows how an aquastat with circulator and high and low limits would operate on an oil-fired boiler. If the water temperature is too low, contacts R-B remain closed, allowing the water to heat up but with no circulation. As the water heats up, R-W closes and R-B opens. If there is a call for heat, power from the switching relay feeds through the closed high limit switch to keep the burner running and through R-W to power the circulator pump.

4.6.2 Reset Controller

To achieve maximum comfort from a hydronic heating system, some systems incorporate an aquastat device called a reset controller. This device monitors the outdoor temperature and adjusts the boiler water temperature for maximize energy savings while still providing comfort. It allows the boiler water temperature to rise on cold days to supply more heat, but limits the boiler water temperature to lower levels on milder days when less heat is required. In addition to a temperature sensor in the boiler jacket, the reset controller contains an outside temperature sensor (*Figure 77*) that is typically mounted on the north side of the structure away from direct sunlight.

03314-13_F77.EPS

Figure 77 Typical reset controller outside temperature sensor mounting.

4.6.3 Low Water Cutoff

Loss of water in a boiler can have catastrophic results. The boiler may produce large amounts of steam, overheat, and build up explosive pressures. To prevent this, a low water cutoff device may be installed in the boiler. The simplest devices are nothing more than a float that activates a switch if the water level drops. The switch shuts off burner operation to prevent overheating and/or activates an alarm or warning light. Electronic versions (*Figure 78*) have a probe that is inserted in the boiler to monitor the water level. Poor water quality caused

03314-13_F76.EPS

Figure 76 Aquastat used for low limit and circulator control in an oil-fired hydronic system.

MINIMUM SAFE
WATER LEVEL

03314-13_F78.EPS

Figure 78 Low water cutoff control.

by contamination will affect the operation of electronic low-water cutoffs. These devices also shut off burner operation and/or sound an alarm when the water level drops too far.

4.6.4 Circulator Pump

The circulator pump in a hydronic system performs a job similar to the blower in a forced air furnace. The major difference is that the pump moves water instead of air. Circulating pumps can be mounted on the floor or in the supply or return water line. However, the preferred location for the pump is in the supply line leaving the boiler. *Figure 79* shows a typical circulator pump. Many pump motors are direct coupled to the pump's centrifugal impeller through a shaft seal in the impeller housing. However, many smaller

pumps are magnetic and have no shaft seal that could leak. They drive the impeller through magnetic force. Circulator operation can be controlled by the aquastat or by the zone valves (if so equipped). Most residential hydronic systems have one circulator pump. In some zoned installations, each zone has its own pump. Some older circulator pumps require periodic lubrication of the motor and pump, while newer models are sealed units with little or no maintenance required.

4.6.5 Zone Valves

Zoned control is easily accomplished with hydronic heating systems due to the simplicity of the zone valve. This device is simply an electrically operated valve. It is placed in the supply line to the zone and opens when the thermostat in the zone calls for heat. *Figure 80* shows a typical zone valve.

03314-13_F79.EPS

Figure 79 Circulator pump.

03314-13_F80.EPS

Figure 80 Zone valve.

In residential applications, zone valves are powered by 24VAC. In commercial applications, higher voltages may be used. Most valves are motorized so that on a call for heating from the zone, the valve opens. Some valves are equipped with an end-switch that closes as the valve opens. This switch closure starts the circulator pump so that water can flow to the zone that is calling for heat. There are a number of ways to control burner and circulator operation on a gas or oil-fired boiler. Much of it depends on the aquastat used but the installer does have some control over the sequence of operation.

4.6.6 Zoned Hydronic System Operating Sequence

In many ways, the operating sequence of a hydronic system is similar to the operating sequence of a forced-air heating system. The gas and oil burners used in both systems operate in a similar manner and often use identical components. For example, the hot surface ignition system used in a furnace is identical to the one used in a residential boiler. As with any HVAC system, knowing its operating sequence is important when seeking to find the source of a problem. The following sequence of operation assumes a simple residential packaged boiler (gas-fired) equipped with an aquastat with low-and high-temperature limits. The structure is divided into three heating zones, each controlled by its own 24V room thermostat.

When power is first applied to a cold boiler with no call-for-heat from any of the zones, the low-limit contacts of the aquastat are closed. The closed contacts allow the burner to operate. This warms the boiler water to a preset temperature. When a call-for-heat signal does come, the boiler water is warm enough to provide immediate heat. The burner will cycle on and off to maintain this minimum water temperature as long as power is available to the boiler.

Assume there is a call-for-heat from the smallest of the three zones (a master-suite bathroom). The room thermostat in the zone signals the zone valve for the bathroom to open. As the motorized zone valve opens, it closes a set of end-switch contacts on the zone valve which provide a 24V signal to energize the circulating pump. The pump moves heated water to the terminal device in the bathroom. At this point, the boiler can react in the following ways:

- Depending on the thermostat setpoint in the bathroom and boiler water temperature, there may be enough heat available in the boiler water to satisfy the room thermostat in that zone without burner operation.
- If the thermostat setpoint in the bathroom is set high and/or additional zone thermostats also call for heat, boiler water temperature will drop and the aquastat low-limit will start burner operation to maintain boiler water temperature.
- If the thermostat setpoint in the bathroom is set high and that is the only zone calling for heat after the burner has come on, that small zone may not be able to rapidly dissipate all the heat the burner is providing. This will cause the boiler water temperature to rise to the point that the aquastat high-limit will shut the burner off.

In any system with multiple zones, the burner will cycle on and off based on water temperature within the boiler and the fluctuating zone heating capacity demands. Troubleshooting the components in a hydronic system is similar to troubleshooting components in a forced-air system. Energized temperature-actuated switches can be checked for correct operation using a voltmeter. Check for switch continuity with power off using an ohmmeter. The circulating pump motor can be checked like any other HVAC motor.

Additional Resources

Electrical Motor Controls for Integrated Systems. 4th Edition, 2009. Gary Rockis and Glen A. Mazur. Homewood, IL: American Technical Publishers, Inc.

Electric Motors and Drives: Fundamentals, Types and Applications. 3rd Edition, 2005. Austin Hughes. Newnes.

4.0.0 Section Review

1. All employees on a job site are authorized to perform lockout/tagout.

    a. True
    b. False

2. The purpose of the legend on a label diagram is to _____.

    a. isolate electrical problems
    b. identify the meaning of symbols
    c. locate components in the equipment
    d. isolate section of a ladder diagram

3. With equipment that operates on three-phase power, the current imbalance in any one leg should not exceed _____.

    a. 2 percent
    b. 5 percent
    c. 8 percent
    d. 10 percent

4. A furnace provides airflow in both heat and cool modes. The furnace operates normally in the heat mode, but the blower does not run in cooling. Which of the components below is most likely at fault?

    a. Blower motor capacitor
    b. Furnace control board
    c. Cooling fan relay
    d. Low-pressure switch

5. When measuring the resistance of motor windings, always _____.

    a. use the R × 10Ω scale
    b. use the R × 10,000Ω scale
    c. zero the meter
    d. ground the meter

6. What is the purpose of the low-limit in an aquastat?

    a. Maintain boiler water temperature
    b. Maintain a safe water level
    c. Prevent boiler freeze-up
    d. Monitor outdoor temperature

5.0.0 VARIABLE FREQUENCY DRIVES

Objectives

Describe the operation of variable frequency drives (VFD) and their selection considerations.

   a. Describe the operation of a VFD.
   b. Identify VFD parameters that can be programmed.
   c. Describe the important considerations for the selection of a VFD.
   d. Explain dynamic motor braking processes.

Trade Terms

Base speed: The rating given on a motor's nameplate where the motor will develop rated horsepower at rated load and voltage.

Insulated gate bipolar transistor (IGBT): A type of transistor that has low losses and low gate drive requirements. This allows it to be operated at higher switching frequencies.

A variable frequency drive (VFD), sometimes referred to as an adjustable frequency drive (AFD), converts three-phase 60Hz input power to an adjustable frequency and voltage output to control the speed of an AC motor. The use of VFD-controlled systems provides energy savings by eliminating most of the losses associated with mechanical or electromechanical methods of adjusting speed used in the past. Many other benefits are derived from using a VFD. These include the following:

- *Controlled starting* – Provides for limited starting current, reduction of power line disturbance on starting, and low power demand on starting.
- *Controlled acceleration* – Provides for soft-start, adjustable acceleration based on time or load, and reduced motor size for pure inertial load acceleration.
- *Adjustable operating speed* – Allows process to be optimized or changed, provides energy savings, allows process start at reduced speed by programmable controller.
- *Adjustable torque limiting* – Current limit is available for quick and accurate torque control, protecting the machinery, product, or process from damage.

- *Controlled stopping* – Soft stop, timed stopping, and fast reversal with much less stress on an AC motor than plug reverse.

5.1.0 Basic VFD Operation

VFDs change the speed of AC squirrel cage induction motors to meet system requirements by converting three-phase 60Hz input power to an adjustable frequency and voltage. The speed of the motor is determined by the formula:

$$\text{Motor speed} = 120 \times \text{frequency} \div \text{number of motor poles}$$

For a specific motor, the number of poles is a constant since they are built into the motor. Examination of the formula shows that the speed of the motor is proportional to the applied frequency. As the frequency increases, the motor speed increases; as the frequency decreases, the motor speed decreases. To maintain a constant motor torque, VFDs are designed to automatically maintain a constant relationship between the voltage and frequency of the excitation output applied from the VFD to the motor terminals. This relationship is called the volts per hertz ratio (V/Hz). If this process is done properly, the speed of the motor can be controlled over a wide variation in shaft speed (0 rpm through twice nameplate) with the proper torque characteristics for the application.

VFDs are made in a wide variety of sizes and designs. Motor drive technology has changed greatly over the last decade. Today, digitally controlled, programmable, microprocessor- based, pulse-width-modulated (PWM) variable frequency drives are commonly used. Their use provides a significant improvement in the elimination of dirty power problems such as harmonics and electro-magnetic interference (EMI). These problems were generated by the analog SCR rectifier VFD units that were the state-of-the-art only a few years ago. For this reason, the remainder of this section will focus on the newer PWM type of VFD device.

Figure 81 shows a block diagram of a typical PWM VFD system. As shown, it consists of three basic parts: the operator control panel, the VFD controller, and an AC motor. The control panel provides an interface between the operator and the drive system. At the control panel, the operator can input commands for starting, stopping, and changing the speed and/or direction of the motor. He or she can also monitor various VFD and motor status signals. Typically, the control panel is removable and can be mounted

LINE-TO-NEUTRAL VOLTAGE

LINE CURRENT

BRAKE RESISTOR

BRAKE CHOPPER

AC CHOKE

CURRENT SENSORS

L1
L2
L3

3-Ø DIODE-BRIDGE RECTIFIER

IGBT INVERTER

MOTOR

POWER SUPPLY

MEASURE-MENTS

MOTOR CONTROL AND GATE DRIVERS

OPERATOR CONTROL PANEL

RS 232

MOTOR AND APPLICATION CONTROL

CONTROL I/O

REMOTE COMMUNICATION

ADJUSTABLE FREQUENCY DRIVE CONTROLLER

03314-13_F81.EPS

Figure 81 Pulse-width modulated variable-frequency drive system.

externally and connected via a cable to the drive. Most VFDs have an input/output (I/O) interface board that allows communication with personal computers (PCs) or programmable logic controllers. The motor used with an VFD is typically a standard NEMA Design A or Design B, squirrel cage induction motor rated for 240V or 480V, three-phase, 60Hz operation. In response to the AC power waveform applied from the VFD, the motor drives the load.

The VFD controller converts three-phase, 60Hz input power to an adjustable frequency and voltage output for controlling the speed of the AC mo-

tor. It typically receives 240V or 480V input power and provides adjustable frequency output power to the motor. The actual range of frequencies used is programmable and depends on the motor application. It also regulates the output voltage in proportion to the output frequency to provide a constant voltage-to-frequency ratio as required by the characteristics of the motor.

The three-phase AC-choke and the DC-link capacitor form an LC filter that, together with the three-phase, diode bridge-type full-wave rectifier, produces a regulated DC voltage for application to the inverter circuit. Note that the circuitry between the rectifier output and the inverter circuit input is sometimes referred to as the DC intermediate bus. The AC choke levels out any high-frequency (HF) disturbances from the utility to the drive and HF disturbances caused by the drive to the utility. It also improves the waveform of the input current to the drive.

Since the inverter section is powered by a fixed DC voltage, the maximum amplitude of its output waveform is fixed. The **insulated gate bipolar transistor (IGBT)** switches in the inverter circuit convert the DC voltage input into a three-phase, PWM, symmetrical AC voltage output for application to the AC motor. The frequency and amplitude of the waveform that is produced are determined by the duration and precise timing intervals of the gates applied to the IGBT switches from the gate driver circuit. The effective value of the output voltage applied to the motor is determined by the width of the zero voltage intervals in the output waveform. Use of IGBTs provides the high switching speed necessary for PWM inverter operation. They are capable of switching on and off several thousand times a second.

VFDs typically have one range of output frequencies over which the ratio of volts to hertz remains constant and another range of output frequencies over which the voltage remains constant while the frequency varies (*Figure 82*). The frequency at which the transition from one range to the other occurs is called the base frequency. Between the base frequency and the maximum frequency points, the rated output voltage remains constant. Below the base frequency point, the output voltage is less than the rated voltage and is based on the volts per hertz ratio (V/Hz). Depending on motor characteristics, these two ranges can correspond to constant torque (constant V/Hz) and constant power (constant voltage with an adjustable frequency).

The motor and application control circuit consists of a programmable microprocessor and related circuits. The microprocessor controls the motor according to measured feedback signals and in accordance with parameter settings and commands. These commands may be applied from the control panel or from a remote device via the control input/output (I/O) circuits. The motor and application control circuit sends signals to and receives feedback signals from the motor control/gate driver circuits. The motor control/gate driver circuits calculate, then output switching position signals to the IGBT switches in the inverter.

When required, dynamic braking can be used with a VFD. This can be done by using a brake chopper circuit and external dynamic brake resistor that operate to absorb motor regenerative energy for stopping the load and to limit the energy flowing back to the drive.

Dynamic braking is accomplished by continuing to excite the motor from the drive after the motor turns off. This causes a regenerative current at the drive's DC intermediate bus circuit. The dynamic brake resistor is then placed across the DC bus to dissipate the returned power. The brake resistors are switched into the circuit under the control of the brake chopper. Dynamic braking cannot occur in an induction AC drive unless the power circuit and control circuit are operational. For this reason, a mechanical brake, actuated in case of a power failure, is often required to stop the motor. The resistor(s) used for dynamic braking must be capable of absorbing at least six times the stored rotational energy of the motor at **base speed**.

5.2.0 Programmable VFD Parameters

The parameters that can be programmed and/or monitored when using a VFD can vary widely based on the specific design and complexity of the VFD, the VFD manufacturer, and the application for which the VFD is to be used. Some of the parameters that can be programmed and/or

Adjustable Frequency Drives

GOING GREEN

Adjustable frequency drives are available for use with a wide range of motors and applications. These drives save energy by eliminating the losses associated with traditional speed control.

03314-13_SA04.EPS

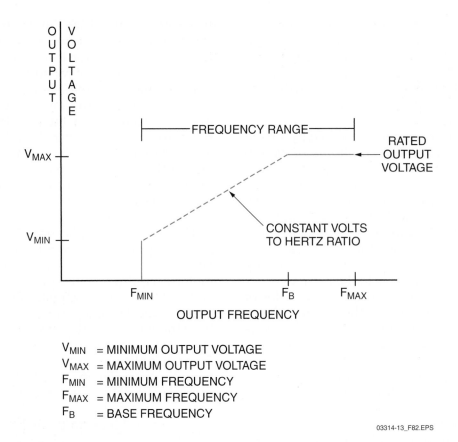

V_{MIN} = MINIMUM OUTPUT VOLTAGE
V_{MAX} = MAXIMUM OUTPUT VOLTAGE
F_{MIN} = MINIMUM FREQUENCY
F_{MAX} = MAXIMUM FREQUENCY
F_B = BASE FREQUENCY

03314-13_F82.EPS

Figure 82 VFD operating frequency range.

monitored are briefly described here. Follow the instructions in the manufacturer's user manual for the specific VFD unit being used.

- *Minimum and maximum frequency* – This defines the frequency limits for the VFD operation.
- *Acceleration and deceleration time* – Acceleration time defines the time required for the output frequency to accelerate from the set minimum frequency to the set maximum frequency. Deceleration time defines the time required to decelerate from the set maximum frequency to the set minimum frequency.
- *Current limit* – This determines the maximum motor current the VFD will provide on a short-term basis.
- *V/Hz ratio* – Typically there are two settings: linear and squared. The linear setting is used with applications requiring constant torque. In this mode, the voltage supplied to the motor changes linearly with the frequency from zero hertz to the nominal frequency of the motor. With the squared setting, the voltage to the motor changes following a squared curve from zero hertz to the nominal frequency of the motor. In this mode, the motor runs under-magnetized (below the normal frequency) and produces less torque and electromechanical noise. This mode is used in applications where

the torque demand from the load is proportional to the square of the speed, such as with centrifugal fans and pumps.

- *V/Hz optimization* – This mode provides an automatic torque boost. The voltage to the motor changes automatically, which allows the motor to produce sufficient torque to start and run at low frequencies. The voltage increase depends on the motor type and horsepower. Automatic torque boost can be used in applications where starting torque due to starting friction is high, such as with a conveyor.
- *Nominal voltage of the motor* – This is the rated voltage from the nameplate of the motor.
- *Nominal frequency of the motor* – This is the frequency from the nameplate of the motor.
- *Nominal speed of the motor* – This is the speed from the nameplate of the motor.
- *Nominal current of the motor* – This is the current value from the nameplate of the motor.
- *Supply voltage* – This is set to a range compatible with the nominal voltage of the supply.
- *Basic frequency reference* – This selects the basis for the frequency reference such as a voltage reference supplied by a potentiometer or a current reference supplied by a transducer or other source.

5.3.0 VFD Selection Considerations

Selecting a VFD for a particular application requires an understanding of the capabilities of various drive units and the characteristics and/or problems encountered with different types of loads. For this reason, evaluating the needs of a drive system and selecting the components for use in the system is a task normally done by an electrical engineer, a drive manufacturer application engineer, or other trained and qualified persons. Most VFD manufacturers offer training that is specific to their product line to help both technicians and engineers understand how to make proper selections and service their product when necessary. *Figure 83* shows a typical checklist used by one manufacturer to gather the data needed to evaluate the requirements for a VFD system. The remainder of this section briefly describes some of the other factors that can affect the VFD selection process.

5.3.1 VFD-Operated Motors

In most applications, the motors used with VFDs are standard NEMA Design A or Design B motors. The use of a motor other than a Design A or Design B usually requires the addition of special drive circuitry. The design letter assigned to a motor by NEMA identifies certain characteristics such as its slip factor and locked rotor torque. Operating self-ventilated motors at reduced speeds may cause abnormal temperature rises. As the speed of a motor is reduced below its 60Hz base speed, motor cooling becomes less effective because of the reduced speed of the self-cooling fan. This limitation determines the minimum allowable motor speed for continuous torque operation. Derating or forced ventilation may be necessary to achieve the rated motor torque output at reduced speeds.

Using VFDs with older existing motors can cause an extra heating effect on the motor and subject the motor winding insulation to high-frequency-induced voltage stress. The load capacity of older motors should be derated to at least the next lower horsepower rating, particularly if the motor has a Class F temperature rise or a 1.0 service factor.

5.3.2 Selecting the Drive

The basic requirement of VFD sizing is to match the torque-versus-speed capability of the VFD to the torque-versus-speed requirement of the driven load.

If the load torque exceeds 150 percent for constant torque drives or 110 percent for variable torque drives during starting or intermittently while running the drive, oversizing of the drive may be required. The minimum and maximum motor speeds for the application will determine the drive's base speed.

The full-load current ratings of typical VFDs are matched to typical full-load, three-phase motor current ratings in current *NEC*® tables. Generally, a VFD of a given horsepower rating will be adequate for a motor of the same rating, but the actual motor current required under operating conditions is the final determining factor for VFD sizing. If the motor will be run at full load, the VFD output current rating must be equal to or greater than the motor nameplate current. If the motor is oversized to provide a wide speed range, the VFD should be sized to provide the current required by the motor at the maximum operating torque. Motor oversizing should generally be limited to an increase of one horsepower. Gross oversizing is not recommended and will result in wasted energy as well.

5.3.3 Environment

The environment in which the motor operates must be examined in order to properly select the equipment and enclosure. The ambient temperature, cooling air supply, and the presence of gas, moisture, and dust should all be considered when choosing a drive, its enclosures, and any protective features.

PWM VFD Outputs to a Motor over Long Cable Runs

When a PWM VFD drive output is applied to a motor over long cable runs, the difference in the motor and cable impedance can cause voltage overshoots to be present at the motor terminals. These overshoots can greatly exceed 1,000V and have the potential to damage the motor's insulation. For small motors, this typically occurs with cable lengths that exceed 33', and for larger motors, with cable lengths that exceed 100'. For situations involving the use of long cable runs, install a reflective wave trap or similar filter at the motor terminals to prevent overshoot.

VFD APPLICATION CHECKLIST

Motor

New_____ Existing _____ Horsepower: _____ Base Speed: _____ Voltage: _____

FLA: _____ LRA: _____ NEMA Design: _____ Gearbox/Pulley Ratio: _____

Service Factor:_____

Load

Application: _____

Load Type: Constant Torque _____ Variable Torque _____ Constant Horsepower _____

Load inertia reflected to motor: _____

Required breakaway torque from motor: _____

Running load on motor: _____

Peak torques (above 100% running): _____

Shortest/longest required accel. time: _____ / _____ secs up to _____ Hz from zero speed

Shortest/longest required decel. time: _____ / _____ secs down to _____ Hz from max. speed

Operating speed range: _____ Hz to _____ Hz

Time for motor/load to coast to stop: _____ secs

VFD

Source of start/stop commands: _____

Source of speed adjustment: _____

Other operating requirements: _____

Will the motor ever be spinning when the VFD is started? _____

Is the load considered to be high inertia? _____

Is the load considered to be hard to start? _____

Distance from VFD to the motor: _____feet

Type of VFD (V/Hz, Flux Vector, Closed Loop Vector): _____

Options desired: _____

_____

_____

Other special requirements/conditions: _____

03314-13_F83A.EPS

Figure 83 VFD application checklist. (1 of 2)

Power Supply

Supply Transformer: _____ kVA and _____ % Z or short circuit current at drive input: _____ amps

(If the drive does not include a built-in line reactor and the available feeder short circuit current is more than 100 times the drive FLA rating, a 1% line reactor or a drive isolation transformer is required.)

Total horsepower of all drives connected to supply transformer or feeder: _____ hp

Is a drive transformer or line reactor desired? _____

Any harmonic requirements? _____ % Voltage THD: _____ % Current THD: _____ IEEE 519: _____

Total non-drive load connected to the same feeder as drive(s): _____ amps

Service

Start-up Assistance: _____   Customer Training: _____

Preventive Maintenance: _____   Spare Parts: _____

Additional Issues

Will the VFD operate more than one motor? _____

Will the power supply source ever be switched with the VFD running? _____

Is starting or stopping time critical? _____

Are there any peak torques or impact loads? _____

Will user-supplied contactors be used on the input or output of the VFD? _____

Does the user or utility system have PF capacitors that are being switched? _____

Will the VFD be in a harsh environment or high altitude? _____

Does the utility system experience surges, spikes, or other fluctuations? _____

03314-13_F83B.EPS

Figure 83 VFD application checklist. (2 of 2)

5.3.4 Torque Requirements

The starting, peak, and running torques should be considered when selecting a drive. Starting torque requirements can vary from a small percentage of the full load to a value several times that of full-load torque. The peak torque varies because of a change in load conditions or the mechanical nature of the machine. The motor torque available to the driven machine must be more than that required by the machine from start to full speed. The greater the excess torque, the more rapid the acceleration potential.

5.3.5 Duty Cycle

Selecting the proper drive depends on whether the load is steady, varies, follows a repetitive cycle of variation, or has pulsating torques. Certain applications may require continuous reversals, long acceleration times at high torque due to inertia loads, frequent high rate acceleration, or cyclic overloads, which may result in severe motor heating if not considered in the selection of the drive. The duty cycle, which is defined as a fixed repeti-

VFD Environment

Heat rejection is a major concern with VFDs. The heat produced in the VFD cabinet can be substantial and can cause failure of the VFDs, SCRs, and IGBTs if their operating temperature limits are exceeded. For this reason, operating VFDs at higher altitudes often requires that the unit capacity be derated.

tive load pattern over a given period of time, is expressed as the ratio of on-time to the cycle period. When the operating cycle is such that the drive operates at idle, or at reduced load for more than 25 percent of the time, the duty cycle becomes an even more important factor in selecting the proper drive.

5.4.0 VFD Dynamic Braking

Braking provides a means of quickly stopping an AC motor when it is necessary to stop it faster than is normally possible by removing the power and letting the motor wind down. Braking can be accomplished in several ways. The method used depends on the application, available power, and circuit requirements. Some common methods of motor braking are:

- Dynamic braking (DC electric braking)
- Dynamic braking (AC drives)
- Electromechanical braking

Electromechanical braking is accomplished using solenoid-operated friction brakes. Friction brakes are used for applications such as printing presses, small cranes, overhead doors, hoisting equipment, and machine tool control. Since electromechanical braking is not used in HVAC applications, only dynamic braking is discussed in this section. Note that braking is not required in every VFD application.

5.4.1 Dynamic DC Braking of an AC Motor

One common method of braking an alternating current induction motor is by using a DC electric braking circuit. This is done by momentarily connecting a DC voltage to the stator winding (*Figure 84*). When DC is applied to the stator winding of an AC motor, the stator poles become electromagnets. Current is induced into the windings of the rotor as the rotor continues to spin through the magnetic field. This induced current produces a magnetic field around the rotor. The magnetic field of the rotor is attracted to the magnetic field produced in the stator. The attraction of these two magnetic fields produces a braking action

in the motor. An advantage of using this method of braking is that motors can be stopped rapidly without having to use brake linings or drums. DC braking cannot be used to hold a suspended load, however. Mechanical brakes must be employed when a load must be suspended, as with a crane or hoist.

The DC brake circuit in *Figure 84* operates as follows:

- Magnetic starter (M) is controlled by a standard Stop/Start pushbutton station with memory. Timing relay (TR) is an off-delay timer. Its normally open, timed-open (NOTO) contact is used to apply power to the braking relay (BR) after the Stop pushbutton is pressed.
- The off-delay of timing relay TR must be adjusted so that its contacts will remain closed until after the motor comes to a stop. At this time, the contacts open, disconnecting the DC power to the motor winding.
- When the STOP pushbutton is pressed, start contactor M de-energizes, thus removing three-phase input power from the motor. Also, braking relay BR is energized through the NC contact M and NOTO contact TR. This causes its normally-open contacts to close, energizing a DC supply consisting of a step-down transformer and full-wave bridge rectifier circuit. It also connects two of the motor leads (T1 and T3) to the output of the DC supply, initiating the motor braking action.
- A transformer with tapped windings is often used to adjust the amount of braking torque applied to the motor. Current-limiting resistors could be used for the same purpose. This allows for a low- or high- braking action depending on the application. The higher the applied DC voltage, the greater the braking force.
- An electrical interlock is provided by the NC contacts of starter M and the NC contacts of braking relay BR. This interlock prevents the motor starter and braking relay from being energized at the same time. This is required because the AC and DC power supplies must never be connected to the motor simultaneously. Total interlocking should always be used on electrical braking circuits. Total interlocking is the use of mechanical, electrical, and pushbutton interlocking.

5.4.2 Dynamic Braking with an AC Drive

Dynamic braking can be used with AC motors controlled by VFDs. Torque will always act to cause the motor's rotor to run towards synchronous speed. If the synchronous speed is suddenly

Figure 84 Circuit for DC braking of an AC motor.

reduced, negative torque is developed in the motor. When this occurs, the motor acts like a generator by converting mechanical power from the shaft into electrical power that is returned to the AC drive unit. This is similar to driving a car downhill in a lower gear so that the car's engine acts as a brake. In order for an AC drive to operate without damage during this condition, a means must exist to deal with the electrical energy returned to the drive by the motor.

Figure 85 shows a simplified diagram of an AC drive unit that uses dynamic braking. Electrical energy returned by the motor can cause voltage in the drive DC bus to become excessively high when added to the existing supply voltage. This excess voltage can damage the drive compo-

nents. To alleviate this problem, a braking resistor is connected across the DC bus through an IGBT switching transistor. The braking resistor is added and removed from the circuit by the switching action of the IGBT switch. Energy returned by the motor is impressed on the DC bus. When this energy reaches a predetermined level, the IGBT switch is turned on by the control logic. This places the braking resistor across the DC bus, causing the excess energy to be dissipated by the resistor, thus reducing the DC bus voltage. When the DC bus voltage is reduced to a safe level, the IGBT is turned off, removing the resistor from the DC bus. This process allows the motor to act as a brake, slowing the connected load.

RECTIFIER   DC BUS   INVERTER

03314-13_F85.EPS

Figure 85 Simplified schematic of an AC drive using dynamic braking.

Additional Resources

Adjustable Frequency Drives, Application Guide, Latest Edition. Milwaukee, WI: Cutler-Hammer.
Electric Motors and Drives: Fundamentals, Types and Applications. 3rd Edition, 2005. Austin Hughes. Newnes.

5.0.0 Section Review

1. If done properly, the volts per hertz (V/Hz) ratio in a VFD can allow a motor to run at twice its nameplate speed (rpm).

    a. True
    b. False

2. When a VFD is in the volts per hertz (V/Hz) optimization mode, the mode provides _____.

    a. an automatic torque boost
    b. a 10 percent reduction in torque
    c. linear voltage increases
    d. time for motor acceleration

3. Oversize the VFD drive used with variable-torque drives if the load torque exceeds _____.

    a. 100 percent
    b. 110 percent
    c. 125 percent
    d. 150 percent

4. How is DC electric braking applied to slow an AC motor?

    a. DC voltage is applied to electric brake calipers.
    b. DC voltage is applied to the stator of an AC motor.
    c. DC voltage is applied to the rotor of an AC motor.
    d. AC motor power is replaced with DC power.

6.0.0 ELECTRONICALLY COMMUTATED MOTORS

Objectives

Identify and describe how to service electronically commutated motors (ECMs).

   a. Identify and describe the operation of ECMs.
   b. Describe how to install and set-up an ECM.
   c. Describe how to troubleshoot an ECM.

Electronically commuted motors (ECMs) are now widely used in residential and light commercial HVAC equipment. These motors offer several benefits to the end user including the following:

- *Increased energy savings* – These motors consume a fraction of the power of a conventional PSC blower motor.
- *Improved comfort* – When coupled with modulating or multi-stage heating and cooling products, they adjust fan speed to match heating or cooling capacity. Fan speeds can be slowed down to allow for better dehumidification without overcooling the conditioned space.
- *Quieter operation* – The motors start at low speed and slowly ramp up to the required fan speed. When the thermostat is satisfied, they slowly ramp down the speed before shutting off. Continuous fan operation can be at a very low speed, allowing for good air circulation without excess air noise.

6.1.0 ECM Characteristics and Operation

ECMs are three-phase, direct-current (DC) motors that are brushless. They also have a permanent magnet rotor. Most ECM motors consist of two sections; the motor section and the motor

---

GOING GREEN

Electronically Commutated Motors

ECMs are true energy savers. An ECM installed in a typical residential furnace or fan coil consumes about 100 watts when running on low speed.

---

control (*Figure 86*). The motor control section has two functions. It first takes single-phase 120V or 240VAC power and converts it to DC. The motor can operate at either of these AC voltages simply by adding or removing a jumper on the five-pin connector in the motor control section. The five-pin connector provides the motor with its primary power supply. The control section then converts the DC power to three-phase power to actually turn the motor. The motor control section houses the electronics that control the speed of the motor.

The permanent magnet rotor is turned by sequentially energizing the three sets of windings in the stator to move the rotor in a clockwise direction. The signal applied to each stator is a square-wave voltage produced from rectified and smoothed DC. To advance the rotor, the square-wave voltage is first applied to the A-B windings (*Figure 87*) that moves the rotor clockwise. Power is then applied to the B-C windings, and then to the A-C windings to complete a full revolution of the rotor.

ECM motors are easy to identify. They look like conventional PSC motors with a section added on to the motor on the end opposite the shaft (*Figure 88*). The added-on section is the motor control section. The size of the motor control section has shrunk over time as more advance designs were introduced. With one popular brand of ECMs, all versions contain a five-pin connector where power is applied. The addition or deletion of a jumper on the five-pin connector determines the motor's AC operating voltage. With that same manufacturer, the connector used to carry the various speed control signals between the motor and the HVAC unit can be either a 16-pin connector (earlier versions) or a four-pin connector (later versions). Unlike conventional PSC motors, power is always available to the motor through the five-pin connector. It is only after additional speed control inputs are received through the 16-pin or four-pin connectors that the motor will start and

Figure 86 ECM major components.

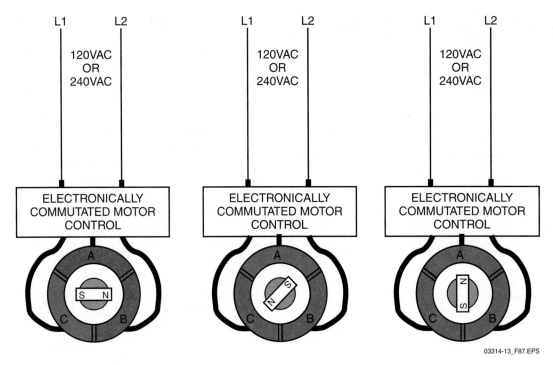

03314-13_F87.EPS

Figure 87 ECM motor rotation.

03314-13_F88.EPS

Figure 88 ECM identification.

run. Although power is applied to the motor system, it cannot function without instructions in the form of electrical signals.

ECMs have another characteristic that can help identify them. When turning the motor shaft by hand, points of high and low resistance can be felt. That is caused by the magnetism in the rotor.

6.1.1 ECM Speed Control

The speed of an ECM is controlled by the electronics in the motor control section and the input signals it receives from the electronic control within the furnace or air conditioner in which the ECM is installed. For example, a furnace control

ICM vs. ECM

Earlier HVAC systems literature and manufacturers referred to these motors as ICM (Integrated Control Module) motors. The words themselves indicate that a module was incorporated into a motor for its control. Today they are referred to as ECMs (Electronically Commutated Motors). However, the ICM acronym will still be seen on some products and wiring diagrams.

Rock and Roll!

When an ECM motor is first started, it may rock back and forth a few times. This is considered normal operation, as it decides which direction the motor should be turning.

board provides different speed control signals to the motor speed control based on the furnace's operating mode. If the furnace has staged heating capacity (high- and low-fire modes), it must provide a signal for a lower fan speed during low-fire operation. During installation, the furnace control board must be configured so that the correct fan speeds are available for the furnace's various operating modes. Different manufacturers accomplish this in different ways.

Figure 89 shows how an ECM installed in a two-stage furnace responds to a call-for-heat. As room temperature drops, the first stage of the room thermostat sends a call-for-heat signal to the furnace control board. Since the call is for first-stage heat, the furnace control board responds by sending a signal to the ECM requesting operation at low speed. These signals are typically in the form of a series of pulses. The frequency of the pulses tells the motor at what speed to operate. If the room thermostat calls for second-stage heat, the furnace control board would send a new signal to the motor to increase its speed.

Figure 89 ECM speed control.

6.2.0 ECM Installation and Setup

Other than replacement scenarios, ECMs are factory-installed by the manufacturer. There is no setup required for the motor itself. Any setup is done on the equipment that contains the ECM. The setup is very important, as they are installed in a variety of systems with varying airflow needs.

For example, a 100,000 Btuh gas furnace with an ECM may be installed with a 2, 2½, or 3-ton cooling unit. Each cooling unit capacity has its own airflow requirements based on cooling capacity (350 to 450 cfm/ton). If that same gas furnace has staged heating, each stage of heating has its own airflow requirements as well. If the system has a dehumidification mode, airflow requirements for that mode are much lower than needed for heating or cooling. Continuous fan operation also can have a dedicated and low fan speed. The control that sends signals to the ECM must be setup to ensure that all the necessary functions are accounted for.

The furnace control board on a deluxe furnace or air handling unit should have the capability to accommodate all of those diverse airflow requirements. How that is accomplished varies by manufacturer. Airflow setup can take the form of rotary selection switches, dip switches, or jumper wires. Some high-end systems may even allow setup to be done with a user interface such as a laptop computer, smart phone or tablet-style device. *Figure 90* shows an interface board used on a packaged heat/cool unit equipped with an ECM. On the board shown, equipment capacity and the size of auxiliary electric heaters are factors that help determine blower speed. The amplitude of the voltage signal and the terminal on the motor to which they are applied are based on the signals received from the room thermostat and/or other input devices.

> **CAUTION**
>
> Always remove power from the equipment before connecting or disconnecting ECM electrical connectors. Do not mate motor connectors with the power applied.

If an ECM has to be replaced, use an exact replacement. The only motor setup required then is to check the five-pin connector jumper to ensure the voltage is correct for the application. The jumper must be installed for 120V operation and removed for 240V operation. Note, however, that the two connectors—the power connector and the communications connector—must be compatible with the connectors provided on the equip-

ICM PRINTED CIRCUIT BOARD

Figure 90 ECM speed control interface board.

03314-13_F90.EPS

ment. The connectors are keyed such that they can only be mated one way and with the opposite gender. This helps eliminate any error in connections. Forcing the connectors together incorrectly will likely result in motor and/or controller damage once they are energized. If an alternate motor must be used, the manufacturer's instructions must be followed carefully to ensure that the motor and the unit controller are compatible and communicating with each other. Note that there are no universal replacements in the ECM world. Even when an exact replacement is provided, it is important to read and apply the instructions provided with the motor. Changes in design or wiring may have taken place since the original, now-failed motor was placed in service.

When a new motor is installed, be sure to position the motor to allow the wiring connections to hang down between the 4 o'clock and 8 o'clock positions. This helps prevent any moisture from entering the connectors. Once the new motor is installed, check the operation thoroughly in all operating modes to ensure the motors response to the conditions is correct. With furnaces, ensure that the temperature rise in the heating mode remains within the allowable range.

6.3.0 ECM Troubleshooting

Troubleshooting an ECM is likely to happen as part of a routine service call, especially one where comfort is an issue. Where comfort and airflow are involved, always check for obvious problems first, such as the following:

- System airflow problems such as dirty air filters, closed duct dampers or blocked supply registers or return grilles. It may be necessary to measure system airflow to confirm that the proper volume of air is being delivered.
- Low refrigerant charge.
- Room thermostat setpoint.
- Connections on the ECM are tight and secure. Both connectors are keyed to ensure a correct connection.
- If the equipment control board has a self-diagnostic feature or displays fault codes, use those features to help identify the problem.

If no obvious problems are found, check the heat/cool equipment to confirm the control board has been properly configured for all airflow contingencies. It is common for some manufacturers to have default settings on their controls as they come from the factory. Those default settings may not be appropriate for a specific equipment application. An incorrect setup in an ECM-equipped system can result in poor system performance.

6.3.1 Motor Troubleshooting

The following troubleshooting procedures were developed by one major manufacturer of ECMs. Other motor manufacturers have similar procedures. These procedures are general in nature and are not meant to replace specific troubleshooting procedures provided by the equipment manufacturer.

> **CAUTION**
>
> Always remove power from the equipment before connecting or disconnecting ECM electrical connectors. Do not mate motor connectors with the power applied.

If the motor is not running, disconnect power from the air handler. Then remove the five-pin connector and restore power. Check for the correct voltage at the connector pins (*Figure 91*). When using a 120VAC power supply, the polarity of the voltage must also be correct. The applied voltage should be ±15 percent of the desired voltage (120V or 240V). Note that the motor must always be well grounded for proper operation.

Equipment manufacturers often supply charts listing the voltages that should be read across the 16-pin or four-pin connectors of the motor. Voltages can be AC or DC. They also may supply the conditions under which the various voltages

should be read. These signals and voltages can be tested when specific information is available. This process determines whether the equipment control module is functional and sending the expected signals.

Devices that simulate communication input signals are available to test the motor (*Figure 92*). The tester shown can be used with motors with 16-pin or four-pin connectors. The tester simulates signals from the control board in the air handler.

Here is how the tester is used to test an ECM.

- Turn off power to the air handler. Disconnect the 16-pin or four-pin connector from the motor but leave the five-pin power supply connector attached to it. Turn off the tester.
- Connect the alligator leads on the tester to the R and C terminals on the control transformer or terminal board to obtain 24VAC. This powers the tester.
- Connect the 16-pin or the four-pin connector on the tester to the matching connector on the motor control section. Both connectors may be provided on the tester to allow it to work with different motors.
- Turn the tester on.

If the motor runs when the tester is turned on, the problem lies elsewhere. Potential causes include a failed air handler control board, or a loose connection. If the motor does not run, the problem may be within the motor control section or the motor itself. Remove the tester and restore all normal connections when done.

If the motor did not run, testing must proceed to the motor and its integrated control. Since the motor control section is bolted to the motor, those bolts can be removed and both sections can be checked separately. The motor section can usually remain attached to its mount in the blower assembly during this procedure. The motor section is tested as follows:

- Turn off power to the air handler. Disconnect the 16-pin or four-pin communications connector from the motor as well as the five-pin power supply connector. Then wait at least 5 minutes before proceeding to the next step.

> **WARNING!**
>
> After disconnecting power and/or communications connectors from an ECM motor, wait at least five minutes before accessing the control section. This allows time for the capacitors to discharge. If the capacitors are not allowed sufficient time to discharge first, an electrical shock may result.

120

AC Volts

Ṽ   Com

5   4   3   2   1

AC LINE   AC LINE   GROUND   POWER JUMPER

240

AC Volts

Ṽ   Com

120VAC motors must
have a jumper between
terminals 1 and 2 for
proper operation.

240VAC motors should
not have a jumper
between terminals 1 and
2 to prevent damage to
the motor.

03314-13_F91.EPS

Figure 91 Input voltage check.

- Unbolt the two sections and disconnect the three-pin connector between the two sections. Place the motor control section aside.
- The three pins on the connector attach to each of the three-phase motor leads. Use an ohmmeter set on its lowest scale to check the resistance of each motor winding (*Figure 93*). Each winding should measure less than 20 ohms. There should be no more than a ±10 percent difference between the three resistance readings.
- Set the ohmmeter on its highest scale to test for grounded motor windings. Place one lead on

the frame of the motor (ground) and the other lead on each of the pins of the three-pin connector. Ideally, all readings to ground will show no continuity (open circuit). Replace any motor that reads less than 100,000 ohms to ground.

If the motor windings are intact and not grounded, the motor control section is likely defective. The motor can be reused and a new control section installed. The motor can remain bolted to its mount in the blower section during the replacement. Of course, the entire ECM must be removed if the motor section is defective.

Figure 93 ECM winding resistance check.

Figure 92 ECM tester.

Additional Resources

The ECM Textbook. Fort Wayne, IN: GE-ECM by REGAL-BELOIT, 2007.

6.0.0 Section Review

1. The signal applied to the stator in an ECM is a _____.

   a. pure sine wave
   b. modulated AC signal
   c. current-limited AC signal
   d. square-wave voltage

2. Remove the jumper on the five-pin connector to enable 120VAC motor operation.

   a. True
   b. False

3. When measuring the motor winding resistance of an ECM, the readings should not vary any more than _____.

   a. ±1 percent
   b. ±10 percent
   c. ±15 percent
   d. ±20 percent

Summary

HVAC systems contain various types of electric motors. These motors are controlled by different devices and different control schemes. The ability to analyze HVAC control systems is a critical skill for the service technician because a large percentage of the problems that occur are control circuit faults. In the past, control circuit devices were electromechanical. Modern equipment relies more and more on electronic devices. Variable frequency drives are now used to efficiently change the speed of large AC motors. Air handlers and furnaces used in residential and light commercial applications use electronically commutated motors (ECMs) for more efficient operation. No matter how complex the control system, it consists of individual components that are combined to form control functions such as heating, cooling, fan control, and defrost. If you know how the components work, you can figure out how they fit together and how the system functions as a whole. Once you have these skills, you can troubleshoot any system.

Effective troubleshooting is a process by which the HVAC technician listens to a customer's complaint, performs an independent analysis of a problem, and then initiates and performs a systematic step-by-step approach to troubleshooting that results in the correction of the problem.

Review Questions

1. All relay contacts shown on schematic diagrams are shown with the relay in the de-energized position.
   a. True
   b. False

2. The purpose of a compressor short-cycle timer is to _____.
   a. prevent compressor restart before system pressures equalize
   b. ensure the compressor runs for at least five minutes
   c. keep track of how often the compressor turns on
   d. run the compressor every five minutes to prevent freeze-up

3. The purpose of a lockout relay is to _____.
   a. prevent unauthorized equipment use
   b. prevent compressor operation until the fan runs
   c. protect the compressor motor from current surges
   d. prevent restart of equipment when a safety control resets

4. The purpose of a motor starter is to _____.
   a. jump-start the compressor if power fails
   b. vary the speed of a motor as conditions change
   c. check motor current when the motor is running
   d. power the motor and provide overload protection

5. Which of the following *is not* an application for an outdoor temperature thermostat?
   a. Control compressor capacity.
   b. Dual-fuel heat pump switchover.
   c. Chiller protection.
   d. Control crankcase heater operation.

6. A limit control is provided on a furnace to _____.
   a. ensure the furnace provides adequate heat
   b. shut off the furnace fan motor
   c. prevent the furnace from overheating
   d. sense the presence of carbon monoxide

7. Low-voltage (24V) control circuits are used in HVAC systems because _____.
   a. step-down transformers produce only 24V output
   b. 24V power is safer and cheaper
   c. 24V power is readily available from the electric company
   d. it is the same voltage used to operate the compressor and fan motors

8. Which of the following is *not* true regarding a cooling compensator?
   a. It compensates for the lag between the call for cooling and when cooling actually begins.
   b. It is a variable resistor just like the heat anticipator.
   c. It causes the cooling thermostat to close sooner than it otherwise would.
   d. It is usually placed in parallel with the cooling thermostat switch.

9. The differential in a thermostat is _____.
   a. the difference between a thermostat's cut-in and cut-out points
   b. the difference between the cooling and heating setpoints
   c. normally set apart at least 12°F
   d. the difference between heat anticipator and heating setpoints

10. Which of the following is *not* a common feature on an electronic thermostat?
    a. Battery backup to prevent loss of programs
    b. Motor-driven heat anticipator
    c. Program override
    d. Digital readouts

11. A good location for a thermostat is _____.
    a. anywhere that it will receive direct sunlight
    b. about five feet above the floor on an outside wall
    c. about five feet above the floor on an inside wall
    d. in a corner where there is minimum air circulation

12. In a standard thermostat wiring scheme, the R terminal is connected to _____.

   a. the fan control
   b. 24V power
   c. the reversing valve
   d. the cooling circuits

13. The best way to obtain the correct setting for a thermostat heat anticipator is to _____.

   a. check the thermostat label
   b. check the thermostat subbase
   c. take current measurements on the heating control circuit
   d. consult the furnace literature

14. What is the purpose of the heat anticipator in heating thermostats?

   a. It shortens or lengthens the burner cycle to provide greater accuracy.
   b. It keeps the thermostat warm in cold weather.
   c. It turns on a light when the furnace is about start.
   d. It directly controls the start time of the burner.

15. In the electrical control circuit of a combined heating/cooling system, the outdoor fan motor _____.

   a. usually has a continuous-on mode for ventilation
   b. runs at low speed in the heating mode
   c. is on whenever the compressor is on
   d. is off whenever the compressor is on

Refer to *Review Question Figure 1* for Review Questions 16 through 20.

16. Which component prevents the compressor from starting immediately after there is a call-for-cooling?

   a. Time-delay relay
   b. Indoor fan relay
   c. Contactor
   d. Short-cycle timer

17. Which component allows the indoor fan to cycle on and off with the compressor?

   a. Time-delay relay
   b. Fan function switch
   c. Contactor
   d. Short-cycle timer

18. Which component prevents the simultaneous energizing of the low- and high-speed windings of the indoor fan motor?

   a. TDR
   b. IFR
   c. PS
   d. IDR

19. Which component prevents the IFR from being energized during a call-for-heat?

   a. Time-delay relay
   b. Flame rollout switch
   c. Pressure switch
   d. Room thermostat

20. When is the time-delay relay energized?

   a. On a call-for-cooling
   b. When the gas valve energizes
   c. On a call-for-heating
   d. When the compressor starts

Figure 1

LEGEND

C	Compressor contactor	OL	Overload
COMP	Compressor motor	PRI	Primary
CH	Crankcase heater	PS	Pressure switch
GV	Gas valve	RS	Flame rollout switch
IDM	Induced-draft motor	SEC	Secondary
IDR	Induced-draft relay	TC	Cooling thermostat
IFM	Indoor fan motor	TDR	Time delay relay
IFR	Indoor fan relay	TH	Heating thermostat
LS	Limit switch	TRAN	Transformer

03314-13_RQ01.EPS

21. A pneumatic-electric relay is one that _____.

    a. uses air pressure to open and close electrical contacts
    b. uses a solenoid to close a pneumatic valve
    c. supplies power to the air compressor
    d. uses fluid pressure to open and close relay contacts

22. A troubleshooting aid normally given on a label diagram is the _____.

    a. troubleshooting tree
    b. troubleshooting table
    c. wiring diagram
    d. fault isolation diagram

23. When testing a relay coil, the multimeter indicates a measurable resistance. This means that the coil is _____.

    a. shorted
    b. usually good
    c. open
    d. partially shorted

24. When checking a 440V motor for grounded windings, the minimum resistance that should be measured between any lead and ground is _____.

    a. 1,000 ohms
    b. 230,000 ohms
    c. 440,000 ohms
    d. infinity

25. When troubleshooting the operation of a compressor start relay circuit by measuring the start winding current draw with a clamp-on ammeter, no current is measured on the clamp-on ammeter when the compressor is first energized. This likely means that the _____.

    a. start relay contacts are open
    b. start relay contacts are closed
    c. start relay coil is open
    d. start capacitor is shorted

26. With a variable-frequency drive (VFD) motor, its speed is proportional to the _____.

    a. applied frequency
    b. voltage amplitude
    c. power factor
    d. motor winding resistance

27. Why are insulated gate bipolar transistors (IGBT) used in VFD controllers?

    a. Low cost
    b. Rapid heat dissipation
    c. High switching speed
    d. Low power consumption

28. An ECM can be described as a _____.

    a. single-phase motor
    b. direct-current motor
    c. three-phase AC motor
    d. split-phase AC motor

29. Power is always available to an ECM motor through the four-pin connector.

    a. True
    b. False

30. Voltages measured at the four-pin or 16-pin connector on an ECM are _____.

    a. AC only
    b. DC only
    c. filtered AC only
    d. AC or DC

Trade Terms Introduced in This Module

Actuator: The portion of a regulating valve that converts one type of energy, such as pneumatic pressure, into mechanical energy (for example, opening or closing a valve).

Base speed: The rating given on a motor's nameplate where the motor will develop rated horsepower at rated load and voltage.

Bleed control: A valve with a tiny opening that permits a small amount of fluid to pass.

Cooling compensator: A fixed resistor installed in a thermostat to act as a cooling anticipator.

Deadband: A temperature band, usually 3°F, that separates heating and cooling in an automatic changeover thermostat.

Differential: The difference between the cut-in and cut-out points of a thermostat.

Droop: A mechanical condition caused by heat that affects the accuracy of a bimetal thermostat.

Electric-pneumatic (E-P) relay: A three-way pneumatic solenoid valve.

Fault isolation diagram: A troubleshooting aid usually contained in the manufacturer's installation, startup, and service instructions for a particular product. Fault isolation diagrams are also called troubleshooting trees. Normally, fault isolation diagrams begin with a failure symptom then guide the technician through a logical decision-action process to isolate the cause of the failure.

Insulated gate bipolar transistor (IGBT): A type of transistor that has low losses and low gate drive requirements. This allows it to be operated at higher switching frequencies.

Invar®: An alloy of steel containing 36 percent nickel. It is one of the two metals in a bimetal device.

Label diagram: A troubleshooting aid usually placed in a convenient location inside the equipment. It normally depicts a wiring diagram, a component arrangement diagram, a legend, and notes pertaining to the equipment.

Ladder diagram: A troubleshooting aid that depicts a simplified schematic diagram of the equipment. The load lines are arranged like the rungs of a ladder between vertical lines representing the voltage source. Normally, all the wire color and physical connection information is eliminated from the diagram to make it easier to use by focusing on the functional, not the physical, aspects of the equipment.

Pneumatic-electric (P-E) relay: A pressure switch in which a pneumatic signal causes an electrical change.

Wiring diagram: A troubleshooting aid, sometimes called a schematic, that provides a picture of what the unit does electrically and shows the actual external and internal wiring of the unit.

COMMON SCHEMATIC SYMBOLS

SWITCHES						

DISCONNECT	MAGNETIC CIRCUIT BREAKER	THERMAL CIRCUIT BREAKER	LIMIT			MAINTAINED POSITION
			SPRING RETURN			
			NORMALLY OPEN	NORMALLY CLOSED	NEUTRAL	
			DOUBLE THROW	DOUBLE POLE	TRIPLE POLE	

Liquid Level		Vacuum & Pressure		Temp. Actuated		Air or Water Flow	
NORMALLY OPEN (1)	NORMALLY CLOSED (2)	NORMALLY OPEN (1)	NORMALLY CLOSED (2)	NORMALLY OPEN (1)	NORMALLY CLOSED (2)	NORMALLY OPEN (1)	NORMALLY CLOSED (2)

CONDUCTORS		FUSES	COILS			
NOT CONNECTED	CONNECTED	OR	RELAYS, TIMERS, ETC.	OVERLOAD THERMAL	SOLENOID	TRANSFORMER

PUSHBUTTONS				
SINGLE CIRCUIT		DOUBLE CIRCUIT	MUSHROOM CIRCUIT	MAINTAINED CONTACT
NORMALLY OPEN	NORMALLY CLOSED			

TIMER CONTACTS				GENERAL CONTACTS		
CONTACT ACTION IS RETARDED WHEN COIL IS:				STARTERS, RELAYS, ETC.		
ENERGIZED		DE-ENERGIZED		OVERLOAD THERMAL	NORMALLY OPEN	NORMALLY CLOSED
NORMALLY OPEN	NORMALLY CLOSED	NORMALLY OPEN	NORMALLY CLOSED			

(1) **Make on rise**
(2) **Make on fall**

03314-13_A01A.EPS

03314-13_A01B.EPS

Additional Resources

This module presents thorough resources for task training. The following resource material is suggested for further study.

Adjustable Frequency Drives, Application Guide, Latest Edition. Milwaukee, WI: Cutler-Hammer.

The ECM Textbook. Fort Wayne, IN: GE-ECM by REGAL-BELOIT, 2007.

Electrical Motor Controls for Integrated Systems. 4th Edition, 2009. Gary Rockis and Glen A. Mazur. Homewood, IL: American Technical Publishers, Inc.

Electric Motors and Drives: Fundamentals, Types and Applications. 3rd Edition, 2005. Austin Hughes. Newnes.

Fundamentals of HVAC/R. Latest Edition. Carter Stanfield and David Skaves. Prentice Hall. Upper Saddle River, NJ.

Refrigeration and Air Conditioning: An Introduction to HVAC/R. Latest Edition. Larry Jeffus. Pearson/Prentice Hall. Upper Saddle River, NJ.

Figure Credits

Topaz Publications, Inc., Module opener, Figures 1, 2, 9, 12B, 13, 16A, 22, 46, SA03

Courtesy of Crydom, Inc., Figure 8

Courtesy of Mara Industrial Supply, SA01

Courtesy of Schneider Electric, Figure 10

Courtesy of Danfoss, Figures 12A, 12C

Courtesy of White-Rodgers/Emerson Climate Technologies, Figures 16B, 23, 39

Courtesy of Honeywell International, Figures 24, 75

Measurement Specialties, Inc., Temperature Products Group, Figure 26

Carrier Corporation, Figures 27, 52, 53, 60, 63, 90

Courtesy of Pinnacle Heating and Cooling, Figure 51

Xylem Applied Water Systems, Figure 78

Courtesy of Taco, Inc., Figure 79, 80

Square D/Schneider Electric, SA04

Courtesy of **www.thedealertoolbox.com**, Figures 86, 88, 91, 93

Section Review Answer Key

Answer	Section Reference	Objective
Section One		
1. d	1.1.7	1a
2. c	1.2.5	1b
Section Two		
1. c	2.1.4	2a
2. a	2.2.0	2b
3. d	2.3.1	2c
4 .a	2.4.0	2d
Section Three		
1. a	3.1.0	3a
2. c	3.2.2	3b
3. a	3.3.1	3c
Section Four		
1. b	4.1.4	4a
2. b	4.2.4	4b
3. d	4.3.2	4c
4. c	4.4.1	4d
5. c	4.5.3	4e
6. a	4.6.1	4f
Section Five		
1. a	5.1.0	5a
2. a	5.2.0	5b
3. b	5.3.2	5c
4. b	5.4.1	5d
Section Six		
1. d	6.1.0	6a
2. b	6.2.0	6b
3. b	6.3.1	6c

NCCER CURRICULA — USER UPDATE

NCCER makes every effort to keep its textbooks up-to-date and free of technical errors. We appreciate your help in this process. If you find an error, a typographical mistake, or an inaccuracy in NCCER's curricula, please fill out this form (or a photocopy), or complete the online form at **www.nccer.org/olf**. Be sure to include the exact module ID number, page number, a detailed description, and your recommended correction. Your input will be brought to the attention of the Authoring Team. Thank you for your assistance.

Instructors – If you have an idea for improving this textbook, or have found that additional materials were necessary to teach this module effectively, please let us know so that we may present your suggestions to the Authoring Team.

NCCER Product Development and Revision

13614 Progress Blvd., Alachua, FL 32615

Email: curriculum@nccer.org
Online: www.nccer.org/olf

❏ Trainee Guide   ❏ Lesson Plans   ❏ Exam   ❏ PowerPoints   Other _____

Craft / Level: _____   Copyright Date: _____

Module ID Number / Title: _____

Section Number(s): _____

Description: _____

_____

_____

_____

_____

Recommended Correction: _____

_____

_____

_____

_____

Your Name: _____

Address: _____

_____

Email: _____   Phone: _____

03210-13

Troubleshooting Cooling

OVERVIEW

The refrigeration system is a closed-loop system, so it is not always easy to tell where in the system a problem has occurred. Since the operation of the refrigerant circuit is always affected by external conditions, it can be challenging to determine if a problem is within the system or the result of an external influence. A thorough understanding of the refrigeration circuit provides the basis for matching specific symptoms to possible causes. An experienced service technician learns to recognize refrigeration cycle failure patterns quickly and efficiently, and to provide complete solutions.

Module Three

Trainees with successful module completions may be eligible for credentialing through NCCER's National Registry. To learn more, go to **www.nccer.org** or contact us at **1.888.622.3720**. Our website has information on the latest product releases and training, as well as online versions of our *Cornerstone* magazine and Pearson's product catalog.

Your feedback is welcome. You may email your comments to **curriculum@nccer.org**, send general comments and inquiries to **info@nccer.org**, or fill in the User Update form at the back of this module.

This information is general in nature and intended for training purposes only. Actual performance of activities described in this manual requires compliance with all applicable operating, service, maintenance, and safety procedures under the direction of qualified personnel. References in this manual to patented or proprietary devices do not constitute a recommendation of their use.

Objectives

When you have completed this module, you will be able to do the following:

1. Describe the operation of the refrigeration cycle and identify problems that can occur.
   a. Identify the primary components of the refrigeration cycle and explain their individual function.
   b. Describe the operation of a typical refrigeration cycle.
   c. Explain the steps to analyzing refrigeration circuit operating conditions.
   d. Describe possible causes for specific abnormal pressures and temperatures.
   e. Explain how condenser and evaporator airflow problems affect the refrigeration cycle.
   f. Identify and describe problems related to fixed metering devices.
   g. Identify and describe problems related to TXVs and distributors.
   h. Identify and describe other problems related to the refrigerant circuit.
2. Explain how to troubleshoot and replace a cooling system compressor.
   a. Identify common problems that can result in compressor failure.
   b. Explain how to troubleshoot compressor mechanical problems.
   c. Explain how to replace a hermetic compressor following a failure.
   d. Describe the additional steps that may be required to replace a compressor following an electrical failure.

Performance Task

Under the supervision of your instructor, you should be able to do the following:

1. Demonstrate the ability to isolate and determine the solution for at least four of the following types of malfunctions:
   - Compressor failures
   - System-related compressor problems
   - Refrigerant undercharge or overcharge
   - Evaporator and condenser problems
   - Metering device problems
   - Refrigerant lines and accessories
   - Noncondensables and refrigerant circuit contamination

Trade Terms

Burnout

Capacitance boost

Industry-Recognized Credentials

If you are training through an NCCER-accredited sponsor, you may be eligible for credentials from NCCER's Registry. The ID number for this module is 03210-13. Note that this module may have been used in other NCCER curricula and may apply to other level completions. Contact NCCER's Registry at 888.622.3720 or go to **www.nccer.org** for more information.

Contents

Figures and Tables

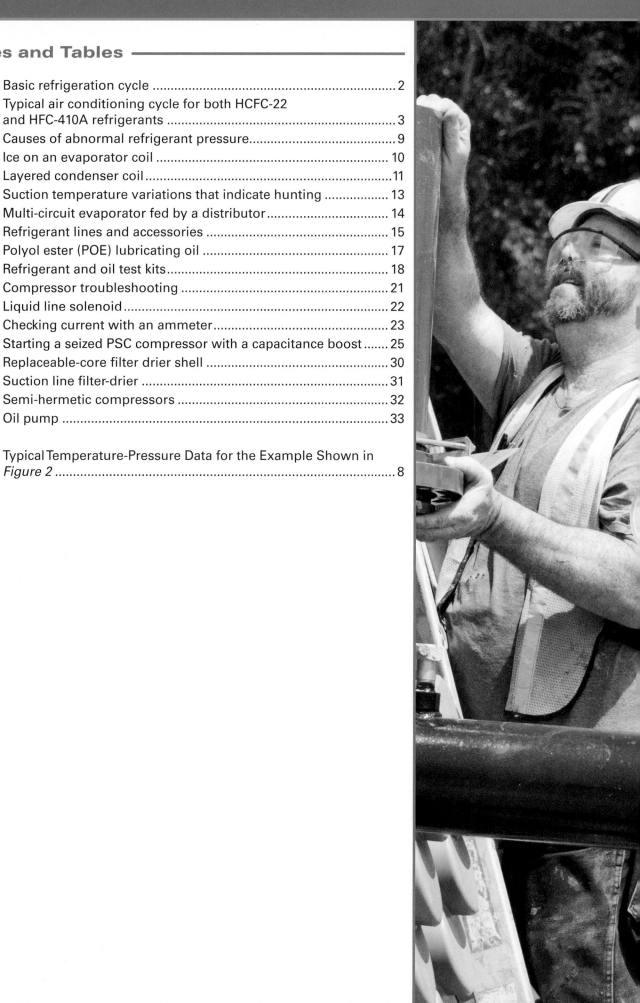

1.0.0 REFRIGERATION CYCLE OPERATION AND TROUBLESHOOTING

Objective

Describe the operation of the refrigeration cycle and identify problems that can occur.

a. Identify the primary components of the refrigeration cycle and explain their individual function.
b. Describe the operation of a typical refrigeration cycle.
c. Identify steps to analyzing refrigeration circuit operating conditions.
d. Describe possible causes for specific abnormal pressures and temperatures.
e. Explain how condenser and evaporator airflow problems affect the refrigeration cycle.
f. Identify and describe problems related to fixed metering devices.
g. Identify and describe problems related to TXVs and distributors.
h. Identify and describe other problems related to the refrigerant circuit.

Trade Terms

Burnout: The condition in which the breakdown of the motor winding insulation causes the motor to short out or ground electrically.

An understanding of the basic refrigeration cycle is critical to the task of troubleshooting cooling systems. There are many types of systems used to provide cooling for personal comfort, food preservation, and industrial processes. Each of these systems uses a mechanical refrigeration system to produce the cooling. The operation of the mechanical refrigeration system is basically the same for most all vapor compression systems. Varying from system to system are the type of refrigerant used, the size and style of the components, pressures and temperatures, and the installed locations of the components and refrigerant lines.

Although the basic refrigeration cycle and its components are covered in previous modules, it is reviewed again here to prepare trainees for troubleshooting. Note that the following discussion is based primarily on an air-cooled system.

1.1.0 Refrigeration Cycle Components and Functions

A basic refrigeration system (*Figure 1*) includes the following components:

- *Evaporator* – A heat exchanger in which the heat from the area or item being cooled is absorbed and transferred to the refrigerant.
- *Compressor* – Creates the pressure differential needed to make the refrigeration cycle work. The compressor is referred to as the heart of the system.
- *Condenser* – A heat exchanger in which the heat absorbed by the refrigerant is transferred to the cooler outdoor air, or another cooler substance.
- *Metering device* – Controls the flow of refrigerant to the evaporator. It provides a pressure drop that lowers the boiling point of the refrigerant just before it enters the evaporator. In *Figure 1*, the metering device is a thermostatic expansion valve, commonly called a TXV or a TEV.

Also shown in *Figure 1* is the refrigerant piping used to connect the basic components in order to provide the path for refrigerant flow. Together, the components and lines form a closed refrigeration system. The arrows show the direction of flow for the refrigerant through the closed system. The lines are as follows:

- *Suction line* – The tubing that carries the low-temperature refrigerant vapor from the evaporator to the compressor.
- *Hot gas line* – The tubing that carries hot refrigerant vapor from the compressor to the condenser. This is also called the discharge line.
- *Liquid line* – The tubing that carries liquid refrigerant, formed in the condenser, to the metering device.

Upon a call for cooling, the evaporator receives a low-temperature, low-pressure liquid/vapor refrigerant mix from the metering device. The evaporator is mainly a series of tubing coils that exposes the cooler refrigerant to the warmer air passing over them. Heat from the warmer air is transferred through the evaporator tubing into the cooler refrigerant, causing it to boil and vaporize. Remember that even though it has just boiled, it is still not considered hot because refrigerants boil at such low temperatures. So it is a cool, low-temperature, low-pressure refrigerant vapor that finally leaves the evaporator. Any increase in vapor temperature beyond saturation is called superheat.

The compressor receives this vapor and compresses it. It then becomes a hot, high-temperature, high-pressure vapor. This vapor travels to

Figure 1 Basic refrigeration cycle.

the condenser via the hot gas line, also known as the discharge line.

Like the evaporator, the condenser is a series of tubing coils through which the refrigerant flows. As cooler air moves across the condenser tubing, the hot refrigerant vapor cools. As it gives up heat to the outside air, it cools to the condensation point, where it begins to change from a vapor into a liquid. Eventually all refrigerant vapor condenses to a liquid. Any decrease in temperature below the saturation point is referred to as subcooling. This warm, high-pressure liquid travels through the liquid line to the input of the metering device.

The metering device regulates the flow of refrigerant into the evaporator. It also decreases the refrigerant pressure significantly. This results in a dramatic drop in refrigerant temperature as well. By restricting refrigerant flow, the metering device converts the warm, high-pressure refrigerant from the condenser into the low-temperature, low-pressure refrigerant needed to absorb heat in the evaporator.

1.2.0 Refrigeration Cycle for a Typical Air Conditioning System

Figure 2 shows a basic air conditioning system with the components divided into two sections.

The high-pressure side includes all the components in which the pressure of the refrigerant is at or above the condensing pressure. This is often referred to as head pressure, discharge pressure, or high-side pressure. The low-pressure side includes all the components in which the pressure is at or below the evaporating pressure. This is often called the suction pressure or low-side pressure. The vertical dashed line represents the division between the two sides of the system. The major pressure changes occur at the compressor and the metering device.*Figure 2*

The following describes a typical air conditioner operating on either HCFC-22 or HFC-410A. These are the two most commonly used refrigerants in today's air conditioning systems. This example demonstrates the temperature/pressure relationships that are typical of those that exist at key points in the system during operation. For our example, assume an indoor air temperature of 75°F and an outdoor air temperature of 95°F. The temperatures and pressures used in the example are representative values only. They will vary due to equipment and load conditions. Load conditions vary from system to system and are based on climatic conditions and the heat load generated within the space. The humidity level of the indoor air also plays a key role in the load.

Note: Lower pressure is HCFC-22/higher pressure is HFC-410A.

TEMPERATURE °F	PRESSURE (IN PSIG)	
	HCFC-22	HFC-410A
27	51.2	91.0
28	52.4	92.9
29	53.6	94.9
30	54.9	96.8
31	56.2	98.8
32	57.5	100.9
33	58.8	102.9
34	60.1	105.0
35	61.5	107.1
36	62.8	109.2
37	64.2	111.4
38	65.6	113.6
39	67.1	115.8
40	68.5	118.1
41	70.0	120.3
42	71.5	122.7
43	73.0	125.0
44	74.5	127.4
45	76.0	129.8
46	77.6	132.2
47	79.2	134.7
48	80.8	137.2
49	82.4	139.7

TEMPERATURE °F	PRESSURE (IN PSIG)	
	HCFC-22	HFC-410A
50	84.0	142.2
55	92.6	155.5
60	101.6	169.6
65	111.2	184.5
70	121.4	200.4
75	132.2	217.1
80	143.6	234.9
85	155.7	253.7
90	168.4	273.5
95	181.8	294.4
100	195.9	316.4
105	210.8	339.6
110	226.4	364.1
115	242.7	389.9
120	259.9	416.9
125	277.9	445.4
130	296.8	475.4
135	316.6	506.9
140	337.3	540.1
145	358.9	575.1
150	381.8	611.9

PRESSURE-TEMPERATURE CHART

03210-13_F02.EPS

Figure 2 Typical air conditioning cycle for both HCFC-22 and HFC-410A refrigerants.

Since a great deal of energy is required to condense airborne moisture, high humidity levels indoors increase the refrigeration load. Condensing moisture on the evaporator coil surface represents a latent load, which becomes a significant part of the total refrigeration load.

The numbers given in the description below correspond to the numbers shown in *Figure 2*. Follow along on the figure as the system is described. Note that the temperature at each point remains the same for both refrigerants. The first pressure shown represents the pressure of HCFC-22 at these conditions; the second pressure represents that of HFC-410A.

1. Refrigerant (roughly 80 percent liquid, 20 percent vapor) is supplied from the metering device to the evaporator. The pressures of 69/118 psig correspond to the 40°F saturation point for each refrigerant. (See PT chart in *Figure 2*.) The 40°F saturation point used here is typical of the temperatures normally used for evaporators in air conditioning systems. Since there is a mixture of refrigerant and liquid present here, the refrigerant is in saturation at this moment.

2. Because the refrigerant flowing through the evaporator is cooler (at 40°F) than the warmer inside room air (at 75°F) passing over the evaporator, it absorbs heat, causing the liquid refrigerant to boil and turn into a vapor. After traveling about 90 percent of the way through the evaporator tubing, all of the refrigerant has boiled into its vapor state. It can now begin to warm up.

3. During the remaining 10 percent of travel through the evaporator, the saturated vapor continues to absorb heat from the warmer air, thus raising its temperature to 50°F. In other words, the saturated vapor is superheated 10°F (50°F – 40°F = 10°F). This superheated vapor flows through the suction line and is drawn into the low-pressure side of the compressor. The cooled inside room air is re-circulated by the evaporator fan back into the room at a temperature of about 55°F.

4. The superheated vapor applied at the suction input of the compressor picks up an additional 3°F to 5°F of superheat because the vapor in the suction line absorbs more heat from the warmer surrounding air as it travels from the evaporator to the compressor. This can be a significant variable from one system to another.

5. After compression, the now highly superheated gas from the compressor flows through the hot gas line to the condenser. The hot gas may be close to 200°F at 297/475 psig, and the refrigerant superheat would be around 70°F as a result (200°F – 130°F = 70°F). Typically, discharge superheat will range between 50°F and 150°F. All of this superheat must be removed before the refrigerant vapor can be condensed into a liquid. The 200°F refrigerant in the hot gas line easily gives up some of its superheat to the surrounding 95°F air. The hot gas line is not normally insulated, and the tubing is a good conductor of heat.

6. Because the refrigerant in the condenser is still hotter than the warmer outside air passing over the condenser, it easily gives up the remaining superheat as it flows through. Its temperature eventually falls to 130°F. In this example, 130°F represents the saturation temperature of the refrigerant. At this temperature, and only at this temperature for the pressures shown in *Figure 2*, liquid and vapor can coexist. As heat continues to be transferred from the vapor to the cooler outside air, the vapor begins to condense into a liquid. After the refrigerant has traveled about three quarters of the way through the condenser, all of the refrigerant has condensed into a liquid.

7. During the remaining travel through the condenser, the liquid refrigerant continues to drop in temperature (subcools). This lowers its temperature to about 115°F. In other words, the liquid refrigerant is subcooled 15°F (130°F – 115°F = 15°F).

8. Subcooled liquid refrigerant from the condenser flows through the liquid line to the metering device. Since the liquid line is often uninsulated, the subcooling may increase or decrease due to length, location, and elevation change. In the case shown in *Figure 2*, 5°F of additional subcooling occurred. The refrigerant temperature is now at 110°F.

9. The metering device controls the flow of liquid refrigerant to the evaporator. Subcooled liquid from the condenser enters at a temperature of 120°F and a pressure of 297/475 psig. A very significant pressure drop occurs here. This pressure drop changes the pressure of the liquid refrigerant from 297/475 psig to 69/118 psig. At this new pressure condition, the refrigerant temperature should be 40°F in its saturated state. This causes some of the refrigerant to flash into the vapor state instantly, absorbing a great deal of heat in the process of cooling the remaining refrigerant liquid. This results in a refrigerant mixture of about

80 percent liquid and 20 percent vapor. The refrigerant leaves the metering device at the reduced pressure of 69/118 psig. All of the subcooling has now been used in the process of cooling the refrigerant from 110°F to 40°F, and the refrigerant is now at saturation (no superheat and no subcooling exists). The low-temperature, low-pressure mixture from the metering device then enters the evaporator coil.

10. The refrigerant has now completed the refrigeration cycle and is ready to start the cycle again.

The pressures and temperatures in this circuit walk-through are just one example of possible system operating characteristics. There are many possible combinations of temperatures and pressures that can represent proper operation for every set of conditions.

1.3.0 Analyzing Operating Conditions

When problems arise in the refrigeration cycle, it is sometimes difficult to determine the cause. Many times, what seems to be the problem is only a symptom. Also, many problems produce the same or similar symptoms. Isolating the problem requires a logical, systematic approach, which generally begins with a preliminary inspection. The preliminary inspection is then followed by an analysis of the operating characteristics.

1.3.1 Preliminary Inspection

A preliminary inspection of the system is performed, by using the senses of sight, hearing, touch, and smell, to identify possible problems with the system. The preliminary inspection is performed as follows:

WARNING!	Be sure all electrical power to the equipment is turned off. Open, lock, and tag disconnects. Watch out for pressurized or hot components. Follow all safety instructions labeled on the equipment and given in the manufacturer's service manual for the equipment.

Step 1 Turn off the equipment. Lockout and tag equipment so it cannot be energized.

Step 2 Look for an evaporator or condenser mounted above the compressor that might dump liquid refrigerant into the compressor.

Step 3 Look for the following piping problems:
- Oversized or undersized refrigerant lines or excessive elbows or fittings
- Long or uninsulated suction line that might develop excessive superheat
- Liquid line running through an unconditioned space (hot or cold) that might affect subcooling
- Buried lines that might cause refrigerant to condense
- Extremely long liquid line that might hold an excessive amount of refrigerant

Step 4 At the evaporator and condenser, check for the following:
- Fin collars corroded
- Fins or coils dirty or damaged
- Supply plenum dirty
- Filters dirty or missing
- Fan belts at improper tension
- Blowers and fans dirty
- Incorrect blower speed
- Evaporator shows signs of freezing up
- For future reference, determine if a variable-speed motor is being used on the indoor blower

Step 5 As applicable, at the compressor:
- Check that the service valves are fully open.
- Check that the hold-down bolts are loosened or unloosened per manufacturer's instructions.
- Check that the crankcase heater is working.
- On open or semi-hermetic compressors, inspect cylinder heads to see if they are scorched or blistered from excessive heat.
- Inspect for rust streaks indicating condensation from cold return gas.
- Check that the oil level is at the proper height in the sight glass.

The preliminary inspection may immediately reveal at least one or more problems with the system. A large puddle of oil under a fitting, for example, shows that at least one significant leak must be repaired. Some items noted may need to be resolved before more in-depth testing begins. However, it should never be assumed that a system has only a single problem. The preliminary inspection may not reveal the most serious problem, but it may reveal one or more issues for concern. In addition, it provides the technician with information to be factored in as more detailed testing begins.

1.3.2 Analyzing System Operating Conditions

In order to tell if the system or compressor has a functional problem, the actual conditions that exist in the system must be known. If the compressor is operable, system operation should be monitored and the critical parameters measured. This is necessary so that the actual conditions can be compared against a set of normal system operating parameters in order to determine if there is a problem, and where it is. Check the system operating conditions and record the values for system parameters as described here.

> **WARNING!**
> Watch out for rotating, pressurized, or hot components. Follow all safety instructions labeled on the equipment and given in the manufacturer's service manual for the equipment.

Step 1 Connect a gauge manifold set to the system gauge ports. Be prepared to go to several different system locations to take air and refrigerant line temperature measurements.

Step 2 Check and record the voltage applied to the compressor contactor.

Step 3 Provide power to the system and adjust the thermostat to call for cooling.

Step 4 Start the system and listen for abnormal sounds.
- Check for excessive vibration of the compressor, piping, motors, and fans.
- Note any compressor knocks or rattles, which may indicate liquid refrigerant is being drawn into the cylinder(s). Such noise may also indicate mechanical damage to the compressor has already occurred.

Step 5 Visually inspect the operating compressor for abnormalities. Check the compressor oil sight glass (if equipped). Heavy foaming at the sight glass should clear 5 to 10 minutes after startup. If not, there may be excessive refrigerant in the oil.

> **WARNING!**
> Danger exists if the compressor terminals are damaged and the system is pressurized. Disturbing the terminals to take measurements could cause them to blow out, causing injury. When making voltage, current, or continuity checks on a hermetic or semi-hermetic compressor in a pressurized system, always take measurements at contactors and other test points inside the unit.

Step 6 Measure and record the input voltage and current at the compressor contactor. Note that the voltage should be compared to the voltage recorded before startup to see if there is any significant voltage drop under load.
- The measured voltage should be within ±10 percent of the motor nameplate value.
- In a three-phase motor, the voltage imbalance between any two phases should not exceed 2 percent. Any current imbalance between phases should not exceed 10 percent.

Step 7 Check that all fans, motors, and pumps are operational and moving the proper amounts of air or water.

Step 8 As applicable, measure and record or calculate the following operating parameters:
- Indoor bulb and wet-bulb temperatures
- Outdoor dry-bulb temperature
- Suction line pressure and actual temperature
- Saturated suction temperature
- Superheat at both the outlet of the evaporator, and the inlet of the compressor
- Discharge pressure
- Saturated discharge temperature
- Discharge line temperature
- Liquid line pressure and actual temperature
- Subcooling at both the condenser outlet, and the metering device inlet
- Liquid line temperature entering the metering device

- Oil pressure (semi-hermetic compressors)
- Compressor temperature at the bottom of the cylinder heads (semi-hermetic compressors)
- Compressor temperature at the top and bottom of the motor barrel (semi-hermetic compressors)
- Crankcase or shell temperature
- Compression ratio

Once the actual conditions that exist in a system are known, they can be compared against a set of normal system operating parameters to determine if there is a problem. *Table 1* shows an example of the types of system parameters that should be recorded for the purpose of comparison. For this example, the data shown reflect the normal operation parameters for the system described in *Figure 2* at a 95°F outdoor temperature. It is important to remember that readings vary widely among systems because of equipment application, ambient conditions, and type of refrigerant used. For the specific equipment you are servicing, refer to any existing operating logs or the manufacturer's service manual to find typical readings. Better yet, if the same kinds of systems are serviced for various customers, or the same system is periodically serviced for one customer, it is recommended that you capture the data listed in *Table 1* when the system is operating properly. This data will prove to be a valuable reference for later use if it becomes necessary to troubleshoot the system.

1.4.0 Abnormal Pressures and Temperatures

Based on the information revealed by your analysis of system operation, first consult the manufacturer's troubleshooting aids to identify the likely causes of the problem. Most refrigeration cycle problems result in abnormal system pressures and/or temperatures. *Figure 3* summarizes the common system problems that relate to high and low pressures. Note that a low head pressure in combination with a high suction pressure and low compressor current normally indicates that the compressor has leaky valves or worn piston rings and should be replaced.

The information given in manufacturers' troubleshooting aids helps isolate the problem to the specific failed component, especially when the problem is related to the evaporator, condenser, and the air or water flow circuits. However, for more complex problems or where multiple problems exist, the troubleshooting aid may only provide some clues for further investigation. For example, the

cause may be listed as a loss of charge, restriction in the refrigeration system, or noncondensables in the system. For this type of problem, more system analysis using a process of elimination is usually needed to isolate the problem.

1.4.1 Low and High Refrigerant Charges

An incorrect refrigerant charge is a common problem. It is important for technicians to understand how an incorrect charge affects system operation so that the problem is more easily recognized.

Loss of refrigerant is a common problem encountered when servicing cooling systems. The customer usually complains of steadily deteriorating cooling performance. Low refrigerant usually results from leaks that occur as a result of poor installation practices, physical damage, or a factory defect. However, it can also be the result of incorrect charging techniques at startup or during a previous visit. For example, if a system with a TXV metering device is charged to only a 5°F subcooling value on a very cool day with a low evaporator load, it will not likely have enough refrigerant to operate properly in higher temperatures and evaporator loads.

Suction and discharge pressures will be lower than normal with a high superheat, especially with a fixed metering device. A fixed metering device is very sensitive to the level of refrigerant charge. If the system has a sight glass, it will likely show bubbles. However, remember that some refrigerants, like HFC-410A, may show bubbles in a sight glass at any time. Systems with an automatic expansion valve may not have low suction pressure because the valve may be operating wide open, trying to raise the evaporator pressure. In some cases, condenser subcooling will be very low because there is little refrigerant to subcool.

Once you determine the cause of a system problem is a low charge, the first thing to do is to find and repair any leaks that may exist in the system. The methods used for leak detection, and the subsequent evacuation/dehydration and charging of the system, are covered in other modules. Never simply add refrigerant to a system without seeking out and repairing a leak, unless there is reason to believe that the undercharge condition is the result of a previous error. It is important to be certain however. Repeated service calls to recharge a leaking system, and the failure to repair the problem, will result in customer frustration and dissatisfaction.

The symptoms of a refrigerant overcharge are similar from system to system. Both the suction and discharge pressures are typically high. The superheat may be low or nonexistent on a system

Table 1 Typical Temperature-Pressure Data for the Example Shown in Figure 2

System Parameter	Typical Operational HCFC-22 Air Conditioning System	Typical Operational HFC-410A Air Conditioning System	Air Conditioning System Under Test
Suction Pressure (psig)	69 psig	118 psig	
Suction Line Temperature (°F)	52°F	52°F	
Saturated Suction Temperature (°F)	40°F	40°F	
Superheat (°F) – TXV	12°F	12°F	
Discharge Pressure (psig)	280 psig	460 psig	
Discharge Line Temperature (°F)	200°F	200°F	
Saturated Discharge Temperature (°F)	130°F	130°F	
Liquid Temperature Leaving Condenser (°F)	125°F	125°F	
Liquid Line Temperature Entering Expansion Device (°F)	120°F	120°F	
Subcooling (°F)	10°F	10°F	
Air Temperature Entering Evaporator (DB°F)	75°F	75°F	
Air Temperature Leaving Evaporator (DB°F)	55°F	55°F	
Air Temperature Drop Across Evaporator (DB°F)	20°F	20°F	
Outdoor Temperature (DB°F)	95°F	95°F	
Outdoor Temperature Rise Across Condenser (DB°F)	20°F	20°F	
Compressor Compression Ratio (Absolute Discharge Pressure/Absolute Suction Pressure)	3.76 to 1	3.72 to 1	
Compressor Oil Pressure (psig)	86 psig	135 psig	

with a fixed metering device, possibly causing floodback. A system with an automatic expansion valve will not have a high suction pressure because it is designed to maintain a constant suction pressure. A high discharge temperature from the compressor may also be a problem, due to a high compression ratio.

Once it has been determined that excess refrigerant charge is the cause of a system problem, recover the excess refrigerant as needed to achieve a correct system charge. When checking the refrigerant charge, remember to focus attention on subcooling for TXV-based systems and on superheat for systems with a fixed metering device.

Refrigerant Charging

When charging a system with HFC-410A refrigerant, always charge with liquid refrigerant. When adding a charge to an operating system, charge with liquid into the suction side of the system through a flow restrictor installed in the hose of the gauge manifold. Alternatively, use the gauge manifold valve as a metering device.

Also remember that many blends tend to fractionate. As a result, even when subcooling is present, bubbles may appear in a liquid line sight glass. This is especially true when the system is lightly loaded and the refrigerant is moving through the system rather slowly. Sight glasses are of little value in charging blends, but most do provide a moisture indicator that remains helpful.

HIGH HEAD PRESSURE

1. Air or noncondensables in the system
2. Obstructions in the condenser coil fins, such as dirt, etc.
3. Overcharge of refrigerant
4. Recirculation of condenser air
5. Higher-than-ambient temperature air entering condenser
6. Wrong rotation of condenser fan blade

LOW HEAD PRESSURE

1. Low refrigerant charge
2. Defective compressor valves
3. Low ambient temperature

HIGH SUCTION PRESSURE

1. Overcharge of refrigerant
2. Damaged or incorrect metering device
3. Damaged or leaking compressor valves
4. Worn compressor piston rings
5. Leaking oil separator

LOW SUCTION PRESSURE

1. Loose or broken blower belt
2. Obstructed or dirty evaporators
3. Dirty air filters
4. Low refrigerant charge
5. Dirty or faulty expansion valve (leaking around push rod)
6. Recirculation of evaporator air (compare return air temperature with conditioned space temperature)
7. Restriction in refrigerant system
8. Restricted or undersized duct work
9. Wrong rotation of evaporator blower

03210-13_F03.EPS

Figure 3 Causes of abnormal refrigerant pressure.

1.5.0 Evaporator and Condenser Problems

Both the evaporator and the condenser often experience airflow problems. These must be corrected for the system to operate properly.

1.5.1 Evaporator Airflow Problems

To perform properly, a blower in a cooling system must move about 350 to 450 cubic feet per minute of air per ton of cooling across the evaporator coil. Multi-speed and variable-speed motors should be checked and properly setup during the initial installation, and then checked during servicing. Too much or too little air can cause comfort problems.

For example, too much air across the evaporator coil usually results in poor humidity control. Decreasing the blower speed usually will correct this problem.

Incorrect airflow can also cause operational problems. For example, a dirty evaporator coil will cause low suction pressure and low superheat. The problem of too little airflow is far more common than too much airflow. Low airflow is generally indicated in refrigerant circuit performance by low suction pressures and, especially with fixed metering devices, a low superheat condition. The airflow can be increased by increasing blower speed, providing that none of the following conditions are the real cause of the reduced airflow:

- The system air filter, blower wheel, or evaporator coil is dirty.
- The blower is rotating in the wrong direction.
- The belts on belt-driven blowers are loose or worn or the belt tension is incorrect.
- There is loose insulation inside the ductwork.
- The air distribution system is poorly designed and undersized, causing excessive resistance to air movement.

A problem with reduced airflow can often appear as the result of a low refrigerant charge. Many technicians who fail to fully investigate a problem are fooled by some of the most common symptoms they encounter when they connect the gauges. The first symptoms of a low refrigerant charge are a low suction pressure coupled with a low head pressure. In a hurried attempt to complete the work, refrigerant may be added immediately. But there are other signs that will indicate a very different problem.

In the situation described above, the technician can simply touch the base of a hermetic compressor. If it is very cool or cold, it is unlikely that an undercharge is the problem. This cool crankcase temperature indicates that floodback is occurring. A check of the superheat will show that it is very low or even non-existent. This confirms floodback is occurring. At this point, it is time to take a look at the evaporator and airflow. Very often in this scenario, poor evaporator flow from a dirty filter will be found. In others, the coil may be frozen over (*Figure 4*). As the low superheat and low suction pressure condition persists, the evaporator temperature drops below freezing, and ice begins to form, Once it begins, it progresses more quickly as the evaporator surface is increasingly blocked. Since the system is not able to satisfy the demand for cooling, the system continues to run, further increasing the thickness of the ice. Although the evaporator coil cannot be seen easily in many installations, the ice often begins to show itself on the outside of the coil cabinet or on the refrigerant lines. Eventually, it can form around the inlet of the compressor, and even engulf a compressor over time.

When looking for low airflow and/or low evaporator load problems, remember that electronically controlled variable-speed motors are being used more and more with indoor blowers. When checking airflow in a system using a variable-speed motor, always ensure that the blower is operating at the expected speed before checking airflow. During some normal operating modes,

03210-13_F04.EPS

Figure 4 Ice on an evaporator coil.

such as dehumidification, variable-speed motors do not operate at their maximum speed. The presence of a variable speed motor should be determined during the preliminary inspection of a system.

1.5.2 Condenser Airflow Problems

For the system to operate properly, the condenser coil must be able to reject heat absorbed by the refrigerant in the evaporator. In addition, the condenser must reject the heat of compression. Therefore, the condenser must be capable of rejecting more heat than the evaporator is able to absorb. Roughly 15 percent of the condenser's heat rejection capacity is dedicated to rejecting the heat of compression.

When sufficient heat rejection does not happen, the system pressures and temperatures build up. If the problem is severe enough, protective devices will shut the compressor off. Some possible causes for poor condenser performance in air-cooled applications are as follows:

- Failure of the condenser fan(s) to operate due to motor or control circuit failure.
- The condenser fan is rotating in the wrong direction.
- The fan blade is not positioned properly in the fan orifice.
- The wrong fan blade has been used, or the pitch of a replacement blade is incorrect.
- Hot condenser discharge is being recirculated through the coil instead of being carried away. This is usually caused by the unit being installed too close to buildings, or under decks or overhangs.
- Higher-than-ambient air temperature is entering the condenser coil from other warm air

sources, such as exhaust vents, dryer vents, and adjacent condensing units.

- Grass clippings or other airborne debris are present on the surface of the coil or embedded deeply within, reducing airflow.
- Airflow is blocked by shrubs and plants.
- A replacement motor has the wrong rpm.

Note that many commercial condenser coil assemblies are constructed of two separate coils in layers (*Figure 5*). The coil fins of any condenser coil are set close together, but dust and dirt can easily make its way in between the coil fins where they either pass through or become trapped. With layered coil construction, a great deal of debris that would normally find its way through the coil fins and exit the opposite side becomes trapped against the face of the rear coil. This happens easily, as the fins of the two coils are not aligned. Cleaning only the face of the coil often results in little or no improvement. To properly clean these layered coils, the two layers must be very carefully separated enough for washing in between.

1.6.0 Fixed Metering Device Problems

Two primary problems occur with fixed metering devices: they are either the wrong size for the application, or they become partially or completely restricted. On systems with a single fixed-orifice

LAYERS MUST
BE SEPARATED
FOR CLEANING

03210-13_F05.EPS

Figure 5 Layered condenser coil.

device, clogging is rare due to the size of the opening. However, some commercial systems that use several fixed-orifice metering devices on the same circuit are usually smaller in size.

Normally, the sizes of both fixed-orifice and capillary tube devices are factory-selected for use with a system. If a capillary tube is too long or the diameter too small, or a fixed-orifice device is too small, the evaporator will be starved for refrigerant. Excess liquid will then build up in the

Coils Using Microgroove Technology

Microgroove technology is a new way to get more from less. Coil manufacturers are beginning to use smaller diameter tubing in both condenser and evaporator coils. The new tubing is internally grooved. Using this type of tubing enhances heat transfer. In addition, smaller tubes are stronger and can withstand higher pressures than conventional coil tubing of an equal thickness.

This new coil technology allows coils to be smaller without sacrificing capacity. This in turn decreases the size of the unit and the total weight of the materials used. Any technology that reduces the use of materials while increasing both heat transfer and energy efficiency is good for the environment. Look for more evaporator and condenser coils to be constructed using this and other innovative approaches.

03210-13_SA01.EPS

condenser. The effect initially will be high head pressure and inadequate cooling. If the unit runs for a while, the liquid trapped in the condenser will cool and the pressure will drop to normal. If a capillary tube is too short or the diameter too great, excess liquid will be fed to the evaporator. The same is true for an oversized fixed-orifice device. This could cause floodback at the compressor.

For any given pressure difference across a fixed metering device, it can pass a larger volume of liquid refrigerant than vapor because the liquid has a much greater density. This makes a fixed metering device very sensitive to excess oil and poor subcooling. Excess oil in circulation displaces some of the liquid refrigerant and reduces the actual weight of refrigerant flowing through the tube. This can starve the evaporator. Excess flashing due to a shortage of refrigerant or a liquid line restriction will cause both liquid and vapor to enter the metering device, also resulting in a starved evaporator coil.

1.7.0 Thermostatic Expansion Valve Problems

TXVs are often blamed for system problems caused by a number of other faults. This is because TXVs are often misunderstood by technicians. Troubleshooting TXV problems should always begin with a measurement of superheat. If too little refrigerant is fed to the evaporator, the superheat will be high. If too much refrigerant is fed to the evaporator, the superheat is low. Although you may think these symptoms are caused by improper TXV control, often the problem is elsewhere. Problems often blamed on TXVs are the following:

- Overfeeding
- Underfeeding
- Erratic operation, called hunting

Flooding can be caused by the valve sticking in an open position. It may also be caused by the following valve or system problems:

- Oversized valve
- Improper superheat setting on the valve
- Improper location of the sensing bulb on the suction line
- Poor thermal contact between the sensing bulb and the suction line
- A very light system load (well below design)
- Wrong type of valve power element for the system refrigerant

Starving can be caused by the TXV sticking closed or by a loss of charge in the power element. The valve may also be restricted, either at the inlet strainer (if equipped) or internally. An improper superheat setting or an undersized valve can also cause the evaporator to be starved for refrigerant. However, neither of these two conditions represents a valve failure; they are the failures of installers and service technicians. The evaporator coil can be starved for the following additional reasons:

- Shortage of refrigerant
- Plugged filter-drier
- Kinked or obstructed liquid line
- Excessively tall liquid line riser
- Undersized or excessively long liquid line
- Plugged refrigerant distributor
- Improper valve location
- Plugged equalizer line

Hunting describes consistent changes in refrigerant flow rates as the TXV attempts to find the proper valve position to maintain the superheat setting. All TXVs may hunt to some degree, but not in an easily noticeable way. Hunting is charac-

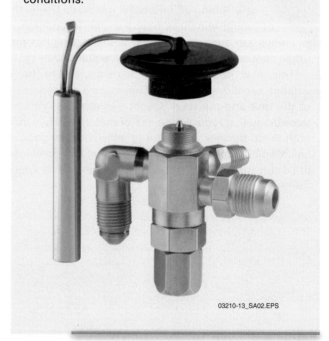

Metering Devices

Because of the need to meet higher energy efficiency standards, some manufacturers are abandoning the use of fixed metering devices on the indoor coils of some products, switching entirely to TXVs. TXVs also provide much better performance in a wider variety of operating conditions.

03210-13_SA02.EPS

terized by swings in superheat of several degrees up and down, rather rapidly.

Hunting can only be verified by making several suction line temperature or pressure measurements over a period of time. If there is a repetitive pattern such as that shown in *Figure 6*, the valve is hunting. Excessive hunting can be caused by sticking or binding of the TXV internal parts. It can also occur if the system load on the TXV drops below 30 percent of its capacity. Other than faults in the valve itself, the following can cause excessive hunting:

- Oversized expansion valve
- Very light load
- Long liquid refrigerant line
- Rapid changes in condensing pressure or temperature
- Rapid load changes
- Intermittent flashing in the liquid line

It is very important to note that TXVs, by design, will try to maintain the superheat setting under any operating condition. When the system charge is low, the superheat will likely be low as well. However, the TXV may be able to mask this symptom somewhat, by simply opening wider and trying to feed enough refrigerant to reduce the superheat. A small loss of refrigerant charge therefore, may not be obvious in the superheat reading. As long as some subcooling exists, and possibly even longer, the TXV can likely compensate for the shortage. If the system is overcharged, the TXV is even better at masking this problem. It will only open as far as necessary to maintain the superheat; any excess refrigerant simply backs up in the condenser or the receiver (if equipped). It may take a considerable overcharge to cause the high-side pressure to rise to the point of being noticed. In this case, the subcooling tells the story. An excessive amount of

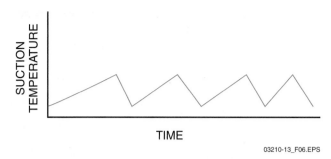

03210-13_F06.EPS

Figure 6 Suction temperature variations that indicate hunting.

subcooling will indicate that there is more refrigerant in the system than necessary. No other obvious symptoms may be noted, unless the overcharge is massive.

Some system problems caused by TXVs are the result of poor installation. Make sure the sensing bulb is securely fastened on the top of a clean, straight section of the suction line, close to the evaporator outlet. On all but the smallest of suction lines, the bulb should be placed at the 10 or 2 o'clock position on a horizontal run of pipe. Also, check the insulation on the bulb to make sure outside air temperatures are not affecting it. Over time, some light corrosion can take place between the bulb and the line, slowing heat transfer. The contact surfaces should be clean.

A distributor is installed downstream of the metering device (usually a TXV) and is designed to ensure that refrigerant leaving the TXV gets evenly distributed to the various circuits in the evaporator coil. Distributors (*Figure 7*) do not function properly at loads well below their design flow capacity. If the distributor flow rate falls below 50 percent of its capacity, the liquid and vapor leaving the TXV will not be properly mixed. As it then passes through the distributor, the liquid will go to the bottom circuits and the vapor will go to the top circuits. A way to check for this

Ice on Suction Lines and Compressors

For technicians working with comfort cooling systems, the appearance of ice on suction lines and compressor inlets indicates trouble. Most all comfort systems are designed to operate at evaporator temperatures well above freezing. However, in the refrigeration world, ice on suction lines and compressor inlets is quite normal.

For a freezer to maintain a box temperature of 10°F, a common evaporator temperature would be around 0°F. Even if the system is operating with a superheat value of 10°F, the suction line temperature will still be 22°F below freezing. No doubt some ice is going to form somewhere. In fact, if a freezer system operates for a significant amount of time each day, the absence of ice on the suction lines indicates that a problem is likely.

This is a perfect example of why technicians must understand their trade and the importance of taking superheat readings, as well as many other readings. To evaluate a system, one must first see the big picture.

When working on a freezer, make sure to scrape off any ice in the immediate area from the pipe surface before measuring the temperature. Otherwise, the reading will be inaccurate.

03210-13_F07.EPS

Figure 7 Multi-circuit evaporator fed by a distributor.

problem is to look for high temperatures at the top of the evaporator and low temperatures at the bottom.

1.8.0 Other Refrigerant Circuit Problems

There are several other problems that are common to the refrigerant circuit. These problems are related to the following:

- Interconnecting refrigerant lines
- Refrigerant circuit accessories
- Contaminants and noncondensables

1.8.1 Refrigerant Line and Accessory Problems

A restriction in a refrigerant liquid line or related accessory (*Figure 8*) reduces or prevents the system refrigerant from getting to (or through) the evaporator. This results in low system capacity. Problems with lines or accessories are rarely, if ever, identified by too much refrigerant flow.

The cause of problems in a system with a restriction is sometimes diagnosed as the system being low on refrigerant charge. This is because the symptoms of low suction and discharge pressures and high superheat are similar. The subcooling temperature can be used to help determine if the problem is a restriction rather than a low charge. Systems with a major restriction anywhere in the liquid line usually show normal or increased subcooling at the condenser, with little or no subcooling at the metering device inlet. This is because

the mass flow of refrigerant in the system slows down. Because of the restriction, most of the system liquid refrigerant remains in the condenser and/or receiver and subcools further. The liquid refrigerant stays in the condenser longer, resulting in greater subcooling because it has a longer time to be cooled.

Refrigerant tubing can become restricted as a result of outside physical damage that causes the tubing to be pinched or bent. This often occurs on systems that have been running normally for a long time. Physical damage to the tubing often goes undetected because it can be hidden under insulation or behind a fixture.

When a partial restriction occurs in a liquid line, it acts like a metering device and drops the pressure and temperature of the refrigerant at the point of the restriction. It is a good practice to consider any accessories that are installed in a system as a potential point of restriction. A restriction could be detected by measuring the pressure drop across the accessory or tubing suspected to have the restriction; however, this is not always practical because gauge ports are not usually available. An easier way is to measure the temperature drop on each side of the accessory or tubing suspected to contain the restriction instead. Split HVAC systems are more susceptible to refrigerant line restrictions and damage because the field-installed interconnecting tubing may be routed through areas where they are subject to accidents, abuse, or vandalism. Packaged HVAC products have all refrigerant lines contained within the security of

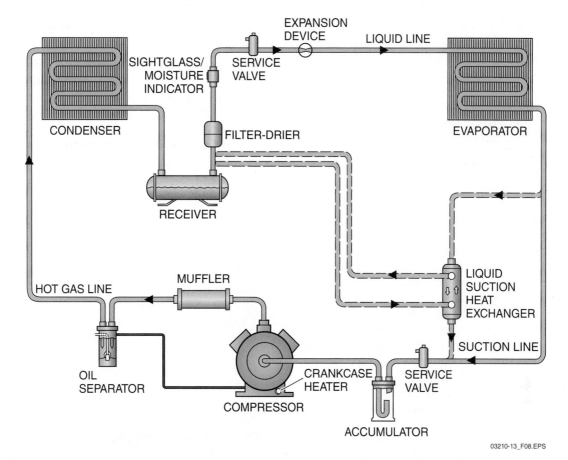

Figure 8 Refrigerant lines and accessories.

the package so they are less susceptible to damage. Restrictions in packaged systems are far more likely to be found at an accessory.

Depending on the severity of the restriction, a technician can sometimes feel a temperature change in the line, or even observe frost or sweat just downstream of the problem area. If the restriction is very slight, temperature measurement with a thermometer is usually the only way to find the location of the restriction.

A plugged or undersized filter-drier is a common cause of a restriction. On a newly installed system, the restriction may occur shortly after startup as a result of being plugged by solid contaminants that entered the system during installation. If a system is equipped with a sight glass installed downstream of the filter-drier, a restricted filter-drier will cause flash gas (bubbles) to appear in the sight glass. To confirm that a filter-drier is restricted, always measure the temperature drop across it, since a restricted filter-drier can go undetected using the touch method. The amount of pressure drop across a filter-drier can also be an indication of a restriction. The pressure drop should not exceed 5 psi.

Restrictions can be caused by expansion valves that do not open all the way. Normally, TXVs either work or do not work. However, a partial restriction can occur in the valve if it loses part of the charge in the thermal bulb, or the inlet strainer becomes dirty. Also, if moisture is present and circulating in a system that operates at a saturation temperature below freezing, it can freeze at the metering device and restrict refrigerant flow. This is a problem that progresses as the system operates. Once the system shuts down for a short period, the frozen moisture melts, remaining in the circuit to progressively freeze again during the next cycle.

Test for loss of charge in the TXV bulb by removing it from the suction line and holding it tightly in the hand. If the valve fails to feed the evaporator more refrigerant, reducing the measured superheat, the bulb has probably lost some charge.

Another potential restriction point in split-system products is the service valves. During normal operation, ensure that both the liquid line and suction line service valves are 100 percent open. During installation, technicians sometimes fail to open these valves fully, causing a restriction. During certain service procedures these valves must be closed, but make sure to open one or both valves 100 percent after the service procedure is complete.

Beyond restrictions, refrigerant lines can be affected by their environment. For example, a liquid line significantly exposed to a very hot environment can lose a lot of subcooling. Conversely, if it is routed through a cold environment, it may gain some beneficial subcooling. Liquid lines experience a 1 psi pressure loss for every 2.3 feet that they rise in height. A riser that is 23 feet in height loses 10 psi. This significant change can erase any subcooling that was gained in the condenser.

Suction lines can also experience these positive or negative changes. However, they are less likely to change due to the insulation that is normally applied. When troubleshooting the refrigerant circuits of split systems, it is extremely important to make superheat and subcooling measurements at each end of the line. Superheat should be checked at the evaporator outlet and at the compressor, and the two values compared. Subcooling should be checked at the condenser outlet and at the metering device inlet for comparison. When pressure readings cannot be taken at both locations, be sure to compensate by considering any normal pressure drop in the lines in the calculations.

1.8.2 Air and Moisture Contamination

Refrigeration systems should contain only refrigerant and oil. Anything else must be considered a contaminant. Air and moisture are the most common contaminants in a refrigeration or air conditioning system. Air is a noncondensable gas, which means that it does not change into a liquid at common operating temperatures and pressures. Air entering the system can also contain moisture in the form of water vapor. When air is in a system, it accumulates in the top of the condenser, taking up valuable condensing surface area. This results in high head pressure and an increase in the condensing temperature.

Under the heat of compression, moisture will react with the refrigerant to form acids. These acids cause corrosion of metals and breakdown of the insulation on the motor windings and wiring. Moisture in the refrigerant can also cause oil sludge, which reduces the lubricating properties of the oil and plugs oil passages and screens in the compressor. Moisture can also freeze at the metering device as mentioned previously.

Air and moisture can enter a system in the following ways:

- Improper or no evacuation/dehydration process at installation or after system servicing
- Refrigerant circuit leaks while under a vacuum
- Failure to install a filter-drier at installation or after servicing
- Failure to purge hoses prior to charging refrigerant

Symptoms of system contamination include the following:

- Sealed-tube refrigerant test shows acid and/or moisture
- Refrigeration oil acid test shows acid
- Discoloration of the moisture indicator in a liquid-line sight glass
- System valves stick
- Clogged metering device
- Poor compressor operation
- Compressor failures
- Overheating of compressor
- High compressor current draw

Capillary Tubes

Capillary tubes, like the one shown here, were once widely used in systems up to five tons. With the introduction of the fixed-orifice metering device, the capillary tube is now mainly used in room air conditioners and packaged terminal air conditioners (PTACs). They also continue to be used in domestic refrigerators and freezers.

03210-13_SA03.EPS

If a system using extremely hygroscopic lubricants, such as POE (*Figure 9*) and PAG types, is left open and exposed to the air, the oil can absorb large amounts of moisture. This moisture cannot be removed with a vacuum pump. Since these oils attract moisture, it is essential that refrigeration circuits not be left open to the atmosphere for any significant length of time.

03210-13_F09.EPS

Figure 9 Polyol ester (POE) lubricating oil.

In a system that uses a TXV, the problem of noncondensables in the system is often diagnosed as an overcharge of refrigerant because the symptoms are somewhat similar.

1.8.3 Acid Contamination

Acid is not introduced into a system. It is formed inside an operating system by the reaction of air or moisture with the refrigerant. It can also be formed during a compressor failure, as the insulation on motor windings breaks down in the extreme heat. Acid creates sludge and varnish, which can plug oil passages and metering devices, and it can restrict the strainers in the compressor and filter-driers, as well as other refrigeration system components.

Acid most likely exists in a system after the compressor has failed as the result of shorted or grounded motor windings, which are called **burnouts**. Both of these conditions are fatal. Burnouts are classified as either mild or severe based on the amount of acid present in the system. A mild burnout tends to yield little acid because it occurs suddenly, causing the motor to be de-energized and stop before the contaminants created by the burnout leave the compressor. A severe burnout usually occurs over a longer period of time, allowing considerable acid and other contaminants to be produced and pumped through the system while the compressor is still running. These situations may also be referred to as a starting burn or a running burn. If the compressor is trying to start, but not pumping, as the motor fails, little acid will be circulated. However, if it actively pumping during the burning process, a great deal of acid can be released into the circuit.

Preventing Noncondensables in a System

Want to prevent noncondensables in an HVAC system? The answer is simple. If the sealed refrigeration system must be opened for any reason, close it as quickly as possible. Always use a vacuum pump and evacuate the system to 500 microns before recharging. Evacuation removes noncondensables most of the time. Also, when installing refrigerant tubing, make sure to purge the tubing with nitrogen before use to remove any moisture caused by condensation trapped within the tubing.

One piece of advice from an industry veteran: Treat cans of lubricant, as well as the refrigerant circuit, as if it is full of angry bees that you do not want to release! If it must be opened, do it quickly and close it just as quickly.

> **WARNING!**
>
> When working on any system containing refrigerant, eye and hand protection must be worn at all times to prevent personal injury. This is especially true when working with acid-contaminated systems. Acid in contact with eyes or skin can cause severe burns.

Before replacing a compressor that has failed because of a burnout, the system must be tested for acid contamination. A sealed-tube acid/moisture test kit (*Figure 10A*) is connected to a system service port to obtain a sample of the refrigerant. Oil acid test kits (*Figure 10B*) require that the test be performed on a sample of the compressor oil. In either case, follow the test kit manufacturer's instructions for using the kit and for determining the level of contamination.

OIL TEST KIT

03210-13_F10B.EPS

Figure 10 Refrigerant and oil test kits. (2 of 2)

BEFORE

AFTER
(POSITIVE RESULT)

SEALED-TUBE TEST KIT

03210-13_F10A.EPS

Figure 10 Refrigerant and oil test kits. (1 of 2)

Refrigerant Smell Can Indicate a Compressor Burnout

An acid test kit can be used to tell if acid is present. The nose can also tell. Refrigerant from a system that has had a moderate to severe compressor motor burnout will have a very strong, pungent odor. Once you've smelled it, you will always be able to recognize it. The burning of some refrigerants can create a material known as phosgene gas, once used in warfare to sicken enemy troops. It can be formed inside the circuit during a severe compressor burnout. It can also be formed when using a torch in the presence of escaping refrigerant gas. This can easily happen while replacing a compressor.

Additional Resources

Refrigeration and Air Conditioning: An Introduction to HVAC/R. Latest Edition. Prentice-Hall.

System Diagnostics and Troubleshooting Procedures. 2003. John Tomczyk. Esco Press.

1.0.0 Section Review

1. Which of the following components takes high-pressure, high-temperature refrigerant vapor and cools it to allow the formation of liquid refrigerant?
   a. Compressor
   b. Evaporator
   c. Metering Device
   d. Condenser

2. The operating pressures for HFC-410A are roughly equal to those of HCFC-22.
   a. True
   b. False

3. During a preliminary inspection of the evaporator and condenser coils, which of the following would logically be done?
   a. Check that the crankcase heater is working.
   b. Look for corroded fin collars.
   c. Inspect for scorching or blistering indicating excessive heat.
   d. Inspect for an excessive number of elbows and other fittings.

4. One of the possible causes of a low suction pressure would be _____.
   a. a loose or broken blower belt
   b. defective compressor valves
   c. an excessive refrigerant charge
   d. the recirculation of condenser air

5. On a unit with a fixed metering device, low airflow is generally indicated by a _____.
   a. high superheat condition
   b. low superheat condition
   c. high subcooling condition
   d. low subcooling condition

6. A capillary tube that is too short in length will result in _____.
   a. high pressures
   b. low subcooling
   c. high superheat
   d. low superheat

7. On a unit equipped with a TXV, overfeeding refrigerant can be caused by _____.
   a. a plugged filter-drier
   b. a shortage of refrigerant
   c. poor thermal contact between the bulb and suction line
   d. an excessively long liquid line riser

8. One advantage packaged cooling units have over split systems is that _____.
   a. they can operate at a lower superheat
   b. they never need a filter-drier
   c. the refrigerant lines are more protected
   d. they can operate properly with little or no subcooling

2.0.0 COMPRESSOR TROUBLESHOOTING AND REPLACEMENT

Objective

Explain how to troubleshoot and replace a cooling system compressor.

   a. Identify common problems that can result in compressor failure.

   b. Explain how to troubleshoot compressor mechanical problems.

   c. Explain how to replace a hermetic compressor following a failure.

   d. Describe the additional steps that may be required to replace a compressor following an electrical failure.

Performance Task 1

Demonstrate the ability to isolate and determine the solution for at least four of the following types of malfunctions:

- Compressor failures
- System-related compressor problems
- Refrigerant undercharge or overcharge
- Evaporator and condenser problems
- Metering device problems
- Refrigerant lines and accessories
- Noncondensables and refrigerant circuit contamination

Trade Terms

Capacitance boost: A procedure used to start a stuck PSC compressor. It involves momentarily connecting a start (boost) capacitor across the run capacitor of the stuck compressor in an attempt to start it by increasing the starting torque.

If not corrected, many cooling system problems will result in compressor failure. If the compressor fails and the cause has not been found and eliminated, a replacement compressor will probably also fail. *Figure 11* summarizes some of the possible causes of symptoms exhibited by the compressor for some common electrical and refrigeration cycle problems.

The measurements and checks used to electrically test compressors motors are provided in the *HVAC Level Two* module, *Compressors*. Electrically troubleshooting the compressor motor is not significantly different than troubleshooting other motors, although it cannot be seen. As a result, this section discusses other types of compressor problems that are frequently encountered.

2.1.0 Problems for Compressors

Compressor problems include slugging, floodback, flooded starts, and overheating. Although overheating can be a symptom of an internal problem, it can also result from an external problem. Slugging, floodback, and flooded starts are problems that are not the fault of the compressor.

2.1.1 Slugging

Slugging occurs when a compressor tries to compress liquid refrigerant, oil, or both, instead of superheated gas. Slugging is generally considered to involve a large single dose of liquid returning to the compressor at once. If slugging occurs, it will usually occur at startup or during a rapid change in system operating conditions.

Slugging can usually be detected by a loud knocking noise at the compressor. In severe cases, it can result in the cylinder cracking, bolts shearing, or parts actually separating from the compressor. Due to the extreme stress it causes, it often results in blown gaskets, broken valves, piston rods, or crankshafts. This type of damage can happen in a single slugging event.

Common causes of slugging include the following:

- An overcharge of refrigerant.
- An oversized, damaged, or failed metering device.
- An overcharge of oil.
- A failed crankcase heater.
- Condensed refrigerant is present in any cold part of the system, such as the evaporator, during the off cycle. Buried refrigerant lines or those passing through cold spots can allow the refrigerant to condense back into a liquid at shutdown.
- Slugs of oil are trapped in the suction line because the suction gas does not have enough velocity to return the oil to the compressor. Normally, the oil and refrigerant mix together effectively. The oil is circulated through the system in very small drops as it is being swept along by the velocity of the refrigerant vapor. If it gets trapped in the system piping and then returns all at once, it can cause slugging. This condition also occurs in systems that use compressors equipped with capacity control devices, especially when after the compressor runs unloaded for long periods of time.

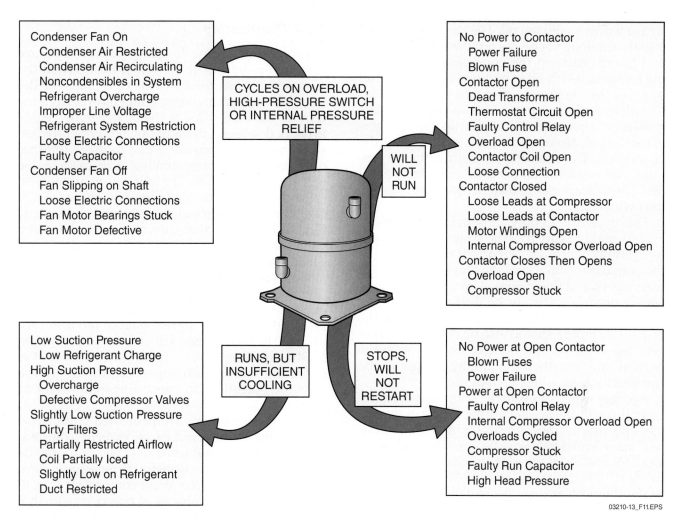

Figure 11 Compressor troubleshooting.

- Poor piping design allows liquid refrigerant to return to the compressor during the off cycle.

2.1.2 Floodback

Floodback is the continuous return of liquid refrigerant or significant liquid droplets in the suction vapor to the compressor during operation. Compared to slugging, it is a more progressive situation that can eventually develop into slugging. Floodback dilutes the oil, resulting in crank-

Slugging

Burying refrigerant lines may sound like a good idea, but it is not. Since it is usually cooler underground, refrigerant in buried lines will condense during shutdown and cause slugging when the compressor starts. For this reason, most HVAC equipment manufacturers prohibit or severely restrict the use of buried refrigerant lines with their equipment.

case foaming and overheating of bearing surfaces. If bad enough, it can damage the pistons, rings, and valves because liquid refrigerant washes the oil off the bearing surfaces. Flooding is most often caused by the uncontrolled flow of refrigerant that results from one or more of the following conditions:

- An oversized thermostatic expansion valve.
- A broken, improperly located, improperly mounted, or poorly insulated TXV power element and/or sensing bulb.
- A low superheat setting.
- Low evaporator load.
- An overcharge of refrigerant, especially on systems using a fixed metering device.

2.1.3 Flooded Starts

Flooded starts are caused by the oil in the compressor crankcase absorbing refrigerant. This is more relevant during long compressor shutdowns at low temperature. Refrigerant will migrate to and condense in a cool or cold compressor shell, where it mixes with the oil. The amount absorbed into the oil depends on the temperature of the oil and the pressure in the crankcase. On startup, the refrigerant-rich oil is pumped through the oil pump of the compressor, resulting in marginal lubrication of the bearings. Also, as the crankcase pressure drops suddenly at startup, the refrigerant will flash from a liquid to a gas, causing the oil to foam. This foaming can restrict the oil passages and cause the oil pressure to build. A great deal of this foam can be also carried over into piston cylinders.

The problem of flooded starts can be minimized by first making sure that the system has a proper refrigerant charge and the correct amount of oil in the crankcase. Make sure the crankcase heater is working, or add one if necessary. Crankcase heaters are used to warm the oil during the off cycle and prevent the condensation of refrigerant in the crankcase. Piping should also be designed to prevent liquid refrigerant from accumulating in the suction line, and then flowing by gravity to the compressor inlet while it remains off.

Another way to prevent the migration of refrigerant through the system during the off cycle is to use a liquid line solenoid valve (*Figure 12*). The liquid line solenoid valve can be wired to operate on a call for cooling along with the compressor. Once the need for cooling is satisfied, the valve closes as the compressor shuts down. On some systems, the solenoid may close before the compressor is shut down. This allows the compressor to pump some or all of the refrigerant remaining in the evaporator and suction line out, and into the condenser and/or receiver, before it shuts down. This prevents any significant amount of refrigerant from condensing in these areas. The compressor eventually shuts down when it reaches the setting of a pressure switch installed for this purpose. When a solenoid valve is used in this way, it is often referred to as a pump-down solenoid.

03210-13_F12.EPS

Figure 12 Liquid line solenoid.

2.1.4 Overheating

Compressors normally generate heat and are designed to handle this normal heat. When the compressor is overheating, the cause must be determined. When discharge temperatures from the compressor exceed 250°F, the oil generally begins to break down and the situation becomes progressively worse. Overheating can damage internal components and cause moving parts to seize due to thermal expansion. The lubricating oil can also be burned and form a residue known as coke. A sample of oil can provide some visual clues. Dark, black oil is a sign that the compressor has been running too hot for a significant period of time. The oil will also show this dark, black color following a severe burnout.

It is important to understand that most compressor types depend on the cool refrigerant gas returning to absorb a great deal of the heat produced. The heat is then rejected from the system by the condenser. As a result, a system that is operating with a high superheat, for any of the possible reasons, can cause the compressor to overheat.

Some compressors are protected from overheating by an internal or an external thermal limit switch. Internal limit switches also provide some back-up protection for three-phase compressors. When a three-phase compressor tries to start with only two phases of power provided, it will become hot very quickly. The thermal limit switch, also referred to as a thermal overload, will then break the electrical circuit to protect it. Without other forms of protection against single-phasing, this cycle can repeat itself many times. The limit switch closes; the compressor tries to start; it does not rotate, but heat builds quickly; the limit switch opens. After several cycles like this, the entire mass of the compressor becomes very warm. It can take some time for it to cool after a technician arrives to stop the cycle.

If a compressor overheats, check for a high superheat. This may be coupled with a low suction pressure. A high compression ratio can also cause overheating. For example, a system with a TXV may have a condenser coil that is badly clogged with grass clippings or other debris. Although the TXV will try to compensate and the superheat may not be excessively high, the compression ratio may be extremely high due to the increase in high-side pressure. The compression ratio increases as the differential between the high and low pressure increases.

2.2.0 Compressor Mechanical Problems

Compressors, of course, do have their own set of problems. Mechanical failures do occur. Some may have been initiated or caused by external problems. Others may be the result of mechanical flaws or normal wear. When a compressor does fail due to external problems, such as a failed metering device, the technician will be unable to discover this until an operable compressor is installed. It is very important that systems with replacement compressors be thoroughly tested and any problems solved to ensure the health of the new compressor.

Again, a visual examination of an oil sample may provide some clues regarding the compressors condition. As noted previously, dark, black oil indicates significant overheating, assuming the compressor has not already experienced a burnout. However, oil with gray or silver coloration, or oil that appears to have tiny pieces of glitter in it, indicates mechanical wear. The minute metal particles that have separated from piston rings, cylinder walls, or bearings will cause this change in color.

2.2.1 Valves and Rings

In a reciprocating compressor, valves and rings provide a seal between the high-pressure and low-pressure sides. If they are damaged, the compressor must be replaced or the valves repaired. Valves can often be field-repaired or replaced in semi-hermetic and open-drive models. Piston rings on the other hand, cannot be field repaired or replaced. Fortunately they are far less likely to be damaged than the valves. Suction and discharge pressure checks can be used to test for this condition. Bad valves or rings may exist if the suction pressure will not pull down or discharge pressure will not build up when the system is properly charged and under normal load. Suction valves are more often damaged than discharge valves.

Another check can be made by measuring the compressor's running current with an ammeter (*Figure 13*). If, under loaded conditions, the running current is considerably lower than normal and the suction/discharge pressures are abnormal (high suction pressures and/or low discharge pressures), faulty valves or rings should be suspected. During this type of test, ensure that no

03210-13_F13.EPS

Figure 13 Checking current with an ammeter.

A Real Situation

In one incident regarding compressor failures, a number of new 20 ton packaged rooftop units were experiencing rapid-fire compressor failures. It did not take long to determine why.

In this industrial plant, a great deal of heat was being generated in a building with a 28-foot roof height. Included in the operation was an industrial oven that added significantly to the heat. The heat collected in the top of the building, as there were no fans or other devices to mix the building air and keep it from stratifying. Temperatures at the roof level exceeded 125°F, as measured in the return air duct by the technician.

The return air grilles for each unit were positioned at the roof level, to save on the air distribution system cost. The technician, upon seeing the return air temperatures, checked the superheat at each compressor of multiple units. The lowest superheat reading found was still in excess of 50°F. Several exceeded 60°F, due to the extreme heat load.

In industrial buildings like this one, there is little or no reason to try and cool the entire volume of the building. The people and equipment are on the ground. Although it was very costly, the technician convinced the client to lower both the return and supply air ducts down to the 16-foot level. Heat that rose above this level was allowed to stratify there intentionally. Not only did the compressors begin to survive and produce more effective cooling, the energy cost fell as well.

compressor capacity control devices are in operation that could cause misleading results.

Many equipment manufacturers specify that power must be applied to their equipment for up to 24 hours before the initial startup or after a prolonged winter shutdown. This is done to allow the crankcase heater enough time to drive any accumulated refrigerant out of the oil in the compressor shell to prevent a flooded start. If this is not done, the compressor may be damaged or destroyed the first time it is started.

A scroll compressor may exhibit the same combination of symptoms that indicate valve problems in a reciprocating compressor. However, a scroll compressor has no valves at all. When the same symptoms are encountered in a scroll compressor, it could indicate a crack or similar damage in the orbiting scroll plate.

2.2.2 Seized Compressors

A seized, or stuck, compressor hums but will not start, and typically draws locked rotor amperage (LRA) for at least a few seconds. Normally, the LRA, which can be four to six times the normal running amperage, lasts for a fraction of a second when the motor first starts. Once the motor begins to rotate, it's starting current drops quickly. When a compressor is mechanically seized, the current will remain at or near its LRA value until a safety device intervenes.

A seized compressor is a mechanical condition. However, there are also electrical problems that can prevent the compressor motor from rotating. Before condemning a seized compressor, check for the following conditions:

- Three-phase compressor contactor not making good contact on all poles (single-phasing)
- Defective or open start relay, preventing the start capacitor from assisting
- Start and/or run capacitor open or weak (single-phase compressors only)

- Unequal system pressures (especially in units using PSC compressor motors)
- Low supply voltage

If the supply voltage is within ±10 percent of the motor's nameplate rating, but a single-phase PSC motor fails to start, a start kit may have to be installed to increase the motor starting torque to correct the problem. This is more often required on systems using a TXV metering device than a fixed restrictor.

If electrical tests prove a PSC compressor has no open, shorted, or grounded windings, and if all external factors check out to be good, a technician can attempt to start a seized PSC compressor with a capacitance boost (*Figure 14*). Capacitance boosting involves momentarily connecting a start capacitor (200 to 300 MFD range) across the compressor's existing run capacitor when power to the compressor is turned on. Since the start capacitor is placed in parallel with the run capacitor, the values of the capacitors are combined, providing increased capacitance and torque to start the PSC motor. Once started, the compressor should continue to run at full speed with the start capacitor removed from the circuit. It is important that the start capacitor be removed because leaving it in the circuit for more than a few seconds will cause it to overheat and burn up. Allow the compressor to run for about ten minutes.

> **WARNING!**
> To avoid the possibility of electric shock, do not handle live wires. Switch the power off and lock it out before connecting the wires. Also, bridge the start capacitor to ensure that it is discharged before connecting wires to it.

After the operating pressures equalize, try restarting the compressor again without the capacitance boost. If the compressor starts the first time and starts on subsequent attempts, a permanently installed start capacitor and start relay are prob-

---

How Much Oil?

One concern for technicians during a compressor change is related to the amount of existing oil in the system. A system may have been operating under unusual conditions, such as undercharged, for a period of time before the compressor failed. As a result, a great deal of the oil may have been trapped in the system rather than returning to the compressor crankcase. Some of the oil remains out in the piping, coils, and accessories at all times. With a semi-hermetic compressor, a sight glass helps to determine how much oil may be missing. However, with hermetic compressors, it is much more difficult to tell. They can be drained after removal to compare their holding capacity to the contents, but all of the oil may not come out. On systems that experience multiple compressor failures, the volume of oil in the system must become a matter of concern. Unfortunately, there is no foolproof way of determining exactly how much is in the circuit.

Figure 14 Starting a seized PSC compressor with a capacitance boost.

ably not needed. If the compressor will not start without a capacitance boost, permanently install a correctly sized start capacitor and start relay. Start kits designed for use with most compressors are available from HVAC distributors.

> **CAUTION**
>
> Always use the exact start kit specified for the compressor. Failure to use the appropriate start kit may cause damage to the compressor motor. Also, never use a megohmeter to check motor windings, or attempt to start a compressor, while the refrigerant circuit is in a vacuum.

Crankcase Heaters

Crankcase heaters come in a variety of sizes (wattage) and shapes. A bellyband heater clamps around the bottom of a welded hermetic compressor shell. Other types of heaters are inserted in a well in the base of the compressor shell that allows the heater to be immersed in the oil. Some equipment is designed so that power is applied to the crankcase heater at all times while others have various thermostatically controlled schemes that apply power to the heater only when temperatures drop below a certain point.

If a compressor has been changed, a different run or start capacitor rating may be required. This is sometimes overlooked when compressors are replaced. The new compressor may be equipped with a different motor than the previous model, and capacitors of a different microfarad value may be needed. Every time a compressor or motor is changed, even when there is no change in capacitance required, the capacitor is best changed with it. The terms of many warranties require installing a new capacitor when a compressor or motor is replaced.

Three-phase motors can sometimes be unstuck by temporarily reversing or interchanging any two of the leads. After the motor has started, always turn off the power, then reconnect the compressor leads as shown on the unit wiring diagram.

2.2.3 No Pressure Change

Occasionally, a compressor will be encountered that does what appears to be absolutely nothing when it is energized. There may be very little audible sound coming from it. A check of the refrigeration gauges shows no activity at all. Yet voltage is being applied to the compressor terminals, and no safety device is, or has been activated.

Capacitors

Use a capacitance checker to determine the capacitance of run and start capacitors. The needle deflection of an analog or digital ohmmeter will not indicate the correct capacitance. A capacitor should operate within 10 percent of its microfarad rating.

Checking for a Grounded Condition

A motor can be grounded by a motor winding that is touching the case. In this case the resistance reading from a winding to ground will be very low. However, resistance readings can be several thousand ohms and still be considered shorted. A good rule of thumb is that the resistance to ground should be greater than 1,000V per operating volt. For example, a motor that operates on 240V should have a minimum resistance of 240,000V to ground. For better accuracy, consult the manufacturer to get the expected resistance values. Both refrigerant and oil contamination can also create low resistance-to-ground readings. To improve these readings, change the refrigerant oil and driers.

These are the symptoms of a compressor with a broken drive or crankshaft. An ammeter placed on the compressor leads will show that some, but very little, power is being consumed. The current reading will be very low, but present. This is strictly the power required to spin the motor without any applied load. When the shaft is broken, the motor is free to spin, and there is no movement in any other part of the compressor.

2.2.4 Troubleshooting Scroll Compressors

Basic troubleshooting for scroll compressors is similar to that of other compressors. Expect a pressure differential between the suction and discharge sides of the compressor.

Like reciprocating compressors, scroll compressors sometimes do require start-assist components, especially in the low-voltage starting conditions that often occur in the heat of summer. Because scroll compressors run backwards momentarily at shutdown, it is important to use short-cycle timers, either those provided by most digital thermostats or separate delay-on-break time-delay relays. Short-cycle timers provide the scroll compressor with the time required for pressures to equalize and rotation to stop completely, allowing the compressor to start without a load and rotate in the proper direction.

If the compressor is restarted too quickly, rotation may continue in the opposite direction. Unlike reciprocating compressors, scroll compressors cannot pump properly when their rotation is reversed. If rotating in the wrong direction, they will pump poorly and overheat. It is important to note that three-phase models must be properly phased to ensure proper rotation.

2.2.5 Troubleshooting Rotary Compressors

Rotary compressors sometimes become oil-locked. This is a condition in which the rotor will not turn due to the tight clearances in the compressor. Oil lock can occur in new compressor motors. To fix an oil-locked compressor, try to start the motor several times. If this does not work, install a soft- or hard-start kit. If the kit does not fix the problem, the compressor must be replaced.

Another problem experienced by rotary compressors is that of stuck vanes. When a vane sticks, there is no pumping action and the motor draws low amperage. The rotor is turning, but the vane(s) is not moving in and out to create compression pockets. Before you replace the compressor, try removing the suction line from the compressor shell. Use a heat gun to warm the tubing and suction area on the compressor. The heat may cause the stuck vane to release. Before reconnecting the suction line, momentarily start the compressor with the palm of your gloved hand over the suction fitting. If the vane is no longer stuck, the compressor will pull on the palm of your hand. This will indicate that the vane is now working and creating suction pressure.

A worn, leaking vane will cause pressure readings that are lower than normal on the high side and higher than normal on the suction side.

2.3.0 Compressor Replacement

The procedures used to change out a compressor differ somewhat depending on whether the cause of the compressor failure is electrical or mechanical. There are many types and variations of compressors. The procedures in this section describe the methods used to replace a welded hermetic-type compressor. Regardless of the type of compressor being replaced, the guidelines given in the procedures normally apply. Always consult the system and replacement compressor manufacturer's service literature for the specific system being serviced and the compressor being used.

When replacing a compressor because of a mechanical problem, use the following basic guidelines:

Step 1 Test the new compressor motor for open windings, grounds, or shorts before removing the failed compressor. Also ensure that it is the correct voltage for the application.

Step 2 Turn off the equipment. Lockout and tag equipment so that no one can start it.

Step 3 Recover the refrigerant, regardless of its condition. Recycle or reclaim the refrigerant as its condition dictates.

Step 4 Disconnect and tag all wiring from the compressor.

Step 5 Use a tubing cutter to cut the system refrigerant tubing connected to the compressor. Do not use a hacksaw; it can introduce harmful chips into the system. Tape the open lines in the equipment to prevent dirt or moisture from entering.

Step 6 Remove the hold-down bolts, and then remove the old compressor. Be sure to get help to avoid injury caused by heavy lifting.

Step 7 Unpack the new compressor and read any manufacturer's literature that accompanies it. Replacement compressors often have new features or devices that should be used. Compare nameplates between the old and new compressors to be sure the new compressor is the correct type and voltage.

Step 8 Remove the vibration-isolating mounts from the old compressor and install them on the feet of the new one.

Step 9 Mount the new compressor in the equipment and bolt it down. Be sure to get adequate help to avoid injury caused by heavy lifting. Remember that the mounting bolts are not tightened all the way, except when the equipment is being prepared for shipment. The bolts should be tightened until they nearly contact the rubber vibration isolators. This allows the isolators to function properly.

Step 10 Measure, cut, and fit new piping components for the compressor suction and discharge lines.

Step 11 Braze the system refrigerant lines to the compressor. Clean any flux from the joints and paint them for protection. Ensure that any flux used is not allowed to enter the refrigerant lines. A very low-pressure nitrogen purge should be maintained at all times to prevent copper oxides and the possibility of oil burning inside the refrigerant lines.

Step 12 Remove the existing liquid line filter-drier and replace it with one that is one size larger. If the system is not equipped with a filter-drier, install one. Make sure that only one liquid-line drier is installed in the system. A suction-line drier may need to be added temporarily if contaminants are expected. The use of suction line driers is discussed more fully in upcoming sections.

Step 13 Leak test the system, then use the proper evacuation techniques.

Step 14 While the evacuation process is ongoing, use the time to consult the equipment wiring diagram and connect all wiring to the new compressor. Replace the contactor if it has been damaged, and replace the capacitor(s) if present.

Step 15 Recharge the system with the correct type and weight of refrigerant. If the charge is not weighed in, ensure a sufficient amount of refrigerant is added to the system for operation while the charge is adjusted at startup.

Step 16 Start the system and allow it to run in order to stabilize the system pressures. Refer to the equipment manufacturer's service instructions and follow them to make any adjustments in the refrigerant charge as deemed necessary. Check and record all of the important operating information.

2.4.0 Compressor Replacement Following an Electrical Failure

Compressor electrical failures are quite common, for a number of reasons. Problems with the incoming power, such as surges from lightning or low voltage conditions, place a great deal of stress on the motor windings. As a general rule, electrical failures related to open windings do not create any more problems than a mechanical failure. This is also true when an internal wiring device such as an overload fails to close. However, when an internal path to ground is formed, lots of action can take place inside the compressor. Arcing occurs, and an astonishing amount of contaminants can be produced very quickly.

Electrical failures involving windings that are shorted to ground are often called burnouts. A burnout refers to the breakdown of the motor winding insulation due to the arcing and high temperatures that are generated at the moment of failure. Burnouts are classified either as mild or severe.

A mild burnout usually occurs very suddenly, causing the motor to stop before the contaminants created by the burnout leave the compressor. Few contaminants, if any, are produced and little or no chemical reaction of the refrigerant and oil has occurred. Mild burnouts often occur as the motor tries to start. If the compressor does not start and is not pumping when the event occurs, little contamination escapes the compressor shell. Another factor in a mild burnout is the overcurrent protection device(s) in the circuit. If a fuse or circuit breaker opens quickly, the amount of time power is applied and damage is occurring is significantly reduced.

Severe burnouts usually occur through a longer period of time. The worst tend to occur while the compressor in operation and pumping. If the process begins slowly and the overcurrent protection is slow to respond, considerable contaminants are produced and are pumped throughout the refrigerant circuit. This is sometimes referred to as a running burn.

> **WARNING!**
> When working on a system suspected of having a compressor burnout, wear appropriate personal protective equipment, including rubber gloves and eye protection. Contaminated refrigerant oil may contain heavy concentrations of acid. Do not allow contact with the skin or eyes as severe burns may result.

Before replacing a compressor that has failed because of a burnout, it must first be determined if the burnout was mild or severe. To do this, check the system for acid, using one of several acid/moisture

---

Liquid Refrigerant Charging

When adding a full refrigerant charge following a compressor replacement, charging with liquid from the cylinder will generally be necessary. If vapor is used, it is highly unlikely that a sufficient amount of refrigerant can be installed before the refrigerant cylinder pressure equalizes with the system. With blend refrigerants, liquid charging is always a necessity. Using a device in the charging hose like the liquid charger shown here from YELLOW JACKET®, between the gauge manifold and the system connection, ensures that liquid does not enter at a higher rate than the compressor can handle. It acts as a simple metering device.

03210-13_SA05.EPS

---

refrigerant test kits or oil test kits that are available. Acid/moisture refrigerant test kits can typically be connected to a system service port to obtain a sample of the refrigerant. Oil test kits require that a sample of the system oil be taken. For systems with hermetic compressors, an oil sample may be very difficult to obtain. If an oil sample trap has been installed in the suction line for this purpose, then a sample can be easily obtained. However, as practical as it may seem, it is very rare. In either case, follow the test kit manufacturer's instructions for using the kit and for determining the amount of contamination in the system.

> **WARNING!**
> Do not cut into or use a torch on a line with any system pressure in it. There is a potential for explosion, as well as production of deadly phosgene gas. The system pressure must be 0 psig before repairs are attempted.

Another way to check out the type of burnout in a hermetic compressor is to cut the suction and discharge lines and check for carbon by running a clean, lint-free swab inside the lines. In a semi-hermetic compressor, the easiest way to tell the type of burnout is to remove the cylinder head. If you find carbon using either method, it is a good indication that a severe burnout has occurred. With a severe burnout, the refrigerant will also have a very strong rotten-egg smell.

2.4.1 Mild Burnout Procedures

Since most contaminants produced by a mild burnout remain within the compressor itself, the procedure for changing the compressor is the same as that used with a mechanical failure. An oversized liquid-line filter-drier may be chosen to provide extra surface area to capture any particles

and contaminants. During the compressor replacement process, also install an oil sample trap for future oil testing. The liquid-line filter-drier should be replaced again after several hours of operating time.

After the compressor installation is completed, the system should be triple-evacuated before charging it with refrigerant. The triple-evacuation method is described in a previous module dedicated to the subject.

2.4.2 Severe Burnout Procedures

Servicing a system after a severe burnout requires not only compressor replacement, but also a thorough cleanup of the entire system to remove all harmful contaminants left behind. In many cases, the compressor replacement is only the beginning of the work to be done. Accomplishing a proper system cleanup can be a time-consuming and very costly process.

Successive failures on the same system can often be traced to improper cleanup. Unless the cleanup is performed correctly, a repeat failure usually occurs in a short period of time. The following is an overview of the procedure for replacing a compressor after a severe burnout. Note that it is essential to follow the system and/or compressor manufacturer guidelines for replacing compressors under these conditions. The warranty usually depends on following specific steps that may differ from this example.

Step 1 Recover the system refrigerant and remove the compressor. This refrigerant should be sent away for reclamation, as it is very unlikely that any field recycling process can successfully remove the contaminants.

Step 2 Remove the liquid line filter-drier.

Oil Sample Traps

Although they can be very handy, oil sample traps are very rarely encountered. They are one of those items that, to some technicians, are too much trouble to install when an installation is being done. However, at the point in time when an oil sample is needed, it is much too late to install a sample trap.

A simple oil sample trap can be constructed in the suction line near the compressor. Locate the trap in a horizontal section of line, preferably at least several inches away from a vertical section. Braze in a tee fitting, with the branch looking straight down. Using the necessary bushings and adapters, adapt the branch of the tee to a ⅜" or ½" female flare fitting. Then install a short, flared ball valve, flared sight glass, and a second short, flared ball valve. Cap the final ball valve.

The sight glass provides a handy reservoir to capture oil as if flows by in the suction line, and a way to see the contents. The two valves can be used isolate the reservoir from the system and collect the sample. After a sample is collected, the valves can also be used to evacuate the reservoir before reopening it to the system to collect another sample.

Step 3 Purge the system piping with dry nitrogen in the direction opposite to normal refrigerant flow.

Step 4 Remove, clean, and/or replace the metering devices, receiver, and accumulator, if equipped.

Step 5 Install the new compressor.

Step 6 Add or replace the liquid line filter-drier. Note that a reversible filter-drier is required in the liquid line of a heat pump. For sealed-core driers, use the next larger size than recommended. For larger systems, it may be best to install a replaceable-core filter-drier shell (*Figure 15*) so that the core can be replaced one or more times once the system is returned to service. Install isolation ball valves on each side of the filter-drier with an access valve to recover refrigerant from this section of the line. The access valve will also be used to evacuate this section and the new filter-drier when it is replaced. This step eliminates the need to recover the entire system charge to change the filter-drier later. Note that most replaceable-core filter-drier shells have a fitting in the cap for the installation of an access valve.

Step 7 Add a suction line filter-drier with input and output pressure taps (*Figure 16*). The taps allow for the pressure drop to be tested while the system is in operation. A replaceable-core type filter-drier may be required to reach the necessary capacity for the system. Per *ANSI/AHRI Standard 730-2005*, suction line filter-driers should be selected for a maximum pressure drop of 3 psi in a permanent installation, and 8 psi for temporary use.

> **NOTE**
>
> Even in the case of a running burn, a limited amount of contamination typically reaches the suction line. This is because the refrigerant is being pumped out of the compressor through the discharge line. Suction line filter-driers can often be removed completely after several hours of system operating time. If a replaceable-core shell is used, the cores can eventually be removed and replaced with simple filters, or they can be left empty. Suction line filter-driers must be larger than liquid-line units due to the extreme pressure drop they can create in the vapor stream. They should not be left in the system indefinitely. If they are, they should be checked periodically to ensure the pressure drop never exceeds 3 psi.

Step 8 Triple-evacuate the system.

Step 9 Recharge the system with new virgin or reclaimed refrigerant.

Step 10 Run the system for one hour. Immediately check the pressure drop through the suction line filter-drier on startup to provide a baseline. Stop the system and change the suction line filter-drier any time the pressure drop shows a significant increase above the baseline pressure drop.

Drier Cores

A variety of drier cores are offered by various manufacturers for specific system uses. Some are designed to capture and hold more water than a normal core; others are designed specifically for POE oils; still others are designed to do a better job of capturing acids. Always consult filter-drier manufacturer literature to determine the proper size and best core type for your specific needs.

PIPE FITTING FOR ACCESS VALVE INSTALLATION

FILTER-DRIER SHELL

REPLACEABLE CORES

03210-13_F15.EPS

Figure 15 Replaceable-core filter drier shell.

Step 11 After one hour has elapsed, stop the system and change the liquid and suction line filter-driers or drier cores. If changing a semi-hermetic compressor, replace the oil.

Step 12 Run the system for two more hours.

Step 13 After two hours have elapsed, stop the system and test for acid or moisture contamination to make sure the system is clean. If the test is negative, change the liquid line filter-drier. Remove the suction line filter-drier or drier cores from the system. Never leave a suction line filter-drier in a system for an extended period of time. For an extra measure of caution, a suction line filter-drier or core can be left in the circuit, but only if the pressure loss through the clean filter is below 3 psi. Even then, note that this additional pressure drop has an effect on system energy use. Liquid-line filter-driers may be permanently left in a clean system.

Step 14 If acid or moisture is still present in the system at Step 13, change the oil (semi-hermetic compressors only) and repeat Steps 12 and 13 as necessary to achieve a clean system.

2.4.3 Changing Oil

After a semi-hermetic compressor has been changed due to a severe burnout, the oil will likely need to be changed one or more times. Although it would be good to change the oil in a hermetic compressor as well, it is impractical to do. The refrigeration oil, since it is always in a liquid state, tends to capture particles, acids, and other contaminants that enter the system.

Before removing oil from a compressor, test a sample of the oil for the presence of contamina-

tion. Use an acid test kit and follow the test kit manufacturer's instructions to determine the amount of acid contamination in the system, if any. The test provides a baseline for future test comparison as the cleanup process continues.

Depending on the compressor model, the oil can be removed one of two ways: through an oil drain plug, or through an oil fill plug hole (*Figure 17*). Some compressors do not have an accessible drain plug. In these cases, the oil fill plug hole must be used.

Remove the oil from a semi-hermetic compressor with a drain plug using the following general guidelines:

> **WARNING!**
>
> When removing oil, be sure to wear rubber gloves and eye protection to prevent possible injury. Contaminated refrigerant oil may contain heavy concentrations of acid. Do not allow contact with your skin or eyes as severe burns may result. Also ensure that the crankcase pressure has been relieved.

Step 1 If possible, run the compressor fully loaded. Then close the suction service valve and reduce the crankcase pressure to 0 psig as the compressor runs. Note that the low-pressure switch may have to be bypassed temporarily, depending on where it is connected. Otherwise, a refrigerant recovery process may be required to reduce the crankcase pressure to 0 psig.

Step 2 Stop the compressor and completely isolate it from the system by closing the discharge service valve.

Step 3 Loosen the drain plug until the oil begins to seep out around the plug threads. Removing the plug too quickly may allow the oil rapidly drain out in an uncontrolled manner. Drain the oil into a suitable container so that it is properly captured and can be properly disposed of. After the oil has been completely drained, reinstall the plug.

When there is no accessible drain plug provided, the oil must be removed through an oil fill plug hole. Remove the oil from a compressor with only an oil fill plug hole using the following general guidelines:

Step 1 If possible, run the compressor fully loaded. Then close the suction service valve and reduce the crankcase pressure to 0 psig. Note that the low-pressure switch may have to be bypassed.

03210-13_F16.EPS

Figure 16 Suction line filter-drier.

OIL DRAIN
PLUG

OIL SIGHT
GLASS

OIL FILL
PLUG

OIL PUMP

CRANKCASE
HEATER

03210-13_F17.EPS

Figure 17 Semi-hermetic compressors.

Step 2 Stop the compressor and isolate it from the system by closing the discharge service valve.

Step 3 Remove the oil fill plug. Use a refrigeration oil pump (*Figure 18*) equipped with a flexible siphon kit to remove the oil from the compressor. Insert the siphon hose into the oil fill opening and pump the oil into a suitable container. Continue to remove the oil in this manner until all the oil is removed.

Oil is charged into a compressor after it has been drained or when oil must be added to make up for some loss. Use a refrigeration oil pump such as the one shown in *Figure 18* to charge oil. This particular pump can also be used to add oil into an operating compressor, since it is able to develop enough pressure to overcome common crankcase pressures.

Before charging oil into the compressor, always refer to the manufacturer's instructions for the compressor being serviced. Make sure to use the correct type of oil and the amount specified by the manufacturer. It is very important to use the exact oil specified by the manufacturer. Do not use substitutes unless authorized to do so. This is especially true with the synthetic oils that are used with the newer refrigerants. Their characteristics can vary widely from one manufacturer to another. Charging oil into an empty compressor is outlined in the following procedure:

Step 1 If the compressor has an oil drain plug, it should now be reinstalled and tight. The oil fill plug should be removed. Some compressors may have a special access valve that can connect directly to the pump.

> **CAUTION**
>
> POE oil in a closed container is free of moisture, but will readily absorb moisture if exposed to air. POE oil should not be used from a container that has been open for any significant length of time or from one that contains used oil. POE oil should never be stored in an open container.

Step 2 Mount and secure the refrigeration oil pump on or next to (depending on the pump style) the container of new oil. Lay the end of the charging hose in a suitable container to capture purged oil. Purge the oil pump line by operating the pump until oil appears without air bubbles. Then quickly connect the end of the hose to the compressor fill opening.

Step 3 Open the service valve (if applicable). While watching the oil level in the compressor sight glass, pump oil into the compressor crankcase from the container of oil as needed.

Step 4 Close the service valve and disconnect the pump.

03210-13_F18.EPS

Figure 18 Oil pump.

Step 5 To remove any moisture that may have entered the compressor, evacuate the compressor to 500 microns.

Step 6 Open the suction and discharge service valves, and then restart the compressor.

Step 7 Run the system for about 20 minutes. Full-load operation is necessary to ensure that refrigerant velocities in the piping system are sufficient to entrain and move oil back to the compressor. Then recheck the oil level at the sight glass. Add additional oil if necessary, using the pump instructions for an operating compressor if possible.

> **NOTE**
>
> The correct oil sight glass level varies from one compressor manufacturer to another. Check the manufacturer's instructions.

Additional Resources

Refrigeration and Air Conditioning: An Introduction to HVAC/R. Latest Edition. Prentice-Hall.

System Diagnostics and Troubleshooting Procedures. 2003. John Tomczyk. Esco Press.

2.0.0 Section Review

1. If a compressor runs but does not provide sufficient cooling, one likely cause is _____.

   a. the thermostat control circuit is open
   b. a faulty run capacitor
   c. defective compressor valves
   d. the motor windings are open

2. Piston rings can easily be changed on a semi-hermetic compressor.

   a. True
   b. False

3. Before putting a new compressor into position, _____.

   a. compare the nameplate information between the two compressors
   b. cut and fit new piping components for the compressor
   c. leak test the system
   d. connect the wiring to the new compressor

4. The worst burnouts, producing the most contaminants, typically occur _____.

   a. when the compressor has not yet started
   b. while the compressor is shut down
   c. when a fuse or circuit opens very quickly
   d. when the compressor is running and pumping

SUMMARY

Effective troubleshooting of cooling systems is truly a process that requires patience, experience, and careful thought. A systematic, step-by-step approach that begins with listening carefully to the complaint and the events that led up to the failure always yields the best results. The refrigerant can, at times, be confusing. It important to remember that more than one issue can affect the operation of the system at any one time. For example, a poorly maintained system may suffer from both a clogged condenser and evaporator coil, but it remained reasonably functional until the TXV power element failed. This set of conditions will not be that unusual for a technician to encounter. A technician must be able to tell whether a given device is functioning properly and to recognize the symptoms arising from the improper operation of any part of the equipment.

Troubleshooting the compressor motor is no different than troubleshooting any other motor, except that it is hidden from view. Like any other motor, it has a difficult time starting under a load. Although mechanical problems with compressors are not uncommon, external issues often cause compressors to fail. When this happens, it is crucial that the cause be found and resolved once the new compressor is installed.

1. The condition of the refrigerant entering an evaporator can best be described as _____.
   a. low-temperature, low pressure
   b. low-temperature, high-pressure
   c. high-temperature, low-pressure
   d. high-temperature, high-pressure

2. The refrigerant entering the evaporator, after passing through a metering device, is usually about _____.
   a. 20 percent liquid
   b. 50 percent liquid
   c. 20 percent vapor
   d. 50 percent vapor

3. Air in a refrigeration circuit will likely show a symptom of a _____.
   a. high suction pressure
   b. low suction pressure
   c. low head pressure
   d. high head pressure

4. A system using a fixed metering device is especially sensitive to the level of refrigerant.
   a. True
   b. False

5. When both the suction pressure and head pressure are low, how might a technician determine that the problem is low evaporator airflow, and *not* the refrigerant charge?
   a. The compressor will likely be very hot.
   b. The subcooling will be normal.
   c. The condenser coil will be frozen.
   d. The compressor crankcase will likely feel cold.

6. Hunting can be caused by _____.
   a. an undersized TXV
   b. a very light load
   c. subcooling below 15°F
   d. a clogged filter-drier

7. If a refrigerant liquid line is routed through an unusually hot area on its way to the metering device, how might this affect it?
   a. It can increase its subcooling.
   b. It can decrease its subcooling.
   c. It can increase its superheat.
   d. It can decrease its superheat.

8. A 1 psi pressure loss occurs in a liquid line riser for every _____.
   a. inch of rise
   b. foot of rise
   c. 2.3 feet of rise
   d. 3.5 feet of rise

9. One problem that may not be evident when a system first starts, but develops progressively over time, is _____.
   a. moisture freezing in the metering device
   b. a plugged filter-drier
   c. a low refrigerant charge
   d. a failed condenser fan motor

10. A running burn is indicated by _____.
    a. a failed start capacitor and/or relay
    b. a failed metering device
    c. a significant amount of acid in the system
    d. only a limited amount of acid in the system

11. Slugging is defined as _____.
    a. an overcharge of refrigerant
    b. an overcharge of oil
    c. the consistent return of liquid refrigerant in small amounts
    d. liquid refrigerant and/or oil entering the compressor cylinder

12. What causes oil to foam in a compressor crankcase?
    a. The flashing of liquid refrigerant to vapor as the pressure suddenly decreases
    b. The flashing of liquid refrigerant to vapor as the pressure suddenly increases
    c. The condensing of refrigerant as the pressure suddenly increases
    d. The condensing of refrigerant as the pressure suddenly decreases

13. Discharge valves are more often damaged than suction valves.

    a. True
    b. False

14. Capacitance boosting involves _____.

    a. placing an additional run capacitor is series with the first
    b. placing a run capacitor in series with the start winding
    c. temporarily replacing the run capacitor with a start capacitor
    d. placing a start capacitor in parallel with the run capacitor

15. When sizing a suction line filter-drier, a permanent installation should be sized for no more than a _____.

    a. 1 psi pressure drop
    b. 2 psi pressure drop
    c. 3 psi pressure drop
    d. 4 psi pressure drop

Trade Terms Introduced in This Module

Burnout: The condition in which the breakdown of the motor winding insulation causes the motor to short out or ground electrically.

Capacitance boost: A procedure used to start a stuck PSC compressor. It involves momentarily connecting a start (boost) capacitor across the run capacitor of the stuck compressor in an attempt to start the stuck compressor by increasing the starting torque.

Additional Resources

This module presents thorough resources for task training. The following resource material is suggested for further study.

Refrigeration and Air Conditioning: An Introduction to HVAC/R. Latest Edition. Prentice-Hall.
System Diagnostics and Troubleshooting Procedures. 2003. John Tomczyk. Esco Press.

Figure Credits

Topaz Publications, Inc., Module opener, Figure 9

Courtesy of Joshua West, Figure 4

Courtesy of Super Radiator Coils, SA01

Courtesy of Parker Hannifin, Sporlan Division, SA02

Courtesy of Henry Technologies, Inc., SA03

Courtesy of QuikProducts by Mainstream Engineering Corp., Figure 10A

Courtesy of Highside Chemicals, Inc., Figure 10B

Courtesy of Emerson Climate Technologies, Figures 12, 15–17

Courtesy of Sealed Unit Parts Co., Inc., SA04

Courtesy of Ritchie Engineering Company, Inc., YELLOW JACKET Products Division, SA05, Figure 18

Section Review Answer Key

Answer	Section Reference	Objective
Section One		
1.d	1.1.0	1a
2.b	1.2.0	1b
3.b	1.3.1	1c
4.a	1.4.0; Figure 3	1d
5.b	1.5.1	1e
6.d	1.6.0	1f
7.c	1.7.0	1g
8.c	1.8.1	1h
Section Two		
1.c	2.0.0	2a
2.b	2.2.1	2b
3.a	2.3.0	2c
4.d	2.4.0	2d

NCCER CURRICULA — USER UPDATE

NCCER makes every effort to keep its textbooks up-to-date and free of technical errors. We appreciate your help in this process. If you find an error, a typographical mistake, or an inaccuracy in NCCER's curricula, please fill out this form (or a photocopy), or complete the online form at **www.nccer.org/olf**. Be sure to include the exact module ID number, page number, a detailed description, and your recommended correction. Your input will be brought to the attention of the Authoring Team. Thank you for your assistance.

Instructors – If you have an idea for improving this textbook, or have found that additional materials were necessary to teach this module effectively, please let us know so that we may present your suggestions to the Authoring Team.

NCCER Product Development and Revision

13614 Progress Blvd., Alachua, FL 32615

Email: curriculum@nccer.org
Online: www.nccer.org/olf

❏ Trainee Guide   ❏ Lesson Plans   ❏ Exam   ❏ PowerPoints   Other _____

Craft / Level: _____   Copyright Date: _____

Module ID Number / Title: _____

Section Number(s): _____

Description: _____

_____

_____

_____

_____

Recommended Correction: _____

_____

_____

_____

_____

Your Name: _____

Address: _____

_____

Email: _____   Phone: _____

Troubleshooting Heat Pumps

OVERVIEW

This module covers the operation and specialized troubleshooting methods for heat pumps. Heat pumps contain more control devices and have a more complex wiring diagram than a cooling-only system. The cooling and heating functions are usually examined separately, depending upon the malfunction. Some problems affect both modes of operation. Since heat pumps are unique, some troubleshooting tasks are unique as well.

Module Four

Trainees with successful module completions may be eligible for credentialing through NCCER's National Registry. To learn more, go to **www.nccer.org** or contact us at **1.888.622.3720**. Our website has information on the latest product releases and training, as well as online versions of our *Cornerstone* magazine and Pearson's product catalog.

Your feedback is welcome. You may email your comments to **curriculum@nccer.org**, send general comments and inquiries to **info@nccer.org**, or fill in the User Update form at the back of this module.

This information is general in nature and intended for training purposes only. Actual performance of activities described in this manual requires compliance with all applicable operating, service, maintenance, and safety procedures under the direction of qualified personnel. References in this manual to patented or proprietary devices do not constitute a recommendation of their use.

Objectives

When you have completed this module, you will be able to do the following:

1. Compare heat pumps to standard cooling systems and describe their operating cycles.
   a. Compare heat pump systems to standard cooling systems and identify the different types.
   b. Describe the three operating cycles of common heat pumps.
2. Describe the sequence of operation for the common operating modes.
   a. Describe the sequence of operation for the cooling mode.
   b. Describe the sequence of operation for the three heating modes.
   c. Describe the sequence of operation for the defrost mode.
   d. Describe the sequence of operation of dual-fuel systems.
   e. Describe the use of microprocessor controls in heat pump systems.
3. Explain how to check and/or troubleshoot various functions and components of heat pump systems.
   a. Explain how to check field and factory wiring.
   b. Explain how to check and troubleshoot heat pump thermostats.
   c. Explain how to test thermistors.
   d. Explain how to check the various types of valves found in heat pumps.
   e. Explain how to check defrost control circuits.

Performance Tasks

Under the supervision of your instructor, you should be able to do the following:

1. Demonstrate the ability to isolate and determine the solution for various electrical and mechanical malfunctions in heat pumps.
2. Initiate the defrost cycle of a heat pump.

Trade Terms

Demand defrost
Direct-acting valve
Electronically commutated motors

Field wiring
Pilot-operated valve

Industry-Recognized Credentials

If you are training through an NCCER-accredited sponsor, you may be eligible for credentials from NCCER's Registry. The ID number for this module is 03311-13. Note that this module may have been used in other NCCER curricula and may apply to other level completions. Contact NCCER's Registry at 888.622.3720 or go to **www.nccer.org** for more information.

Contents

Topics to be presented in this module include:

Figures and Tables

1.0.0 HEAT PUMP OPERATING CYCLES

Objectives

Compare heat pumps to standard cooling systems and describe their operating cycles.

   a. Compare heat pump systems to standard cooling systems and identify the different types.

   b. Describe the three operating cycles of common heat pumps.

A service call for a heat pump is likely to be more complicated than one for a cooling-only system or a furnace. Heat pump control circuits are more complex because they control heating and cooling cycles as well as a defrost cycle.

Additionally, there are several types of heat pumps, most of which are air-to-air split systems or packaged units. There are also types that use water as a heat transfer medium. Heat pumps can also be used to heat domestic hot water.

Ultimately, a heat pump is still a cooling unit, so many of the same troubleshooting principles covered in earlier modules also apply to heat pumps.

1.1.0 Heat Pump Systems

A heat pump is a combination heating and cooling unit. It produces cooling in the same manner as a conventional air conditioner, and then reverses the cycle to produce heat (*Figure 1*). Both systems have these same components:

- Compressor
- Condenser and evaporator coils
- Indoor blower fan
- Condenser fan
- Service valves or service ports
- Pressure and temperature controls
- Refrigerant and refrigerant lines

In addition to the typical air conditioning components, the heat pump requires a reversing valve and additional metering devices and control circuits. Because heat pumps perform both heating and cooling functions, the coils are identified by location. The condensing and evaporator coils on an air conditioner are identified as the outdoor and indoor coils on a heat pump.

The main advantage of a heat pump is that it is less expensive to operate than electric resistance heat. Depending on the operating conditions, a heat pump may be able to provide three times as much heat for the same electrical cost. Depending on the climate and fuel costs, a heat pump may or may not be less costly to operate than gas or oil heat. It is, however, more likely to be cost effective for combined heating and cooling than other methods.

1.1.1 Heat Pump Operation

Even at very cold outdoor temperatures, there is some heat in the air. The air-to-air heat pump operates on the principle of extracting this heat from the outdoor air and transferring it to the indoor air. Of course, heat can be extracted from water as well (water-to-air heat pump).

The reversing valve is critical to heat pump operation, and sets heat pumps apart from cooling-only units. It is responsible for reversing the flow of refrigerant through the system. The outdoor coil then becomes the evaporator and the indoor coil becomes the condenser. The control circuits are almost always arranged so that the reversing valve is de-energized in the heating mode. If there is a reversing valve failure, the heating function will continue to operate. Since the defrost cycle is based on the cooling mode, a reversing valve failure will however, result in a frozen outdoor coil. A reversing valve failure in the summer will leave the structure without cooling.

Another key difference between cooling-only systems and heat pumps is the metering device arrangement. In *Figure 2*, the heat pump has two metering devices. In the arrangement shown, the devices are fixed-restrictor metering devices. Built-in or separate check valves are used to block or allow refrigerant flow, depending on the mode

GOING GREEN

Heat Pump Swimming Pool Heaters

In mild climates where swimming pools can be used year-round, heat pumps are used as pool heaters. Heat is extracted from the air and transferred to the swimming pool by a refrigerant-to-water heat exchanger. This method of heating swimming pools is an energy-efficient alternative to conventional fossil-fuel pool heaters.

HEAT IS
EXTRACTED
FROM THE
OUTSIDE AIR

THE REFRIGERANT
CARRIES THE HEAT
TO THE INDOOR COIL

COOL REFRIGERANT
RETURNS TO THE
OUTDOOR COIL

HEAT IS
TRANSFERRED
FROM THE
REFRIGERANT TO
THE INDOOR AIR

HEAT PUMP

03311-13_F01.EPS

Figure 1 Heat pump operation in the heating mode.

of operation. In this unique case, the restrictor is truly fixed; it cannot move inside the housing. In other applications, fixed-orifice, piston-type metering devices are used. When this type is used, check valves are not required because the piston acts as its own check valve. Check valves may be used with TXVs, although some TXVs also provide this function internally.

1.1.2 Heat Pump Classifications

Heat pumps are classified according to their heat source and the medium to which the heat is transferred (heat sink). For example, a water-to-air heat pump picks up heat from water flowing through a heat exchanger and transfers it to the air flowing through another coil.

The more common types of heat pumps are: air-to-air, air-to-water, water-to-water, and water-to-air.

In the air-to-air system, air is the external heat source and the medium (heat sink) that comes in contact with the indoor refrigerant coil. Some larger heat pump systems may have more than one heat source and may also use both water and air as heat sinks.

An air-to-water system also uses air as the heat source, but the heating and cooling within the structure are provided by a water distribution system. The air-to-water system uses convection devices to transfer the heat between the water and the air. These systems are used primarily to heat water.

Ground water, from the standpoint of temperature, is an excellent heat source but is not available in all areas. Ground water can also be in limited supply during certain periods. Because of these limitations, air is the predominant heat source in residential and small commercial installations.

Because the Earth's temperature is very stable, heat pump systems may use a buried coil to extract that heat (ground source or geothermal heat pump). In this system, fluid is circulated through the buried coil (*Figure 3*) and the Earth is used as a heat sink to absorb or dissipate heat. The fluid (typically a glycol-based anti-freeze solution) is circulated to a fluid-to-refrigerant heat exchanger located in the air handler inside the structure. The refrigerant then flows to the indoor coil in the air handler where the heat is rejected (heating mode) or absorbed (cooling mode).

SAT. DISCH. TEMP. 130°F
ENT. AIR TEMP. 95°F

LIQUID LINE

SAT. SUCT. TEMP. 40°F
ENT. AIR TEMP. 75°F

OUTDOOR FAN

CHECK VALVE CLOSED

METERING DEVICE

OUTDOOR COIL

WARM OUTDOOR AIR

HOT AIR

INDOOR COIL

CHECK VALVE OPEN

COOL SUPPLY AIR

ACCUMULATOR

WARM ROOM RETURN AIR

INDOOR FAN

ENERGIZED

AUX. HEATER

SUCTION PRESSURE 69 PSIG

SERVICE VALVE

SUBCOOLING 15°F

SUCTION TEMP. 50°F

VAPOR LINE

SUPERHEAT 10°F

REVERSING VALVE

DISCHARGE PRESSURE 300 PSIG

VIEW A – COOLING MODE

SAT. SUCT. TEMP. 19.5°F
ENT. AIR TEMP. 45°F

LIQUID LINE

SAT. DISCH. TEMP. 103°F
ENT. AIR TEMP. 70°F

OUTDOOR FAN

CHECK VALVE OPEN

METERING DEVICE

OUTDOOR COIL

COLD OUTDOOR AIR

COLD AIR

INDOOR COIL

CHECK VALVE CLOSED

HEATED SUPPLY AIR

ACCUMULATOR

COLD ROOM RETURN AIR

INDOOR FAN

DEENERGIZED

AUX. HEATER

SUCTION PRESSURE 42 PSIG

SERVICE VALVE

SUBCOOLING 15°F

SUCTION TEMP 34°F

VAPOR LINE

REVERSING VALVE

DISCHARGE PRESSURE 210 PSIG

VIEW B – HEATING MODE

03311-13_F02.EPS

Figure 2 Refrigerant flow in an air-to-air heat pump.

Figure 3 Geothermal heat pump system.

A typical geothermal heat pump is a packaged product that contains the compressor, refrigerant-to-fluid heat exchanger, indoor coil, and blower motor. The pump that circulates the fluid through the ground coil is often a separate component.

In addition to air and water, other heat sources are used. Examples are waste heat from selected industrial processes, exhaust air from ventilation, solar energy, and heat extracted from refrigerated spaces. These sources are commonly used in addition to, rather than as replacements for, the basic heat source.

1.2.0 Heat Pump Operating Modes

A heat pump has three primary operating cycles or modes: cooling, heating (reverse-cycle or compression heat), and defrost.

In addition to the three primary modes, compression heat may operate with supplemental electric heaters and an emergency heat mode. When selected, the emergency heat mode locks out compressor operation. It will then enable supplemental electric heaters to heat the structure. The emergency heat mode would be used if a component failure that hinders compression heating is suspected or identified.

1.2.1 Cooling Mode

In most heat pumps, the reversing valve is energized when the thermostat calls for cooling and de-energized when it calls for heating. This guar-antees that heat is available even if there is a failure in the reversing valve control circuits. When the unit is in the cooling mode, the refrigerant flow (see arrows) is the same as that of any cooling unit. Refer to *Figure 2, View A*. Note that the pressures and temperatures shown are based on HCFC-22.

Note the path through the reversing valve. The check valve used with the indoor coil metering device is closed, forcing the refrigerant to be metered. The cold, low-pressure refrigerant flowing through the indoor coil (evaporator) absorbs heat from the conditioned space and is boiled into a superheated vapor.

Hot, high-pressure refrigerant gas leaving the compressor is pumped through the outdoor coil (condenser), where the heat is rejected. The check valve used with the outdoor coil metering device is open, allowing refrigerant to bypass the metering device. The refrigerant pressures and temperatures shown are typical of those in a normally operating air conditioning system.

1.2.2 Heating Mode

In the heating mode (*Figure 2, View B*), the reversing valve changes position. The refrigerant leaving the compressor is routed in the opposite direction from that of the cooling cycle. Instead of flowing through to the outdoor coil, the hot, high-pressure refrigerant vapor leaving the compressor flows to the indoor coil. When this happens, the indoor coil is acting as a condenser. The check valve used with the indoor coil metering device is pushed open, allowing refrigerant to bypass the metering device.

Heat is extracted from the refrigerant because the air in the conditioned space is cooler than the refrigerant in the coil. The cooled refrigerant then flows to the outdoor coil. At this point, the check valve used with the outdoor coil metering device is pushed closed, forcing the refrigerant to flow through the metering device. Acting as an evaporator, the outdoor coil absorbs heat from the relatively warmer outdoor air.

Heat provided in this way is known as reverse-cycle heat or compression heat. Note the significant differences in the pressures and temperatures at key points in the system as compared with the cooling mode.

An accumulator and crankcase heater, which are optional accessories on cooling units, are essential on most heat pumps. The crankcase heater keeps liquid refrigerant from condensing in the crankcase of the compressor, where it could cause slugging at startup. The accumulator captures liquid refrigerant that gets into

Geothermal Heat Pumps

Geothermal systems can be installed a number of different ways, depending upon the available property and its geology. The illustration on the left shows a vertical closed loop system. Closed loop systems can be installed horizontally or vertically in the ground, or in a body of water. Open loop systems use two separate water wells, as shown in the diagram on the right.

CLOSED VERTICAL LOOP SYSTEM

OPEN LOOP SYSTEM
(TWO SEPARATE WELLS)

03311-13_SA01.EPS

the suction line during operation in the heating mode. The two coils—indoor and outdoor—have different internal volumes. As a result, less refrigerant is needed in the circuit during operation in the heating mode, when the indoor coil becomes the condenser. This excess refrigerant charge makes its way to the suction line and into the accumulator. The liquid refrigerant slowly vaporizes in the accumulator and returns to the compressor.

1.2.3 Defrost Mode

At outdoor temperatures below 50°F, frost will likely form on the outdoor coil of the heat pump while it is running in the reverse cycle heating mode (*Figure 4*). The amount of frost that collects is directly related to moisture content of the air. As the frost builds up, it will restrict airflow and significantly reduce heat transfer. The unit in the figure shows a heat pump that has gone beyond the normal level of frost accumulation.

The frost is usually melted by reversing the flow of refrigerant so that the hot, high-pressure refrigerant from the compressor discharge flows through the outdoor coil. The defrost cycle is au-

tomatic. The start and termination of the cycle may be based on elapsed time, coil temperature, or a combination of the two.

Note that the defrost mode is basically the same as the cooling mode. The unit is operating as a cooling unit, sending the hot refrigerant to the outdoor coil. The primary difference is that

03311-13_F04.EPS

Figure 4 Frost on a heat pump outdoor coil.

the condenser fan remains off during a defrost cycle. Since the object is to transfer heat to the coil and melt frost and ice, the condenser fan would work against the cycle. During a defrost cycle, it is common for steam to be seen rising from the melting frost.

Since the unit is actually operating in the cooling mode, the discharge air temperature can be quite cold. During the defrost mode, a supplementary electric heater or the burner of a fossil-fuel furnace is usually turned on to prevent the occupied areas from becoming cold.

Additional Resources

Troubleshooting & Servicing Heat Pump Systems, 2002. Mt. Prospect, IL: ESCO Press.

1.0.0 Section Review

1. Each of the two metering devices used with a heat pump may be equipped with a(n) _____.
   a. reversing valve
   b. check valve
   c. gate valve
   d. pressure reducing valve

2. What happens when frost builds up on a heat pump's outdoor coil?
   a. Faster outdoor fan speed
   b. Increased coil efficiency
   c. Heat pump shuts off
   d. Heat transfer drops

SECTION TWO

2.0.0 HEAT PUMP OPERATING SEQUENCES

Objectives

Describe the sequence of operation for the common operating modes.

   a. Describe the sequence of operation for the cooling mode.
   b. Describe the sequence of operation for the three heating modes.
   c. Describe the sequence of operation for the defrost mode.
   d. Describe the sequence of operation of dual-fuel systems.
   e. Describe the use of microprocessor controls in heat pump systems.

Trade Terms

Electronically commutated motors: A variable-speed electric motor based primarily on a DC-driven permanent magnet design that responds to microprocessor inputs to achieve the desired speed.

Understanding the operating sequence in all operating modes is the key to troubleshooting any heat pump. The control circuits for a heat pump are generally more complicated than those of a cooling-only or heating-only system because there are more controls and more operating modes.

For most air-to-air heat pumps, there are five different operating modes:

- Cooling mode
- Heating mode with compression heat only
- Heating mode with compression heat and supplemental heat
- Defrost mode
- Emergency heating (compression heat disabled)

2.1.0 Cooling Sequence of Operation

Use the wiring diagram in *Figure 5* to follow the sequence of operation in cooling. When the room thermostat function switch is placed in the cooling mode, the reversing valve (1) is energized. In this position, refrigerant will be directed to the outdoor coil which acts as a condenser, and to the indoor coil which acts as an evaporator. As a general rule, the reversing valve will remain ener-

gized and in this position, even when the system is not actively cooling.

As room temperature rises, room thermostat contacts (2) close to generate a call-for-cooling signal. Assuming the discharge temperature switch and high-pressure switch (3) are closed, 24-volt control circuit power is supplied from the Y terminal to energize the compressor contactor (4). With the contactor contacts closed, power is supplied to start the compressor and outdoor fan motor (5).

At the same time the compressor is energized, a 24-volt signal from the G terminal on the room thermostat energizes the indoor blower relay (6) to energize the indoor blower motor (7).

The unit will continue to operate until the room thermostat is satisfied. In the cooling mode, a heat pump acts like a conventional air conditioner.

2.2.0 Heating Sequence of Operation

Use the wiring diagram in *Figure 6* to follow the sequence of operation in heating.

Unlike the cooling mode, the reversing valve is de-energized in the heating mode. In the de-energized position, refrigerant is directed to the outdoor coil which acts as an evaporator, and to the indoor coil which acts as a condenser.

As room temperature drops, room thermostat contacts (1) close to generate a call-for-heat signal. Assuming the discharge temperature switch and high-pressure switch (2) are closed, 24-volt control circuit power is supplied from the Y terminal to energize the compressor contactor (3). With the contactor contacts closed, power is supplied to energize the compressor and outdoor fan motor (4). At the same time as the compressor is energized, a 24-volt signal from the G terminal on the room thermostat energizes the indoor blower relay (5) to energize the indoor blower motor (6).

At this point, the unit is supplying compression heat only. If the heating load is low and the outdoor temperature is above the balance point, the unit should be able to satisfy the room thermostat on compression heat alone. However, in the event that compression heat alone is unable to satisfy the room thermostat, the second stage of the room thermostat (7), closes as the room temperature continues to drop. This energizes the W1 terminal and supplemental electric heat by first energizing heat relay No. 1 (8).

Assuming the outdoor thermostat is closed, heat relay No. 2 (9) is energized through the auxiliary contacts of heat relay No. 1. The outdoor thermostat prevents the second stage of electric heat from operating above a preset outdoor temperature. The compressor and supplemental electric

Figure 5 Cooling sequence of operation.

COOLING SEQUENCE OF OPERATION

1 – The reversing valve is energized when the thermostat is switched to the cooling mode.
2 – Thermostat contacts close on a cooling demand.
3 – 24-Volt control circuit through normally closed discharge temperature switch and high-pressure switch.
4 – The compressor contactor is energized and its contacts close.
5 – Both the compressor and the outdoor fan motor are energized.
6 – As the compressor circuits are energized, the indoor blower relay is energized through the G terminal.
7 – The indoor blower relay closes its N.O. contacts to start the indoor blower motor.

03311-13_F05.EPS

Figure 6 Compression heat with supplemental heat sequence of operation.

HEATING SEQUENCE OF OPERATION

HEAT PUMP ONLY

1 – Thermostat controls close on a heating demand.

2 – 24-Volt control circuit through normally closed discharge temperature switch and high-pressure switch.

3 – The compressor contactor is energized and its contacts close.

4 – Both the compressor and the outdoor fan motor are energized.

5 – As the compressor circuit is energized, the indoor blower relay is energized through the G terminal.

6 – The indoor blower relay closes its N.O. contacts to start the indoor blower motor.

SUPPLEMENTAL HEAT

7 – If additional heat is needed, the thermostat closes its second stage heating contacts.

8 – Heat relay no. 1 is energized to power elements 1 and 2.

9 – Heat relay no. 1's auxiliary contacts also close to energize heat relay no. 2, providing the outdoor thermostat is closed.

03311-13_F06.EPS

heater(s) will continue to operate until the room thermostat is satisfied.

Use the wiring diagram in *Figure 7* to follow the sequence of operation in emergency heat.

If the compressor fails to operate, no compression heat will be available. On some systems, a service light on the room thermostat (1) will come on to alert the occupants that there is a problem. This light is usually responding to a lock-out relay or a similar device.

To provide heat until the compressor problem is resolved, the room thermostat function switch can be placed in the emergency heat mode (2) to enable emergency heat. Once in this mode, an emergency heat signal light (3) is energized on the room thermostat and the emergency heat relay (4) is also energized from the E terminal of the thermostat. The light reminds the owner that the system is operating in emergency heat, which provides a lesser amount of heat and is more costly to operate. Once the thermostat is switched to Emergency Heat, only the supplemental or auxiliary heat source and the blower operate. The compressor does not function at all.

One set of contacts (4) of the emergency heat relay bypasses the outdoor thermostat for the electric heat, enabling all available stages of supplemental heat to be energized. Without compression heat, both stages of electric heat will likely be needed to maintain comfort. Once in the emergency heat mode, heat relay No. 1 (5) can energize all available stages of electric heat with the help of heat relay No. 2.

The unit will operate until the room thermostat is satisfied. Some thermostats may make the electric heat respond as the first stage of heating in place of compression heat. On thermostats that do not provide this function, the electric heat continues to respond to the second stage. The result is that the room temperature lags behind the thermostat setting by 2°F to 3°F. Once in emergency heat, it is important to find and correct the problem that is preventing compressor operation.

2.3.0 Defrost Mode Sequence of Operation

Use the wiring diagram in *Figure 8* to follow the sequence of operation during defrost.

During normal heating operation, an air-to-air heat pump will accumulate frost on the outdoor coil. If allowed to build up, the frost restricts airflow across the coil, reducing heat transfer. To remove the accumulated frost, the heat pump must periodically defrost the outdoor coil.

Defrost is initiated and controlled by a defrost control (1). Before a defrost cycle can be initiated, the unit must be operating in the heating mode. An idle heat pump does not start up with the intention of defrosting itself. The unit must be running with the compressor contactor closed in order for this circuit to be energized. This means that both the compressor and indoor fan are running as the defrost cycle begins. Both components—the indoor fan and the compressor—must continue running throughout the cycle.

There are various strategies used to initiate defrost. Some systems check the outdoor coil temperature on a timed basis to see if a defrost cycle is required. Very old systems may actually enter defrost on a timed basis only, without monitoring temperatures. Once the defrost mode is initiated by the control, a set of contacts changes position. Contacts between terminals 2 and 3 open, de-energizing the outdoor fan motor. Contacts between terminals 2 and 1 close, energizing the defrost control relay.

Once energized, the defrost control relay (3) energizes the reversing valve and a second defrost relay. This second relay is normally located at the indoor unit to enable the electric heat. To prevent cold air from being blown into the conditioned space, this defrost relay energizes heat relay No. 1 (4). With the heat relay energized, line-voltage power is available to the supplemental electric heat (5), which warm and temper the air being discharged into the conditioned space. Without the supplemental heat, the system will deliver very cold air to the structure as long as the defrost mode is active. Note that the outdoor thermostat may still be open, leaving only one stage of electric heat energized during the defrost cycle.

With the reversing valve now energized, the unit has changed to its cooling mode, sending hot refrigerant vapor to the outdoor coil to melt accumulated frost. The compressor and indoor blower motor (2) continue to operate non-stop throughout the defrost cycle.

Once it has been determined that the coil is free of frost, defrost is terminated by the defrost control. Termination is usually based on both time and temperature. The defrost cycle will terminate when a specific temperature has been reached at the outdoor coil (monitored by a thermostat or thermistor), or when a maximum amount of time has elapsed. Strategies for defrost termination may vary among manufacturers. Once the defrost cycle terminates, it returns to operation in the heating mode.

EMERGENCY HEAT SEQUENCE OF OPERATION

1 – Should a heat pump malfunction occur, the service light comes on.
2 – Place the thermostat in the emergency heat position.
3 – Emergency heat signal light illuminates.
4 – Emergency heat relay is energized from terminal E.
5 – Heat relay no. 1 remains energized as before and elements no. 1 and no. 2 are energized.

03311-13_F07.EPS

Figure 7 Emergency heat sequence of operation.

DEFROST CYCLE SEQUENCE OF OPERATION

1 – The defrost control switches position under defrost conditions. This turns off the outdoor fan and energizes the defrost control relay.
2 – The compressor and indoor blower motor continue to operate.
3 – The defrost control relay closes its N.O. contacts to energize the reversing valve and a second defrost relay.
4 – The defrost relay closes its contacts to energize heat relay no. 1.
5 – Heat relay no. 1 closes its contacts to power elements 1 & 2.

Figure 8 Defrost sequence of operation.

03311-13_F08.EPS

2.4.0 Dual-Fuel Heat Pump Operation

A popular, energy-efficient use of air-to-air split-system heat pumps is as an add-on to a fossil-fuel furnace. In cooling, the heat pump operates like any other split-system air conditioner with the furnace acting as the air handler. It pushes air through the indoor coil mounted in the supply duct above the furnace.

In the heating mode, the heat pump operates under some conditions, while the fossil-fuel furnace will maintain comfort when other conditions are present. The primary factor for changeover is related to the balance point. Above the balance point, a properly sized heat pump can handle the heating needs of the structure without supplemental heat. The balance point is typically above freezing as well. As a result, only a limited amount of frost tends to build on the outdoor coil, especially if the heat pump is cycling off some. By operating the fossil-fuel heating equipment below the balance point, the heat pump only operates when it is most efficient and with a very limited need for energy wasting defrost cycles.

In *Figure 9*, an outdoor thermostat or thermistor is used to switch the system between compression heat and fossil-fuel heat. At low outdoor temperatures (generally below the balance point) where the heat pump is less efficient and lacks heating capacity, the gas or oil furnace heats the structure. Above the balance point, the more efficient heat pump heats the structure.

The temperature at which the changeover takes place between the two methods of heating can vary. The balance point is only one possible factor, although it is the primary one. Another factor that may affect the chosen point for changeover is the cost of electricity compared to the cost of the fossil fuel used. In most cases though, a temperature slightly above the actual balance point of the structure is chosen as the point where compression heat is locked out and fossil fuel is used to

Figure 9 Dual-fuel heat pump operation.

heat the structure. This ensures that indoor comfort is maintained at lower outdoor temperatures.

Unlike a conventional all-electric heat pump where compression and supplemental electric heat are often supplied at the same time, only one form of heat is available with dual-fuel systems; compression heat or fossil fuel heat. Since the heat pump indoor coil is downstream of the furnace, the heat pump compressor would quickly overload with the furnace blowing hot air across its already-hot coil. Reversing the order of the coil and furnace would not help. The furnace would exceed its high-temperature limits if hot air from the heat pump coil were provided to its inlet. The one exception to this situation applies during the defrost cycle. Instead of using supplemental electric heaters to temper the indoor air, the system's gas or oil burner comes on to supply heat. The indoor coil is usually mounted downstream of the burner. That results in very hot air passing over the indoor coil while it is serving as the evaporator. In summer conditions, this would quickly overload the compressor. However, with frost/

ice on the outdoor coil, there is an enormous heat sink available, preventing the discharge pressures from rising too high. The furnace heat is transferred to the refrigerant where it moves to the outdoor coil. The large amount of heat carried by the refrigerant results in a very fast and thorough defrost cycle.

For many years, this dual-fuel arrangement was only available in a split-system configuration. Today, packaged heat pumps are available that contain gas burners, allowing users of packaged products to benefit from this unique energy-saving combination. Aside from the unique differences just described, an air-to-air heat pump installed with a fossil fuel furnace is no different in its operation than any other air-to-air heat pump.

2.5.0 Microprocessor Controls in Heat Pumps

The operating sequences just described are typical for heat pumps using electromechanical controls. They are ideal for understanding basic electrical operating sequences. While millions of electromechanical heat pumps are still in use, they are rapidly being replaced with heat pumps that use electronic controls and variable-speed motor technology. Electronic controls can do everything that can be done with electromechanical controls, but better and more reliably. Electronic controls have few, if any, moving parts which are prone to fail after repeated use. Compare a motor-driven defrost timer turning cams that activate switches to an electronic timer with no moving parts.

Figure 10 is the schematic diagram of a modern packaged heat pump that contains electronic controls and a variable-speed blower motor. Even though what is going on inside the control boards cannot be seen, it is possible to determine a lot about the sequence and timing of operations. The unit contains two control boards, a defrost board (DB) and an integrated control motor (ICM) board that controls the operation of the variable-speed indoor blower motor.

Looking at the defrost board helps reveal how it functions. A timing (logic) circuit is in series with the compressor contactor coil and the low- and high-pressure switches in the Y circuit. This indicates that the defrost board contains a compressor start-delay timer. Another timing (logic) circuit is in series with the defrost thermostat. This indicates that defrost is based on elapsed time and outdoor coil temperature. Elapsed time is field-selectable and is done based on the position of dip switches, as shown on the schematic.

A direct connection between the room thermostat cooling contacts and O on the defrost board indicates the reversing valve is energized in cooling. Open relay contacts within the defrost board imply that during defrost, they close to energize the reversing valve and apply power to the electric heat relays (HR) by providing a current path from R to W2. Closed relay contacts across OF1 and OF2 open during defrost to stop outdoor fan operation. The same contacts can be seen in series with the outdoor fan motor (OFM) in the high-voltage side (upper part) of the schematic.

The ICM board is used to control the variable-speed indoor fan motor (IFM). Looking at the schematic, one could conclude that there are four electric motors in the unit: three in the high-voltage side (outdoor fan motor, compressor motor, and indoor fan motor), and one in the low-voltage side, labeled as the ICM. In fact, the indoor fan motor and the ICM motor are one and the same. The indoor fan motor has line-voltage power applied at all times, even when the contactor is de-energized. This is a characteristic of electronically commutated motors. In order to run, it must also receive various low-voltage control signals from the ICM board through the 16-pin connector on the motor. This connector is labeled as the ICM in the lower part of the diagram. This results in the indoor fan motor having two physical connections, one for line-voltage (upper part of schematic), and another that carries the low-voltage input signals (lower part of schematic). The ICM board generates the low-voltage signals based on the inputs it receives from the room thermostat.

During cooling operation, signals from the thermostat generate output signals from the ICM board that tell the fan to run at its highest speed. If continuous fan operation is required, the lack of a Y signal (cooling) and the presence of a G signal (fan operation only) cause the ICM board to signal the fan to run at a much lower speed for ventilation purposes only. The ICM board generates the appropriate speed control signals based on the position of jumper wires across pins on the board. These selections are made during the initial system start-up.

The startup selections made are partially based on system capacity. The cooling fan speed of a 3-ton system, for example, would be higher than the selected cooling fan speed of a 2-ton system. The electric heater size in kilowatts (kW) also determines fan speed in the heating mode. Larger heaters require more airflow. When properly configured, the ICM board will allow the fan to operate at a number of different speeds based on the equipment's operating mode. Configuration pro-

Figure 10 Packaged heat pump schematic.

cedures must be obtained by reading the manufacturer's installation instructions.

For the special features offered by an electronic control, always read the manufacturer's literature. The training programs available from manufacturers are also very helpful in learning about the specifics of a particular product or family of products. Technicians are often intimidated when they see the number of different components found on electronic control boards. In fact, HVAC technicians do not need to know how each component on a control functions, nor do they have to troubleshoot electronic controls to the component level. What they do need to know is if the control is functioning the way it is supposed to. If it is not, the control must be replaced.

Most electronic controls respond to input signals of some sort. If it has the correct input signals, it responds by generating appropriate output signals. So if a technician can confirm that input signals are correct, but the control is not responding with the correct output signals, there is a good chance the electronic control is defective. On the other hand, if inputs and outputs from a control are what they should be, and the device being controlled is not responding, the problem is likely with the device being controlled and not the electronic control. For example, if the control board is sending power to the reversing valve to change its position, but the position does not change, then the reversing valve coil should be tested next; the control board is not at fault.

Always check the device in question before replacing it. Something as simple as a loose connection may be preventing normal operation.

Manufacturers often provide the input and output signals expected from an electronic control. One example of that type of information is shown in *Figure 11*.

ECMs vs. ICM Motors

ECMs and ICM motors are actually the same thing. Earlier versions of HVAC equipment that used these motors referred to them in literature and wiring diagrams as Integrated Control Module (ICM) motors. The ICM actually referred to the electronic module that drove the motor, which is (and was) often integrated into the motor itself as shown here. In later years, the term ECM was commonly adopted to describe these motors.

03311-13_SA02.EPS

BONNET SENSOR (INPUT) J C S
BONNET SENSOR (INPUT)
THERMOSTAT 1ST STAGE (INPUT) Y
24V (INPUT) R
24V COMMON (INPUT) B
REVERSING VALVE (INPUT) O
THERMOSTAT 2ND STAGE (INPUT) W
EMERGENCY HEAT LIGHT (OUTPUT) X
1ST STAGE SUPPLEMENTAL (OUTPUT) 66
STANDBY ELECTRIC HEAT (OUTPUT) 53
2ND STAGE SUPPLEMENTAL (OUTPUT) 60

MODULE-CONTROLLED CONTACTS

1K 3K 4K 5K 2K

3K 5K-2 5K-1 4K 2K

R D P M O B

1t/LL 2t/DIS 3t/OD
SENSOR (INPUT)

24V (OUTPUT)
DEFROST (OUTPUT)
HIGH-PRESSURE LOCKOUT CIRCUIT (INPUT)
M CONTACTOR (OUTPUT)
24V COMMON (OUTPUT)
SOLENOID (OUTPUT)

03311-13_F11.EPS

Figure 11 Input and output signals.

2.0.0 Section Review

1. When is the indoor blower motor usually energized during a call-for-cooling?

   a. Before the compressor is energized
   b. After the compressor is energized
   c. When the compressor is energized
   d. After the outdoor fan is energized

2. As room temperature rises, a call-for-heat signal is generated by the room thermostat.

   a. True
   b. False

3. Without the assistance of supplemental heat while in the defrost mode, a heat pump _____.

   a. adds heat to a structure
   b. removes heat from a structure
   c. must stop compressor operation
   d. would de-energize the reversing valve

4. In a dual-fuel heat pump, the indoor coil is usually mounted _____.

   a. in a basement
   b. near the air handler
   c. in the return duct
   d. above the furnace

5. Electronic controls are more reliable than electromechanical controls because they _____.

   a. are hermetically sealed
   b. have few moving parts
   c. operate on DC voltage
   d. draw very low current

SECTION THREE

3.0.0 TESTING AND TROUBLESHOOTING HEAT PUMPS

Objectives

Explain how to check and/or troubleshoot various functions and components of heat pump systems.

   a. Explain how to check field and factory wiring.
   b. Explain how to check and troubleshoot heat pump thermostats.
   c. Explain how to test thermistors.
   d. Explain how to check the various types of valves found in heat pumps.
   e. Explain how to check defrost control circuits.

Performance Tasks

1. Demonstrate the ability to isolate and determine the solution for various electrical and mechanical malfunctions in heat pumps.
2. Initiate the defrost cycle of a heat pump.

Trade Terms

Demand defrost: A defrost control strategy that typically does not depend on time to determine when a defrost cycle is needed. Instead, the system responds directly to current conditions, such as temperature or pressure drop through the outdoor coil, to determine when frost is present.

Direct-acting valve: A valve in which the solenoid operates the valve port.

Field wiring: Wiring that is installed during installation of the system; for example, such as a thermostat hookup to unit.

Pilot-operated valve: A valve in which the solenoid operates a plunger, which in turn creates or releases pressure that operates the valve.

Although heat pumps are more complex than cooling units, they have several operating modes that share common functions. If a problem such as a disabled compressor is observed in the cooling mode, check to see if the same problem occurs in the heating or defrost mode. If it does, the problem is most likely in the compressor or one of its control or protective de-

vices. If it does not, then the problem is most likely in some device that is unique to the cooling mode. Once the operating sequence of a particular heat pump is known, a process of elimination can often isolate a failed component merely by observing how the unit operates in each mode. These observations can often be done without using test equipment. Some examples of this process follow.

Refer again to *Figure 5.* If the compressor does not operate in the cooling mode, but works in the reverse cycle heat and defrost modes, the malfunction would likely be found in the room thermostat. All safety controls must be closed, or the compressor would not run at all. The portion of the thermostat that initiates a call for cooling may be faulty. This would be a good place to begin troubleshooting.

What is the likely cause if the unit runs in heating but does not defrost and refuses to switch to the cooling mode of operation? A defective reversing valve is a strong possibility, because heating takes place with the valve de-energized. To operate in cooling, the reversing valve must be energized. If the unit does not go into defrost while in the heating mode, it is a further indication that the reversing valve is defective because the reversing valve must be energized during defrost. Note that the valve itself may be fine. The solenoid coil is more likely to fail than the valve itself. Testing and/or changing the solenoid coil is very simple. Replacing an entire reversing valve is a lengthy and tedious job.

Many manufacturers provide troubleshooting guidance in their product literature. *Figures 12* and *13* are examples of a troubleshooting road map provided by one manufacturer. Additional troubleshooting guides are provided in the *Appendix* for trainee review. Troubleshooting aids like these can be helpful, but some are either too general or too complicated. During troubleshooting, there is no substitute for knowing and understanding the operating sequence of a product.

3.1.0 Checking Wiring

The wiring for heat pumps and similar equipment can usually be broken down between **field wiring** and factory wiring. Factory wiring is the line-voltage and low-voltage wiring that is installed and terminated at the point of manufacture. Field wiring is the wiring installed and terminated by the installer.

3.1.1 Field Wiring

Control circuit field wiring must be connected by at least 18 AWG copper wire, or the relays and contactors may chatter or fail to operate due

Figure 12 Troubleshooting procedures.

to voltage drop. As a general rule, if the wiring circuits are longer than 100', wire heavier than 18 AWG copper will have to be used in order to prevent excessive voltage drop. 16 AWG wire is the next available size. It is very rare that control circuit loads require a larger wire size than this.

Low primary voltage at the transformer also may cause relays and contactors to chatter or fail to operate. Therefore, the primary voltage applied to the transformer must be checked. The secondary voltage, typically 24VAC, is dependent upon the primary voltage applied. If the primary voltage is low, the secondary voltage will be low as well. If the primary voltage is below 10 percent of the rating stamped on the nameplate, erratic control operation may result. This situation will usually cause the secondary control voltage to fall below 22VAC.

When looking for wiring problems, check for loose connections at the terminals and wire nut connections. The additional resistance caused by a poor connection can cause erratic control device

Restricted Airflow

The supply air temperature of a heat pump in the heating mode is typically lower than that of electric, gas, and oil heating systems. If the air being discharged into the conditioned space from a heat pump is warmer than 120°F, it could indicate restricted airflow. Otherwise, the supply air temperature will rarely approach this temperature.

A heat pump is typically designed to provide airflow of 450 to 500 cfm per ton. Undersized ductwork, a dirty air filter, or a dirty indoor coil can cause reduced airflow, which in turn will cause an increase in compressor discharge temperature. High temperatures can cause the compressor to overheat and trip on the overload protection.

Figure 13 Troubleshooting diagram.

03311-13_F13.EPS

operation. Control circuit fuses or breakers (if any) should also be checked before checking the field wiring.

To check field wiring, first check the terminals for proper connection. Disconnect each end of the wire in a specific circuit and measure the resistance with an ohmmeter. Excessive resistance will indicate loose connections or corroded terminals. Open circuits will be indicated by a reading of in-finity on the ohmmeter. To check thermostat wiring in the wall, join two wires with a wire nut and check continuity at the other end.

3.1.2 Factory Wiring

Factory wiring in the unit can be checked in the same manner as field wiring. However, field wiring should be checked before factory wiring. When factory wiring errors occur, the malfunction should show up when the unit is checked out during installation. Factory wiring problems are less likely to occur after the system has been in operation over the short term. However, over time, all terminations tend to deteriorate and need to be checked. Factory connections are generally of better quality since they are completed under ideal conditions using precise and consistent components.

Check the equipment wiring using the wiring diagram attached to the unit. Trace the circuits by observing the actual location of the electrical devices, terminal designations, and wire color codes as designated on the wiring diagram.

3.2.0 Checking Heat Pump Thermostats

Heat pump thermostats (*Figure 14*) should be checked to ensure that the thermostat is the correct model number to match the system. Thermostats should not be mounted where extraneous cooling or heating sources will affect the sensing bulb. They should be mounted level and the wiring hole in the wall behind them should be plugged. A perfectly level installation is not necessary for performance, as digital thermostats are not affected by level like older electromechanical models. However, a level thermostat does indicate the proficiency of the installer. The wiring hole should be plugged to prevent air from inside the wall from affecting the thermostat's accuracy is it reads room temperature. Today's heat pump thermostats are far more likely to be digital electronic controls rather than electromechanical mercury-bulb thermostats.

There are several reasons for this. Most importantly, many older thermostats contained mercury—a toxic substance. If the glass capsule containing the mercury breaks, people could be exposed to the mercury. In replacement situations, old mercury-bulb thermostats must be treated as hazardous waste and disposed of properly.

Digital electronic room thermostats offer more precise temperature control. This allows the full range of features designed into modern heat pumps to be fully utilized. The digital room thermostat shown in *Figure 14* is typical of the newest generation of controls. It offers the versatility of being used for heating, cooling, and heat pump operation. The installer configures the thermostat for the desired mode of operation. Within the heat pump mode, some thermostats offer the versatility of programming for heat pump operation with auxiliary electric heaters or with a dual-fuel system. Also note the signal strength indicator in the upper right-hand corner of the display; this thermostat communicates wirelessly with the rest of the system.

Often, the traditional mode selection subbase switches are removed and replaced with a mode selection button or touch-sensitive screens. Adjustable heat anticipators have been replaced by a method to select burner cycle rates directly. Within the thermostat itself, the traditional hard-wired electromechanical circuits are replaced with electronic circuits. Miniature relays provide the switching that was previously done by the subbase switches.

Because the thermostats are electronic devices, they must be powered to operate. This can be done with batteries and/or the 24VAC available from the control circuit. If control circuit power is used, a 24VAC common lead must be provided and connected to the common (C) terminal on the thermostat. Most models have batteries that maintain programming for an extended period of time if the power is interrupted.

To check a heat pump room thermostat, it is best to begin by determining the function controlled by each terminal. These terminal desig-

Figure 14 Heat pump thermostat.

Field Wiring Problems

Field wiring problems are most often found at connection points. This is because HVAC installers may fail to use the correct devices or procedures to join wires. For example, it is common to encounter low-voltage thermostat wires that have been simply twisted together and taped. A correctly sized wire nut should be used to make this type of connection.

nations are fairly standardized in industry, but occasional variations will be found. In the case of the unit shown in *Figure 14*, the subbase terminals are related to the following circuits or functions:

- *Terminal C* – Wire color is often blue or brown; it is the 24VAC common lead from the control transformer secondary circuit.
- *Terminal R* – Wire color is red; it is the 24VAC power lead from the control transformer secondary circuit. There may be two R terminals on a thermostat, marked as R_C and R_H. Some systems, especially older systems, may have two control transformers in operation; one serves the heating equipment and one serves the cooling equipment. Separating the R terminals keeps the two transformer circuits apart. When a single transformer is used (very common), a jumper is placed across the two R terminals.
- *Terminal Y* – Wire color is yellow; it energizes the compressor control circuit. The Y terminal may be followed by a number (Y1, Y2, etc.) to indicate the cooling stage it controls.
- *Terminal G* – Wire color is green; it energizes the indoor fan relay.
- *Terminal O or B* – Wire color is usually orange or blue; these terminals are energized when modes are selected. When cooling is the selected mode of operation, O is energized and remains so. B is energized when the heating mode is selected; O becomes de-energized. The O terminal is commonly used to power the heat pump reversing valve. Some manufacturers, however, do use terminal B and energize the reversing valve in the heating mode.
- *Terminal W* – Wire color is white: it controls the heat source relay for supplementary heat. The W terminal is often followed by a number (W1, W2, etc.) to show the different stage of heat each terminal controls.
- *Terminal S* – Wire color is optional; may be red and white when two-conductor thermostat wire is used. The S terminals are used to connect external temperature sensors.
- *Terminal E or X* – Any color wire; it connects to the emergency heat circuit.
- *Terminal L* – Any color wire; provides power for the Emergency Heat light on the thermostat. The light is powered constantly when the thermostat is switched to Emergency Heat. Otherwise, it can be powered by heat pump control circuits that indicate a fault has occurred.

Note that this wiring description applies to the specific thermostat shown in *Figure 14*. It does not apply to all thermostats.

Aluminum to Copper Wire Connections

Joining aluminum and copper wires requires special techniques and connectors. If correct procedures are not followed, the aluminum-copper connection can overheat or fail. For this reason, many HVAC equipment manufacturers prohibit the connection of aluminum wires to their products.

In most heat pumps the reversing valve is energized in cooling and defrost, and de-energized in heating. It is done that way so heat can be available if the reversing valve solenoid or something in the circuit fails while in the heating mode. However, there are some heat pumps that were designed to operate with the reversing valve energized in heating. To determine if the reversing valve is energized during cooling, use the following procedure:

- If the reversing valve is wired directly to the room thermostat through the O terminal, it is energized in cooling. The O terminal is always a terminal energized immediately when the cooling mode is selected.
- If the reversing valve is wired directly to the room thermostat through the B terminal, it is energized in heating. The B terminal is historically energized immediately when the heating mode is selected.
- Check for normally open contacts of a defrost relay in the reversing valve solenoid circuit. The relay contacts close during defrost in the heating mode to energize the reversing valve solenoid. Being in defrost is like being in the cooling mode as far as reversing valve operation is concerned.

Figure 15 is a detailed schematic of an older electromechanical mercury-bulb room thermostat similar to that shown in the *Figure 13* wiring diagram. Although these thermostats are no longer manufactured, there are still millions of them in use.

3.2.1 Room Thermostat Troubleshooting Sequence

Electromechanical room thermostats are nothing more than a series of manual switches that are placed in various combinations of positions to get a desired result. Electronic thermostats provide the same service, but in a different way. The easiest way to determine if a room thermostat is faulty is to remove it from the subbase and

03311-13_F15.EPS

Figure 15 Thermostat wiring diagram.

use jumper wires at the terminals to simulate the various circuit closures. The R terminal is the hot terminal, so directly connecting it to each of the functional terminals will energize the appropriate component or function. With the thermostat removed from the subbase and 24VAC power on, place jumpers across the following terminals or thermostat wires to check for operation:

- R to G will energize the indoor fan.
- R to Y will energize the compressor and outdoor fan motor.
- R to O should energize the reversing valve.
- R to W and G will energize the resistance auxiliary heat if the indoor fan relay contacts are closed.

- R to X2 and G will energize the emergency heat controls (heat pump) if the indoor fan relay contacts are closed.

If a component fails to function with the jumper in place, the problem is not in the thermostat. If the component does function with the jumper, but fails to function with the thermostat installed, the problem is somewhere in the thermostat. Problems with electronic room thermostats can be isolated in a similar manner.

Thermostat troubleshooting can also take place at the furnace or fan coil unit. If the thermostat is calling for cooling, for example, there should be 24VAC at the furnace C and Y terminals. Before condemning any thermostat, verify that all switch positions are correct and that any installer programming has been properly completed. In many cases, incorrect setup of an electronic thermostat will make it appear to be malfunctioning, when in fact it is operating as instructed.

3.2.2 Heat Pump Outdoor Thermostats

Electromechanical outdoor thermostats are used with air-to-air heat pumps in several ways:

- Lock out supplementary electric heaters above an outdoor temperature
- Lock out compression heat at very low outdoor temperatures
- Switch between compression heat and a furnace in dual-fuel applications
- Stage banks of supplementary electric heaters as outdoor temperature drops

Electromechanical outdoor thermostats are still in widespread use, but thermistors have replaced them in newer equipment. Check the operation and calibration of an outdoor thermostat as follows:

Step 1   Connect the ohmmeter as shown in *Figure 16*. Set the sensing bulb in a container with a slushy ice and water mixture. The mixture, after a few minutes, should be at or very near 32°F. This can be verified with an electronic thermometer. Slowly

Special Thermostats

There is a growing number of thermostats on the market now that operate using a communication bus, rather than through common 24VAC wiring terminals. The Carrier Infinity™ system, for example, uses a four wire approach to control wiring. The wires are designated as A, B, C, and D. The A and B wires are used for communication signals to travel back and forth, while the C and D terminals provide 24VAC power. The primary equipment must also be configured to accept this type of control arrangement. Troubleshooting these units requires a different strategy than troubleshooting standard controls. When problems are encountered, be sure to acquire the service information for the specific model in use.

OHMMETER

RX1

ICE AND WATER

03311-13_F16.EPS

Figure 16 Outdoor thermostat check.

rotate the outdoor thermostat dial until the ohmmeter indicates that the contacts are closed. The dial should indicate about 32°F.

Step 2 By rotating the dial one way and then the other several times, while watching the ohmmeter (with many thermostats, the open and close action of the contacts are audible), a technician can determine if the contacts are operating in the 32°F range with reasonable accuracy. Note that laboratory-level accuracy is not required of an outdoor thermostat. However, if more than a ±5°F variation is found to exist between the ice solution and the dial, calibrate or replace the outdoor thermostat.

3.3.0 Testing Thermistors

Thermistors (*Figure 17*) are used in modern heat pumps to sense outdoor air temperature, coil temperature, and other temperature values that are helpful in controlling a modern heat pump. Compared to a conventional electromechanical thermostat that typically senses a single temperature and responds accordingly by opening or closing a switch, a thermistor is capable of measuring a wide range of temperatures and sharing the information with a controller. This makes the thermistor much more versatile than electromechanical outdoor temperature thermostats.

Thermistor calibration can be checked in a manner similar to that used to check conventional outdoor temperature thermostats.

Step 1 Turn off power to the unit, disconnect the thermistor, and immerse it in an ice water solution.

03311-13_F17.EPS

Figure 17 Thermistors.

Step 2 Place an accurate electronic thermometer sensor in the ice water and stir the solution until 32°F registers on the thermometer.

Step 3 Place ohmmeter leads across the thermistor leads and read the resistance. Compare the resistance at 32°F to the resistance that the thermistor should read at that temperature by consulting a resistance versus temperature chart (*Figure 18*).

In the example shown, the resistance measured at that temperature should be about 33,000 ohms. If the temperature that the thermistor is attempting to measure is accurately known (air temperature or the surface temperature of a refrigerant line), enter that value into the chart to find the resistance at that temperature. An ice bath is not always necessary.

THERMISTOR CURVE

Figure 18 Resistance versus temperature chart.

03311-13_F18.EPS

Then check the manufacturer's literature to determine if the resistance measured is within acceptable limits.

3.4.0 Troubleshooting Heat Pump Valves

Depending on its operating mode, refrigerant in a heat pump flows to different components in different directions. To direct the refrigerant to where it needs to be and to prevent the refrigerant from being where it should not be, various types of flow-control valves are used.

3.4.1 Solenoid Valves

When the coil is energized on a typical solenoid valve, a magnetic field is produced, which pulls the valve stem on positions the valve open or closed. There are two types of solenoid valves:

- **Direct-acting valves** – The solenoid coil pulls the valve port open (or draws it closed) directly by lifting the pin. Power limitations of the solenoid coil restrict the port size of a given valve however. A large valve, especially one that must operate against significant pressure differentials, may be impossible to open with electromagnetic force applied to a coil alone.
- **Pilot-operated valves** – The solenoid coil operates a plunger covering a small pilot port hole. When the port hole is open, system pressure trapped between a larger piston and the port

hole escapes. The resulting pressure differential on the two sides of the piston causes it to move toward (or away from) the pilot port hole; this opens (or closes) the main port hole on the other side of the piston. Heat pump reversing valves are usually pilot-operated valves.

When troubleshooting solenoid valves, look for the following major categories of problems:

- Valve does not open
- Valve does not close
- Valve produces excess noise

When checking or testing solenoid valve coils, perform the following:

Step 1 Inspect for loose or improperly installed wires.

Step 2 Test supply voltage to the coil with a voltmeter. The voltage should fall within +10 percent and –15 percent of the rated voltage.

Step 3 Place the tip of a steel screwdriver on the solenoid valve plunger when it is energized. The magnetic force of the solenoid should attract the screwdriver, indicating the magnetic attraction of the coil is active.

Step 4 Inspect the coil for burning, which can result from too high of an ambient temperature, too much moisture, or too much humidity. Energizing the solenoid coil for a period of time without a magnetic load imposed on it can also damage the coil or cause it to fail. This occurs whenever the solenoid coil assembly is removed from the valve. If the coil needs to be energized while removed from its solenoid valve stem, place a steel screwdriver or similar object through the opening to impose a magnetic load on it.

Step 5 Always check the coil resistance while it is disconnected from the circuit.

Assuming the solenoid coil is functional, but the valve refuses to move, internal sticking or seizing may be the problem. Before condemning or disassembling a valve for internal inspection, check the pressure differential across the valve. Solenoid valves have a maximum pressure differential specification. If the differential is too high, the valve cannot move. Pilot-operated valves usually have both a minimum and a maximum pressure differential specification, since they depend on system pressure differences to physically change position.

3.4.2 Reversing Valves

A reversing valve (*Figure 19*) has a four-way slide mechanism that directs the refrigerant in one of four ways, depending on the system needs. The following is one way to test a reversing valve:

Step 1 Inspect the valve and solenoid coil for dents, deep scratches, or cracks. These may indicate failure.

Step 2 Check the electrical system to be sure the solenoid has power and can be energized. This can be determined by the sound of the plunger moving in and out on some units.

03311-13_F19.EPS

Figure 19 Reversing valves.

Solenoid Valves

Solenoid valves may be used in the liquid lines of split-system heat pumps and air conditioners. These valves are used to prevent the migration of refrigerant into the outdoor unit during the compressor off cycle.

03311-13_SA03.EPS

Step 3 If the coil cannot be energized, check other components in the electrical system for the source of the problem. Check the coil resistance with the wiring disconnected to be sure it is not open or shorted.

Step 4 Check the system pressure differential. Remember that a reversing valve is a pilot-operated valve. For a reversing valve to operate, a minimum operating pressure differential must exist. While some valves will shift when the differential is as low as 10 psig, others may require over 100 psig. Most reversing valve manufacturers do not guarantee the movement of the valve unless a minimum 100-psig differential exists. Also check the system for the proper amount of charge. If the charge is too low, the pressure differential may be insufficient.

The reversing valve may also be tested for internal leakage by measuring temperatures. *Figure 20* shows the temperature measurement points for testing a reversing valve while it is in the heating position. To avoid error, temperature readings should be made at least five inches from the valve body. The temperature probes should be insulated. Infrared thermometers may not provide the needed accuracy to determine the valve condition; therefore, it is recommended that only contact thermometers be used.

Step 1 Set the thermostat to turn on cooling.

Step 2 Attach the thermometers at T1, T2, and T3.

Figure 20 Reversing valve temperature test points.

Step 3 If the temperature difference between thermometers T2 and T3 is greater than 3°F, the valve is leaking internally. Replace the reversing valve.

Step 4 Set the thermostat to turn on heating.

Step 5 If the difference between T1 and T3 is greater than 3°F, the valve is leaking internally and requires replacement.

> **CAUTION**
> While reversing valves appear rugged, they can still suffer from mishandling. What appear to be small dents or dings can result in sticking and/or internal leakage. Handle reversing valves with care.

The current refrigerant (HFC-410A) used in residential heat pumps produces much higher pressures than the refrigerant it replaced (HCFC-22). Components used with HFC-410A heat pumps are rated for those higher pressures. If a heat pump that uses HFC-410A requires a replacement reversing valve, make sure to install a proper replacement as specified by the manufacturer. If a generic valve is installed, it is crucial that it be properly rated for the application and the refrigerant in use.

3.4.3 Check Valves

If a unit has two sets of capillary tubes for metering devices there will be two check valves. Some units, especially older models, may also have thermostatic expansion valves that also require external check valves. During the heating cycle, one check valve is used to bypass the indoor metering device. The check valve on the outdoor coil is forced closed, sending the refrigerant through the metering device. During the cooling cycle, the outdoor check valve allows refrigerant to bypass the outdoor metering device. The indoor check valve closes, forcing refrigerant to flow through the metering device. Modern heat pump expansion valves have the check valve and bypass built into the valve.

Check valves will open to bypass refrigerant and close to force refrigerant to be metered. Note that these check valves have no solenoid coil or other operating device. They are simple ball-check devices. The force of refrigerant trying to flow in one direction or another positions the ball. In one direction, the ball is forced against a seat that does not allow refrigerant to pass (closed). In the other, the ball is forced against a surface that allows refrigerant to pass around it on all sides (open). In its open position, the valve offers a low resistance (bypass) path for the refrigerant to travel.

When a check valve is suspected to be defective, test the heat pump equipment in both the heating and cooling cycles. If the equipment maintains normal head and suction pressures during either the heating or cooling cycle, but pressures are abnormal in the other cycle, a check valve is probably faulty.

When testing check valves during cold weather, cover the outdoor coil surface area with paper or plastic while conducting cooling cycle tests. By

---

Think About It

Troubleshooting Exercises

Refer to the heat pump wiring diagram in *Figure 13* to solve the following problems. A list of common HVAC abbreviations associated with wiring diagrams is provided in *Appendix B*.

1 Is the reversing valve energized or de-energized in the heating mode? What is the energizing path?
2 If the outdoor fan (ODF) continues to run during defrost, what component should be checked? What symptoms would likely be seen if the fan continues running during defrost?

---

blocking the outdoor coil until the head pressure is around 250 psig, cooling cycle conditions can be reasonably simulated.

Figure 21 illustrates typical check valve construction. When a check valve becomes faulty, a number of predictable problems will usually be encountered. *Table 1* provides a review of the problems that are associated with specific check valve failures.

One very easy test should be conducted to see if a check valve ball is stuck in place. See *Figure 22*:

Step 1 Shut down the equipment, allowing head and suction pressures to equalize. Once the pressures are equal, slide a strong magnet back and forth along the check valve body.

Step 2 If the ball valve is stuck, nothing will be heard. If the ball valve is free, a clear clicking sound will be heard as the magnet moves the ball back and forth in its housing.

If it is determined that the check valve must be replaced, follow these steps:

Step 1 Properly recover or isolate the refrigerant.

Step 2 Remove the faulty check valve.

Step 3 Avoid bending or deforming the valve body when installing the replacement. Use care when handling the valve.

Step 4 Avoid overheating the check valve or applying excessive brazing material when brazing. Wrap the valve body with a wet rag to keep the valve from overheating during brazing.

Figure 21 Check valve construction.

3.5.0 Troubleshooting Defrost Controls

Most heat pump defrost controls encountered today will be solid-state packaged controls that will be replaced if they are determined to be defective. These controls usually rely on a combination of compressor run time and outdoor coil temperature. However, there are some that use only outdoor coil temperature to both initiate and terminate defrost.

Most electronic defrost controls provide a feature that allows the technician to initiate the defrost cycle in order to avoid delays while troubleshooting. Otherwise, a technician would have to wait until the system initiates a defrost cycle on its own, which could take hours. Ways to initiate this self-test feature vary from manufacturer to manufacturer, but most are very similar. The most popular method is to use a coin or the tip of a screwdriver blade to jumper across the two specific pins or terminals on a defrost control board. The two pins must be continually shorted together for around five seconds in most cases. In *Figure 23*, the speed-up terminals, as they are commonly called, are labeled TEST. Other boards may have them labeled as SPEED-UP. Once the self-testing feature is initiated, the system enters the defrost mode, allowing the technician to observe the sequence of operation.

If an electromechanical type of defrost control is used, technicians can also manipulate the unit into starting a defrost cycle. The mechanical timer usually has a manual dial that allows a technician to advance the time. The action of the timer to initiate a defrost cycle is usually audible as the internal contacts switch and the cycle begins. The

Metering Device Advances

Check valves in heat pumps have become far less common in recent years. Fixed-orifice metering pistons have been used for years, combining a metering device and check valve in a single component. Special heat pump thermal expansion valves (TXVs) were also developed that have the check valve and bypass features built into the valve body.

Note that these advances are more about reducing manufacturing costs and increasing the flexibility of components than about eliminating the check valve. Although they can be fouled after compressor burnouts and other events that add undesirable materials to the circuit, TXVs can be fouled the same way.

Table 1 Check Valve Troubleshooting

CHECK VALVE PROBLEMS	DESCRIPTION
Outdoor check valve sticks closed during a cooling cycle	There is low suction pressure, high discharge pressure, and high superheat at the indoor coil. The compressor is hot and may trip the internal overload protector. If the indoor unit has a thermostatic expansion valve, a defective thermal element on this valve can lead to similar symptoms.
Outdoor check valve sticks closed during the cooling mode	The equipment operates at normal pressures in the heating cycle.
Indoor check valve sticks closed during a heating cycle	There is low suction pressure, high discharge pressure, and high superheat at the outdoor coil. The compressor is hot and may trip the internal overload protector.
Outdoor unit has a thermostatic expansion valve	A defective thermal element can lead to similar symptoms.
Indoor check valve sticks closed during the heating cycle	The equipment operates at normal pressures in the cooling cycle.
Indoor check valve sticks open during the cooling cycle	There is high suction pressure and liquid flooding back to the compressor. During the heating cycle, the equipment operates at normal pressures.
Outdoor check valve sticks open during a heating cycle	There is high suction pressure, low superheat, and liquid flooding back to the compressor. The compressor is cool or cold. During the cooling cycle, the equipment operates at normal pressures.
Indoor check valve leaks during the cooling cycle	There is a slightly high suction pressure and a slightly low head pressure. Depending upon the rate of the leak, there could be cool or normal temperatures in the compressor. In the heating cycle, the equipment operates at normal pressures.
Outdoor check valve leaks during the heating cycle	There is slightly high suction pressure and slightly low head pressure. The compressor will be near normal temperatures or slightly cool, and the equipment operates at normal pressures during the cooling cycle.

03311-13_T01.EPS

temperature switch is built-in to most of these older controls, using a refrigerant-filled remote bulb and a capillary tube connected to the timer.

Ensuring that the defrost controls are properly set-up before troubleshooting a problem is important. Most time-temperature defrost controls allow the installer to select the time between defrosts based on the local climate. As an example, in areas of freezing fog or drizzle and near lakes or wetlands, a shorter interval would be selected because frost would build up quickly under these conditions. In drier climates, a longer interval would be more suitable because frost would accumulate more slowly. *Figure 24* shows a jumper wire permanently attached to the board, which is placed over one of three pins to complete a circuit. The pins on this board are labeled 30, 60, and 90 minutes. Most heat pump units and replacement boards ship with the setting at 30 minutes, as shown here. What is being set is the amount

Figure 22 Magnet test.

03311-13_F22.EPS

DEFROST TEST
SPEED-UP TERMINALS

03311-13_F23.EPS

Figure 23 Defrost speed-up terminals.

of compressor run-time required before a defrost cycle initiates. On warmer days, a heat pump may go for hours between defrost attempts before it accumulates 30 minutes of run time.

The time-temperature defrost method is used in most heat pumps, but other methods may be used. With a time-temperature cycle, the selected period of time must pass before a defrost cycle can begin. At the end of the time period, the system will enter defrost if the coil temperature is below a fixed value. The temperature setting of the coil thermostat or thermistor is usually around 30°F. If the coil is not cold enough to indicate the presence of frost/ice, the unit continues operating in the heating mode, and the time period begins again. If it is below the temperature, a defrost cycle begins and continues until the coil temperature rises above a set temperature, or a certain amount of time has elapsed (usually 10 minutes).

Other methods include the use of a pressure switch that monitors the pressure drop across the

COMPRESSOR RUN-TIME
SELECTION JUMPER

03311-13_F24.EPS

Figure 24 Compressor run-time selection for defrost control.

outdoor coil. As frost builds on the coil and less air flows through it, the pressure drop increases until it reaches a pre-set point. At that point, the defrost cycle is triggered. Defrost control that is based solely on need, without time involved, is called a **demand defrost** system.

Another demand defrost strategy is to use two separate thermistors. One monitors the coil temperature, while the other monitors the ambient temperature. When the outdoor temperature is above a preset level, 40°F for example, a defrost cycle is never initiated. Below that temperature, the coil temperature is monitored and evaluated with the outdoor temperature. The defrost control board decides when the conditions indicate the likely presence of frost and initiates a defrost cycle.

3.5.1 *Checking Inputs and Outputs*

Any electronic control, including defrost controls, responds to input signals. If an electronic defrost control has the correct input signals, it responds by generating appropriate output signals or relay contact closures. If it is confirmed that input signals are correct but the control is not responding with the correct output signals, there is a good chance the defrost control is defective. However, if the inputs and outputs from a defrost control are normal and the device being controlled (such as a reversing valve) is not responding, the problem is most likely with the reversing valve or its coil and not the defrost control.

The defrost control manages the various input/output signals using miniature relays that are part of the control. Some typical defrost control board inputs include the following:

- *24V power* – If the defrost control does not have power, it will not operate.
- *Timing input* – With defrost controls that are time/temperature based, the timing function is built-in to the board.
- *Temperature input (binary/switch closure)* – With time/temperature defrost systems, this input takes the form of an open or closed set of outdoor coil thermostat contacts. Closed contacts provide an input indicating that defrost is needed. Open contacts indicate outdoor coil temperature is not low enough to initiate defrost. When those same contacts open *during* defrost, it signifies the outdoor coil temperature is warm enough to terminate defrost. The presence of open/closed contacts can be verified with a voltmeter (power on) or an ohmmeter (power off). Note that the switch will not close at the same temperature that it opens. A differential is built-in to the temperature switch. A defrost thermostat that closes at 30°F, for example, may re-open at 40°F. The same switch is typically used to terminate the defrost cycle as well as initiate it.
- *Temperature input (electronic)* – If a heat pump uses one or more thermistors to sense system temperatures, their input signal(s) would be in the form of a resistance that varies based on the temperature being monitored. Thermistor inputs can be checked with an ohmmeter and the manufacturer's resistance versus temperature chart like the one in *Figure 18*. These temperature inputs are used to terminate the defrost cycle as well as initiate it.
- *Pressure input* – The pressure drop across an outdoor coil (indicating a frosted coil) may be used to initiate and terminate defrost. The pressure switch opens and closes a set of contacts,

so the input to the board is binary. The pressure switch is used to terminate the defrost cycle as well as initiate it.

Some typical defrost control output signals include the following:

- *24V power* – The reversing valve solenoid is typically powered by 24V. Once defrost is initiated, this output voltage is supplied to the solenoid valve to energize it, and removed to terminate defrost. 24V is also provided to energize supplemental electric heaters used to temper the indoor air during defrost. These voltages are can typically be measured at output terminals on the defrost control board.
- *Line-voltage power* – The outdoor fan motor is cycled off during defrost. Often there are terminals on the defrost control where the outdoor fan motor is connected. Voltage can be measured at the terminals.

Keep in mind that different manufacturers use different methods to initiate and terminate defrost. However, they all depend on the defrost control receiving the correct input signals, whatever form they may take. With the correct input signals, a properly functioning electronic defrost control will deliver the correct output signals.

3.5.2 Defrost Troubleshooting Tips

There are a number of things that can interfere with the successful completion of a defrost cycle. Even when defrosting is effectively completed, it may create discomfort for the occupants if the supplemental heat does not work. Below are some common problems encountered and suggestions for their resolution. However, as always, it is important to review the manufacturer's literature

Wind Baffles

If a heat pump is installed in an area that is subject to high winds, such as a roof-mounted unit, it may be necessary to install baffles around the outdoor coil. Baffles will shield the coil from the wind. Wind blowing across a defrosting coil has the same effect as leaving the outdoor fan running during defrost.

WIND BAFFLE

03311-13_SA04.EPS

for the specific heat pump model being repaired. The sequence of operation and construction of heat pumps has varied widely over the years.

- *The defrost cycle activates, but does not readily melt the frost buildup.* The problem could be that the outdoor fan is not being turned off during defrost. If that is the case, check the switch or relay contacts that control the outdoor fan. A strong wind blowing across the outdoor coil could also be a factor. However, this would likely be an intermittent problem rather than a consistent one. Also remember that an incorrect refrigerant charge can also affect defrost performance. Check the refrigerant charge according to the manufacturer's instructions.
- *Cold air is being blown into the conditioned space during defrost.* This means that the electric heater or furnace burner is not turning on when it should. In a system with supplemental electric heat, if the heater works when staged on by the thermostat, it means the problem is likely related to the defrost relay or the defrost control board, depending on the design. If the electric heater does not work in either situation, check the heating element and its associated thermal protection.

Some dual-fuel systems do allow the furnace burner to operate during defrost. However, many of these systems do not operate the heat pump far below the balance point. Therefore, defrost cycles are infrequent, and are usually conducted at fairly warm temperatures outdoors. If cold air is blowing into the conditioned space with a dual-fuel system, check to be sure the furnace operates properly with the outdoor temperature below the balance point, or by switching the system to Emergency Heat. If so, the problem is likely related to the defrost relay, defrost control board, or the interconnecting wiring. Of course, if the furnace does not operate in these modes, then attention should be focused on the furnace controls.

- *The defrost cycle activates when there is no frost or ice on the coil, or the cycle will not terminate even when all the frost is clearly gone.* With a time/temperature defrost scheme, this proves that there is nothing wrong with the timing cycle. Otherwise, the cycle would never start. This problem is often caused by a failed defrost thermostat that has remained closed. The system will then initiate a defrost cycle at the end of every timing cycle, and the cycle will continue for the full amount of time allowed by the defrost control (usually 10 minutes). A defective thermistor, or a failed or out-of-calibration pressure switch can cause the same symptoms.
- *Ice tends to develop and remain on the lower part of the outdoor coil.* The operation of all the defrost controls may appear to be normal with this problem. However, it can be caused by problems with the defrost thermostat or thermistors. If they are not in their proper location or not making good surface contact, the defrost control board may terminate the cycle before the task is done. One of the most likely causes though, is an incorrect refrigerant charge. If the refrigerant charge is low, the hot refrigerant entering the outdoor coil will condense very

Defrost Board Replacement

When replacing a defrost board, it is important to use a board that is designed for the unit being serviced. Boards differ in programming, so using an incorrect board can result in immediate defrost problems. A new board can be ordered from the equipment distributor by heat pump model and serial number to ensure the board features are correct. Some component manufacturers specialize in building boards that are also specifically designed for certain heat pump models, but can be purchased directly from HVAC/R parts houses. These after-market boards usually perform well and offer competitive pricing. Technicians must be sure though, that the correct replacement board is selected. The after-market defrost board shown here is designed to replace several specific Trane heat pump boards.

03311-13_SA05.EPS

quickly, delivering all of its latent heat to the top half of the coil. Since the vast majority of the heat that melts the ice is the latent heat from the change of state, there is not enough heat remaining in the refrigerant to thaw the remainder of the coil. Check the refrigerant charge first when this problem is encountered.

Depending on how the outdoor coil is circuited, the lower portion may always be the last part of the coil to thaw. In some cases, it may be that the defrost cycles are simply not occurring frequently enough. If all aspects of defrost cycle operation are normal and the refrigerant charge is correct, try increasing the frequency of the cycles by adjusting the compressor-run-time on the defrost control board.

There are a number of other troubleshooting guidelines related to heat pumps in the *Appendix* of this module.

Additional Resources

Troubleshooting & Servicing Heat Pump Systems, 2002. Mt. Prospect, IL: ESCO Press.

3.0.0 Section Review

1. If a heat pump outdoor unit is located 110′ from the air handler containing the 24V transformer, the correct control circuit copper wire size should be _____.

   a. 24 AWG
   b. 22 AWG
   c. 18 AWG
   d. 16 AWG

2. What feature is likely to be found on a modern digital heat pump room thermostat?

   a. Mercury bulb switch contacts
   b. Subbase function switches
   c. Heat anticipator
   d. Mode selection button

3. How are thermistors calibrated?

   a. Immersion in ice water
   b. Heated with a heat gun
   c. Immersion in hot water
   d. Immersion in anti-freeze

4. Reversing valves can be checked for internal leakage using a _____.

   a. magnet
   b. thermometer(s)
   c. stethoscope
   d. leak detector

5. The self-test feature on a heat pump defrost control typically speeds up timing functions of the control.

   a. True
   b. False

Summary

The control circuits for a heat pump are more complex than those of a cooling-only system because the heat pump performs more functions: cooling, heating, and defrost. If you consider that many heat pumps also have supplementary electric heaters or dual-fuel capability in addition to emergency heat functions, it makes the control circuits seem even more complex. The added complexity is often an advantage to a troubleshooter, however, because the ability to switch between functions makes it easy to eliminate large portions of the control circuits when isolating a problem.

Technicians troubleshooting heat pumps should have a good understanding of the techniques and tools used in troubleshooting electrical devices and cooling equipment. For effective troubleshooting of a heat pump, it is important to learn the electrical operating sequence. If an electronic defrost control is used, it is important to know the input signals it must receive in order to provide output signals to operate the various heat pump components.

Mechanical troubleshooting is similar to that of a cooling system. Temperature and pressure readings will be different in the heating mode than they are in cooling, however. A heat pump is also much less able to tolerate a variation in refrigerant charge, so a refrigerant leak or a slight charging error can have a significant impact on the heat pump's performance.

1. If a heat pump works normally in the heating mode but will not provide cooling or a defrost mode, which component is most likely defective?

   a. Compressor
   b. Room thermostat
   c. Reversing valve
   d. Compressor contactor

2. Air-to-air heat pumps produce heat by _____.

   a. circulating indoor air through an electric heating element
   b. circulating low-pressure, low-temperature refrigerant through the indoor coil
   c. extracting heat from the outdoor air and transferring it to the indoor air
   d. circulating hot, high-pressure refrigerant through the outdoor coil

3. In addition to the reversing valve, the other major difference between a heat pump and a cooling-only system is that _____.

   a. the heat pump requires a head pressure control
   b. the cooling unit requires a filter-drier
   c. heat pumps use only fixed-orifice metering devices
   d. heat pumps require two metering devices

4. In the cooling mode, the reversing valve is normally _____.

   a. energized
   b. de-energized
   c. bypassed
   d. back-seated

5. When a heat pump is operating in the heating mode, the refrigerant bypasses the _____.

   a. indoor coil metering device
   b. outdoor coil metering device
   c. indoor coil
   d. outdoor coil

6. In the reverse cycle heating mode, _____.

   a. the indoor coil acts as an evaporator
   b. the indoor coil acts as a condenser
   c. the outdoor coil acts as a condenser
   d. defrost is only needed in very cold climates

7. Refer to *Review Question Figure 1*. What happens in the cooling mode if the discharge temperature switch fails to close?

   a. Indoor fan does not run
   b. Indoor blower relay de-energized
   c. Outdoor fan does not run
   d. Contactor will chatter

8. Refer to *Review Question Figure 2*. If the outdoor thermostat is not closed and this system is in the heating mode, a likely symptom would be _____.

   a. no compressor operation
   b. less supplemental heat available
   c. no supplemental heat available
   d. no defrost operation

9. Which of the following conditions is *not* common and proper during the defrost mode?

   a. The outdoor fan shuts off
   b. The indoor fan runs
   c. An electric heater is energized
   d. The compressor shuts off

10. Refer to *Review Question Figure 3*. Defrost is initiated by the _____.

    a. defrost control
    b. outdoor thermostat
    c. defrost relay
    d. room thermostat

11. If a heat pump occasionally blows cold air into the conditioned space during the heating season, a likely cause is _____.

    a. a low refrigerant charge
    b. that the supplemental electric heaters are not on during defrost
    c. the outdoor fan is not turning off during defrost
    d. the indoor thermostat is defective

Figure 1

03311-13_RQ01.EPS

Figure 2

03311-13_RQ02.EPS

Figure 3

03311-13_RQ03.EPS

12. During dual-fuel heat pump operation, the compressor and furnace burner can operate together _____.

a. any time there is a call for heat
b. only if the outdoor temperature is above 40°F
c. only if the outdoor temperature is below 40°F
d. during periods of defrost

13. Refer to *Review Question Figure 4*. The variable-speed indoor fan motor receives low-voltage input signals by way of _____.

a. a 16-pin connector
b. IFM terminal PL-4
c. ICM terminal PL-2
d. the 24V splice box

14. If a heat pump will not switch into defrost, and you suspect the reversing valve, the quickest way to verify your suspicion is to _____.

a. see if the reversing valve switches when there is a call for cooling
b. see if the reversing valve switches when there is a call for heating
c. bypass the reversing valve and see if the unit switches into defrost
d. wait for the next defrost cycle

15. Low primary voltage at the control circuit transformer may cause _____.

a. thermostat droop
b. relay chatter
c. fuse failure

d. temperature swings

16. Why must a 24V common lead be attached to a digital room thermostat?

a. To power the thermostat.
b. To power the transformer.
c. To charge the thermostat battery.
d. To maintain battery polarity.

17. Which heat pump component is controlled through room thermostat terminal O?

a. Indoor fan relay.
b. Compressor contactor.
c. Electric heat relay.
d. Reversing valve.

18. Which of the following is true of a pilot-operated valve?

a. It is non-magnetic.
b. The solenoid coil opens the main valve port.
c. The solenoid coil operates a plunger which opens a smaller valve port.
d. It must be manually operated.

19. The function of a reversing valve will be affected by all of the following, *except* _____.

a. a defective solenoid
b. low system refrigerant charge
c. a crack or dent in the valve body
d. high superheat settings

20. If the outdoor check valve sticks closed during the cooling cycle, there will be _____.

a. a high suction pressure and high superheat at the outdoor coil
b. a low suction pressure and low superheat at the outdoor coil
c. a low suction pressure and high superheat at the indoor coil
d. abnormal temperatures and pressures in the heating cycle

Figure 4

03311-13_RQ04.EPS

Trade Terms Introduced in This Module

Demand defrost: A defrost control strategy that typically does not depend on time to determine when a defrost cycle is needed. Instead, the system responds directly to current conditions, such as temperature or pressure drop through the outdoor coil, to determine when frost is present.

Direct-acting valve: A valve in which the solenoid operates the valve port.

Electronically commutated motors: A variable-speed electric motor based primarily on a DC-driven permanent magnet design that responds to microprocessor inputs to achieve the desired speed.

Field wiring: Wiring that is installed during installation of the system; for example, such as a thermostat hookup to unit.

Pilot-operated valve: A valve in which the solenoid operates a plunger, which in turn creates or releases pressure that operates the valve.

Appendix A

TROUBLESHOOTING CHART

GENERALLY THE CAUSE Make these checks first.	OCCASIONALLY THE CAUSE Make these checks only if the first checks failed to locate the trouble.	RARELY THE CAUSE Make these checks only if other checks failed to locate the trouble.
PROBLEM: Compressor and outdoor fan motor do not start		
Power failure Blown fuse Faulty wiring – Motor to line side of contactor – Load side, contactor to motor terminal – Control circuit Loose terminals – Motor to line side of contactor – Load side, contactor to motor terminal – Control circuit Compressor overload Contactor coil Pressure control – Internal overload in compressor Refrigerant charge low High head pressure Low suction pressure Run capacitor Start capacitor – Compressor motor – Compressor stuck	Low voltage – Motor to line side of contactor – Load side, contactor to motor terminal – Control circuit – Defective contacts in contactor Control circuit thermostat	Unbalanced power supply Outdoor fan motor Indoor fan motor Refrigerant overcharge Noncondensibles Defective power element – Outdoor expansion valve – Indoor expansion valve Thermostat setting
PROBLEM: Outdoor fan motor will not start		
Faulty wiring – Load side, contactor to motor terminal Loose terminals – Load side, contactor to motor terminal Defrost relay Outdoor fan motor Outdoor fan motor capacitor	Defective control, timer, or relay	
PROBLEM: Compressor hums but will not start		
Faulty wiring – Motor to line side of contactor – Load side, contactor to motor terminal Loose terminals – Motor to line side of contactor – Load side, contactor to motor terminal Compressor overload Potential relay Start capacitor	Unbalanced power supply Low voltage – Load side, contactor to motor terminal Defective contacts in contactor Compressor motor Compressor bearing defective Compressor stuck	Refrigerant overcharge High head pressure Noncondensibles Restrictions

03311-13_A01A.EPS

GENERALLY THE CAUSE Make these checks first.	**OCCASIONALLY THE CAUSE** Make these checks only if the first checks failed to locate the trouble.	**RARELY THE CAUSE** Make these checks only if other checks failed to locate the trouble.
PROBLEM: Compressor cycles on overload		
Faulty wiring – Motor to line side of contactor – Load side, contactor to motor terminal Loose terminals – Motor to line side of contactor – Load side, contactor to motor terminal Low voltage – Motor to line side of contactor – Load side, contactor to motor terminal Unbalanced power supply Potential relay Run capacitor Start capacitor	Defective contacts in contactor Compressor overload Compressor motor Refrigerant charge low Refrigerant overcharge High head pressure High suction pressure Noncondensibles Excessive load cooling Defective reversing valve	Compressor valve defective Compressor oil level Fins dirty or plugged – Indoor section of system – Outdoor section of system Fan belt slipping Coil air short circuiting Air volume low (for cooling) Blower belt slipping Air volume low Air filters dirty
PROBLEM: Compressor off or cycling on low-pressure control		
Refrigerant charge low Low suction pressure Defective power element Push rod packing leak – Outdoor expansion valve Indoor fan motor Outdoor fan motor Air volume low Air filters dirty Small or restricted ductwork Restrictions	Indoor fan relay Indoor fan motor Sensing element loose or poorly located Cycle too long (clock timer) Defective power element Fins dirty or plugged Fan belt slipping Low-temperature coil air (for cooling)	Pressure control Out of adjustment (de-ice control) Dirty internally Push rod packing leak – Indoor expansion valve Coil air short circuiting Air stratification in conditioned space
PROBLEM: Compressor off on high-pressure control		
Defrost relay Outdoor fan motor Indoor fan relay Indoor fan motor Refrigerant overcharge Cycle too long (clock timer) Coil Defective control timer or relay Fins dirty or plugged Fan belt slipping Coil air short circuiting Blower belt slipping Air volume low Air filters dirty	Out of adjustment (de-ice control) Defective reversing valve Air volume low (for cooling) Fins dirty or plugged Small or restricted ductwork	Pressure control Temperatures Noncondensibles Air stratification in conditioned space Auxiliary heat upstream from indoor fan

GENERALLY THE CAUSE Make these checks first.	OCCASIONALLY THE CAUSE Make these checks only if the first checks failed to locate the trouble.	RARELY THE CAUSE Make these checks only if other checks failed to locate the trouble.
PROBLEM: Compressor noisy		
Hold down bolts Compressor valves defective Compressor oil level Refrigerant overcharge	Run capacitor Leaking or defective check valve – Indoor section of system – Outdoor section of system Superheat setting TXV stuck open Loose thermal bulb – Indoor section of system	Compressor bearings defective
PROBLEM: Compressor loses oil		
Refrigerant leak Low suction pressure Refrigeration piping incorrect	TXV stuck open Restrictions	Superheat setting Refrigeration lines too long
PROBLEM: Unit runs in heat cycle – pumps down in cooling or defrost cycle		
Defective reversing valve Defective power element Restrictions	Check valve sticking closed Dirty internally	Push rod packing leak – Indoor section of system
PROBLEM: Unit runs normal in one cycle – high suction pressure in other cycle		
Leaking reversing valve Leaking or defective check valve – Outdoor section of system Loose thermal bulb – Outdoor section of system	Superheat setting Leaking or defective check valve – Indoor section of system Loose thermal bulb – Indoor section of system	TXV stuck open – Outdoor section of system – Indoor section of system
PROBLEM: Head pressure too high		
Refrigerant overcharge Temperatures Noncondensibles Fins dirty or plugged Fan belt slipping Air volume too low (for cooling) – Outdoor coil Blower belt slipping Air volume too low – Indoor coil Air filters dirty	Excessive load cooling Coil air short circuiting Fins dirty or plugged Refrigeration piping incorrect	Compressor oil level Small or restricted ductwork Thermostat setting
PROBLEM: Head pressure too low		
Refrigerant charge low Low suction pressure Leaking reversing valve Low-temperature coil air (for cooling)	Compressor valves defective	

03311-13_A01C.EPS

GENERALLY THE CAUSE Make these checks first.	OCCASIONALLY THE CAUSE Make these checks only if the first checks failed to locate the trouble.	RARELY THE CAUSE Make these checks only if other checks failed to locate the trouble.
PROBLEM: Suction pressure too high		
Compressor valves defective High head pressure Excessive load cooling Reversing valve leaking	Refrigerant overcharge Temperatures Leaking or defective check valve – Indoor section of system – Outdoor section of system Loose thermal bulb – Outdoor section of system TXV stuck open – Indoor section of system	TXV stuck open – Outdoor section of system Superheat setting Loose thermal bulb – Indoor section of system
PROBLEM: Suction pressure too low		
Refrigerant charge low Fan belt slipping Blower belt slipping Air volume low Air filters dirty Indoor fan motor Indoor fan relay	Temperatures Leaking or defective check valve – Outdoor section of system – Indoor section of system Superheat setting Dirty internally Defective power element Fins dirty or plugged Coil air short circuiting Small or restricted ductwork Check valve sticking closed Restrictions	Low head pressure Push rod packing leak – Outdoor section of system – Indoor section of system Air stratification in conditioned space Refrigeration piping incorrect
PROBLEM: Blower will not start		
Power failure Blown fuse Run capacitor Faulty wiring – Motor to line side of contactor – Load side of contactor to motor – Control circuit Indoor fan motor	Indoor fan relay	Low voltage – Motor to line side of contactor – Load side, contactor to motor terminal – Control circuit Thermostat
PROBLEM: Indoor coil frosting or icing		
Low suction pressure Blower belt slipping Air volume low Air filters dirty – Indoor section of system Indoor fan motor Indoor relay Small or restricted ductwork	Refrigerant charge low Temperatures Reversing valve sticking closed Dirty internally	

03311-13_A01D.EPS

GENERALLY THE CAUSE	OCCASIONALLY THE CAUSE	RARELY THE CAUSE
Make these checks first.	Make these checks only if the first checks failed to locate the trouble.	Make these checks only if other checks failed to locate the trouble.
PROBLEM: Compressor runs continuously – no cooling		
Compressor valve defective Refrigerant charge low Reversing valve leaking Reversing valve defective Fins dirty or plugged (outdoor) Blower belt slipping Air volume low Air filters dirty Small or restricted ductwork – Indoor section of system	High suction pressure Noncondensibles Superheat setting Dirty internally Push rod packing leak – Indoor section of system Restrictions	Fins dirty or plugged (indoor) Coil air short circuiting Air volume low (for cooling) Low-temperature coil air Defective power element Check valve sticking closed – Indoor section of system Check valve leaking or defective Air stratification in conditioned space
PROBLEM: Compressor runs continuously – cooling		
Faulty wiring (control circuit) Thermostat (control circuit)	Thermostat location	Thermostat setting Air duct not insulated
PROBLEM: Liquid refrigerant flooding back to compressor in cooling cycle – TXV system		
Superheat setting – Indoor section of system Loose thermal bulb – Indoor section of system	Refrigerant overcharge Temperatures TXV stuck open – Indoor section of system Check valve leaking or defective – Indoor section of system	Fins dirty or plugged Blower belt slipping Air volume low Air filters dirty Small or restricted ductwork Air stratification in conditioned space
PROBLEM: Liquid refrigerant flooding back to compressor in cooling cycle – cap tube system		
Refrigerant overcharge Indoor fan motor Indoor fan relay Blower belt slipping Air volume low Air filters dirty	High head pressure Fins dirty or plugged Small or restricted ductwork Check valve leaking or defective	Temperatures Air stratification in conditioned space
PROBLEM: Compressor runs continuously – no heating		
Refrigerant charge low Reversing valve leaking Contactor stuck	Defective control, timer, or relay	Superheat setting Dirty internally Push rod packing leak – Outdoor section of system Air stratification in conditioned space
PROBLEM: Compressor runs continuously – heating		
Faulty wiring (control circuit) Thermostat Thermostat location Contactor stuck	Thermostat setting	Air duct not insulated

03311-13_A01E.EPS

GENERALLY THE CAUSE	OCCASIONALLY THE CAUSE	RARELY THE CAUSE
Make these checks first.	Make these checks only if the first checks failed to locate the trouble.	Make these checks only if other checks failed to locate the trouble.
PROBLEM: Unit runs in cooling cycle – pumps down in heating cycle		
Defective reversing valve Defective power element	Dirty internally Fins dirty or plugged Fan belt slipping Coil air short circuiting Check valve sticking closed Restrictions	Push rod packing leak – Outdoor section of system
PROBLEM: Unit cycles on low-pressure control when switching cycle or at end of defrost cycle		
Defective low pressure control Refrigerant charge low Defective reversing valve Defective power element Push rod packing leak – Outdoor section of system – Indoor section of system		
PROBLEM: Defrost cycle initiates – no ice on coil		
Refrigerant charge low De-ice control out of adjustment Sensing element loose or poorly located Defective control, timer, or relay	Faulty wiring (control circuit) Fins dirty or plugged Fan belt slipping Coil air short circuiting Air volume low (for cooling)	Cycle too long (clock timer) Defective reversing valve Superheat setting
PROBLEM: Defrost cycle initiates but won't terminate defrost relay		
Defrost relay Refrigerant charge low De-ice control out of adjustment Defective control, timer, or relay	Faulty wiring (control circuit) Compressor valves defective Reversing valve defective	
PROBLEM: Reversing valve will not shift		
Faulty wiring (control circuit) Loose terminal (control circuit) Defective reversing valve Refrigerant charge	Low voltage (control circuit) Defrost relay Compressor stuck	
PROBLEM: Ice buildup on lower part of outdoor coil		
Defrost relay Refrigerant charge low De-ice control out of adjustment Sensing element loose or poorly located	Compressor valves defective Cycle too long (clock timer) Reversing valve leaking Reversing valve defective Superheat setting – Outdoor section of system	Loose thermal bulb – Outdoor section of system
PROBLEM: Liquid refrigerant flooding back to compressor in heating cycle – TXV system		
Superheat setting – Outdoor section of system Loose thermal bulb – Outdoor section of system	Leaking or defective check valve TXV stuck open – Outdoor section of system	Refrigerant overcharge Temperatures

03311-13_A01F.EPS

GENERALLY THE CAUSE	OCCASIONALLY THE CAUSE	RARELY THE CAUSE
Make these checks first.	Make these checks only if the first checks failed to locate the trouble.	Make these checks only if other checks failed to locate the trouble.

PROBLEM: Liquid refrigerant flooding back to compressor in heating cycle – cap tube system

Refrigerant overcharge High head pressure	Leaking or defective check valve	Temperatures

PROBLEM: Auxiliary heat on – indoor blower off

Faulty wiring – Load side, contactor to motor terminal – Control circuit Loose terminals – Load side, contactor to motor terminal – Control circuit Fan relay Indoor fan motor	Thermostat	

PROBLEM: Excessive operating costs

Compressor stuck Refrigerant charge low De-ice control out of adjustment Reversing valve leaking Reversing valve defective Blower belt slipping Air volume low Air filters dirty Air duct not insulated No outdoor thermostat Outdoor thermostat adjustment	Refrigerant overcharge Fins dirty or plugged Fan belt slipping Coil air short circuiting Thermostat location Air stratification in conditioned space Refrigeration piping incorrect Unit incorrect size for application	Faulty wiring – Motor to line side of contactor Loose terminals – Motor to line side of contactor Noncondensibles Superheat setting TXV stuck open Dirty internally Defective power element Thermostat setting

PROBLEM: Unit short cycles on defrost control

Refrigerant charge low De-ice control out of adjustment Defective control, timer, or relay Fan belt slipping	Defective power element Fins dirty or plugged	

PROBLEM: Outdoor blower does not stop in defrost cycle

Fan relay IOD or ID section Defrost relay		

03311-13_A01G.EPS

HVAC COMMON WIRING DIAGRAM ABBREVIATIONS

AH	Supplementary Heater		HVTB	High-Voltage Terminal Board
AR	Auxiliary Relay		IOL	Internal Overload Protector
BH	Supplementary Heater Contactor		1K	Economizer Relay
BHA	Supplementary Heater Contactor – Auxiliary		2K	Compressor Lock-out Relay
BLR	Balancing Relay		LA	Lock-in Relay
CA	Cooling Anticipator		LAS	Low Airflow Sensing
CB	Circuit Breaker		LB	Lock-in Relay
CF	Fan Capacitor		LPCO	Low-Pressure Cutout
CFA	Outdoor Fan Capacitor		LT	Light
CH	Supplementary Htr. Contactor		LTS	Low-Temperature Switch
CL	Contactor Relay		LVTB	Low-Voltage Terminal Board
CN	Wire Connector		MAC	Mixed Air Control
COC	Changeover Control		MPM	Motor Protection Module
CR	Run Capacitor		MPP	Min. Position Potentiometer
CPR	Compressor		MPS	Motor Protection Sensor
CSR	Capacitor Switching Relay		MS	Compressor Motor Contactor
D	Defrost Relay		MSA	Compressor Motor Contactor – Auxiliary
DFC	Defrost Control		MTR	Motor
DFT	Defrost Timer		ODA	Outdoor Temperature Anticipator
DH	Supplementary Htr. Contactor		ODF	Outdoor Fan Relay
DR	Defrost Relay		ODS	Outdoor Temperature Sensor
DS	Diaphragm Switch		ODT	Outdoor Thermostat
DT	Defrost Termination Switch		OFT	Outdoor Fan Thermostat
EDC	Evaporator Defrost Control		RH	Emergency Heat Relay
EDR	Evaporator Defrost Relay		RHS	Resistance Heat Switch
ETB	Economizer Terminal Board		S-FU	1/16 Amp Fuse
ER	Electric Heater Relay		SEN	Sensor
F	Indoor Fan Relay		SC	Switchover Valve Solenoid
FL	Flexible Link		SM	System Switch (Room Thermostat)
FP	Feed-Back Potentiometer		T	Thermistor
FR	Contactor Relay		TCO	Temperature Limit Switch
FA	Contactor Relay		TDL	Discharge Line Thermostat
FM	Manual Fan Switch		TDR	Time Delay Relay
FS	Fan Signal Relay		TM	Compressor Motor Thermostat
FST	Fan Switch Terminal		TNS	Transformer
FTB	Fan Terminal Board		TS	Heating & Cooling Thermostat
FU	Fuse		TSC	Cooling Thermostat
FUA	Compressor Fuse		TSH	Heating Thermostat
FUB	Transformer Fuse			
H	Heating Relay			
HA	Heating Anticipator			
HPCO	High-Pressure Cutout			
HTR	Heater			

03311-13_A02.EPS

Additional Resources

This module presents thorough resources for task training. The following resource material is suggested for further study.

Troubleshooting & Servicing Heat Pump Systems, 2002. Mt. Prospect, IL: ESCO Press.

Figure Credits

Answer	Section Reference	Objective
Section One		
1.b	1.1.1	1a
2.d	1.2.3	1b
Section Two		
1.c	2.1.0	2a
2.b	2.2.0	2b
3.b	2.3.0	2c
4.d	2.4.0	2d
5.b	2.5.0	2e
Section Three		
1.d	3.1.1	3a
2.d	3.2.0	3b
3.a	3.3.0	3c
4.b	3.4.2	3d
5.a	3.5.0	3e

NCCER CURRICULA — USER UPDATE

NCCER makes every effort to keep its textbooks up-to-date and free of technical errors. We appreciate your help in this process. If you find an error, a typographical mistake, or an inaccuracy in NCCER's curricula, please fill out this form (or a photocopy), or complete the online form at **www.nccer.org/olf**. Be sure to include the exact module ID number, page number, a detailed description, and your recommended correction. Your input will be brought to the attention of the Authoring Team. Thank you for your assistance.

Instructors – If you have an idea for improving this textbook, or have found that additional materials were necessary to teach this module effectively, please let us know so that we may present your suggestions to the Authoring Team.

NCCER Product Development and Revision

13614 Progress Blvd., Alachua, FL 32615

Email: curriculum@nccer.org
Online: www.nccer.org/olf

❏ Trainee Guide   ❏ Lesson Plans   ❏ Exam   ❏ PowerPoints   Other _____

Craft / Level: _____   Copyright Date: _____

Module ID Number / Title: _____

Section Number(s): _____

Description: _____

_____

_____

_____

_____

Recommended Correction: _____

_____

_____

_____

_____

Your Name: _____

Address: _____

_____

Email: _____   Phone: _____

Troubleshooting Gas Heating

OVERVIEW

Gas furnaces and other gas heating devices contain a variety of controls. Most of these controls are safety devices that either shut the system down in the event of a problem or prevent it from operating if certain conditions are not met in a specified sequence. Service technicians must be able to determine how control devices interact to control the system. They must also learn to recognize combustion-related problems and be able to properly test the combustion system using specialized test instruments.

Module Five

Trainees with successful module completions may be eligible for credentialing through NCCER's National Registry. To learn more, go to **www.nccer.org** or contact us at **1.888.622.3720**. Our website has information on the latest product releases and training, as well as online versions of our *Cornerstone* newsletter and Pearson's product catalog.

Your feedback is welcome. You may email your comments to **curriculum@nccer.org**, send general comments and inquiries to **info@nccer.org**, or fill in the User Update form at the back of this module.

This information is general in nature and intended for training purposes only. Actual performance of activities described in this manual requires compliance with all applicable operating, service, maintenance, and safety procedures under the direction of qualified personnel. References in this manual to patented or proprietary devices do not constitute a recommendation of their use.

Objectives

When you have completed this module, you will be able to do the following:

1. Describe how to troubleshoot the components related to gas heating.
   a. Describe the control circuits and typical sequence of operation of various gas heating units.
   b. Describe the operation and troubleshooting process for thermocouples.
   c. Describe the operation and troubleshooting process for ignition devices.
   d. Describe the operation and troubleshooting process for flame sensors.
   e. Identify common problems associated with system airflow.
2. Identify infrared gas heaters and describe how they operate.
   a. Identify various types of infrared gas heaters.
   b. Describe the operating characteristics of infrared gas heaters.
3. Explain how to conduct a combustion analysis on a gas furnace.
   a. Identify combustion analysis equipment and the combustion byproducts that are of importance to the analysis.
   b. Describe the combustion analysis process and how to interpret basic results.

Performance Tasks

Under the supervision of your instructor, you should be able to do the following:

1. Using the proper tools, instruments, and control circuit diagrams, isolate and correct malfunctions in a gas heating system.
2. Complete a combustion analysis on a gas furnace or boiler.

Trade Terms

Flame rectification
Modulating gas valve
Net stack temperature

Reverberator
Submerged flame

Industry-Recognized Credentials

If you are training through an NCCER-accredited sponsor, you may be eligible for credentials from NCCER's Registry. The ID number for this module is 03209-13. Note that this module may have been used in other NCCER curricula and may apply to other level completions. Contact NCCER's Registry at 888.622.3720 or go to **www.nccer.org** for more information.

Contents

Topics to be presented in this module include:

Figures and Tables

SECTION ONE

1.0.0 TROUBLESHOOTING GAS HEATING COMPONENTS

Objectives

Describe how to troubleshoot the components related to gas heating.

a. Describe the control circuits and typical sequence of operation of various gas heating units.
b. Describe the operation and troubleshooting process for thermocouples.
c. Describe the operation and troubleshooting process for ignition devices.
d. Describe the operation and troubleshooting process for flame sensors.
e. Identify common problems associated with system airflow.

Trade Terms

Flame rectification: A process in which exposure of a sensing rod to a flame produces a tiny current that can be sensed and used to control the gas valve.

Modulating gas valve: A gas valve that precisely controls burner capacity by adjusting the flow of gas to the burners in small increments.

To properly troubleshoot a gas heating system, the technician must consider three separate but related subsystems: controls, air or water flow, and gas combustion. The ability to read circuit diagrams, use electrical test equipment, recognize and correct combustion-related problems, and measure and adjust airflow all come together and play a role in troubleshooting gas-fired heating equipment.

Control circuits for gas-fired systems come in many varieties. It is rare to find two furnace models that are precisely the same electrically, even if they are made by the same company. Although packaged electronic controls have allowed manufacturers to standardize control circuits, variations from model to model will be found, if only in the programming of the microprocessor. Small changes are often made from one furnace series to another.

Included here are examples of circuit diagrams from different furnace manufacturers, along with a description of their operating sequences. By studying these circuits, trainees will reinforce and expand their knowledge of furnace controls. This will help to analyze the control circuit of an unfamiliar product to locate the cause of a malfunction. It is important to note however, that the circuits discussed here represent only a few possible examples of those found in today's furnaces. It is important to develop the skill of evaluating all circuits, rather than memorizing how to read only the examples found here.

In addition to gas furnaces and boilers, gas heating can also be provided by packaged heating and cooling units. One major difference is that packaged units are installed outdoors and need no vent piping. The control circuits of packaged systems and gas furnaces are roughly the same.

If a gas heating system is not working at all, the ignition system and/or gas valve may be malfunctioning. In older furnaces with standing pilots, thermocouples are used to indicate the presence of a pilot flame to the main gas valve. Thermocouples develop a voltage in the range of 30mV from the heat of the pilot flame. Modern furnaces use flame rectification sensors in conjunction with intermittent pilot spark igniters or direct ignition devices such as hot surface igniters (HSIs) (*Figure 1*). Flame rectifiers typically develop a small current from the heat of the flame, rather than a voltage. HSIs may wear out or crack after several years of use. Older HSIs are constructed of a ma-

---

Indoor Comfort

What is indoor comfort? To many consumers, comfort means hot air coming from the registers in winter and cold air coming from them in summer. True indoor comfort is more than just hot or cold. Today's equipment manufacturers are taking advantage of new technologies that allow them to provide improved temperature and humidity control, improved air quality, and energy efficiency levels that were only dreamed of a few years ago. These include the widespread use of electronic controls, capacity control using staged gas valves and new compressor technology, and the use of variable-speed blower motors. Technicians must be prepared to deal with these new technologies when servicing modern gas heating systems.

---

terial called silicon carbide. When replacing a hot surface igniter, consider replacing it with a more durable and reliable type made of silicon nitride igniter.

This section focuses primarily on furnace controls. Troubleshooting the combustion system in gas heating equipment consists primarily of observing the flame and the mechanical operation of the unit. On the basis of these observations, it may be possible to identify the source of a problem right away or with minimal testing. The troubleshooting guide example in *Appendix A* outlines common problems that occur in gas heating systems, along with their probable causes.

1.1.1 Electronic Controls in Heating Systems

Microprocessor-based electronic controls are now common across all HVAC product lines. Although older technologies exist in legacy systems, it is rare to find a modern heating unit that does not contain some kind of electronic control. Manufacturers put these controls in their equipment to increase reliability and energy efficiency. Electronic controls also allow engineers to design more versatility into the products than was ever possible with electromechanical controls.

Electronic control boards are checked for correct operation by making sure that the board is receiving correct data, such as voltage signals from the room thermostat and open/closed circuit indications from devices such as limit switches. If all inputs are correct, the control should provide appropriate outputs. For example, if the control is receiving a room thermostat input to W (the typical 24V control terminal used for heating) on the control, and all other inputs are within an acceptable range, the control will provide output signals to energize the draft inducer, igniter, gas valve, and blower motor. If the correct input and output signals are present and the furnace does not run, the problem is likely related to the control board. Various input and output signals are typically measured with a multimeter.

Figure 2 shows a printed circuit board containing the microprocessor control circuits for a

(A) GAS SPARK IGNITION
(RELATIVELY COMMON)

(B) GLOW COIL
(RELATIVELY OBSOLETE)

(C) SAFETY PILOT

(D) HOT SURFACE IGNITOR

(E) FLAME RECTIFICATION SENSOR

(F) NITRIDE IGNITOR

(MOST COMMON IN NEWER FURNACES)

03209-13_F01.EPS

Figure 1 Gas furnace ignition devices.

packaged gas heating/electric cooling unit. *Figure 3* shows the wiring diagram for a gas forced-air furnace. Note that *Figure 3* is a combination diagram, with a connection diagram on the left and a schematic diagram on the right. Having both types of diagrams helps the troubleshooting process by allowing the technician to see the control structure from two points of view. Highlights of the operating sequence for an induced-draft gas forced-air furnace are as follows:

> **WARNING!**
>
> LP gas, also known as propane, is heavier than air. While natural gas tends to rise and dissipate easily, LP gas tends to collect in low places and remain. Use extreme caution when dealing with LP gas leaks or purging the gas lines of equipment operating.

- As in all furnaces, the sequence is initiated by a call for heat from the room thermostat in the form of a 24V signal. The control board often starts the combustion process by starting the draft inducer through a board-mounted relay.
- Before the microprocessor sends a signal to the igniter to initiate combustion, it will first verify that each safety control is in its correct position. For example, the microprocessor might look for answers to the following questions before sending a signal to energize the igniter and a signal to open the gas valve:

    Is there continuity through the rollout and limit switches, indicating they are closed?

    Is the pressure switch closed to indicate that the induced-draft motor is running?

If these conditions are met, the microprocessor will energize the igniter and signal the gas valve to open.

03209-13_F02.EPS

Figure 2 Gas unit control board.

- After the igniter and gas valve have been energized, the microprocessor looks for a current signal from the flame sensor to verify that there is a steady, strong flame. If so, the ignition process ends. If not, the microprocessor may be programmed to retry for ignition a number of times over a period of several minutes. After each attempt, the microprocessor will impose a delay that gives the induced-draft fan time to purge the heat exchangers, preventing gas buildup. One electronic furnace control on the market will retry for ignition 33 times over a 15-minute period before it stops. At that point, the microprocessor locks out ignition and a fault indicator begins to flash. The system is then disabled until the power is manually reset. Note that systems using LP gas have different lockout schemes, since LP vapors are heavier than air and tend to fall rather than rise easily out of the heat exchanger area.
- Once the burner flame is proven, the microprocessor fan logic will be notified and the blower time-delay will begin. This delay gives the heat exchangers time to warm up before the blower starts circulating air.

The operating sequence just described is common to many different brands of gas furnaces that use electronic controls. Differences result from special features that can be programmed into the microprocessor-controlled system. For example:

- Some systems have a feature to prevent nuisance trips of the high-limit switch. Furnace blowers run for a period of time after the wall thermostat is satisfied and combustion has stopped to extract more heat from the heat exchanger. If the blower stops too quickly, built-up heat in the heat exchanger could trip the high limit switch. The microprocessor will automatically adjust the blower-off delay on the limit switch if the unit is tripping off after the wall thermostat has been satisfied. It will also flash a fault message to notify maintenance that this has happened.
- Variable-speed blowers are common features in high-efficiency furnaces. Some are programmed to automatically adapt to changing conditions such as the opening and closing of zone dampers in a zoned system.
- Most microprocessor-controlled systems can diagnose and isolate common faults. Most automatic diagnostics include a failure message to indicate that the control board/module itself has failed. This must be checked before the board is replaced because other problems, such as an intermittent power loss, might also cause this message.

Figure 3 Gas furnace control board and wiring schematic.

03209-13_F03.EPS

• Some gas heating controls can automatically adjust the length of the high-fire and low-fire cycles of two-stage burners to maintain indoor comfort.

Electronic gas furnace controls (*Figure 4*) are usually equipped with a diagnostic aid in the form of multiple or single light-emitting diodes (LEDs). Controllers with a single LED only light the LED when the ignition sequence is interrupted by a fault. The LED usually flashes in a code corresponding to the type of failure detected. For example, two flashes might mean the pressure switch is stuck closed, while three flashes could mean the pressure switch is stuck open. Controllers with multiple LEDs usually light the LEDs in a sequence corresponding to the ignition sequence of the controller during operation. If the ignition sequence fails, the last LED lighted is usually the last valid action completed. Controllers with LEDs bear labels stating the meaning of the LEDs or listing the flash codes.

To isolate the failure, voltages at the designated (or next) component in the sequence can be monitored by the technician as the ignition sequence is restarted. Consult the manufacturer's literature to verify the actual ignition sequence and components involved.

Older solid-state gas furnace controllers may not have diagnostics, or may have only limited diagnostics. The older controller shown has an LED that flashes to indicate an external failure has occurred. If continuously lit, this indicates an internal controller failure.

LED ON = FAILED CONTROLLER
LED FLASHING = EXTERNAL COMPONENT FAILURE
CONTROLLER WITH MINIMAL DIAGNOSTICS

DIAGNOSTIC INDICATOR FLASH CODES

1. System lockouts (Retries of lockouts exceeded)
2. Pressure switch stuck closed
3. Pressure switch stuck open
4. Open high-temperature limit switch
5. Flame rollout sensed
6. 115VAC power reversed/improper ground
7. Low flame sense signal

Continuous flash – flame sensed 0.5 sec without gas valve

Continuous on – internal control failure

03209-13_F04.EPS

Figure 4 Gas furnace controllers.

Think About It
Circuit Diagram Interpretation Exercises

Test your knowledge of microprocessor-controlled heating systems with these exercises. Use the diagram in *Figure 3* as a reference.

1 Which component on the control board controls blower speed in the heating and cooling modes?
2 If the draft inducer (IDM) runs and the igniter (HSI) energizes, but the gas valve (GV) does not energize, what should you do?
   a. Replace the gas valve.
   b. Replace the control board.
   c. Check to see if 24V is available at the gas valve.
   d. Check to see if there is a call-for-heat signal from the room thermostat.

To determine an ignition-sequence failure, the furnace must be observed during the ignition sequence to visually check where the sequence is halting. Then, voltages representing inputs from, or outputs to, components associated with the sequence failure can usually be monitored at test points on a connection block. This is done to isolate the malfunctioning component while the controller is recycled. With these types of controllers, consult the manufacturer's literature for the ignition sequence and test point information.

1.1.2 Natural-Draft Packaged Units

Government-mandated energy efficiency requirements for gas furnaces cannot easily be met with furnaces using natural-draft venting technology. Induced-draft furnaces allow for the use of more restrictive heat exchanger designs, thus placing the hot air in contact with a larger amount of surface area. Greater surface area results in a higher degree of heat exchange and a more efficient furnace. This has made natural-draft furnaces virtually obsolete. However, natural-draft systems will be in use for some time to come, as they were still being installed in the 1990s.

Figure 5 represents the control circuits and sequence of operation for a packaged air conditioning unit with natural-draft gas heat. *Figure 6* is the wiring diagram for this unit. Because we are dealing with gas heating, only portions of the cooling system are shown.

This type of unit is sometimes called a year-round air conditioner (YAC). It may also be known as a packaged gas-electric unit (gas heat, electric cooling). A packaged electric-electric unit would be an air conditioning package with electric resistance heaters.

The basic operating sequence of any natural-draft unit follows:

1. The wall thermostat calls for heat.

2. The system verifies (proves) that the pilot is lit.

3. Assuming all safety controls are closed, the gas valve is turned on and the gas ignites.

4. The blower comes on after the heat exchanger heats and closes the fan switch.

5. The wall thermostat opens. The blower continues to operate until the heat exchanger cools, opening the fan switch.

The heating side of the unit shown in *Figure 5* has a couple of variations from the basic sequence:

- It uses electronic pilot igniter rather than a standing pilot.
- There is a blower time delay at the beginning and end of the cycle.

1.1.3 Induced-Draft Gas Furnace

A common induced-draft heating system is an 80 percent Annual Fuel Utilization Efficiency (AFUE) gas furnace (*Figure 7*). In this particular unit, the ignition is accomplished with an HSI, the burner flame is monitored by a flame sensor, and the sequencing/monitoring of the furnace is accomplished by a sealed, solid-state controller containing micro-miniature relays.

The furnace shown in *Figure 7* is similar to one represented by the wiring diagram in *Figure 3*. Use *Figure 3* to follow the operating sequence of a typical 80 AFUE gas furnace.

1. The electronic control board (CPU) is powered through terminals Com and R. Power to the R terminal first flows through the 3-amp fuse (FU1). A 24V signal from the room thermostat is applied to the W terminal of the control board. This happens when the thermostat heating contacts close, applying power from the R terminal to the W terminal of the thermostat.

2. The electronic control board (CPU) performs a self-check and verifies that the pressure switch (PRS) contacts are open. If the pressure switch contacts are open, the draft inducer motor (IDM) can then start. It will run for 15 seconds to purge the heat exchanger. This causes the pressure switch contacts to close, proving that the inducer is running and that

Flame Sensors

The flame sensor on a gas furnace is placed in the burner flame and senses a very small electrical current (in the microamp range) that flows through the burner flame. The furnace's electronic control senses this current and allows the burner to continue operating as long as the current is above a certain value. Over time, an oxide coating can build up on the flame sensor, reducing current flow and preventing furnace operation. Cleaning the sensing rod with fine steel wool can remove the oxide coating and restore normal operation.

1. Line potential feeds through the door interlocks (if used). Access panels must be in place to energize machine.
2. Transformer provides 24V control circuit.
3. On a heating demand, the thermostat heating contacts close.
4. The 24V control circuit is complete to the pilot valve through limit controls, W leg of the thermostat, and internal ignitor circuits.
5. After the pilot flame has been proven, the main gas valve is energized. Main burners are ignited by the pilot flame.
6. As the main gas valve is energized, the fan control heater is also activated after a brief delay.
7. After a brief delay, the heater provides sufficient heat to close the fan control contacts.
8. This energizes the blower motor on the selected speed.
9. As heating demand is satisfied, the thermostat heating contacts open. This de-energizes the ignition control, gas valve, and fan control heater.
10. The blower motor continues running until the furnace temperature drops below fan control set point.

03209-13_F05.EPS

Figure 5 Natural-draft gas heating system.

Figure 6 Field wiring diagram.

03209-13_F06.EPS

TEMPERATURE LIMIT SWITCH (PRIMARY LIMIT CONTROL)

EXHAUST VENT

COMBUSTION BLOWER

VENTING PRESSURE SWITCH

COMBUSTION AIR INLET

BURNER BOX (HSI, FLAME SENSOR, AND MANUAL-RESET FLAME ROLLOUT LIMIT SWITCH [PRIMARY ROLLOUT LIMIT] INSIDE)

GAS VALVE

TOP HALF OF FURNACE

BURNER BOX

CONTROLLER

DOOR SWITCH

MANUAL-RESET TEMPERATURE LIMIT SWITCH (AUXILIARY LIMIT CONTROL)

AIR CIRCULATION BLOWER

BOTTOM HALF OF FURNACE

03209-13_F07.EPS

Figure 7 Typical 80-percent AFUE induced-draft gas furnace.

airflow through the heat exchanger is adequate for safe combustion.

3. At the end of the purge period, the HSI is energized. After the igniter warms up, the gas valve relay (GVR) contacts close, applying power to the gas valve (GV), to energize it. The igniter will light the gas that is now flowing to the burners. The flame sensor (FSE) has two seconds to prove that an adequate burner flame is present. The igniter is de-energized as soon as an adequate flame is proven. **Modulating gas valves** can provide very precise levels of indoor comfort. They are often combined with a variable-speed blower motor that matches airflow to burner output. The gas valve can typically modulate between 40 to 100 percent burner capacity in 5 percent increments. A microprocessor-based furnace control tells the gas valve at what level to modulate the burner flame based on various values such as indoor and outdoor temperature and the length of previous burner cycles. The unit represented by *Figure 3* does not use a modulating gas valve.

4. Failure to sense an adequate burner flame will cause the control board to de-energize the gas valve and attempt to re-establish a flame. If these attempts fail, the control board will lock out any more attempts to establish a flame. Interrupting the line voltage to the furnace or the control power to the module will reset the control.

5. When the burner flame is proven, the blower motor (BLWM) is energized 25 seconds after the gas valve opens. The blower motor operates on HEAT speed. The motor speed for heating is field-selectable to meet the needs of the air distribution system.

6. If the room thermostat is satisfied, its contacts open and the 24V signal to W ceases. This de-energizes the gas valve, causing the burner flames to extinguish. The inducer motor will run for a short period to purge any combustion byproducts from the heat exchanger. The blower motor will continue to run for up to 180 seconds. The blower-off delay time is field-selectable.

Electronic Ignition

The development of electronic ignition for gas furnaces was an important step for all furnaces. However, it may have been even bigger for outdoor units. Since these units are exposed to winds and have very short flue vents, standing pilots were far more easily blown out. The pilot then had to be relit manually to restart the unit.

If a flame rollout limit switch has been tripped, the cause of the flame rollout must be corrected and the switch must be manually reset before the gas valve will function. Flames may roll out of the combustion chamber and into the furnace vestibule if the heat exchanger is blocked or the draft inducer blower wheel is loose or deteriorated. If the draft inducer pressure switch has failed closed, the control module will not be aware of these conditions.

If the furnace shuts down repeatedly because of high-limit switch activation, and the supply air filters, ducts, and blower are clean, check the temperature rise of the furnace. If it is normal, the switch may be defective. If it is not normal, too many registers or air terminals may have been closed, or the blower speed may require adjustment. *Appendix B* is an example of an induced-draft gas furnace-troubleshooting flowchart.

1.1.4 Condensing Furnaces

A condensing furnace (*Figure 8*) has both a primary and secondary heat exchanger. Their ability to extract additional heat from flue gases produces AFUEs that typically exceed 90 percent. Because the secondary heat exchanger extracts most of the remaining heat from the vent gases, condensation of some of the combustion products occurs in the secondary heat exchanger.

This acidic condensate is drained from the secondary heat exchanger and routed through a condensate drain trap to an appropriate drain, either directly or by use of a condensate pump. Some local codes require that a replaceable neutralizer cartridge be installed in the condensate drain line before it is routed to a drain. The typical sequence of operation, along with the combustion devices, controller, and monitoring devices, is essentially the same as the 80-percent AFUE gas furnace. However, the condensate pump serving the furnace may be equipped with a float switch that opens if the pump fails to operate and fills. The switch is wired into the furnace control circuit as an additional safety device, and can shut down the furnace. This is an additional control that would not be found in the control circuit of other furnace types.

Most manufacturers do not recommend operating condensing furnaces with return air temperatures below 55°F for prolonged periods. Condensation may occur in the primary heat exchanger, causing it to corrode and fail prematurely. Instruct the owners of such furnaces to set the room temperature thermostat no lower than 60°F if the building will be unoccupied for an extended period.

1.1.5 Gas-Fired Boiler Operation

The operating sequence for a gas-fired boiler is different from a forced-air furnace. A typical boiler control circuit is shown in *Figure 9*.

The operating sequence for a small packaged gas-fired boiler follows:

1. A call for heat from the room thermostat closes the circuit between the two T terminals on the aquastat, energizing the 1K relay. What is not shown on the connection diagram to the right is that there is a small transformer on the aquastat board, and one leg of that power is applied to one of the two T terminals. A thermostat closure completes the circuit and applies that power to the second T terminal. This completes the circuit to the 1K relay coil. Again, having both a connection diagram and a schematic or ladder diagram is extremely helpful.

2. 1K relay contacts 1K1 and 1K2 close. Assuming the high-limit switch is closed, 120V power is applied to the circulator pump and the draft inducer.

3. The draft inducer comes up to speed. If there is no vent blockage, the pressure switch closes, indicating proper airflow through the combustion chamber. This applies 24V power to the intermittent pilot control through the flame rollout switch.

Soot Buildup

If a lot of soot is noticed on the secondary heat exchanger of a condensing furnace, there may be a problem with the gas pressure or combustion air source. Perform a gas pressure check.

OUTDOOR COMBUSTION AIR SUPPLY

GAS VALVE

SEALED BURNER BOX

VENTING PRESSURE SWITCH

AIR CIRCULATION BLOWER

CONDENSATE DRAIN PUMP

EXHAUST VENT

COMBUSTION BLOWER

DOOR SWITCH

CONDENSATE DRAIN TRAP

CONTROLLER

GAS VALVE

TEMPERATURE OVERLIMIT SWITCH

FLAME SENSOR

FLAME ROLLOUT SENSOR SWITCH

03209-13_F08.EPS

Figure 8 Typical 90-percent AFUE condensing gas furnace.

4. The intermittent pilot control opens the pilot gas valve, allowing the spark generated by the pilot control to light the pilot.

5. When an adequate pilot flame is proven to the pilot control, the spark ceases.

6. The main gas valve now opens and the pilot ignites the main burners.

7. The main burner heats water within the boiler. If the water temperature exceeds the high-limit setpoint, burner operation will cease, but the circulator will continue to run until the room thermostat is satisfied. As a general rule, the circulator responds solely to the thermostat and is not otherwise involved in the operation of the boiler. When the water temperature drops below the setpoint, burner operation will resume, assuming the room thermostat is still calling for heat.

8. If the vent system becomes blocked, the pressure switch will open, stopping burner op-

eration. The draft inducer stays powered as long as the room thermostat continues to call for heat and the limit switch contacts remain closed. If the vent blockage clears, burner operation resumes. When a technician approaches a unit with the draft inducer running but no combustion is taking place, it is often related to a blocked vent or a failed pressure switch.

9. When the room thermostat is satisfied, the 1K relay de-energizes, opening the 1K1 and 1K2 contacts to stop the burner, circulator pump, and draft inducer operation.

1.2.0 Troubleshooting Thermocouples

A thermocouple generates only enough voltage to hold open the safety valve disc when it is manually held open by depressing the pilot valve. It is not able to physically open the valve; it is designed to hold it open only after it is opened manually. Thermocouples are typically rated at 30 millivolts,

Door Interlock Switch

Door interlock switches have traditionally been used to disconnect power when the door of a cabinet containing electrical circuits is opened. The purpose is to reduce the hazard of electrical shock. On gas furnaces, the door interlock serves a second purpose: it prevents furnace operation if the blower door is removed. Keeping the door in place during furnace operation prevents harmful products of combustion from being drawn into the duct system by the blower. During troubleshooting procedures, it is acceptable to temporarily tape the door interlock switch closed. A suitable jumper wire may also be used. However, any tape must be removed when the troubleshooting procedure is complete. If a jumper is used, place it so that it hangs outside of the compartment, so that it cannot be forgotten when the repairs are complete.

Figure 9 Gas-fired boiler control circuit.

03209-13_F09.EPS

but may operate at 18 to 25 millivolts with the load of the safety valve imposed on it. The voltage produces current through a holding coil that creates enough of a magnetic field to hold the disc open. Heating up the thermocouple takes a little time, so the pilot valve must be held open manually for as long as 60 seconds.

If the pilot flame goes out after this warm-up period, the thermocouple is probably not generating enough voltage to hold the safety valve open. This can be verified by checking if the thermocouple is producing sufficient voltage. If the thermocouple is operating correctly, the problem is most likely in the gas valve. On many gas valves, the holding coil can be replaced. In some cases, the entire gas valve assembly must be replaced. *Figure 10* shows the method for testing a thermocouple.

When tested without a load, as shown in *Figure 10*, the thermocouple should generate 20 to 30 millivolts DC. An adapter (*Figure 11*) can be used to test the thermocouple under actual operating (loaded) conditions. In a test under load, the voltage will be about half that for an unloaded test.

Here are some other troubleshooting hints to consider when checking a thermocouple:

• The flame must be directed toward the tip of the thermocouple for a voltage to be generated.
• A flickering pilot flame, or one that lifts off the thermocouple, generally indicates too much draft or leakage in the air system. If the flame

appears to be lifting off of the pilot orifice, it usually indicates a gas pressure problem.
• A floating pilot flame may indicate inadequate draft.
• A yellow pilot flame, caused by a partially blocked orifice in the pilot assembly, will not produce enough heat to generate an adequate voltage. Take the pilot assembly apart and clean the orifice to restore the flame to its proper blue color.

There are several ways a thermocouple can fail:

• The tip of the thermocouple may be burned from extended use.
• The sensing element can be misaligned so that the pilot flame does not make adequate contact with it.
• The lead can be damaged or kinked.
• If the connecting nut is too tight, the insulator may have collapsed and shorted out the thermocouple. If it is not tight enough, moisture may have entered and corroded the connection. Finger-tighten the nut, then turn it one-quarter turn with a wrench only. Take care not to overtighten it.

1. Disconnect the thermocouple from the gas valve.
2. Connect the thermocouple lead to the multimeter as shown.
3. Take the multimeter reading with the pilot flame on the thermocouple.

03209-13_F10.EPS

Figure 10 Thermocouple test (no load).

1. Connect the multimeter to the thermocouple lead and adapter as shown.
2. Take the multimeter reading with the pilot flame on the thermocouple.

03209-13_F11.EPS

Figure 11 Thermocouple test under load.

- A dirty contact can block current flow. Do not touch the contact, because oil from your fingers can cause the same problem.

1.3.0 Troubleshooting Pilot Re-Ignition and Direct Ignition Devices

Some systems use spark igniters to light the pilot flame when the thermostat calls for heat. Pilotless systems use either an HSI or a direct spark igniter. Both ignite the gas at the burner.

The spark ignition system will have a flame sensor that sends a signal to the gas valve proving combustion whenever the burners are lighted. The sensor operates through flame rectification, changing an AC signal to DC. When an HSI is used, the flame-sensing rod, or flame rod, may not be part of the igniter assembly. However, it will be located somewhere in the burner area and positioned in the main burner flame. Some HSI devices do incorporate the flame rod into the HSI assembly. Regardless of the type of igniter used, if the flame sensor is not producing an output, the gas supply will be shut off.

When troubleshooting spark igniters or HSIs in an electronically controlled system, keep in mind that the system will lock out the ignition circuit if it fails to establish and maintain a flame after a number of tries or a certain amount of elapsed time. Usually, either the control voltage or the primary power must be reset before the system will operate again once a lockout occurs. Also remember that ignition devices require a good ground to operate. If the ignition device is poorly grounded, it will not create a spark.

Remember that propane-fired equipment has different lockout schemes. Also remember that propane vapors are heavier than air. They tend to fall and collect in low spaces, creating a potentially hazardous situation.

1.3.1 Spark Igniters

You should be able to see and hear whether the spark igniter is working. If you can hear the spark, but the gas is not being ignited, the ceramic insulator may be cracked. This allows the spark to be generated to another surface, outside of the pilot gas flow. If there is no spark at all, the spark gap may need to be readjusted according to the manufacturer's instructions. This is a relatively common problem.

If the gap is correct but there is still no spark or the spark is weak or intermittent, the problem is most likely in the ignition module or the high-voltage wire to the igniter. If there is a spark but no pilot flame, the pilot orifice may be plugged. During the summer, small insects may build webs or cocoons inside the orifice.

> **WARNING!**
> Be very careful when troubleshooting spark igniters because they are operated by a very high voltage pulse (10,000 to 20,000 volts) from an ignition transformer. Do not touch the spark igniter with your fingers or with a screwdriver. Do not attempt to disconnect the igniter without first turning off the power.

1.3.2 Hot Surface Igniters

Older HSIs are made of silicon carbide. The silicon carbide is resistive to current flow. When electrical power is applied through the material, it heats up rapidly, becoming very hot and bright red. The carborundum element of the igniter is very brittle and can be easily cracked or broken if handled carelessly or bumped during equipment cleaning. If a bright white line is visible anywhere on an operating HSI element, the element is cracked, but hasn't completely failed. Contrary to popular belief, any oils transferred to the element during handling will not harm the element. Silicon nitride materials are used to construct many HSIs today, and are considered to be more durable and reliable.

If an HSI fails to glow at the proper time during an ignition sequence and the required preceding actions (inducer motor running, pressure switch closed, and so forth) have occurred, then one or more of the following has occurred:

- The element is open because it is cracked, broken, burned out, or the HSI wiring is defective.
- No power is being supplied to the element.

The Incredible Shrinking Furnace

As late as the 1980s, commonly available upflow gas furnaces were up to 60 inches in height. Today, thanks to improvements in heat exchanger technology, several manufacturers offer upflow furnaces less than 36 inches in height.

Observe the HSI element without removing it from the burner section. If the HSI element is not visibly cracked, broken, or burned out, monitor the 120VAC power applied to the HSI during the ignition sequence at its inline connector inside the equipment (*Figure 12*).

Cycle the equipment through its ignition sequence while monitoring the HSI voltage. If the proper voltage is detected, the HSI has failed and must be replaced. If multiple failures of the same HSI have occurred, make sure that the voltage does not exceed 125VAC. High-operating voltages can significantly reduce the life of an HSI.

In other cases, the HSI may glow but will be faint and fail to reach a temperature hot enough to ignite the gas. This indicates the HSI has lost some of its resistance after many cycles. The resistance can be tested, and should conform to the manufacturer's specifications. If the resistance measured is outside of the proper range, replace the HSI.

If no power is found at the inline connector, the power source (inline fuse, controller, or separate relay) for the HSI has failed and must be replaced. In this case, make sure the HSI leads are not damaged or shorted to any metal part of the furnace before reapplying power to the HSI.

1.4.0 Troubleshooting the Flame Sensor

In direct ignition systems, the flame rod is not an integral part of the igniter. It is installed in the path of the burner flame (*Figure 13*). It uses flame rectification to produce a tiny DC current in the microampere range. This current must be sensed by the gas-fired equipment controller, or the main gas valve will be turned off.

Figure 14 shows the method for testing a flame sensor. A multimeter capable of reading DC current in the low microampere range (less than 10 microamps) is connected in series between the

03209-13_F13.EPS

Figure 13 Flame sensor.

Figure 12 Checking HSI source voltage.

WIRES TO HSI FROM INLINE CONNECTOR

① WITH 120VAC POWER OFF, DISCONNECT HSI LEADS OR PLUG FROM INLINE CONNECTOR

② CONNECT INSULATED ALLIGATOR METER LEADS TO CONNECTOR CONTACTS

HSI INLINE CONNECTOR

READING MUST BE WITHIN MANUFACTURER'S SPECIFIED RANGE

WIRES TO 120VAC POWER SOURCE FOR HSI

③ WITH 120VAC POWER ON, CHECK FOR 120VAC AT INLINE CONNECTOR CONTACTS AS EQUIPMENT CYCLES THROUGH IGNITION SEQUENCE

03209-13_F12.EPS

flame sensor electrode and the flame-proving circuit. Then the power is turned on and the thermostat is set to call for heat. After the burners ignite, the current reading should be within the range specified by the manufacturer. Typically, the normal range is somewhere between 0.5 and 4.5 microamperes. Follow the manufacturer's specifications for this value.

If the current reading is absent or too low, it could be caused by one of several problems:

- The flame-sensing rod is incorrectly positioned.
- The ignition module is not properly grounded.
- The sensor electrode is not making good electrical contact.
- The sensing rod has become oxidized. In this case, clean the rod with fine steel wool.
- The controller or flame sensor insulation/wiring is defective.
- Improper 120V polarity.

Special equipment is available for testing spark generators, flame sensors, gas valves, and other ignition devices. One manufacturer provides optional accessories for a digital multimeter. These accessories permit voltage and current measurements in the very low ranges necessary for testing ignition devices.

A furnace manufacturer might provide a furnace tester for use with specific furnace models. The tester can be plugged into the furnace. Switches on the tester can then be manipulated to simulate various inputs.

1.4.1 Testing a Combined HSI/Flame Sensor

Instead of a separate HSI and flame sensor, many gas-fired systems use a controller that switches the connections to the HSI so that it functions as the igniter for a prescribed time, then as a flame sensor using flame rectification. The controller accomplishes this internally by removing the AC current from the HSI after the ignition sequence, opening one of the HSI leads, then substituting a high-resistance circuit and an AC signal in the other lead to the HSI element. The element can then function as a separate flame sensor.

As the HSI ages, a protective oxide coating slowly forms on the surface of the element. This coating is normal, but it may reduce the strength of the flame signal to the controller. If the gas ignites, but immediately or occasionally goes out before the thermostat is satisfied (nuisance trips), a weakened flame-sensing signal may be the cause. Always check the continuity of the burners and their flame carryover tube to ground, as well as the controller ground connection. If the grounds are satisfactory, visually check the HSI element for contamination. If not visibly contaminated, check that the HSI element is positioned so that ¾" to 1" of the element is continuously immersed in flame.

If desired, the flame signal can be checked using a test circuit as shown in *Figure 15*. The meter used must have a DC microammeter range of about 10 microamps.

Figure 14 Flame sensor test.

Figure 15 Testing a combined HSI/flame sensor.

NCCER – HVAC Level Three 03209-13

After turning off all primary 120V power, connect the test circuit at a convenient point (inline connector) between the controller and the HSI. Depending on the gas equipment manufacturer, the test circuit may require various inline connector plugs to interface to the equipment.

Reapply power and cycle the system through an ignition sequence. When the gas ignites and after the igniter function de-energizes, push and hold the momentary contact switch to read the DC flame signal on the meter. If no reading occurs, the switch may be in the wrong lead from the controller. In that case, turn off the primary 120V power, reverse the test leads to the controller, and repeat the process.

Check that the flame signal is within the prescribed limits for the controller manufacturer. If the flame signal is out of limit, or the flame goes out before a reliable reading can be obtained, replace the HSI and retest. If the problem reoccurs, replace the controller and retest. If the problem is solved with replacement of the controller, reinstall the original HSI and retest.

1.5.0 Airflow-Related Problems

Forced-air gas furnaces can malfunction if there are problems in the air distribution system. There are several ways that airflow problems can create problems for a gas-fired furnace. Proper combustion airflow is also essential to furnace operation and reliability.

1.5.1 Air Distribution System and Blowers

In addition to component failure, three kinds of problems can occur in the supply and return air system:

- *Obstructed airflow* – The most likely cause of an airflow problem is a dirty filter. Dirty heat exchangers or a dirty blower wheel will also reduce airflow and prevent the appliance from providing adequate heating. If internally insulated ductwork is used, it is possible for the insulation to come loose and block airflow. Insufficient airflow can cause nuisance trips of the temperature limit switch because heat is allowed to accumulate around the heat exchangers. Even when the airflow is not sufficiently obstructed to overheat the furnace, it does cause a loss of efficiency. Another possible source of obstructed airflow is a cooling coil installed in the ductwork. If air filters are removed and/or filter maintenance is neglected, dirt can build up on the fins of the coil, blocking airflow. Always access the interior of the ductwork to check the cleanliness of the finned coils when other possible causes are not present.
- *Ductwork* – Problems with the ductwork size are likely to occur only if the replacement appliance has significantly greater or smaller capacity and the ductwork was not resized. Such systems will often be noisy, especially if the ductwork is too small.
- *Blower motor* – If the appliance has previously been operating properly and the blower wheel has been periodically cleaned, it is unlikely that the blower size or speed is the source of a problem. However, a blower motor may have difficulty starting or may be beginning to fail, providing intermittent or short-term service only.

Electronically controlled, variable-speed motors are frequently used as furnace blower motors. Note that these are different from a multispeed motor. While multispeed motors may operate at several different fixed speeds, variable-speed motors modulate the speed through a wide range. When checking airflow in a furnace that uses a variable-speed blower motor, always ensure that the blower is operating at the correct speed for the conditions before checking airflow. For example, if the space temperature is more than 3°F lower than the thermostat setting, the motor may be running faster than when the two temperatures are very close together. One major difference to remember is that the variable-speed motors come up to speed slowly and reduce speed slowly before stopping.

The example troubleshooting chart in *Appendix A* identifies symptoms and likely causes of air system problems.

Heating and cooling generally require different volumes of air. If a unit with a single-speed blower is used for both heating and cooling, it may be inadequate in one mode or the other. For example, a high-speed blower that is suitable for cooling may prevent the heat exchangers from getting warm enough to do their job. In such cases, it may be necessary to replace the motor and its controls with a multispeed motor and controls that automatically choose the correct blower speed based on the mode of operation.

1.5.2 Induced-Draft Blowers

Today's high-efficiency furnaces and boilers use draft inducers to help move products of combustion through the heat exchangers and the vent system. Problems related to air movement through the heat exchangers and vent piping are usually the result of inadequate combustion air, a blockage in the furnace or vent, or an incorrectly sized vent. This assumes that the draft inducer motor is operating properly, the combustion blower has not deteriorated, and devices to prove the airflow are set properly and functional.

Induced-draft furnaces have more restrictive heat exchangers that are designed to extract more heat from the flue gases. The inducer helps the flue gases overcome any restriction so that they can be properly vented. In 80-percent AFUE furnaces and boilers, the flue gas that exits the heat exchanger should be at a neutral pressure. Natural currents established in the vent then act to remove the flue gas from the structure. The draft inducer is designed primarily to overcome the restriction of the heat exchanger only, not the vent system. However, in 90-percent AFUE condensing furnaces, the inducer does impart a positive pressure to the flue gases, forcing them out the vent system, which is sealed and pressurized.

If a blockage occurs in the furnace or vent, safety devices in the appliance will prevent operation until the problem is corrected. In most sequences of operation, the first event that occurs after a call for heat is the start-up of the draft inducer motor. A pressure switch is typically used to prove that the inducer is running. The closure of the pressure switch tells the system that an adequate flow of air through the heat exchanger exists for the rest of the ignition sequence to occur. If the inducer runs but no ignition occurs, look for blockage or restrictions in the air supply, heat exchanger, or vent.

A defective or improperly adjusted pressure switch can also prevent ignition from occurring. In addition to the pressure switch, there are other manual-reset safety switches that open if the vent becomes blocked or flames roll out the front of the burner.

Due to the high level of sensitivity required of the draft inducer pressure switch, it is best to avoid attempting any field adjustments. Although many are adjustable, they are typically factory set and calibrated. This is often noted by a small amount of colored cement placed on the threads of the adjusting screw. Field adjustment of the switch may result in unsafe operation. If a pressure switch is not operating properly, replacement with a factory-provided part is generally the best course of action. Otherwise, very sensitive and properly calibrated pressure measurement instruments that can be tied into the same pressure-sensing line are required.

Condensing furnaces have some unique venting problems. Most condensing furnaces use a sealed combustion system in which outside air is brought in through a dedicated plastic pipe and flue gases are vented out a second plastic pipe. The size of both pipes is critical. Failure to size the pipes according to the manufacturer's instructions may result in a restriction that would prevent the inducer from closing the pressure switch. Improper routing of one or both pipes can also cause inadequate airflow. Dips, sags, or an improper pitch in the vent may cause condensate water to pool in the vent pipe, which causes a restriction. If the combustion air intake and vent are not properly terminated, ice may form at the vent outlet and snow may be drawn into the air intake. Both conditions represent restrictions that will prevent or interfere with furnace operation.

Additional Resources

Refrigeration and Air Conditioning, An Introduction to HVAC/R, Fourth Edition. Larry Jeffus. Air Conditioning and Refrigeration Institute. Upper Saddle River, NJ: Prentice Hall.

1.0.0 Section Review

1. In an induced-draft furnace, on which control board terminal is the heating control voltage applied?
   a. C
   b. G
   c. W
   d. Y

2. Without a load, a normally functioning thermocouple would provide an output voltage of at least _____.
   a. 9 millivolts
   b. 12 millivolts
   c. 15 millivolts
   d. 20 millivolts

3. Hot surface igniters are often made of _____.
   a. silicon carbide
   b. nichrome
   c. sodium nitride
   d. aluminum nitrate

4. The output signal of a flame sensor is typically a(n) _____.
   a. stabilized AC voltage
   b. rectified DC current
   c. variable resistance
   d. oscillating waveform

5. In most sequences of operation, the call-for-heat signal first energizes the _____.
   a. HSI
   b. gas valve
   c. pressure switch
   d. inducer motor

SECTION TWO

2.0.0 INFRARED GAS-FIRED HEATERS

Objectives

Identify infrared gas heaters and describe how they operate.

a. Identify various types of infrared gas heaters.
b. Describe the operating characteristics of infrared gas heaters.

Performance Task 1

Using the proper tools, instruments, and control circuit diagrams, isolate and correct malfunctions in a gas heating system.

Trade Terms

Reverberator: A wire screen used to cover the ceramic grids in a gas-fired heater that operates by submerged flame.

Submerged flame: A method of gas-fired heater operation that uses burning gas within two-layer porous ceramic grids.

Infrared heaters are used in buildings with a high volume of air exchange resulting from numerous external door operations, or in high-bay buildings where forced-air heating is impractical. The basic purpose of infrared heaters is to heat objects, not air. Since air is not being heated, opening a large door does not result in the significant loss of heat. This makes them ideal for commercial applications in factories, foundries, large workshops, vehicle repair garages, warehouses, aircraft hangers, and sports arenas. Smaller versions are available for limited use in residential applications.

It is important to note however, that infrared heat functions on a line-of-sight principle. If the person or object is not within view of the heater, they cannot absorb the infrared rays.

2.1.0 Infrared Heater Types

Infrared heaters are available in both low- and high-intensity types. High-intensity gas-fired heaters have been around since the 1950s. They are usually mounted high on a ceiling, use an open flame to heat a ceramic surface, and are typically unvented. Low-intensity heaters use a steel tube in which the flame is enclosed, often with a draft inducer to overcome the resistance of the tube. This heats the tube and the tube then emits radiant heat. Low-intensity heaters are typically vented.

2.1.1 Low-Intensity Infrared Gas-Fired Heaters

Figure 16 shows a typical gas-fired, tubular, low-intensity infrared heater, also called an indirect radiant heater. They are mounted at a relatively high level in a building and are available in straight-tube and U-tube configurations. A gas flame is directed from a burner assembly through the steel tube, heating the tube. The tube's hot outer surface produces infrared radiant energy focused by a stainless steel reflector onto the floor, occupants, and other surfaces or objects under the heater. The effect is similar to that caused by the sun's infrared rays. Surfaces, occupants, and objects under the heater are warmed, creating a comfortable environment without producing drafts.

03209-13_F16.EPS

Figure 16 Low-intensity infrared tubular heater.

Tubular heaters are available in either vented or non-vented versions. Most models require some form of mechanical assistance for venting, due to the resistance of the long tubular heat exchanger. If more than one heater is connected to a single vent, a vacuum is created by a single induced-draft blower located at the vent's exit to the outdoors. Single-vented or non-vented units may be assisted (power-vented) by a blower located at the burner assembly itself. Venting can be through the roof or horizontally through a wall.

In a non-vented unit, combustion products are discharged into the heated space for additional heat. These heaters are permitted in areas with a high outdoor air infiltration rate, or when exhaust fans change the building air a set number of times over a specific period. The burner assembly operates the same as the burner assembly in an induced draft gas furnace. As noted earlier, an excessive amount of heat is not lost due to the constant air exchange.

Safety devices used in low-intensity infrared gas-fired heaters are similar to those used in induced-draft gas furnaces. The pilot ignition devices used, as well as various electrical and electronic devices including glow coils, are also similar to induced-draft gas furnace equipment.

2.1.2 High-Intensity Infrared Gas-Fired Heaters

These non-vented units (*Figure 17*), also called direct-fired radiant burners, are used at high levels off the floor as spot or space heaters. The units operate by burning gas within two-layer porous ceramic grids, which is referred to as a **submerged flame**. The grids attain a red heat level of about 1,600°F and produce intense infrared radiant energy. In some cases, the grids can also be closely covered by an optional wire screen, called a **reverberator**, that increases the infrared radiation when heated.

High-intensity infrared gas-fired units must be mounted at a specified angle to allow the combustion products to flow upward off each grid and into the room. Reflectors around the perimeter of the grids focus the radiant energy to the desired location. These non-vented units are permitted for use in areas with a high outdoor air infiltration rate or when exhaust fans change the building air a set number of times over a specific period. This ensures that the byproducts of combustion are diluted and eventually removed from the space. The gas burner, located behind the grids in a plenum, operates in the same way as the burner in a natural-draft gas furnace.

Like low-intensity units, the safety devices used in high-intensity infrared gas-fired heaters are similar to those used in induced-draft gas furnaces.

2.2.0 Infrared Heater Operation

Multiple commercial infrared heaters may be controlled from a central multi-zone control panel (*Figure 18*). Panels typically include digitally programmed time switches that can be set up to deal with a variety of different heater control scenarios. If needed, a built-in temperature sensor can also be used in conjunction with the programmed switches for better temperature control.

03209-13_F18.EPS

Figure 18 Multi-zone control panel.

03209-13_F17.EPS

Figure 17 High-intensity infrared heater.

> ## Clean Natural Gas
>
> **GOING GREEN**
>
> With today's concerns about global warming, people are looking at natural gas as the best fuel to burn to reduce carbon emissions. Natural gas produces 30 percent less carbon emissions than oil and 45 percent less carbon emissions than coal. Plus, natural gas does not produce ash particles like coal and oil. Recent advances in drilling technology now allow access to this very clean fuel.

Conditioned-space temperature control with gas-fired infrared heaters is different from temperature control with more conventional heating systems. Infrared heaters are usually used for spot heating in an otherwise unheated area. A simple on/off switch can often provide adequate temperature control. For example, a worker at a desk on an unheated loading dock can turn the heat on when he or she is present, then turn off the heat when he or she leaves the work area. Infrared heaters may use low-voltage or line-voltage thermostats for temperature control.

However, conventional thermostats may not offer the best temperature control for a variety of reasons:

• The spot area being heated is often surrounded by cold air which can affect the thermostat.
• Line-of-sight radiant heat energy can heat the thermostat. A radiant shield may have to be installed to prevent this.
• Thermostats installed near open doors such as found on a loading dock can be exposed to cold drafts.
• Workstation machinery may generate heat that can affect the thermostat.

Figure 19 shows a wiring diagram for a low-intensity tubular heater that is controlled by a thermostat. The components used in the heater and its operating sequence are very similar to that of a conventional gas-fired furnace.

The operating sequence of the unit shown in *Figure 19* is as follows:

• On a call-for-heat, the inducer (blower) starts running. When it reaches operating speed, the air switch closes and provides an input to activate the ignition module.
• After a pre-purge period of about 30 seconds, the ignition module energizes the spark igniter. Simultaneously, the gas valve is energized for about 15 seconds to allow the burner to ignite.
• If burner ignition occurs, the flame-sensing rod sends a signal to the ignition module. This input to the ignition module allows the gas valve to remain energized.

If no burner flame is detected, the ignition module shuts off the gas valve and the inducer runs for 30 seconds to purge any unburned gas from the system. The ignition module will attempt to restart the burners several times on natural gas systems. Units operating on LP gas typically do not retry ignition. Lock-out occurs on the first failure.

If no restart occurs, the ignition module will lock out for 1 hour or until it is manually reset by removing power for at least 5 seconds.

If a 24V thermostat is used for temperature control, the inducer will run for 30 seconds after the burners shut off to purge products of combustion from the system. This feature is de-activated if a line-voltage thermostat is used for temperature control.

> **WARNING!**
>
> Infrared heaters are often suspended from high ceilings. Maintenance or repair of these units may have to be done from a ladder, scaffold, or self-propelled aerial lift. When working at heights above 6 feet, you must wear a personal fall arrest system (PFAS).

WIRING DIAGRAM ALL MODELS SERIES U

24 VAC THERMOSTAT

120 VAC THERMOSTAT

03209-13_F19.EPS

Figure 19 Low-intensity infrared heater wiring diagram.

Additional Resources

Refrigeration and Air Conditioning, An Introduction to HVAC/R, Fourth Edition. Larry Jeffus. Air Conditioning and Refrigeration Institute. Upper Saddle River, NJ: Prentice Hall.

2.0.0 Section Review

1. Ceramic grids in a high-intensity infrared heater operate at a temperature of about _____.
    a. 500°F
    b. 1,200°F
    c. 1,600°F
    d. 2,000°F

2. The simplest method of controlling the operation of an infrared space heater is to use a(n) _____.
    a. line-voltage thermostat
    b. low-voltage thermostat
    c. multi-zone panel
    d. on/off toggle switch

3.0.0 GAS COMBUSTION ANALYSIS

Objectives

Explain how to conduct a combustion analysis on a gas furnace.

   a. Identify combustion analysis equipment and the combustion byproducts that are of importance to the analysis.

   b. Describe the combustion analysis process and how to interpret basic results.

Performance Task 2

Complete a combustion analysis on a gas furnace or boiler.

Trade Terms

Net stack temperature: A value needed to perform a combustion efficiency analysis. It is calculated by subtracting room temperature from the temperature measured in a furnace's vent or vent connector.

Combustion analysis of an operating gas-fired furnace or boiler provides several benefits including the following:

- Reduced fuel consumption. A burner operating at maximum efficiency burns less fuel.
- Undesirable exhaust emissions such as carbon monoxide (CO) and other noxious gases are reduced when a burner is operating properly.
- A furnace or boiler in which the burner is properly adjusted is much safer. For example,

CO is less likely to form and soot is less likely to build up.

Combustion analysis entails measuring the flue gas temperature and sampling the flue gas using specialized test instruments. After the temperature is measured and levels of certain gases in the flue gas are known, combustion efficiency can be determined. Use the listed AFUE of the furnace or boiler as a guide for expected efficiencies. It is common to expect combustion efficiency numbers on new and properly tuned equipment to be slightly lower than the AFUE number, which is determined in laboratory conditions. Also expect lower numbers on older equipment and equipment in need of service or repairs.

Furnace efficiency stated as AFUE and combustion efficiency are terms that are often used interchangeably. While similar, the two terms have slightly different meanings. Using these interchangeably can lead to confusion. When the AFUE is determined for a furnace or boiler, the product is tested in a laboratory where conditions are tightly controlled and the furnace is cycled on and off under part-load conditions.

A field combustion efficiency analysis on the other hand, is done under real-world conditions with all its variables. The furnace or boiler must be operating continuously under steady-state conditions. In fact, some people believe that a steady-state efficiency test is a more accurate description of a combustion efficiency test.

Because of test conditions differences, it is unrealistic for a combustion efficiency test to produce efficiency numbers that match AFUE numbers generated under precise laboratory conditions. Actual field tests have shown that combustion efficiency tests produce efficiency numbers several points lower than the posted AFUE of a product. These number differences should not be of sig-

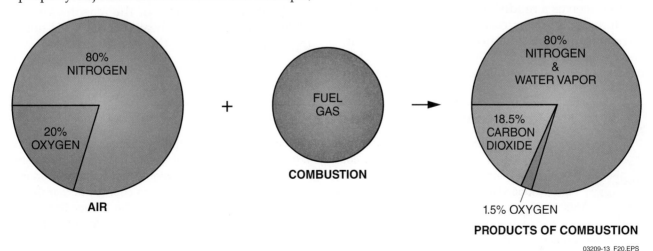

Figure 20 Products of combustion.

nificant concern to service technicians. Just be aware that you are unlikely to match any stated AFUE value when you perform a field combustion analysis.

3.1.0 Gas Combustion Byproducts

Air used for combustion is composed of about 20 percent oxygen, about 80 percent nitrogen, and a very small percentage of other gases. In an ideal situation, the oxygen in the air combines with hydrocarbons in the fuel to produce the combustion products carbon dioxide (CO_2) and water vapor. In fact, the byproducts of combustion (*Figure 20*) also contain some oxygen because more air is supplied to the combustion process than is required. This excess air is also provided to ensure that only insignificant quantities of toxic CO can form.

Byproducts of combustion from a normally operating furnace will contain CO_2, water vapor, and a small amount of oxygen. Nitrogen is not required for combustion so it passes through the burner. Even though nitrogen theoretically passes through the burner unchanged, the heat of the combustion process can form small amounts of nitrous oxides that are considered pollutants. Flue gas temperature also affects combustion efficiency. If the flue gas is too hot, it indicates that heat is going out the vent instead of being used to heat the structure.

Some states and regional pollution control authorities require testing of gas-fired heating equipment to determine levels of CO, CO_2, and nitrous oxides. Nitrous oxide (NO_x) is the generic term for a group of highly reactive gases, all of which contain nitrogen and oxygen in varying amounts. Many of the nitrous oxides are colorless and odorless.

However, one common pollutant, nitrogen dioxide (NO_2), along with particles in the air, can often be seen as a reddish-brown layer over many urban areas. Nitrous oxides form when fuel is burned at high temperatures. The primary man-made sources of NO_x are motor vehicles, electric utilities, and other industrial, commercial, and residential sources that burn fuels. NO_x can contribute to acid rain, as well as global warming, and is known to cause degradation of water quality.

The flue gas of an efficient and normally operating gas furnace or boiler will have the following characteristics:

- Flue gas temperature will be such that there is adequate heat available for proper operation of the vent, but not excessive to the point that too much heat is being lost out the vent.
- CO_2 and oxygen levels in the flue gas will be at acceptable levels indicating that enough excess air is being supplied to the burners.
- Only minute quantities (less than 25 parts per million) of CO should be present in the flue gas. The presence of any significant amount of CO is an indication of incomplete combustion. If significant CO is present, ensure that the furnace is receiving adequate combustion air and check for other problems that may be causing the level to be high.
- Traces of nitrogen and sulfur oxides may be present.

Natural-draft and induced-draft furnaces and boilers must be installed so that they receive adequate air for combustion. Furnaces or boilers installed in confined spaces must have air openings into the confined space that are large enough to supply adequate air. Opening sizes are based on the input rating of the furnace. If the building has tight construction, outside air may need to be ducted into the confined space. Having an adequate supply of combustion air for condensing furnaces (90 AFUE and higher) is not the problem that it is with other furnaces because condensing furnaces use a sealed combustion system that takes in outdoor air for combustion through a dedicated intake pipe. Indoor air is not used for combustion. This is one of the reasons for their added efficiency, since heated room air is not being removed to satisfy the venting process.

GOING GREEN

Low NO$_x$ Furnaces

NO_x byproducts are considered to be an air pollutant and their emissions are subject to regulation in many parts of the country. For example, residential gas furnaces and gas water heaters installed in California must be designed to reduce these emissions. Since NO_x byproducts are produced at higher burner temperatures, one solution to reduce them is to cool the burner flame. This is often done by positioning metal rods, metal screens, or metal tabs into the burner flame. The metal acts as a heat sink, cooling the flames so that fewer pollutants are produced. In the United States, all major furnace manufacturers now offer low NO$_x$ furnaces.

3.1.1 Gas Combustion Analysis Equipment

Both mechanical and electronic gas burner combustion test equipment is available (*Figure 21*). The mechanical equipment shown is available as a tool kit that contains indicators for CO_2, oxygen, and CO in flue gas samples. The kit also contains a thermometer for measuring flue gas or supply air temperatures; a draft gauge to measure the pressure in the vent; a gas pressure manometer to test the gas pressure to the burner manifold; and a slide-rule style combustion efficiency calculator. Blank combustion efficiency calculation forms are often provided with mechanical test kits.

The electronic analyzer shown can be used for both gas- and oil-burning appliances. It is one of several makes of electronic combustion analyzers currently available. Even though both types will produce similar test results, electronic combustion analyzers are far more versatile than mechanical test kits. They can provide a combustion analysis in a fraction of the time that it would take with a mechanical analyzer. They typically contain multi-function or dedicated sampling probes that can be plugged into the analyzer to obtain various readings. Results are available instantly on a digital display. Information gathered can be saved, printed out, or downloaded into a personal computer or smart phone. The basic steps for a combustion analysis described in the next section apply to both mechanical and electronic combustion analyzers.

Before attempting to use any mechanical or electronic combustion analyzer, read the instruction manual and become familiar with the various features of the device.

3.2.0 Conducting a Combustion Analysis

Before conducting a combustion analysis on a gas-fired furnace or boiler, conduct an inspection of the equipment to see if there is any problem that might affect the combustion analysis. Correct any problems found.

Things to look for during a preliminary inspection include:

- Plugged or dirty supply air filters.
- Rust or holes in vent connectors; holes in furnace heat exchangers.
- Soot buildup that could indicate incomplete combustion.
- Restricted or blocked air intakes into an equipment room or a blocked or restricted outside air intake.
- Blocked or restricted air intake and/or vent pipes on condensing furnaces.

ELECTRONIC GAS/OIL
BURNER COMBUSTION ANALYZER

MECHANICAL GAS BURNER
COMBUSTION TEST EQUIPMENT

03209-13_F21.EPS

Figure 21 Gas burner combustion test equipment.

Prepare the equipment for the combustion analysis by drilling a sampling hole in the vent pipe to sample the flue gas and measure flue gas temperature. *Figure 22* shows the location of the sampling hole in the different types of furnaces and boilers.

Each type of appliance has special requirements:

- No sampling hole needs to be drilled when checking a natural-draft furnace. Instead, flue gas and flue gas temperature is sampled at the outlet of each clam-shell heat exchanger, at the draft diverter. For best results, sample each heat exchanger outlet and then average the values of all samples.
- Natural-draft boilers equipped with an external draft diverter should have the sampling hole below the draft diverter and as close to the boiler vent connector collar as possible.
- 80-percent AFUE furnaces and boilers that use induced-draft technology should have the sample hole drilled directly above the vent connector collar on the furnace.
- Condensing furnaces and boilers should have two sample holes; one in the vent pipe as it

leaves the furnace and one on the air intake pipe just before it enters the furnace. If codes or manufacturer's literature prevents drilling holes in the plastic pipes, samples can be taken at the outdoor terminations of the two pipes.

After the sample holes have been drilled, set the room thermostat high enough so that the furnace or boiler will run for the duration of the combustion analysis. Allow the burners to operate for at least five minutes to allow the furnace or boiler to reach stability.

3.2.1 Measuring Flue Gas Temperature

Flue gas temperature is a key component of a combustion analysis. Actual vent or stack temperature cannot be used for the calculation, since the temperature of the flue gas is the sum of the heat created by combustion and the temperature of the air as it enters the burner. Heat created by combustion is the important value. For example, if the flue gas temperature is 400°F, that temperature is actually the sum of the room temperature (70°F, for example) and the heat added by combustion (330°F).

Figure 22 Sampling hole locations.

03209-13_F22.EPS

The temperature needed to calculate burner efficiency is known as **net stack temperature**. Net stack temperature is calculated by subtracting room temperature (or temperature in the air intake pipe as it enters the furnace) from the measured flue gas temperature.

Measured stack temperature –
Room temperature = Net stack temperature

$$400°F - 70°F = 330°F$$

To find net stack temperature, insert the thermometer probe into the sample hole. Many probes have a self-sealing feature that allows for a tight seal. Slide the probe back and forth until the highest temperature is reached. On electronic analyzers, pressing a button will record the value. Otherwise, write down the highest stack temperature.

Measure the temperature in the room that the furnace or boiler is located to obtain the net stack temperature. Electronic analyzers will automatically perform this calculation. Otherwise, do the math. Since condensing furnaces do not use indoor air for combustion, measure the temperature of the air entering the furnace through the combustion air intake pipe to calculate net stack temperature.

If codes prevent drilling sampling holes in the plastic air intake and vent pipes of condensing furnaces, measurements may be taken at the terminations of these two pipes. However, measurements taken at the terminations will not be as accurate as those taken close to the furnace. As the pipes move through the structure to or away from the furnace, there may be some heat exchanged with the air outside of the pipe. For example, if the piping is routed through the attic, combustion air may become warmer while the vented air may become cooler. These gains and losses affect any temperatures measured close to the furnace. Taking them into consideration will result in a more accurate analysis.

3.2.2 Sampling Flue Gas

Flue gas is sampled through the same hole used to measure stack temperature. Follow the combustion analyzer's instructions to obtain a flue gas sample.

Electronic devices will quickly calculate burner efficiency and display it with all other values at the push of a button (*Figure 23*). Mechanical analyzers will require you to manually record the levels of oxygen and CO_2 measured and then make some additional calculations.

03209-13_F23.EPS

Figure 23 Electronic combustion analyzer display.

With net stack temperature and levels of oxygen and CO_2 known, these values can also be entered into a chart or slide rule-style calculator to determine burner efficiency (*Figure 24*). For example, with a net stack temperature of 380°F, a measured CO_2 level of 8.4 percent and a measured oxygen level of 6 percent, the calculated burner efficiency is 79.5 percent.

At the completion of the combustion analysis, make any adjustments to the equipment to improve efficiency (if required) and restore the equipment to its previous operating status. Make a final test to ensure there were positive results.

3.2.3 Analyzing the Combustion Efficiency Test

When the combustion analysis is complete, results should be analyzed and corrective action taken to solve any problems. Here are some problems that the analysis may reveal, along with suggested corrective actions:

% O_2	% CO_2	NET STACK TEMPERATURE DEGREES °F						
		340	360	380	400	420	440	460
5.0	9.0	81.1	80.5	80.0	79.5	78.9	78.4	77.8
5.5	8.7	80.8	80.3	79.7	79.2	78.6	78.1	77.5
6.0	8.4	80.6	80.0	79.5	78.9	78.3	77.8	77.2
6.5	8.2	80.3	79.7	79.2	78.6	78.0	77.4	76.8
7.0	7.9	80.0	79.4	78.8	78.2	77.6	77.0	76.4
7.5	7.6	79.7	79.1	78.5	77.9	77.3	76.7	76.0
8.0	7.3	79.4	78.8	78.1	77.5	76.9	76.2	75.6
8.5	7.0	79.1	78.4	77.8	77.1	76.4	75.8	75.1
9.0	6.7	78.7	78.0	77.3	76.6	76.0	75.3	74.6

03209-13_F24.EPS

Figure 24 Combustion efficiency calculator chart.

- Higher than normal flue gas temperatures may be the result of an over-fired burner or a high temperature rise across the heat exchanger. Check burner input and measure temperature rise to confirm and/or correct the problems.
- Lower than normal CO_2 and oxygen levels may indicate the burner is not receiving adequate combustion air. Check for and correct any combustion air problems.
- Higher than normal CO levels may indicate incomplete combustion caused by a lack of combustion air or a burner that is damaged or not properly adjusted.

3.2.4 Other Combustion Efficiency Information

Both electronic and mechanical combustion analyzers are capable of providing other valuable information including:

- Vent draft can be measured through the sampling hole in the vent of natural-draft and 80-percent AFUE furnaces and boilers. Draft on gas furnaces is a function of vent height and diameter. If draft is excessive, it may cause excess heat to be moved out the vent, decreasing combustion efficiency.
- A smoke spot tester in mechanical combustion analysis kits can detect smoke in the flue gases. This device in used when doing a combustion analysis on oil burners, but is of little use with gas-fired appliances.
- Electronic combustion analyzers have the ability to detect other gases that may be present in flue gas, including NO_x and sulfur, or CO. Many states and localities have limits on oxides of nitrogen (NO_x) emissions. The combustion analysis can tell if those emissions are within acceptable limits.

Paperwork

When you complete a combustion analysis, always present a copy of your before-and-after analysis to the customer. Having a printed copy of your analysis boosts the credibility of your work and shows the customer that you are a professional, even if they do not understand the process completely.

Additional Resources

Refrigeration and Air Conditioning, An Introduction to HVAC/R, Fourth Edition. Larry Jeffus. Air Conditioning and Refrigeration Institute. Upper Saddle River, NJ: Prentice Hall.

3.0.0 Section Review

1. Which substance would you not expect to find in any significant quantity in the flue gas of a normally operating gas furnace?

   a. water vapor
   b. oxygen
   c. CO_2
   d. CO

2. Where is the flue gas sampling hole drilled in a natural-draft boiler?

   a. In a clam-shell heat exchanger.
   b. Directly below the draft diverter.
   c. Directly above the inducer motor.
   d. In the combustion air intake pipe.

Summary

Troubleshooting gas heating appliances requires skill in reading and interpreting electrical diagrams as well as the ability to evaluate and test the ignition, combustion, and air distribution systems. The HVAC technician must have good observation skills and a working knowledge of the various gas heating systems and accessories available. Technicians must also possess a thorough understanding of safety precautions to be taken while working with gas heating equipment.

Although all gas heating controls have elements in common, controls are implemented by various means. Packaged gas heating units include natural-draft and induced-draft units. Gas furnace types include induced-draft furnaces and low- and high-intensity gas-fired heaters. It is important to be able to analyze the various types of control circuits used in these devices.

The HVAC technician must understand the operation of various types of control devices, including igniters, gas valves, and flame sensors. Combustion analysis of an operating gas-fired furnace or boiler provides several benefits including reduced fuel consumption; elimination of undesirable exhaust emissions such as CO and other noxious gases; and a furnace or boiler that is safer to operate.

---

1. The most durable hot surface igniters are made with nitrides.

   a. True
   b. False

2. Natural-draft furnaces are no longer available for new installations because they are _____.

   a. expensive to build
   b. not energy-efficient
   c. too noisy
   d. not code-compliant

3. The ignition device used in the system depicted in *Figure 5* is a _____.

   a. hot surface igniter
   b. glow coil
   c. spark igniter
   d. standing pilot

4. The ignition device used in the system depicted in *Figure 3* is a _____.

   a. hot surface igniter
   b. glow plug
   c. spark igniter
   d. intermittent pilot

5. Pilotless heating systems use either direct spark igniters or _____.

   a. indirect spark igniters
   b. bimetal igniters
   c. thermocouple igniters
   d. hot surface igniters

6. Which of the following is *not* a problem that can occur with a spark igniter?

   a. Faulty cold junction
   b. Incorrect spark gap
   c. Ignition module failure
   d. Cracked ceramic insulator

7. When testing a flame-sensing rod, the reading should be _____.

   a. 12 volts
   b. 30 ohms
   c. 0.5 to 4.5 milliamps
   d. 0.5 to 4.5 microamps

8. Oxidation on a flame-sensing rod protects the rod and improves flame sensing.

   a. True
   b. False

9. The basic purpose of infrared heaters is to heat _____.

   a. objects
   b. air
   c. water
   d. coolant

10. High-intensity infrared gas-fired heaters are also called _____.

    a. vented units
    b. indirect radiant heaters
    c. direct-fired radiant burners
    d. tubular heaters

11. Optional wire screens used to cover the grids of high-intensity infrared gas-fired heaters are called _____.

    a. u-tubes
    b. radiators
    c. vents
    d. reverberators

12. A lack of adequate combustion air would be indicated by the flue gas containing _____.

    a. CO_2
    b. moisture
    c. CO
    d. oxygen

13. Which mathematical function is used to calculate net stack temperature?

    a. Addition
    b. Subtraction
    c. Multiplication
    d. Division

14. What temperature should you seek when the temperature probe is inserted into the vent connector?

    a. Lowest
    b. Average
    c. Highest
    d. Most stable

15. When calculating burner efficiency using a mechanical (non-electronic) combustion analyzer, efficiency is determined by entering values found into a(n) _____.

    a. smart phone
    b. personal computer
    c. digital calculator
    d. slide rule-style calculator

Trade Terms Introduced in This Module

Flame rectification: A process in which exposure of a sensing rod to a flame produces a tiny current that can be sensed and used to control the gas valve.

Modulating gas valve: A gas valve that precisely controls burner capacity by adjusting the flow of gas to burners in small increments.

Net stack temperature: A value needed to perform a combustion efficiency analysis, calculated by subtracting room temperature from the temperature measured in a furnace's vent or vent connector.

Reverberator: A wire screen used to cover the ceramic grids in a gas-fired heater that operates by submerged flame.

Submerged flame: A method of gas-fired heater operation that uses burning gas within two-layer porous ceramic grids.

Appendix A

GAS FURNACE TROUBLESHOOTING GUIDE

GENERALLY THE CAUSE Make these checks first.	OCCASIONALLY THE CAUSE Make these checks only if the first checks failed to locate the trouble.	RARELY THE CAUSE Make these checks only if other checks failed to locate the trouble.
PROBLEM: No heat – Burner fails to start		
Power failure Blown fuses or tripped circuit breaker Open disconnect switch or blown fuse Thermostat set too low Thermostat switch not in proper position Pilot Manual fuel valve Pilot valve	Control transformer Limit controls (manual reset) Pilot safety control Gas valve Thermocouple Fuel lines Gas meter	Faulty wiring Loose terminals Low voltage Thermostat faulty Improper burner adjustment Plugged burner orifices Gas pressure Thermostat not level (mercury bulb thermostat only)
PROBLEM: No heat – Inducer starts but burner fails to ignite		

Combustion air switch
Inducer motor
Primary or safety controls
Gas valve
Flame detector or electrode
Improper burner adjustment
Manual fuel valve
Pilot valve
	Ignition devices
	Regulator
	Plugged vent or air supply
	    (induced draft only)
	Plugged pressure switch line
	    at inducer (induced draft only)
		Low voltage
		Fuel lines
		Gas pressure
		Plugged burner orifice
		Condensate trap broken or empty
		(condensing furnace only)

PROBLEM: Insufficient or no heat – Burner starts and fires, then switches off when safety switch trips		
Safety controls Flame detector or electrode Improper burner adjustment Dirty air filter Dirty or plugged heat exchanger	Combustion air switch Inducer motor	Faulty wiring Loose terminals Fuel lines Gas pressure
PROBLEM: Insufficient or no heat – Burner cycle too short		
Limit controls Thermostat not level (mercury bulb thermostat only) Thermostat location Blower belt broken or slipping Dirty filters Thermostat heat anticipation incorrect	Fan control Blower bearings Blower motor speed Low air volume Dirty blower wheel	Faulty wiring Loose terminals Low voltage Blower wheel Ductwork small or restricted Restrictions
PROBLEM: Insufficient or no heat – Burner runs continuously		
Blower belt slipping Dirty blower wheel Dirty filters Gas pressure	Orifice size Low air volume Improper burner adjustment Plugged burner ports Input too low Undersized furnace	Ductwork small or restricted Displaced or damaged baffles Insufficient insulation Excessive infiltration

03209-12_A01A.EPS

GENERALLY THE CAUSE Make these checks first.	OCCASIONALLY THE CAUSE Make these checks only if the first checks failed to locate the trouble.	RARELY THE CAUSE Make these checks only if other checks failed to locate the trouble.
PROBLEM: Too much heat – Burner cycle too long		
Thermostat heat anticipation incorrect Thermostat out of calibration	Thermostat not level Thermostat location	
PROBLEM: Combustion noise		
Pilot Gas valve Improper burner adjustment Plugged burner orifice Air shutter adjustment (if so equipped)	Inducer motor Primary or safety controls Ignition transformer Flashback to venturi Gas pressure Pressure regulator	Faulty wiring Loose terminals Regulator Dirty or plugged heat exchanger Vent or flue Displaced or damaged baffles Blocked heat exchanger Input too high
PROBLEM: Mechanical noise		
Blower bearings Blower belt cracked or slipping	Inducer motor Blower motor Cabinet	Low voltage Control transformer Control relay or contactor Blower wheel Cracked or ruptured heat exchanger Displaced or damaged baffles Fuel lines
PROBLEM: Air noise		
Blower Cabinet Ductwork small or restricted Air leaks in ductwork	Other air system restrictions	Blower wheel Dirty filters
PROBLEM: Odor		
Vent or flue Flue gas spillage Fuel leaks Cracked or ruptured heat exchanger Blocked heat exchanger	Regulator Low air volume Improper burner adjustment Displaced or damaged baffles Humidifier stagnant water Water or moisture	Faulty wiring Loose terminals Control transformer Dirty filters Dirty or plugged heat exchanger Input too high Input too low Outdoor odors
PROBLEM: Cost of operation		
Combustion motor Blower motor Dirty filters Improper burner adjustment Equipment size	Low air volume Insufficient insulation Excessive infiltration	Low voltage Fan control Orifice Blower belt slipping Dirty blower wheel Ductwork small or restricted Vent or flue Input too high

03209-13_A01B.EPS

GAS FURNACE TROUBLESHOOTING FLOW
CHART

03209-13_A02A.EPS

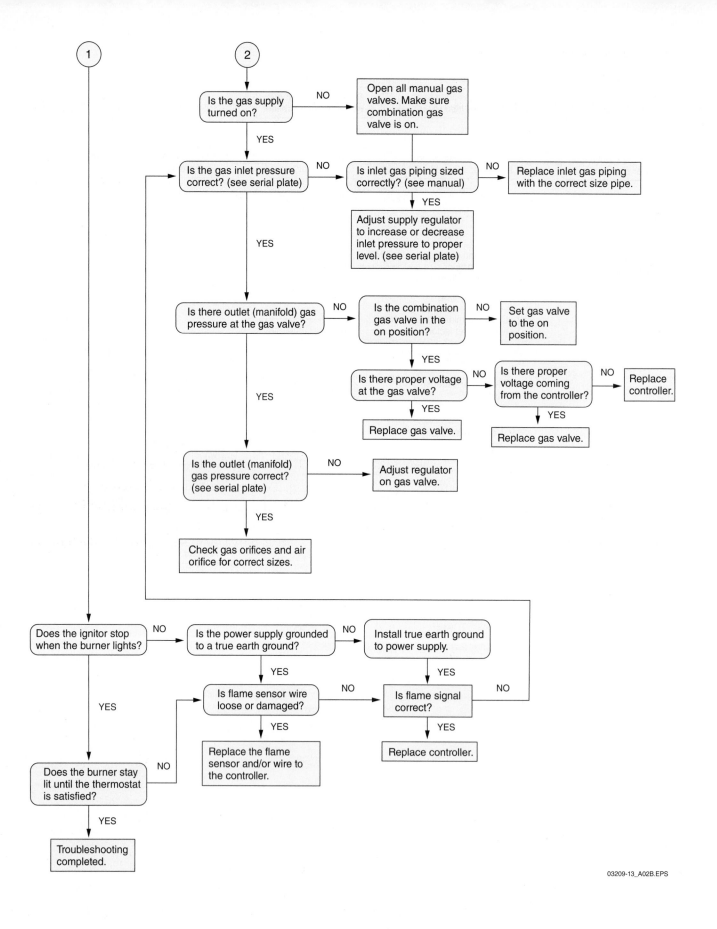

Is the gas supply turned on? — NO → Open all manual gas valves. Make sure combination gas valve is on.

↓ YES

Is the gas inlet pressure correct? (see serial plate) — NO → **Is inlet gas piping sized correctly? (see manual)** — NO → Replace inlet gas piping with the correct size pipe.

↓ YES (from inlet piping) → Adjust supply regulator to increase or decrease inlet pressure to proper level. (see serial plate)

↓ YES

Is there outlet (manifold) gas pressure at the gas valve? — NO → **Is the combination gas valve in the on position?** — NO → Set gas valve to the on position.

↓ YES

Is there proper voltage at the gas valve? — NO → **Is there proper voltage coming from the controller?** — NO → Replace controller.

↓ YES

Replace gas valve.

↓ YES (from controller voltage)

Replace gas valve.

↓ YES

Is the outlet (manifold) gas pressure correct? (see serial plate) — NO → Adjust regulator on gas valve.

↓ YES

Check gas orifices and air orifice for correct sizes.

Does the ignitor stop when the burner lights? — NO → **Is the power supply grounded to a true earth ground?** — NO → Install true earth ground to power supply.

↓ YES

Is flame sensor wire loose or damaged? — NO → **Is flame signal correct?** — NO → Replace controller.

↓ YES (flame sensor)

Replace the flame sensor and/or wire to the controller.

↓ YES (flame signal)

Replace controller.

↓ YES (ignitor stop)

Does the burner stay lit until the thermostat is satisfied? — NO → (to flame sensor wire)

↓ YES

Troubleshooting completed.

03209-13_A02B.EPS

Additional Resources

This module presents thorough resources for task training. The following resource material is suggested for further study.

Refrigeration and Air Conditioning, An Introduction to HVAC/R, Fourth Edition. Larry Jeffus. Air Conditioning and Refrigeration Institute. Upper Saddle River, NJ: Prentice Hall.

Figure Credits

Section Review Answer Key

Answer	Section Reference	Objective
Section One		
1. c	1.1.3	1a
2. d	1.2.0	1b
3. a	1.3.2	1c
4. b	1.4.0	1d
5. d	1.5.2	1e
Section Two		
1.c	2.1.1	2a
2.d	2.2.0	2b
Section Three		
1.d	3.1.0	3a
2. b	3.2.0	3b

NCCER CURRICULA — USER UPDATE

NCCER makes every effort to keep its textbooks up-to-date and free of technical errors. We appreciate your help in this process. If you find an error, a typographical mistake, or an inaccuracy in NCCER's curricula, please fill out this form (or a photocopy), or complete the online form at **www.nccer.org/olf**. Be sure to include the exact module ID number, page number, a detailed description, and your recommended correction. Your input will be brought to the attention of the Authoring Team. Thank you for your assistance.

Instructors – If you have an idea for improving this textbook, or have found that additional materials were necessary to teach this module effectively, please let us know so that we may present your suggestions to the Authoring Team.

NCCER Product Development and Revision
13614 Progress Blvd., Alachua, FL 32615

Email: curriculum@nccer.org
Online: www.nccer.org/olf

❏ Trainee Guide ❏ Lesson Plans ❏ Exam ❏ PowerPoints Other _____

Craft / Level: _____ Copyright Date: _____

Module ID Number / Title: _____

Section Number(s): _____

Description: _____

_____

_____

_____

_____

Recommended Correction: _____

_____

_____

_____

_____

Your Name: _____

Address: _____

_____

Email: _____ Phone: _____

03310-13

Troubleshooting Oil Heating

OVERVIEW

In northern sections of the United States, many homes are heated with oil furnaces. There was a time when oil was a very inexpensive way to heat, so oil furnaces became very popular. In other instances, natural gas just was not available to the home as it was being built, so oil was the best alternative. Although the end result is the same as that of a gas furnace, the method of producing heat is significantly different. One of the main differences is that the oil is stored on site and must be pumped to the point of combustion, whereas natural gas is supplied under pressure through pipelines. Although its use as a home heating fuel is diminishing as time goes on, it is still an important fuel that will likely be encountered in the field.

Module Six

Objectives

When you have completed this module, you will be able to do the following:

1. Identify the primary components of an oil-fired furnace and explain its operation.
   a. Describe a basic oil-fired heating system.
   b. Describe the primary components and operation of a pressure-type oil burner.
   c. Describe the safety controls used on oil furnaces.
   d. Describe the fuel supply system used with oil furnaces.
2. Describe how to perform periodic servicing of a typical oil-fired heating system.
   a. Describe the basic servicing procedures performed on an oil-fired system.
   b. Describe how to perform a combustion efficiency test.
3. Describe how to troubleshoot a typical oil-fired heating system.
   a. Describe troubleshooting procedures for typical oil furnace controls.
   b. Describe troubleshooting procedures for common oil heating problems.

Performance Tasks

Under the supervision of your instructor, you should be able to do the following:

1. Complete two of the following:
   - Remove and reinstall an oil pump in single-pipe and two-pipe systems.
   - Test a cad cell flame detector.
   - Conduct a complete combustion analysis (smoke test and draft included).
   - Remove and replace an oil burner nozzle and set the electrode gap.

Trade Terms

Annual Fuel Utilization Efficiency (AFUE)
Atomize
Cad cell
Coping
Downdraft
Flame-retention burner
Joist
Lockout

Over-firing
Primary control
Refractory
Smoke spot test
Soot
Temperature differential
Weatherized furnace

Industry-Recognized Credentials

If you are training through an NCCER-accredited sponsor, you may be eligible for credentials from NCCER's Registry. The ID number for this module is 03310-13. Note that this module may have been used in other NCCER curricula and may apply to other level completions. Contact NCCER's Registry at 888.622.3720 or go to **www.nccer.org** for more information.

Contents

Topics to be presented in this module include:

Figures and Tables

SECTION ONE

1.0.0 OIL-FIRED HEATING SYSTEMS

Objectives

Identify the primary components of an oil-fired furnace and explain its operation.

   a. Describe a basic oil-fired heating system.
   b. Describe the primary components and operation of a pressure-type oil burner.
   c. Describe the safety controls used on oil furnaces.
   d. Describe the fuel supply system used with oil furnaces.

Trade Terms

Annual Fuel Utilization Efficiency (AFUE): An industry standard used to define furnace efficiency.

Atomize: Reduce to a fine spray.

Cad cell: A light-sensitive device (photocell) in which the resistance reacts to changes in the amount of light. In an oil furnace, it acts as a flame detector.

Flame-retention burner: A high-efficiency oil burner.

Lockout: A safety feature that prevents a furnace from automatically cycling if the flame fails.

Primary control: A combination switching and safety control that turns the oil burner on and off in response to signals from the thermostat and safety controls.

Refractory: Materials such as ceramics designed to withstand extremely high temperatures.

Weatherized furnace: A furnace that has been well insulated and has a weather-resistant case as specified by government standards.

Oil furnaces are designed to burn fuel oil, which is mixed with air, then sprayed as a fine mist into the furnace's combustion chamber, where it is ignited by an electric spark. Heat produced in the combustion chamber heats up the furnace heat exchanger. Air returned from the conditioned space is directed over the heat exchanger by a blower in the furnace unit. Heat is then transferred to that air and recirculated by the duct system.

A typical oil furnace is shown in *Figure 1*. Oil furnaces do not achieve the efficiency levels of gas-fired furnaces. Current high-efficiency oil

furnaces achieve an Annual Fuel Utilization Efficiency (AFUE) of 86.5 percent in comparison with gas furnaces, which report AFUE ratings up to 98.2 percent. The US Department of Energy has established minimum AFUE ratings for oil furnaces of 83 percent for weatherized furnaces and 78 percent for non-weatherized furnaces. These ratings do apply to oil furnaces used in mobile homes, which have lower minimum efficiency ratings.

1.1.0 Oil-Fired Heating System Basics

In a typical oil-fired furnace, the oil burner receives oil from a storage tank on the premises (*Figure 2*). The oil burner shoots a spray of oil into the combustion chamber, where it is mixed with combustion air. Electrodes located at the point where the oil spray enters the combustion chamber provide a high-voltage spark that ignites the oil.

The purpose of the combustion chamber is to protect the heat exchanger from flame damage and to provide reflected heat to the burning oil. The reflected heat warms the tips of the flame, enhancing combustion. No part of the flame should touch the combustion chamber surface. If it does, incomplete combustion could result. The chamber must fit the flame and the nozzle must be located at the correct height above the chamber floor.

Three types of materials are commonly used for combustion chamber construction: metal, typically stainless steel; insulating fire brick; and molded ceramic. In small furnaces such as those used in residential applications, the combustion

03310-13_F01.EPS

Figure 1 Oil furnace.

 03310-13 Troubleshooting Oil Heating   Module Six 1

Figure 2 An oil-fired heating system.

chamber is built into the furnace. For large commercial and industrial applications, the combustion chamber may be constructed on site. Such chambers can be made of fire brick or pre-cast ceramic (**refractory**) material. If pre-cast materials do not fit the job, a combustion chamber may be formed and poured on site using ceramic materials designed for this purpose.

A moldable refractory material, called by the generic name of wet-pack, is used to build or re-line combustion chambers. This material is moist when it is purchased, and is shipped in sealed plastic bags. Because it is moist, it can be easily molded into the desired shape. Continued exposure to air causes the wet-pack to dry into a very hard, rigid structure. The material has very poor heat transfer characteristics, making it a good insulator and chamber construction material.

The heat exchanger of an oil-fired furnace has primary and secondary heating surfaces. The primary heating surface surrounds the flame, whereas the secondary heating surface is often a series of steel sections over which the hot combustion gases pass before leaving the heat exchanger on their way to the chimney flue.

The main elements of an oil-fired system are the following:

- Oil burner assembly
- Controls
- Fuel supply system
- Flue/chimney exhaust system

1.2.0 Pressure-Type Oil Burners

There are two categories of pressure-type oil burners: the low-pressure gun-type and the high-pressure gun-type. In a low-pressure gun-type burner, which may be found in older furnaces, oil and primary air are mixed before going through a nozzle. At a pressure of 1 to 15 psi, the mixture is forced through the nozzle where it **atomizes** as it exits. Secondary combustion air is drawn into the spray mixture after it is released from the nozzle. An electric spark is used to light the combustible mixture.

A high-pressure gun-type burner forces oil through the nozzle under a pressure of 100 to 150 psi. Some burners may even be designed for operation at pressures up to 200 psi. This breaks the oil into much finer, mist-like droplets. The atomized oil spray leaving the nozzle creates a low-pressure area into which the secondary combustion air is drawn in and mixes with the oil. Combustion air is supplied by a vane fan, which is part of the burner assembly. The high-pressure, gun-type burner is the most popular domestic burner. It is simple in construction and efficient in operation. The parts are mass-produced, readily available, and relatively low in cost.

Oil furnaces in current production provide an atomized oil vapor mixed with the correct proportion of air to the combustion chamber. An ignition transformer provides the spark at a set of electrodes to ignite the fuel-air mixture in the combustion chamber. The primary control locks the system out on a flame failure. The lockout must be manually reset. Cad cells are commonly used to sense the burner light in order to verify the presence of a flame for continued operation.

A pressure-type oil burner assembly (*Figure 3*) consists of a pump, an air tube and nozzle assembly, and a fan. Oil is pumped to the nozzle, which converts it to a fine mist. The oil mist is mixed with air in the air tube (also known as a blast tube) and ignited by an electric spark. The burner cycles on or off in accordance with the demands of the room thermostat. The burner mechanism is outside the furnace or boiler, making it easily accessible for servicing. Grade 2 fuel oil is commonly used with pressure-type oil burners.

1.2.1 Power Assembly

The power assembly of the high-pressure burner unit consists of the motor, fan, and oil pump (*Figure 4*). The motor drives both the fan and oil pump through a single shaft. The fan, mounted on the motor shaft, forces air through a blast tube to provide combustion air for the atomized oil. The inlet to the fan has an adjustable opening so that the air volume can be controlled manually. The oil pump, coupled to the motor shaft, draws oil from the storage tank and delivers it to the nozzle. The fuel/air mixture is ignited by an elec-

(A) PRESSURE OIL BURNER COMPONENTS

(B) OIL BURNER ASSEMBLY

03310-13_F03.EPS

Figure 3 Pressure oil burner components.

tric spark that is formed between two electrodes in the nozzle assembly.

A pressure-regulating valve on the oil pump has an adjustable spring that permits adjustment of the oil pressure. The pump delivers more oil than the system can use. The excess oil is re-

Heating With Biodiesel

Biodiesel is a fuel made from new or recovered vegetable oil or animal fat. For several years, people with diesel-powered cars have been using filtered oil from restaurant deep fryers to power their cars. Experiments with blends of heating oil and biodiesel in oil furnaces have been successful and people using a blend containing 20 percent biodiesel have experienced no problems. In addition, no modifications of the furnaces have been necessary. Further experiments by the US Department of Energy have shown that a 70-30 blend can be used with no change in heating performance. At this writing, if a heating fuel blend contains less than 10 percent of biodiesel, it need not be labeled as being a blend. One of the main benefits of using such a blend is a reduction of carbon emissions. In addition, biodiesel, unlike fossil fuel, is a readily renewable resource.

Figure 4 Power assembly.

turned to the tank or bypassed back to the sup-ply line internally. The choice depends on the oil line configuration. A two-pipe system allows for the excess oil to return to the tank, while a single-pipe system does not.

On some burners, a solenoid-controlled oil flow valve stops the flow of oil as soon as power is re-moved, thereby preventing oil from dripping into the combustion chamber. Oil pumps are designed for single-stage or two-stage operation. The sin-gle-stage pump is used where the oil supply is above the level of the burner and the oil is mainly fed to the pump by gravity. The two-stage unit is used where the storage tank is below the burner and the oil must be lifted. The first stage draws the oil to the pump and the second stage provides the pressure required by the nozzle. Pump suc-tion should not exceed 15 inches of vacuum.

1.2.2 Nozzle Assembly

The nozzle assembly (*Figure 5*) consists of the oil feed line, the nozzle, ignition electrodes, and the transformer connections. The nozzle prepares the oil for mixing with the air by atomizing the fuel. Oil passes through a strainer and enters slots

Figure 6 Nozzle patterns.

that direct the oil to the swirl chamber. The swirl chamber gives the oil a rotary motion when it en-ters the nozzle orifice, shaping the spray pattern. The nozzle orifice increases the velocity of the oil. The oil leaves the nozzle in the form of a mist or spray and mixes with the air from the blast tube.

Figure 5 Nozzle assembly.

Nozzles are selected by the manufacturer to provide the correct oil volume and spray pattern for a given oil burner. If it is necessary to replace a nozzle, the replacement should be the same size and type as the original.

Due to the fine tolerances of the nozzle construction, dirty or defective nozzles are usually replaced. Nozzle spray patterns vary with the type of application. The shape of the spray and the angle between the sides of the spray can be varied. There are three basic spray patterns, as shown in *Figure 6*: hollow (H), semi-hollow (SH), and solid (S). The hollow and semi-hollow are most popular on domestic burners as they are more efficient when used with modern combustion chambers.

The angle of the spray must correspond to the type of combustion chamber (*Figure 7*). An angle of 70 to 90 degrees is usually best for square or round chambers; an angle of 30 to 60 degrees is best suited to long, narrow chambers. Again, it is important to use the spray pattern and nozzle angle recommended by the manufacturer for proper performance. Although another spray pattern or nozzle angle may prove functional, it will not provide the best performance and could result in irreparable damage to the furnace.

Oil burner nozzles are sized by the gallons per hour of oil they will pass at the design oil pressure. Information regarding the size and type of an oil burner nozzle is stamped on the nozzle body. Different nozzle manufacturers may use different letters or combinations of letters to identify the spray pattern. When replacing a nozzle, consult the manufacturers catalog information if unfamiliar with the markings.

1.2.3 Ignition System

The ignition system for high-pressure burners consists of a step-up transformer (*Figure 8*) connected to two electrodes. The transformer supplies high voltage that causes a spark to jump between the two electrodes. The force of the air in the blast tube causes the arc to bend into the fuel-air mixture, igniting it.

Ceramic insulators surround the electrodes (*Figure 9*) and serve to position them. The step-up

70° TO 90°

70° TO 90° FOR ROUND OR
SQUARE CHAMBERS

30° TO 60°

30° TO 60° FOR LONG
NARROW CHAMBERS

03310-13_F07.EPS

Figure 7 Spray angles.

03310-13_F08.EPS

Figure 8 Ignition transformer.

Oil Pressure

For many years, it was commonly accepted that the oil-burner pump outlet pressure should be 100 psi. Today, some manufacturers specify higher pressures, and 150 psi has become common. Always check the manufacturer's literature before checking or adjusting the pump outlet pressure.

transformer increases the voltage from 120V to about 10,000V, but reduces the amperage to about 20 milliamps. The ceramic insulators prevent arcing from the electrode to other parts of the burner, since a voltage this high can jump across a significant gap to a grounded surface. The low amperage reduces wear on the electrode tips.

The correct position of the electrodes (*Figure 10*) is important. If the electrodes are not centered on the nozzle orifice, the flame will be one-sided and will cause carbon to form on the nozzle.

The oil burner nozzle flow rate determines burner input. For example, one gallon of Grade 2 fuel oil has a heat content of approximately 140,000 Btus. If using a half-gallon nozzle, the burner input would be 70,000 Btus per hour (Btuh) because it delivers only half as much fuel as the one-gallon nozzle. The nozzle flow rate is stamped on the nozzle body. Note that the flow rate discussed here is related to furnace input and not the output. The output will be lower, determined by the combustion efficiency of the burner.

CERAMIC INSULATORS

03310-13_F09.EPS

Figure 9 Ceramic insulators.

1.2.4 *Flame-Retention Burners*

Flame-retention burners (*Figure 11*) are used in modern oil-burning appliances and are designed to produce significant improvements in overall heating efficiency, while simultaneously reducing the number of burner-related service calls. Flame-retention burners use motors that operate at 3,450 rpm instead of the conventional 1,725 rpm. This higher speed is needed to overcome the greater resistance to air flow that is created by the retention ring.

The advantages of a flame-retention burner include the following:

- Better air-oil mixing and flame retention within the air pattern.
- A higher flame temperature with less excess air.
- More usable energy produced for the amount of oil consumed.
- Less effect on flame from stack draft variations.
- Fewer products of incomplete combustion.
- During the off-cycle, the retention head reduces the flow of air through the burner, over the heat exchanger, and out the stack, thereby removing less heated air from the structure.

1.3.0 Oil Safety Controls

All oil-fired furnaces are equipped with a central electrical control known as the primary control (*Figure 12*), which is connected to the room thermostat. The primary control is usually mounted on the burner assembly. It provides both control and safety functions. As a control device, its job is to turn the burner on and off in response to commands from the thermostat. As a safety control, the primary shuts off power to the burner if the burner does not ignite or if the burner flame goes out. Its response is based on one or more simple

Nozzle Identification

The specific characteristics of oil burner nozzles are clearly identified. For example, a stamp that reads 0.75, 90H means that the nozzle has a flow rate of 0.75 gallons per hour at the design oil pressure (usually 100 psi), a spray angle of 90 degrees, and a hollow spray pattern.

0.75 GALLONS PER HOUR (GPH)

0.75

90 H

90° SPRAY ANGLE

SPRAY PATTERN H* (HOLLOW)

* SPRAY PATTERN CODE MAY VARY AMONG MANUFACTURERS

03310-13_SA01.EPS

ELECTRODE GAP

DISTANCE ELECTRODES
EXTEND BEYOND NOZZLE
FACE (DEPENDENT ON
NOZZLE ANGLE)

DISTANCE BETWEEN
ELECTRODES AND
NOZZLE CENTER

03310-13_F10.EPS

Figure 10 Electrode position.

03310-13_F12.EPS

Figure 12 Primary control for an oil-fired boiler.

03310-13_F11.EPS

Figure 11 Flame-retention oil burner.

(A) CAD CELL

IGNITION
ELECTRODE

IGNITION TRANSFORMER

AIR TUBE

CAD CELL

NOZZLE

LINE OF SIGHT
TO FLAME

(B) CAD CELL LOCATION

03310-13_F13.EPS

Figure 13 Cad cell.

inputs. The primary control may also control the circulating pump in simple hydronic heating systems.

In modern furnaces, a cadmium sulfide cell, or cad cell (*Figure 13*) is used to sense the presence of a flame. The cad cell performs the same basic function as a gas furnace flame detector, but in a very different way. It is photosensitive, and is installed in the burner assembly on a line of sight with the burner flame. The light of the flame causes the cad cell to generate a small amount of voltage, proving that a flame is established. If the cell does not sense a burner flame, the primary control will not sense any voltage and will turn off the burner.

Another safety device is the limit switch, which is actuated by a bimetal element located in the discharge air plenum of the furnace. It is in series with the power source to the primary control. The limit switch is usually calibrated to

open if the discharge air temperature reaches or exceeds 200°F. This limit switch will usually reset when the temperature drops 25°F below the cutout point. A secondary (auxiliary) limit switch is an additional safety control used on some down-flow furnaces to protect the conditioned air blower motor against excessively high temperatures. If a limit switch or other safety device trips more than once, the cause should be determined.

Older installations may have a stack switch (*Figure 14*) in place of a cad cell. The stack switch is installed in the flue vent (stack) ahead of the draft damper. It senses heat and shuts down the oil burner if heat is not detected in the stack. Most stack switches combined the function of a primary control and burner ignition proof into a single component. These are more accurately referred to as stack controls or stack relays.

1.4.0 Oil Supply System

Oil furnaces may be supplied from indoor or outdoor oil tanks. Outdoor tanks may be buried. Because many underground oil tanks have leaked, there are strict regulations governing placement, construction, and alarm systems for these tanks. Aboveground tanks over a certain size may require special containment arrangements to ensure the oil from leaks can be captured. No oil tank installation should be undertaken without a clear understanding of local and national codes.

1.4.1 Fuel Oil

Fuel oils are rated according to their Btu per gallon content and the American Petroleum Institute

Figure 14 Stack switch.

(API) gravity. The API gravity is an index related to the heating values for standard grades of fuel oil. There are six common grades of oil: Nos. 1, 2, 4, 5 (light), 5 (heavy), and 6. The lighter-weight oils have a higher API gravity.

- *Grade 1* – A light-grade distillate for use in vaporizing-type oil burners.
- *Grade 2* – A heavier distillate for use in domestic pressure-type oil burners.
- *Grade 4* – A light residue or heavy distillate used for higher-pressure commercial oil burners.
- *Grade 5 (light)* – A medium-weight, residual-type fuel used in commercial oil burners that are specifically designed for it.
- *Grade 5 (heavy)* – A residual-type fuel for commercial oil burners. Usually requires preheating.
- *Grade 6* – A very heavy residual fuel used for industrial burners. Preheating is necessary in the tank to permit pumping; additional preheating is needed at the burner to convert the fuel into a fine spray or atomize it. Grade 6 is also known as Bunker C.

Grade 1 fuel oil (kerosene) has a slightly lower heat content than Grade 2 fuel oil. In northern climates, Grade 1 may be used if the oil is to be stored in an outside tank. Grade 2 fuel oil has a tendency to thicken when exposed to extreme cold, which makes it hard to pump from the storage tank to the burner. Fuel oil must be properly atomized in order to burn.

1.4.2 Piping Systems

Piping connections to the oil pump are of two configurations: single-pipe and two-pipe. In the single-pipe system, there is only one pipe from the storage tank to the burner. In the two-pipe system, two pipes are routed from the tank to the burner. One carries supply oil; the other carries excess oil returning to the tank from the pump. Oil lines are typically connected using flare fittings; compression fittings should never be used on oil system piping.

Oil piping must be purged of air in order for a solid column of oil to enter the pump. While single-pipe systems generally need to be primed and the air bled from the lines, two-pipe systems prime easily and remain free of air.

Figures 15 through *18* are examples of oil furnace installations that have been used in the past. National, state, and local codes covering oil furnace installations, especially underground tanks, are under constant review and are subject to change. Before beginning any installation, consult the manufacturer's installation instructions, as well as applicable codes. If an underground tank is used, make sure it meets current construction standards. The US Environmental Protection Agency (EPA) has published special requirements for commercial installations with underground tanks that are different from those for residential installations.

Figure 15 shows a single-pipe system with the supply tank in the same space as the burner. The single-pipe system should only be used when the burner is located on the same level or above the bottom of the oil supply tank. Single-pipe systems often depend primarily on gravity to feed oil to the burner. The pump used on pressure-type burners does create some vacuum to assist oil movement, but not enough to lift the oil very far. In gravity-feed installations, a solenoid oil flow valve at the tank outlet may be appropriate to help prevent oil leakage from a damaged oil line.

The piping on this application must be leak-free. No air should be allowed to enter the oil piping and pump after the supply system has been purged. If air does enter through a leak, the system will likely need to be primed again before the burner will operate.

Figure 16 depicts a two-pipe system with the burner located below or on the same level as the supply tank. However, note that the routing of the oil piping fell below the level of the burner. As a result, a two-pipe system was used.

Note that the storage tank is buried outside the structure. An all-brass check valve should be installed in the supply piping as close to the tank as possible. It is also acceptable to install a globe valve in the supply piping so the oil can be shut off for burner service or removal.

Figure 17 shows the two-pipe system with the burner located above the tank when a lift of less than 15′ is required. A check valve and globe shutoff valve should also be included in this piping network.

Figure 18 shows a piping system required where a lift of more than 15′ is necessary and multiple units are tied together with a single oil system. A boost pump is often used in this application.

Figure 15 Single-pipe oil supply system.

AIR VENT (ABOVE SNOW LEVEL)

INSTALL IN ACCORDANCE WITH UNDERWRITERS AND LOCAL REGULATIONS

MINIMUM DISTANCE FROM ANY BUILDING OPENING

2'

5'

1¼" AIR VENT PIPE

OIL FILL CAP

2" FILL PIPE

SHUTOFF VALVE

CHECK VALVE

FILTER

RETURN LINE

SUPPLY LINE (SAME SIZE AS RETURN)

03310-13_F16.EPS

Figure 16 Two-pipe oil supply—burner below tank level.

The schematic in the illustration portrays a series-loop system where all the oil is pumped through individual smaller tanks on each unit in series and then returned directly to the main supply through a separate return line.

Some manufacturers recommend that the piping system use two pipe connections at the boost pump and at each oil rooftop unit. The supply piping from the main tank to the boost pump should include a fusible valve as well as oil filter(s) of adequate size.

To guarantee a trouble-free supply system for multiple units, the following factors should be considered:

- A rooftop system should be limited to a maximum of eight oil-fired units.
- The fittings and lengths of supply and return lines must be accurately calculated to guarantee proper delivery of fuel to all units.
- Supply piping should be run to the farthest burner first and then piped toward the boost

pump so that the last unit is the closest to the boost pump. The manufacturer's recommendations as to supply pump pressure must be followed to eliminate seal failure.

- The return piping should never include smaller pipe sizes than the supply piping. One size larger is sometimes recommended on a vertical return line to permit free flow of fuel back to the supply tank.
- A vacuum breaker should be installed at the highest point in the system. It should be located on the last unit in the supply run, closest to the return piping.
- All piping exposed to low ambient outside temperature must be insulated.
- Local and other applicable codes must be followed and will take precedence over any other design and construction considerations for the fuel oil supply system.

INSTALL IN ACCORDANCE
WITH UNDERWRITERS AND
LOCAL REGULATIONS

Figure 17 Two-pipe oil supply—burner above tank level.

The Bypass Plug

Most oil pumps are designed to operate on either single-pipe or two-pipe systems. A bypass plug is generally inserted into these pumps for two-pipe applications. For single-pipe applications, the bypass plug must be removed for the pump to function properly. Be sure to follow the manufacturer's instructions carefully for the type of piping system in use.

Moisture in a Fuel Storage Tank

How does water get into a fuel storage tank? Most of it gets in through condensation of water vapor on the interior walls of the tank. Since the tank is vented, moisture-laden air can move into the tank. This can happen with all fuel tanks, but it is a bigger problem in outdoor and buried fuel tanks. Keeping the fuel tank full throughout the year can help prevent air migration and condensation. Any accumulated moisture can be removed by inserting a rigid pipe or hose through the top of the tank until it touches the bottom. The water can then be pumped or siphoned out. Since it is heavier than oil, water settles at the bottom of the tank. Of course, this is also where the pick-up for the burner oil supply is located. As a result, a significant accumulation of water in the tank will be picked up by the pump first.

Figure 18 Oil piping system for rooftop units.

Additional Resources

Fundamentals of HVAC/R. Carter Stansfield and David Skaves. Upper Saddle River, NJ: Prentice Hall.

Refrigeration and Air Conditioning, an Introduction to HVAC/R. Larry Jeffus. Air Conditioning and Refrigeration Institute. Upper Saddle River, NJ: Prentice Hall.

1.0.0 Section Review

1. Oil furnaces have the same range of AFUE ratings as gas furnaces.

   a. True
   b. False

2. In an oil furnace, combustion air is supplied by _____.

   a. the vane fan
   b. gravity
   c. the furnace blower
   d. the oil pump

3. The limit switch in an oil furnace is usually found in the _____.

   a. combustion chamber
   b. heat exchanger
   c. discharge plenum
   d. oil burner assembly

4. A two-pipe oil supply system is used when the _____.

   a. oil tank is below the level of the furnace
   b. oil tank is above the level of the furnace
   c. oil tank is located on the roof
   d. oil furnace is located in the basement

SECTION TWO

2.0.0 SERVICING OIL HEATING SYSTEMS

Objectives

Describe how to perform periodic servicing of a typical oil-fired heating system.

   a. Describe the basic servicing procedures performed on an oil-fired system.

   b. Describe how to perform a combustion efficiency test.

Performance Task 1

Remove and reinstall an oil pump in single-pipe and two-pipe systems.

Conduct a complete combustion analysis (smoke test and draft included).

Remove and replace an oil burner nozzle and set the electrode gap.

Trade Terms

Over-firing: A situation caused by an oversized nozzle in which excess oil is pumped into the combustion chamber.

Smoke spot test: A test to determine the level of smoke in the flue gases.

Soot: Fine carbon particles resulting from incomplete combustion. An accumulation of soot on heat exchangers will reduce combustion efficiency and result in increased carbon monoxide production.

Temperature differential: The difference in temperature between the flue gases and the gases above the combustion chamber.

Like the gas furnace, the oil furnace should be checked at the beginning of each heating season. Air filters must be cleaned or replaced, the blower and heat exchangers should be cleaned, and the oil filter should be replaced. Oil burner operation should also be checked. If there is not enough combustion air, the flame may be orange or red instead of yellow. The presence of smoke, soot, or odors could indicate improper oil pressure, poor draft, or an improper mix of oil and air. A pulsating sound could indicate that the flame is touching the combustion chamber. In any of these cases, some cleaning, adjustment, or repair will be needed.

Oil-fired burners typically require more servicing than gas-fired systems because of the continuing need to maintain proper burning efficiency. Inefficient burning will cause soot buildup, which will result in a further loss of efficiency.

The following items are commonly used when servicing oil furnaces:

- AC ammeter
- Bucket
- Carbon dioxide tester (if not part of the combustion test equipment)
- Oil burner combustion test equipment
- Draft gauge, smoke spot tester, CO_2 tester
- Electrode gauge
- Flashlight
- Hand mirror
- Insulated jumper wire
- Multimeter
- Oil pressure gauge kit
- Nozzle wrench (A nozzle tool should always be used to remove the nozzle; pliers and wrenches may cause damage.)
- Stiff brush
- Temperature probes
- Utility knife
- Vacuum cleaner

2.1.0 Servicing Procedures for Oil-Fired Systems

In addition to the specific safety precautions provided in the product literature, the following general safety precautions should be followed when servicing an oil-fired system:

- Unless it is necessary to have the power on for a particular procedure, always turn off and lock out the power when working on the equipment. If it is necessary to check voltage or current, be sure to keep one hand outside the unit and carefully avoid touching bare wires and connections.
- Do not use compression fittings on oil piping. Use flare fittings to prevent air and oil leaks in the fuel lines.
- Wear protective respiratory equipment when cleaning the soot out of a furnace.
- Do not try to clean the oil burner nozzle. It should be replaced.

Before performing any service or maintenance procedure, check with the customer and listen to any complaints. Make note of any comment that would indicate a possible source of trouble or equipment malfunction. Then proceed as follows:

- Check the thermostat for proper operation
- With the furnace disconnect turned off, check and clean the blower compartment and blower motor.
- Check the electrical wiring for loose connections, insulation breakdown, and corrosion or dirt on the terminals.

- Inspect the combustion chamber using a hand mirror and flashlight. Look for carbon buildup, holes in the combustion chamber, and erosion of the oil burner head. When the combustion chamber is lined with refractory to isolate the chamber walls from intense heat, look for holes or other damage.
- Check for soot in the heat exchanger:

    Remove the flue or vent pipe.

    Use a flashlight or trouble light to check inside the heat exchanger flue for carbon buildup. It is not necessary to clean the heat exchanger unless there seems to be an excessive amount of carbon or scale on the heat exchanger surfaces.

    If the heat exchanger needs cleaning, refer to the manufacturer's service manual for the proper cleaning procedure.
- Check the oil piping:

    Check all oil supply lines between the supply tank and burner for dents, rust, kinks, loose connections, cracks, or holes. Wetness around fittings, or oil spots on the equipment or the floor, may indicate a leak.

    Check for proper support of oil piping. Check the routing for areas where potential damage could occur.

Soot Removal

There are vacuum cleaners designed specifically for soot removal. While they look something like a shop vacuum, there are significant differences. Soot vacuums have a metal canister that provides a tight seal to prevent soot from escaping. They also use a special fine filter to capture the soot. Soot vacuums are also, in general, more powerful than shop vacuums.

There are several chemical products available that assist in cleaning the soot-coated surfaces of an oil furnace. Always follow both the furnace and chemical manufacturers' instructions and cautions when using soot-removal products.

03310-13_SA02.EPS

2.1.1 Changing the Oil Filter

The oil filter (*Figure 19*) should be changed during a preventive maintenance visit. The following is a basic procedure for changing the filter:

- Remove the ring gasket at the top of the filter.
- Check the filter bowl for the name and type of filter element.
- Place a container below the filter to catch any spilled fuel oil.
- Close the fuel line valve and unscrew the bowl mounting screw or attaching hardware. Gently pry the bowl loose from the casting, remove the filter, and place it in the waste bowl container.
- Clean the inside of the filter bowl with filter clean, soft cloth.
- Remove the ring gasket at the top of the filter bowl. It will probably stick to the casting. Scrape away any filter gasket particles with a wide-blade putty knife, and then clean the casting and the top of the bowl with a clean cloth.
- Replace the filter with one of proper size and construction and fill the cartridge with fresh oil.
- Install a new ring gasket, replace the bowl, and firmly tighten the mounting hardware. Be careful not to over-tighten it.
- Open the fuel supply valve and bleed air from the filter bowl through the bleed port at the filter as it fills with oil, if the oil tank is above the level of the burner. This can help speed the priming and air removal process. Then connect a plastic hose to the bleed port of the oil pump and operate the burner. Route the end of the hose into a suitable container to capture the oil. The bleed port is cracked open (roughly ¼ turn) with a small wrench. When oil starts to discharge from the port, close the bleed port.

03310-13_F19.EPS

Figure 19 An oil filter element.

- Open the oil supply control valve and then check for leaks.

2.1.2 Check the Safety Controls

Check the cad cell flame detector. While the burner is operating, disconnect the cad cell leads at the primary control. The burner should stop in less than 90 seconds. If it does not, the primary control is likely defective. The cad cell should be cleaned during the maintenance procedure. If the burner locks out when combustion appears to be good, the cad cell needs to be tested. Cad cell testing is covered in the next section related to troubleshooting.

Check the safety limit control. Prevent the blower from operating by removing the belt or disconnecting the common wire terminal from the blower relay. Initiate a call for heat at the thermostat and place a suitable thermometer in the heat exchanger plenum. The safety limit control should shut down the furnace at 200°F or after 2½ to 3 minutes of operation. If the furnace continues to operate after reaching the high-limit temperature, shut it off immediately and replace the safety limit control.

2.1.3 Check the Temperature Rise

Temperature rise is the difference in temperature between the air returned from the conditioned space and the air leaving the furnace. It reflects the amount of heat transferred to the air as it passes over the heat exchanger. The temperature range limits are specified by the manufacturer and it found on the furnace label. If the temperature rise is out of tolerance, it can often be corrected by adjusting the furnace blower speed. Note however, that this is also a symptom of a furnace being over- or under-fired by an incorrect nozzle size.

- Initiate a call for heat at the thermostat. Make sure all the furnace access doors are closed and the filters are replaced or clean.
- Drill holes in the supply and return ductwork (*Figure 20*) and insert a temperature probe into each hole. The thermometer in the supply duct must be located out of the line of sight of the furnace heat exchanger to prevent radiant heat from affecting the thermometer.
- If the temperature differential is below the specified range, reduce the blower speed. If it is above the range, make sure that there are no airflow restrictions and that the burner is not over-firing, then increase the blower speed.

Figure 20 Hole locations for temperature-rise test.

2.1.4 Servicing the Oil Burner Assembly

The oil burner assembly (*Figure 21*) is the primary active component in an oil-fired system. It must be carefully inspected and serviced. The following procedure is a guide to oil burner servicing:

- Turn the power off.
- Disconnect all wiring and piping to the gun and vane fan.
- Remove the assembly, being careful not to damage any of the components of the burner

or ignition system and controls. Do not disturb the slip ring for the air gate or lose track of the nozzle adjustment or the insertion dimensions.

- Clean the vane fan wheel with a stiff bristle brush.
- Clean the burner tube with a soft cloth or disposable towels.
- Remove, clean, and inspect the electrodes. Use fine steel wool for cleaning. Do not use sandpaper.
- Remove and replace the oil burner nozzle. Nozzles are inexpensive and should always be replaced as part of any planned maintenance procedure. Make sure the nozzle installed is the correct one for the equipment by checking the furnace or boiler data plate. Do not refer to the data plate on the burner assembly alone, as the burner will have a wide range of nozzle sizes that it can accommodate. The furnace or boiler is the controlling factor for the nozzle size. A nozzle with too large an orifice will result in over-firing, which, in turn, can cause overheating. A nozzle with an orifice that is too small will result in under-firing. This can cause a reduction of flue gas temperature and may result in condensation forming in the flue.
- Replace the nozzle using the nozzle wrench designed for that purpose (*Figure 22*). Do not overtighten it. The nozzle wrench consists of two sockets, with the outer socket larger than the second socket on the inside. The wrench is slipped over the nozzle, where the outer socket seats on the hexagonal section of the oil tube. The inner socket is then pushed over the nozzle, and the nozzle is loosened with minimal effort by rotating the handle attached to the inner socket.
- Reinstall the electrodes and check the position of the electrode tips. Electrode insulators must be clean and free of cracks and carbon buildup. Clean or replace them as necessary. Adjust the gap and spacing from the nozzle (*Figure 23*) as specified by the oil burner manufacturer.

Accurately setting the position of the electrode tips can be very challenging. The angle of the nozzle spray pattern determines how far forward the electrodes are positioned, and this precise measurement must be known to successfully service the burner. Nozzles with narrow angles require that the electrode tips be positioned further forward to ensure that the spark is at the edge of the oil spray. Wider angles require the electrode tips to be moved further back. The distance between the two tips, and the vertical distance of the tips above the nozzle, are also crucial to proper ignition.

Figure 21 Oil burner assembly.

03310-13_F22.EPS

Figure 22 Nozzle wrench.

The simple device pictured in *Figure 24* eliminates a lot of guesswork and simplifies the process. The electrode-setting tool is slipped over the nozzle tip. On the tip of the tool, the various possible spray angles for nozzles are molded into

the plastic. Simply select the positioning guide for the angle of the nozzle being adjusted, and then lay the electrodes in the appropriate guide. This will properly position the electrodes in all three dimensions. Once the electrode mount clamp is tightened, the tool is slipped off the end of the nozzle. No additional measuring instruments or nozzle angle charts are required. For technicians servicing oil burners with any regularity, this is an indispensable tool.

- Inspect the cad cell flame detector and clean it if necessary. A photocell flame detector must see light to determine that an oil burner is operating properly. If the face of the cad cell becomes coated with soot or a film of oily dirt, the light is obscured and the cad cell may shut down the burner. During every routine service of an oil burner, clean the face of the cad cell with a soft cloth or swab to remove any soot or oil film. Check for proper contact with the mounting bracket and reinstall the unit.
- Replace the gun assembly and reassemble per the heating plant manufacturer's instructions.

"A" = Usable Air Tube Length
"L" = "A" + $^5/_8$"
"R" = "A" + $2^7/_8$"
"S" = $2^5/_{16}$"
"Q" = "A" + $3^1/_4$"
"Z" = $1^1/_8$"

ELECTRODE

5/32" GAP

7/16" ABOVE

NOZZLE

SEE BELOW

ELECTRODE ADJUSTMENTS

NOTE: ELECTRODE POSITION AHEAD OF NOZZLE

MODELS	DIMENSIONS
ALL	0 TO $1^1/_{16}$"

A. BURNER HEAD, SPECIFY TYPE F
B. AIR TUBE
C. ELECTRODE CLAMP, STATIC PLATE, AND NOZZLE LINE SUPPORT ASSEMBLY
D. CENTERING SPIDER
E. PORCELAIN INSULATOR
F. ELECTRODE ROD EXTENSION ADAPTER, AS REQUIRED

G. ELECTRODE ROD EXTENSION, AS REQUIRED
H. NOZZLE LINE AND VENT PLUG
I. ELECTRODE ROD AND TIP
J. NOZZLE
K. NOZZLE ADAPTER – SINGLE
L. STATIC PLATE
M. LOCKNUT BULKHEAD FITTING
N. BULKHEAD FITTING

03310-13_F23.EPS

Figure 23 Electrode adjustment.

Figure 24 Electrode setting tool.

Check for air (blast) tube length (*Figure 25*) and reinstall.

• Check the burner motor coupling; replace or service it as necessary.

• Remove and clean the oil pump screen.

• Perform a combustion efficiency check.

2.2.0 Measuring Combustion Efficiency and Flue Gases

Combustion efficiency is a measure of how well an appliance is burning its fuel. The ideal, of course, is 100 percent efficiency, which means all the energy available in the fuel is being used to produce usable heat. Combustion efficiency can therefore be translated into operating cost. That is, the higher the percentage of usable energy extracted from the fuel, the lower the cost of operating the system.

The highest efficiency gas furnaces, with AFUEs exceeding 95 percent, approach this level of efficiency, but oil furnaces do not. The highest AFUE oil furnace at the time of this writing has an AFUE of 86.5 percent, which means that 13.5 percent of the available heat is being wasted in the flue gases.

THE AIR TUBE LENGTH (DIMENSION A) IS THE DISTANCE FROM THE FRONT OF THE BURNER HOUSING TO THE DRAIN HOLE IN THE BURNER HEAD

NOTE: ADJUSTABLE FLANGE WIDTH - 1⅛"

Figure 25 Typical air tube length.

Combustion efficiency is determined by measuring the temperature of the flue gases and the amount of oxygen and CO_2 in those gases. The lower the flue gas temperature, the higher the combustion efficiency, as it means there is less heat escaping up the chimney. In perfect combustion, all oxygen would be converted to CO_2, so the ratio of oxygen to CO_2 is a measure of combustion efficiency.

Draft is also an important factor in combustion efficiency. Insufficient draft can cause dangerous carbon monoxide (CO) to back up into occupied spaces and can also cause overheating of the combustion chamber. Too much draft will result in lower combustion efficiency.

There are other important factors to consider, however. The most efficient burning produces the least smoke. The smoke produced by burning the fuel contains soot, which will be deposited on the surfaces of the heat exchanger, combustion chamber, and flue gas exhaust system. Heavy soot deposits on the heat exchanger will cause a reduction in heat transfer efficiency and will result in increased operating cost.

The electronic gas/oil combustion analyzer shown in *Figure 26* is a key item of test equipment. It directly measures and displays O_2, CO, stack temperature, draft, differential pressure, combustion air temperature and other optional selected gas emissions. Some testers have optional print-

Disposing of Waste Oil

Waste oil cannot simply be thrown away or poured on the ground because it contaminates the environment. Dispose of waste oil at an appropriate recycling facility or service station.

Module Six   19

Figure 26 Electronic gas/oil combustion analyzer.

03310-13_F26.EPS

03310-13_F27.EPS

Figure 27 Smoke spot tester kit.

ers that can be connected to the device with a USB cable or by wireless connection.

The instrument simultaneously calculates and displays combustion efficiency, excess air (EA), CO_2, NO_X, and O_2 reference values. The analyzer performs combustion calculations for ten fuels, including natural gas and propane, as well as #2, #4, and #6 fuel oil. The only combustion parameter it does not measure is smoke level. Although there are mechanical instruments available to perform the various tests that are required, the electronic device has a number of advantages:

- It consolidates the various tests performed with a number of different mechanical devices into a single device.

- Electronic instruments allow adjustments to be performed in real time. That is, the adjustments can be made while observing the effect on the instrument readings.
- Electronic instruments are easier to use, more accurate, and reduce the chance of human error.
- Tests can be performed more quickly with electronic instruments.

The combustion analyzer shown in *Figure 26* is equipped with a probe and hose assembly. The probe is inserted into an opening in the flue pipe. The hoses in the hose assembly are terminated in sensors that are plugged into connections on the bottom of the tester.

GOING GREEN

Multi-Fuel System

The system shown here is designed burn coal, wood, or oil. In order to conserve cost in fluctuating oil markets, this furnace can burn wood or coal at any time. At times when there is no one around to stoke the furnace, oil can be used.

03310-13_SA03.EPS

2.2.1 Smoke Spot Testing

A smoke spot test should be performed and the air shutter adjusted for proper smoke level before combustion testing is performed. The smoke spot test is performed using a smoke spot tester and comparison chart (*Figure 27*).

- Drill a test hole in the flue vent downstream of the draft diverter and close to the breeching, as shown in *Figure 28*. This opening will also be used for combustion efficiency testing.
- Operate the furnace for 10 to 15 minutes before performing the smoke spot test.
- Before starting the test, insert test paper into the tester in accordance with the manufacturer's instructions.
- Insert the hose of the smoke spot tester into the hole in the vent and draw 10 samples into the tester at 1- to 2-second intervals. The flue gases are drawn in by pulling the handle of the tester back. The gases pass through the filter paper inserted into the chamber, leaving a black circle on the paper. The darkness of the circle indicates the level of smoke in the gases.
- The initial test should show some smoke staining on the paper. If it does not, adjust the air shutter to lower the air until some smoke appears. Then, re-adjust the air shutter for the proper smoke level. For a flame retention head burner, the air shutter should be adjusted for zero smoke. A No. 1 smoke is standard for conventional burners.
- Compare the results to the sample chart provided by the tester manufacturer. For residential equipment that burns No. 2 oil or kerosene, the smoke reading should be between zero and a trace. In a heavy boiler where regular tube cleaning is performed, a smoke reading of two is often acceptable. If it is higher, the air setting at the burner may have to be readjusted.

2.2.2 Combustion Efficiency and Flue Gas Testing

Before starting the test, make sure the inspection door and the blower compartment door are securely closed. The furnace should be operated for at least 10 minutes before testing.

- Set up the combustion analyzer in accordance with the manufacturer's instructions. The hose assembly on the test set typically contains a water trap and filter assembly. Before using the device, make sure the trap is dry and empty, and replace the filter. The test instrument also contains oxygen, CO, and CO_2 sensors, which require periodic replacement. Follow the man-

HORIZONTAL FLUE CONNECTION

VERTICAL FLUE CONNECTION

03310-13_F28.EPS

Figure 28 Test hole locations.

ufacturer's instructions. It should be started up and allowed to stabilize before inserting the probe into the flue.
- Insert the probe so that the tip is halfway into the vent pipe and allow time for the readings to stabilize.
- Record or print the test results and compare them to the manufacturer's requirements. The values will vary, depending on the type of burner and the AFUE rating of the furnace, so it is important to make this comparison. Leave the probe in the flue pipe. Typical values for a flame retention burner are as follows:
  - Oxygen: 3 to 5 percent
  - CO_2: 10 to 12.5 percent
  - CO: <50 ppm (diluted)
  - Stack temperature: 400°F (60-79 AFUE); 450°F (80-plus AFUE)
  - Overfire draft: 0.02 in wc
  - Stack draft: 0.02 to 0.04 in wc
- If the combustion efficiency is not within the required tolerances, adjust the air shutter on the burner to obtain the correct value. If this value cannot be achieved by air adjustment, troubleshooting will have to be performed to isolate the cause.

If the combustion performance specifications cannot be met, check for air leaks into the combustion chamber and repair the leaks with furnace cement or high-temperature sealant.

To check for dilution by leakage into the combustion chamber, measure the carbon dioxide content at as high a point as possible over the fire. Insert a stainless steel tube through the fire door sample hole and compare the reading with the CO_2 measured in the vent pipe. A differential of more than one percent carbon dioxide between the vent pipe and the over-fire readings will usually indicate air entry into the combustion chamber.

If the required carbon dioxide level cannot be reached without exceeding No. 1 smoke, poor mixing of air and fuel is likely. This could be caused by an improper match between the air pattern and nozzle spray pattern. This problem can frequently be corrected by replacing the nozzle with one having a different spray angle and pattern.

If the proper combustion efficiency cannot be achieved, check for an unbalanced flame. If the flame appears unbalanced, the fuel input may be too low. Check the nozzle size and oil pressure. If the nozzle is partially plugged or defective, replace it with another as recommended by the manufacturer or choose one from an interchange chart.

- Take a negative draft reading through the inspection door (over the flame). A difference between this reading and the reading from the flue of more than 0.002 in wc indicates a dirty or restricted heat exchanger.
- Place the system into operation and observe it through at least one complete heating cycle to make certain all controls are functioning properly.
- With the heating plant in operation for five or more minutes, conduct a CO check in the plenum using an appropriate CO detection device and probe. CO levels of less than 50 parts per million are acceptable.
- Clean up the work area, seal test holes that were drilled, and complete the service call.
- Leave a copy of a complete service and maintenance checklist with the owner.

Additional Resources

Fundamentals of HVAC/R. Carter Stansfield and David Skaves. Upper Saddle River, NJ: Prentice Hall.

Refrigeration and Air Conditioning, an Introduction to HVAC/R. Larry Jeffus. Air Conditioning and Refrigeration Institute. Upper Saddle River, NJ: Prentice Hall.

HVAC Maintenance Procedure. Syracuse, NY: Carrier Corporation.

http://www.bacharach-inc.com/PDF/Brochures/4097percent20Combustion.pdf; Instructions for performing and evaluating a combustion efficiency test using an electronic test instrument.

2.0.0 Section Review

1. Which of the following conditions should occur if the cad cell leads are disconnected while the furnace is running?

   a. The temperature limit switch will open.
   b. The burner will shut off.
   c. The furnace will continue running.
   d. The blower fan will turn off.

2. A sample of flue gases for smoke analysis is drawn into the tester by pumping the handle _____.

   a. once
   b. twice
   c. five times
   d. ten times

SECTION THREE

3.0.0 OIL FURNACE TROUBLESHOOTING

Objectives

Describe how to troubleshoot a typical oil-fired heating system.

    a. Describe troubleshooting procedures for typical oil furnace controls.

    b. Describe troubleshooting procedures for common oil heating problems.

Performance Task 1

Test a cad cell flame detector.

Trade Terms

Coping: The upper masonry course of a chimney designed to keep water out.

Downdraft: A situation in which outdoor air is forced down a chimney, preventing proper draft. Downdraft can be caused by not extending the top of the chimney above the roof line.

Joist: Framing member used to support floors and ceilings.

Figure 29 Oil burner control system.

When starting out on a trouble call, it is always a good idea to listen carefully as the customers describe the problem from their perspective. They may not be technically knowledgeable, but what they have experienced can give you a clue to what might be wrong with the oil heating system.

3.1.0 Control Circuit Troubleshooting

The ability to effectively troubleshoot oil furnace problems demands the ability to understand the operating sequence and safety controls for the furnace. This, in turn, requires the ability to interpret the schematic and/or wiring diagrams provided by the manufacturer. The control systems for oil furnaces are generally not as complex as those for gas-fired systems. *Figure 29* provides an overview of an oil furnace control system.

This section begins by providing examples of schematic and wiring diagrams from two furnaces, along with descriptions of their operating sequences.

Figure 30 is the schematic diagram for an oil furnace and add-on air conditioner with a cad cell flame detector. The operating sequence is included on the diagram. *Figure 31* is the wiring diagram for the same furnace. In a troubleshooting situation, the wiring diagram will help determine where to connect the test meter. Although these diagrams represent only one of the many oil-fired systems a technician may encounter, studying them will help gain more insight into the operation of similar systems.

3.1.1 Primary Control Troubleshooting

The primary control shown in *Figure 32* is a combination control and safety device. It contains an electronic switching circuit that turns on the system by transferring line voltage to the burner motor and the ignition transformer when a call for heat is received from the thermostat. Although the low voltage control wiring terminals are visible on top, the line voltage connections enter through the bottom of the control. The primary control is usually mounted on a junction box or similar enclosure that is part of the burner assembly. The safety feature is controlled by the combustion detector, which in this case is a cad

 03310-13 Troubleshooting Oil Heating

Module Six 23

1. Line potential feeds through the secondary limit and limit controls to power primary control.

2. The primary control provides 24-volt control circuit.

3. On a heating demand, the thermostat heating bulb makes.

4. The fan control heater is energized through primary control. After a short period, the heater provides sufficient heat to close the fan contacts. This energizes the blower motor.

5. The primary control simultaneously energizes the burner motor and ignition transformer at the oil burner. The burner motor operates the oil pump and combustion blower to feed air and oil vapor into the combustion chamber. The fuel mixture should ignite with the spark furnished by ignition transformer.

6. If combustion does not take place within the time specified on the primary control, as detected by cad cell, the primary control locks itself out.

7. Should a flame failure occur during the "on" cycle, the primary control locks itself out in response to the cad cell.

8. The secondary limit opens at temperatures above setpoint. This de-energizes the primary control but still allows blower motor operation through the NO contacts.

9. As the heating demand is satisfied, the thermostat heating bulb breaks. This de-energizes the oil burner circuits.

10. The blower motor continues running until furnace temperature drops below fan control setpoint.

03310-13_F30.EPS

Figure 30 Operating sequence number one.

cell flame detector. The flame detector must be matched to the primary control. However, there are very few cad cells to choose from on the market. If the primary control does not receive a continuous signal that combustion is taking place, its relay will be de-energized, turning off the burner motor. The primary control allows a set time to elapse before it shuts off and locks out the burner. The timing of a primary control is not adjustable. Different primary controls have different lockout times. The burner manufacturer selects a primary control that suits the design of the burner. If a flame is established within the prescribed time, burner operation will continue.

The following are operating features of this control circuit:

- The room thermostat senses changes in the space temperature and then signals the primary control to energize or de-energize the burner.

- The primary control acts as the nerve center. It receives signals from the thermostat and flame detector and makes or breaks power to the oil burner.

- The cad cell flame detector has a similar function to that of a thermocouple or flame rod in a gas furnace. In older furnaces, a bimetal detector in the stack was used for this purpose. Either type may be encountered when working on residential and light commercial systems.

- The line voltage limit control de-energizes the burner if overheating occurs.

Figure 31 Field wiring diagram number one.

- Auxiliary controls may include devices to control the blower or fan, low-limit controls, zone valve controls, air cleaner controls, humidifier controls, time delay controls, and motor control relays.

Checking the primary control and cad cell must be done with live circuits; therefore, the troubleshooter must observe all the precautions necessary to avoid the danger of electrical shock and equipment damage.

To troubleshoot a primary control if the burner starts and locks out, proceed as follows:

Step 1 Make sure the limit switches are closed and line voltage at the black and white leads of the primary control is 120 volts.

RESET BUTTON

THERMOSTAT TERMINALS

CAD CELL TERMINALS

(A) PRIMARY CONTROL

BLACK

J-BOX

LINE VOLTAGE THERMOSTAT

THERMOSTAT

SS

1K

1 K2

ORANGE

LIMIT

L1 (HOT)

R1 BILATERAL SWITCH

TRIAC

SAFETY SWITCH HEATER

1 K1

BURNER MOTOR

L2

R2

OIL VALVE (OPTIONAL)

CAD CELL

CAPACITOR

WHITE

IGNITION

⚠ POWER SUPPLY. PROVIDE DISCONNECT MEANS AND OVERLOAD PROTECTION AS REQUIRED.

(B) PRIMARY CONTROL WIRING DIAGRAM

03310-13_F32.EPS

Figure 32 Primary control and wiring diagram.

Switch off the power at the furnace disconnect.

Step 2 Disconnect at least one room thermostat lead from terminals T-T. Place a jumper wire across terminals T-T.

Step 3 Disconnect the two cad cell leads from terminals F-F. Connect one end of a 1,500V resistor to one of the F-F terminals. This resistor will simulate a cad cell that senses light. Make sure that there is enough lead length to quickly connect the other end of the resistor to the other F-F terminal.

Step 4 Close the furnace disconnect, press the reset button on the primary control, and quickly (within 5 to 10 seconds) connect the other end of the 1,500V resistor to the remaining F-F terminal.

Step 5 If the primary control locks out with the resistor across F-F, it indicates that the primary control is defective. When in doubt, wait five minutes and repeat the test, making sure to connect the resistor quickly. If the primary does not lock out with the resistor in place, the control is functioning properly.

3.1.2 Cad Cell Flame Detector Troubleshooting

The cad cell (photocell) flame detector consists of a plug-in, light-sensitive cell and a socket with a mounting bracket and lead wire (one-piece units are also available). It is mounted in the burner unit and wired into the control circuitry as shown in *Figure 32*. The location of the cad cell is critical. It must be positioned so that it can detect the light from the flame in order to function properly.

Before checking the cad cell and primary control, check the following:

• Main power supply (unit disconnect) on

- Electrodes properly positioned and with proper gap
- Contact between ignition transformer and electrodes is good
- Oil pump pressure adequate
- Oil burner nozzle clean and correctly sized
- Oil supply adequate and piping correctly installed

Using a cotton-tipped swab, clean off the soot or oil film that has formed on the face of a cad cell flame detector.

The following is an example of control circuit troubleshooting based on the circuit shown in *Figure 32*.

When checking the cad cell resistance with an ohmmeter, the resistance should range from 300 to 1,000 ohms (Ω) when the burner is operating (light showing). With the cad cell face covered (no light available), the resistance should be roughly $100,000\Omega$. This is high enough to block any current flow in the circuit. If the cad cell resistance is above $1,600\Omega$ during the burner run cycle, the cad cell may need cleaning or aligning, or the burner flame may need to be adjusted. The equipment required for checking the cad cell and primary control includes the following:

- 0 to 150VAC voltmeter
- $1,500\Omega$ resistor
- Insulated jumper wires
- Ohmmeter

To troubleshoot a cad cell, proceed as follows:

Step 1 Switch off the furnace power. Disconnect at least one lead of the cad cell from primary control terminals and check it using an ohmmeter. Resistance should be around $100,000\Omega$.

Step 2 Reconnect the cad cell to the terminals, disconnect at least one lead of the room thermostat from the T-T terminals, and place a jumper wire across the T-T terminals.

Step 3 Press the reset button on the primary control and switch on the furnace power to initiate burner operation.

Step 4 After the burner ignites the fuel, disconnect both leads of the cad cell from the terminals. The burner should stop after a short interval (15 to 45 seconds). If the burner does not shut off, the primary control is probably defective.

Step 5 Switch off the furnace power and connect a jumper wire to one of the cad cell terminals.

Step 6 Press the reset button on the primary control and turn on the primary power. The burner should start and run. As soon as the burner starts, connect the other end of the jumper wire to the other cad cell terminal.

Step 7 With the burner operating, measure the resistance across the two disconnected cad cell leads. The resistance should not exceed $1,600\Omega$. Ideally, it should be between 600Ω and 700Ω.

Step 8 If the resistance is $1,500\Omega$ or less, the cad cell is functioning properly. If the resistance is higher than $1,600\Omega$, the cad cell face may be dirty, cracked, broken, or misaligned such that it does not completely sense the light from the flame.

3.1.3 Ignition Components Troubleshooting

The ignition system consists of the ignition transformer and the spark electrodes. The ignition transformer can be checked as follows:

WARNING!

Follow the manufacturer's instructions and be extremely careful when performing this procedure. The ignition transformer produces a 10,000V pulse.

Perform the following steps to troubleshoot the ignition controls:

Step 1 Turn off power to the unit.

Step 2 Shut off the fuel supply or disconnect the burner motor lead.

Step 3 Connect a voltmeter capable of measuring high voltage across the ignition transformer output terminals, and then turn the power back on. The reading should be at least 10,000V. If this result is not obtained, replace the transformer.

The ignition transformer must make good contact with the contacts of the electrodes. The contacts mounted to each electrode is flexible and can be bent or curved to ensure good contact is made with the transformer high-voltage terminals. If the electrode insulators are cracked or broken, an arc may be established between that electrode and another part of the burner assembly. If the ignition transformer is working properly, yet there is no arc observed at the end of the electrodes, check their position and gap, and inspect for flaws in the insulators.

3.2.0 Troubleshooting Common Oil Heating Problems

This section provides a troubleshooting guide to common causes of the primary complaints associated with oil heating service calls. The following checks should be made on if the customer complains that the appliance is not producing the desired heat, or is not operating at all:

Step 1 Check for power at the main disconnect switch.

Step 2 Check the burner motor fuse.

Step 3 Check the burner on-off switch.

Step 4 Check the oil supply.

Step 5 Make sure all manual oil valves are open.

Step 6 Check the limit switches. They must be closed.

Step 7 Reset the safety switch, set the thermostat to call for heat, and then perform the following checks:

> **WARNING!**
>
> If the switch is off, make sure the combustion chamber is free of oil, or oil vapor, before turning the switch on. Failure to do so could result in an explosion.

- Simulate flame failure by shutting off the oil supply hand valve while the burner is operating. The burner should lock out after the primary control has timed out.
- Simulate ignition or fuel failure by shutting off the power supply while the burner is operating

normally. The system should shut down immediately. Wait a few minutes for the stack to cool, and then restore power. The system should restart immediately if the thermostat is calling for heat.

If the system still does not operate properly, note the point at which it fails. Use the troubleshooting chart in *Appendix A* to determine which parts of the oil heating system are most likely to be at fault. *Appendix B* shows the most likely causes of common oil heating system failures.

3.2.1 Oil Burner Troubleshooting

The components of the burner system include the oil supply components and the burner. The gun-type atomizing burner, shown in *Figure 33*, is the most commonly used unit. This type of burner uses a fuel pump to deliver oil under pressure (typically 100 psig, but sometimes up to 150 psig) to a calibrated nozzle where it is broken into a fine, cone-shaped mist. An electric spark provided by an ignition transformer and electrode ignites the mixture.

The oil burner nozzle assembly is a common source of problems. The following conditions can be traced to a faulty nozzle:

- No flame (partial or complete blockage)
- Pulsating pressure
- Flame changing in size and shape
- Flame impinging on (touching) the sides of the combustion chamber (*Figure 34*)
- Low carbon dioxide reading in flue gases (less than 8 percent)
- Delayed ignition
- Oil odors

If the nozzle appears to be faulty, it should be replaced. In general, it is a good idea to replace it annually. If the nozzle strainer is clogged, determine the cause. The fuel oil line or the oil pump

Figure 33 Gun-type burner.

Figure 34 Improper burner flame.

strainer may be plugged, or there may be water or other contaminants in the fuel tank.

If a varnish-like substance has formed on the nozzle, it indicates overheating. There are four major causes of overheating:

- An over-fired burner (nozzle too large)
- The fire burning too close to the nozzle
- Too small a firebox
- The nozzle being positioned too far forward

If the fire is too close to the nozzle, a small increase in oil pump pressure may correct the problem. When the nozzle is too far forward, it may be moved back by using a short adapter or by shortening the oil line.

3.2.2 Fuel Supply Troubleshooting

If an oil pump is working properly, the needle on its pressure gauge should be steady. If the needle pulsates, check for a partially clogged fuel filter or pump screen and/or air leaks in the fuel supply lines. A vacuum gauge connected to the inlet side of a single-stage pump should read less than 10 inches of vacuum. The vacuum gauge on a two-stage pump should read less than 15 inches of vacuum. If a higher vacuum is detected, check for kinks or restrictions in the fuel supply lines and/or a plugged fuel filter. Also listen for a whine at the pump, which can indicate a high suction and a possible supply restriction. Before checking the pump, valves, and controls, check the fuel supply for proper installation and piping.

Oil supply lines in residential installations generally use ⅜" (9.53mm) OD soft copper tubing. To prevent oil and air leaks in fuel lines, always use flare fittings on copper tubing connections. Compression fittings should never be used on copper fuel supply lines.

Water can form in the fuel tank due to condensation. If it collects to a sufficient depth in the bottom of the tank, it can make its way to the pump. Here it can cause corrosion and eventually ruin the pump, in addition to preventing ignition. Fuel tanks should be periodically checked for the presence of water. If water is found, it must be removed.

3.2.3 Odors

To find the source of a fuel oil odor, proceed as follows:

Step 1 Check for leaks in the oil supply and return piping systems. Tighten or replace connections or ruptured lines as necessary.

Step 2 Check the fuel tank for proper venting to the outdoors if it is a basement tank. Correct venting problems where necessary and securely tighten or seal the vent connections.

GOING GREEN

Disposing of Oil Storage Tanks

Oil storage tanks for heating systems, particularly those for residential systems, may not be covered by the same federal regulations as industrial underground storage tanks, but they still contain potentially hazardous fuel oil residue. Always check and abide by all applicable federal, state, and local codes when disposing of oil storage tanks.

Over-Fired Oil Burners

How can you tell if an oil burner is over-fired? First, check the furnace information plate to find the input rate. Then, inspect the oil burner nozzle and find the flow rate stamped on the nozzle body. The nozzle flow rate determines burner input. For example, one gallon of No. 2 fuel oil has a heat content of approximately 144,000 Btus. A one-gallon-per-hour nozzle would therefore deliver 144,000 Btuh of input. If a half-gallon-per-hour nozzle is used, the burner input is 72,000 Btuh. A nozzle that produces an input above the computed rate, as determined using the nozzle body stamp and the furnace information plate, is over-fired.

Step 3 Check for seepage around the fill or vent connections if there is an oil film on the tank.

Step 4 Check for seepage through the oil pump cover gasket or the pump shaft bushings. If necessary, replace the pump.

Step 5 Check for loose high-pressure oil line connections to the gun assembly, and tighten the connections or replace the oil line as necessary.

Step 6 Check for fuel filter leakage, correct the cause, and replace the filter.

3.2.4 Flue and Chimney Exhaust Troubleshooting

The flue pipe and chimney exhaust can also become a source of odor or improper burning operation. As with any other HVAC systems, the proper design and installation of the flue pipe and chimney exhaust system eliminate problems that could be encountered by the HVAC troubleshooter.

Refer to *Figure 35* and the numbered list in the text below for an explanation of some of the areas that might lead to chimney problems:

1. **Downdraft** might be caused by the top of the chimney being lower than surrounding objects. The correction requires extending the chimney above all objects within 30' or as recommended by local building codes.

2. A restricted opening may be caused by problems with the chimney cap, ventilator, or **coping**.

3. Obstructions in the chimney can be located by using a light and a mirror.

4. A **joist** projecting into the chimney can cause problems.

5. A break in the chimney lining or tile may be present.

6. A collection of soot at a narrow opening or turn may be present.

Figure 35 Common chimney problems.

7. Two or more openings may exist in the same chimney. Refer to local codes for recommendations.

8. Loose or open connections at joints in the vent pipe may be present.

9. A burner smoke pipe that extends into the chimney can cause problems.

Oil Burner Nozzle Wear

Oil burner nozzles can wear out over time. The abrasive action of microscopic particles in the oil, which is forced at high pressures through a relatively small orifice, can erode the inside of the orifice. In time, this abrasion will distort the nozzle spray pattern.

10. A loosely seated vent pipe in the chimney can cause problems.

11. Failure to extend the flue pipe into the flue opening can cause problems.

12. An oversized chimney might not be able to establish an adequate draft.

3.2.5 Draft Regulator

A draft regulator (*Figure 36*) consists of a small door in the side of the flue pipe. It is hinged near the center and controlled by adjustable weights. A draft regulator is necessary to control heat and combustion. If too high a draft exists, it causes undue loss of heat through the chimney. If too little draft is available, incomplete combustion could result. The draft regulator should maintain a constant draft over the fire, usually 0.01 to 0.03 in wc.

3.2.6 Solenoid Oil Flow Valves

Solenoid on-off flow valves can be mounted directly in the pipeline or on support brackets.

These magnetic valves provide fuel oil to the burner only when the burner operates. They close immediately upon a loss of power. Some models have an integral thermistor that delays opening until the burner motor reaches full speed. They are wired in parallel to the line voltage supply to the burner motor (*Figure 37*).

Under normal operation, the valve opens when the thermostat calls for heat and closes immediately when the call for heat is ended. These valves usually offer reliable service and should not be replaced until all other sources of trouble have been eliminated. A failure usually results in the valve not opening, resulting in no fuel supply to the burner. To check solenoid oil flow valve operation, proceed as follows:

Step 1 Check the valve operation by listening for an audible click when it opens and closes.

Step 2 If the valve does not open when the thermostat calls for heat:

- Check for normal fuel pressure.
- Make sure that the bleed line is not obstructed.
- Check the power supply at the valve. If no voltage is available, check the power source and circuit controls (relays and primary control). If the proper voltage is available, replace the coil or the entire valve.

DRAFT REGULATOR

03310-13_F36.EPS

Figure 36 Draft regulator.

OIL FLOW VALVE

BURNER MOTOR

TO RELAY

03310-13_F37.EPS

Figure 37 Oil valve circuit.

Solenoid Oil Flow Valve

A solenoid oil flow valve can help maintain the prime in the oil piping when the tank is below the burner. It also provides a safety feature by preventing oil flow if there are leaks or breaks in the oil line. This is especially true if the valve is used in a gravity-feed unit, where the tank or oil level is higher than the burner.

3.2.7 Electronic Time Delay

A few burners may be fitted with an electronic time delay external to the primary control. The electronic time delay allows the burner motor time to establish draft. It also can be used to establish first-stage flame before the opening of a second-stage fuel valve on an oil burner. It is connected in the control circuit as illustrated in the schematic diagram in *Figure 38*. On a call for heat, the electronic circuit in the time delay device (ST 70) delays the opening of the oil valve for approximately five seconds. Troubleshooting the electronic time delay consists of checking for supply voltage. If the input voltage is present and the delay does not function by passing the voltage through to the output, it must be replaced.

PRIMARY CONTROL

03310-13_F38.EPS

Figure 38 Time delay schematic.

Chimney Liners

Correcting masonry-related chimney problems is much easier today due to advances in technology. If a chimney is oversized or is improperly lined, a flexible chimney liner kit rated for use with fuel oils can be installed to correct the problem.

Additional Resources

Fundamentals of HVAC/R. Carter Stansfield and David Skaves. Upper Saddle River, NJ: Prentice Hall.

Refrigeration and Air Conditioning, an Introduction to HVAC/R. Larry Jeffus. Air Conditioning and Refrigeration Institute. Upper Saddle River, NJ: Prentice Hall.

3.0.0 Section Review

1. In *Figure 30*, the indoor blower will energize when the _____.

   a. thermostat calls for heat
   b. limit control closes
   c. thermostat fan switch is set to ON
   d. cad cell senses burner firing

2. Flare fittings should always be used to connect oil supply piping.

   a. True
   b. False

SUMMARY

Oil furnaces are designed to burn fuel oil, which is mixed with air, then sprayed as a fine mist into the furnace's combustion chamber, where it is ignited. An oil burner system consists of the oil supply and the burner. Gun-type atomizing units are the most common burners used. Air returned from the conditioned space is directed over the heat exchanger by a blower in the furnace unit. Heat is transferred to the air and recirculated by the duct system. The primary component of an oil furnace is the pressure burner, which contains the fuel pump and the nozzle.

Periodic servicing, including inspection, cleaning, and combustion testing are essential for an oil-fire furnace. A critical part of the servicing is measuring combustion efficiency. Oil furnace troubleshooting usually starts with an owner's complaint, which must be analyzed in a logical manner to arrive at the solution to the problem. One of the critical tasks of the HVAC technician is to check a system thoroughly before condemning any control or component. Oil burner nozzle assemblies are a common source of problems. Problems may also occur in oil furnace control systems. Components to check include thermostats, primary controls, flame detectors, limit controls, auxiliary controls, and ignition components. Auxiliary components such as solenoid oil flow valves and electronic time delays, may also cause problems.

At the time of the troubleshooting call, it is a good idea to replace filters and similar devices that will contribute to the efficient operation of the system. The responsible troubleshooter will not leave the site until the reason for the system failure has been corrected. To make sure of this, the system should be cycled several times and then monitored for a reasonable amount of time after startup.

1. The combustion efficiency of a non-weather-ized oil furnace must be at least _____.

   a. 65 percent
   b. 78 percent
   c. 92 percent
   d. 100 percent

2. The vane fan in a high-pressure oil burner is used to _____.

   a. prevent overheating
   b. cool the oil
   c. atomize the oil
   d. provide combustion air

3. The burner input, in Btus, is determined by the _____.

   a. type of fuel
   b. nozzle flow rate
   c. position of the electrodes
   d. size of the ignition transformer

4. An advantage of a flame-retention oil burner is _____.

   a. it provides more oil to the flame
   b. with a flame-retention burner, combustion occurs inside the burner, rather than in a combustion chamber
   c. the flame-retention burner operates at a much lower speed
   d. the flame-retention burner produces a higher temperature flame

5. Which of the following receives a signal from the thermostat and switches power to the burner and other components?

   a. Cad cell
   b. Limit switch
   c. Primary control
   d. Master cylinder

6. The device most commonly used in oil furnaces to verify a continuous flame is the _____.

   a. flame rectifier
   b. thermopile
   c. cad cell
   d. hot surface igniter

7. The return lines in a two-pipe system should _____.

   a. never be smaller than the supply lines
   b. be the same size or larger than the supply lines
   c. always be smaller than the supply lines
   d. always be larger than the supply lines

8. A single-pipe oil system is used when the _____.

   a. oil supply level is above the burner
   b. oil supply level is below the burner
   c. the vertical distance from the tank to the furnace is more than 15'
   d. the vertical distance from the tank to the furnace is more than 30'

9. An oil burner flame should be _____.

   a. blue with a yellow tip
   b. red with a yellow tip
   c. yellow
   d. orange

10. In a residential oil furnace burning No. 2 oil, the smoke reading should be _____.

    a. zero
    b. between zero and a trace
    c. between 1 and 2
    d. no greater than 2

11. The test hole for smoke spot and combustion efficiency testing should be drilled _____.

    a. as close as possible to the furnace
    b. between the draft regulator and the chimney
    c. at least two flue pipe diameters from the furnace breeching
    d. at least one flue pipe diameter from the furnace breeching

12. If the burner flame is burning off to one side, striking the combustion chamber wall, the most likely cause is _____.

    a. electrode spark gap too wide
    b. a damaged or worn nozzle
    c. weak ignition transformer
    d. loose oil pump coupling

13. A pulsating needle on an oil pump pressure gauge can indicate _____.
    a. a defective cad cell
    b. clogged fuel filter
    c. cracked heat exchanger
    d. defective nozzle

14. A downdraft in a chimney could be caused by _____.
    a. the top of the chimney being too high
    b. the chimney being below surrounding surfaces
    c. an obstruction in the chimney
    d. a crack in the chimney lining

15. An electronic time delay mounted on an oil burner external to the primary control is sometimes used to _____.
    a. control the second stage of a two-stage burner
    b. delay burner ignition until oil flow is established
    c. delay burner ignition until airflow is established
    d. limit the amount of time the burner runs on each cycle

Trade Terms Introduced in This Module

Annual Fuel Utilization Efficiency (AFUE): An industry standard used to define furnace efficiency.

Atomize: Reduce to a fine spray.

Cad cell: A light-sensitive device (photocell) in which the resistance reacts to changes in the amount of light. In an oil furnace, it acts as a flame detector.

Coping: The upper masonry course of a chimney designed to keep water out.

Downdraft: A situation in which outdoor air is forced down a chimney, preventing proper draft. Downdraft can be caused by not extending the top of the chimney above the roof line.

Flame-retention burner: A high-efficiency oil burner.

Joist: Framing member used to support floors and ceilings.

Lockout: A safety feature that prevents a furnace from automatically cycling if the flame fails.

Over-firing: A situation caused by an oversized nozzle in which excess oil is pumped into the combustion chamber.

Primary control: A combination switching and safety control that turns the oil burner on and off in response to signals from the thermostat and safety controls.

Refractory: Materials such as ceramics designed to withstand extremely high temperatures.

Smoke spot test: A test to determine the level of smoke in the flue gases.

Soot: Fine carbon particles resulting from incomplete combustion. An accumulation of soot on heat exchangers will reduce combustion efficiency and result in increased carbon monoxide production.

Temperature differential: The difference in temperature between the flue gases and the gases above the combustion chamber.

Weatherized furnace: A furnace that has been well insulated and has a weather-resistant case as specified by government standards.

Appendix A

TROUBLESHOOTING SUMMARY

PROBLEM	POSSIBLE CAUSE	CORRECTIVE ACTION
Burner motor does not start		
Trouble in primary control	Check cad cell	Replace if resistance is incorrect
Trouble in thermostat	Broken wires or loose connections Defective thermostat	Replace wires or tighten connections Replace thermostat
Faulty burner components	Broken wires or loose connections Motor start switch or thermal overload switch open Defective motor Defective pump	See manufacturer's instructions
Burner motor starts – no flame established		
Trouble in primary control	Check cad cell or stack relay Loose connections or broken wires between ignition transformer and primary control	Replace wires or tighten connections
Trouble in ignition system	Defective transformer Ignition electrodes: • Improperly positioned • Spaced too far apart • Loose • Dirty Dirty or damaged ceramic insulators	Replace transformer Check manufacturer's instructions Replace or clean electrode assembly
Faulty burner components	Dirty nozzle Loose, misaligned, or worn nozzle Clogged oil pump strainer Clogged oil line Air leak in suction line Defective pressure regulator valve Defective pump Improper draft Water in oil Oil too heavy	Correct per manufacturer's instructions
Burner motor starts – flame goes on and off after start-up		
Trouble in primary control	Cad cell problems	Check and/or clean cad cell
Trouble in ignition system	Limit settings too low Dirty oil filter	Readjust setting or differential Replace filter
Trouble in thermostat	Broken wires or loose connections Defective anticipator Defective thermostat	Replace broken wires or tighten connections Adjust anticipator Replace thermostat
Faulty burner components	Low oil pressure Defective pump, clogged or dirty oil lines Fluctuating water level (hydronic systems) Weak transformer Cracked electrodes	See manufacturer's instructions Replace transformer Replace porcelain electrodes

03310-13_A01A.EPS

PROBLEM	POSSIBLE CAUSE	CORRECTIVE ACTION
System overheats house		
Trouble in primary control	Defective primary	With burner running, disconnect one low-voltage thermostat lead from primary control terminal; if relay does not drop out, replace primary
Trouble at thermostat	Wiring shorted	Repair or replace wiring
	Thermostat out of calibration	Recalibrate thermostat with an accurate thermometer
	Defective anticipator or anticipator improperly set	Replace or readjust anticipator; replace thermostat, if necessary
	Thermostat stuck in call-for-heat	Repair or replace thermostat
	Thermostat improperly located in area under control, or draft on stat through wall	Relocate thermostat: must be out of drafts, away from radiating surfaces, ducts, steam pipes, and sunny locations
	Thermostat not level	Level thermostat
Trouble in distribution system	Ductwork close to thermostat	Relocate thermostat
	Flow control valve (hot water) stuck in ON position	Repair or replace valve
	Circulator does not stop running	Check circulator circuits, replace switching device
Faulty burner components	Extremely oversized system	Replace system
	System control valves stuck in ON position	Repair or replace control valve
System underheats house		
Trouble at thermostat	Open or loose wiring	Repair or replace wiring
	Thermostat out of calibration	Recalibrate thermostat with an accurate thermometer
	Defective anticipator or anticipator improperly set	Replace or readjust anticipator; replace thermostat, if necessary
	Thermostat stuck in not calling for heat	Repair or replace thermostat
	Thermostat improperly located	Relocate to sense more accurately
	Thermostat not level	Level thermostat
Trouble at high- or low-limit controller	Controller set too low	Raise setpoint
	Slow to return to ON position	Adjust differential
	Controller defective	Replace controller
Trouble at blower (for forced-air system)	Cutting out or cycling on overload	Check air filters
	Burned out blower motor	Replace motor
	Running too slow or inadequate capacity	Check/replace motor
	Fan belt broken or loose	Replace or tighten belt
System defects	Undersized furnace or boiler system (possible if addition has been made to house)	Replace furnace or boiler system
	Distribution system inadequate	Correct problem
	Sooted heat exchanger	Clean heat exchanger
	Dirty warm-air filters	Replace filters
	Dirty boiler water (steam system)	Drain/refill boiler

PROBLEM	POSSIBLE CAUSE	CORRECTIVE ACTION
System underheats house (continued)		
Faulty burner components	Fuel line clogged or strainer plugged	Clear clog
	Air in fuel line or excess air in system	Check fuel lines for leaks
	Dirty or improperly sized nozzle	Replace nozzle
	Low oil pressure or defective pump	Adjust or replace pump
	Motor does not come up to proper speed	Replace motor
	Inoperative or inadequate circulator (hydronic system)	Replace circulator
System cycles too frequently (this problem is outside the primary control; the most common problems are listed below)		
Trouble at thermostat	Poor location	Relocate thermostat
	Poorly adjusted anticipator	Readjust anticipator; replace thermostat, if necessary
	Loose connections or defective wiring	Tighten connections or replace broken wires
Trouble at high limit	Limit set too low	Reset limit
	Differential too narrow	Reset differential; replace control, if necessary
Faulty burner system	Dirty or clogged air filter	Replace filter
	Loose connections	Repair connections
	Fluctuating voltage	Contact utility
	Intermittent shorts	Find short circuit
Relay chatters after pulling in		
Trouble at thermostat	Open contacts	Replace thermostat
Trouble at primary control	Supply voltage too low	Check for loose connections
	Defective load relay	Replace primary

03310-13_A01C.EPS

Troubleshooting Chart

GENERALLY THE CAUSE	OCCASIONALLY THE CAUSE	RARELY THE CAUSE
Make these checks first.	Make these checks only if the first checks failed to locate the trouble.	Make these checks only if other checks failed to locate the trouble.

PROBLEM: No heat – burner fails to start

Power failure Blown fuses or tripped circuit breaker Open disconnect switch or blown fuse Thermostat set too low Thermostat switch in improper position Primary or safety controls Flame detector	Control relay or contactor Limit controls Thermostat not level Fuel pump Burner motor	Faulty wiring Loose terminals Low voltage Thermostat faulty

PROBLEM: No heat – burner starts but fails to ignite

Electrodes Ignition transformer Improper burner adjustment Oil supply and oil pressure Manual fuel valve	Fuel pump Nozzle Burner motor Oil line filter Fuel lines	Low voltage Water in oil

PROBLEM: No heat – burner starts and fires, then loses flame before locking out on safety

Fuel pump Nozzle Oil supply and oil pressure Oil line filters Water in oil Primary or safety controls Flame detector Improper burner adjustment	Fuel lines Nozzle Displaced or damaged baffles	Faulty wiring Loose terminals Vent or flue problem

PROBLEM: Not enough heat – burner cycle too short

Control relay or contactor Thermostat heat anticipator setting Thermostat not level Thermostat location Dirty filters	Limit controls Fan controls Blower bearings Blower motor Defective blower wheel Dirty blower wheel	Faulty wiring Loose terminals Low voltage Low air volume Ductwork small or restricted

PROBLEM: Not enough heat – burner cycle too long

Blower belt broken or slipping Dirty filters Dirty or plugged heat exchanger	Fuel pump Low air volume Dirty blower wheel Oil line filter	Ductwork small or restricted Oil supply and oil pressure

03310-13_A02A.EPS

GENERALLY THE CAUSE	OCCASIONALLY THE CAUSE	RARELY THE CAUSE
Make these checks first.	Make these checks only if the first checks failed to locate the trouble.	Make these checks only if other checks failed to locate the trouble.
PROBLEM: Too much heat – burner cycles too long		
Thermostat heat anticipator setting too high Thermostat out of calibration	Thermostat not level Thermostat location	
PROBLEM: Too much heat – burner runs continuously		
Faulty wiring Control relay or contactor	Thermostat faulty	Thermostat not level Thermostat location
PROBLEM: Combustion noise		
Nozzle Electrode Dirty or plugged heat exchanger Fuel pump Burner air adjustment	Burner motor Ignition transformer	Faulty wiring Loose terminals Vent or flue problem Oil supply
PROBLEM: Air noise		
Blower Ductwork small or restricted	Cabinet	Blower wheel Dirty filters
PROBLEM: Cost of operation		
Burner motor Barometric control Blower motor Dirty filters Improper burner adjustment Inefficient burner	Low air volume Insufficient insulation Excessive infiltration	Low voltage Fan control Nozzle Ignition transformer Blower belt broken or slipping Dirty blower wheel Ductwork small or restricted Vent or flue problem Input too low
PROBLEM: Mechanical noise		
Blower bearings Blower motor Blower belt broken or slipping Blower wheel out of balance	Cabinet	Low voltage Control relay or contactor Fuel pump Burner motor Oil valve Ruptured heat exchanger Displaced or damaged baffles Fuel lines
PROBLEM: Odor		
Barometric control Vent or flue problems Flue gas spillage Fuel lines Oil supply Oil pump shutoff valve	Low air volume Improper burner adjustment Ruptured heat exchanger Displaced or damaged baffles Blocked heat exchanger Stagnant water in humidifier Water or moisture in air system	Faulty wiring or loose terminals Control transformer Fuel pump Dirty filters Dirty heat exchanger Input too high Input too low Outdoor odors

03310-13_A02B.EPS

Additional Resources

This module presents thorough resources for task training. The following resource material is suggested for further study.

Fundamentals of HVAC/R. Carter Stansfield and David Skaves. Upper Saddle River, NJ: Prentice Hall.

Refrigeration and Air Conditioning, an Introduction to HVAC/R. Larry Jeffus. Air Conditioning and Refrigeration Institute. Upper Saddle River, NJ: Prentice Hall.

HVAC Maintenance Procedure. Syracuse, NY: Carrier Corporation.

http://www.bacharach-inc.com/PDF/Brochures/4097percent20Combustion.pdf; Instructions for performing and evaluating a combustion efficiency test using an electronic test instrument.

Figure Credits

Section Review Answers

Answer	Section Reference	Objective
Section One		
1.b	1.0.0	1a
2.a	1.2.0	1b
3.c	1.3.0	1c
4.a	1.4.2	1d
Section Two		
1.b	2.1.2	2a
2.d	2.2.1	2b
Section Three		
1.c	3.1.0	3a
2.a	3.2.2	3b

NCCER CURRICULA — USER UPDATE

NCCER makes every effort to keep its textbooks up-to-date and free of technical errors. We appreciate your help in this process. If you find an error, a typographical mistake, or an inaccuracy in NCCER's curricula, please fill out this form (or a photocopy), or complete the online form at **www.nccer.org/olf**. Be sure to include the exact module ID number, page number, a detailed description, and your recommended correction. Your input will be brought to the attention of the Authoring Team. Thank you for your assistance.

Instructors – If you have an idea for improving this textbook, or have found that additional materials were necessary to teach this module effectively, please let us know so that we may present your suggestions to the Authoring Team.

NCCER Product Development and Revision
13614 Progress Blvd., Alachua, FL 32615

Email: curriculum@nccer.org
Online: www.nccer.org/olf

❏ Trainee Guide   ❏ Lesson Plans   ❏ Exam   ❏ PowerPoints   Other _____

Craft / Level: _____   Copyright Date: _____

Module ID Number / Title: _____

Section Number(s): _____

Description: _____

_____

_____

_____

_____

Recommended Correction: _____

_____

_____

_____

_____

Your Name: _____

Address: _____

_____

Email: _____   Phone: _____

03312-13

Troubleshooting Accessories

OVERVIEW

Residential furnaces, especially in cold, dry northern climates, are often equipped with humidifiers to maintain a healthy moisture level in the indoor air. Electronic air cleaners may also be installed in both residential and commercial environments. Economizers are common in commercial systems. Energy and heat recovery ventilators are another accessory that can be applied in both residential and commercial systems. Servicing these types of devices requires some specialized knowledge. This is especially true of economizers, because there are several types, each with its own specific operating characteristics.

Module Seven

Trainees with successful module completions may be eligible for credentialing through NCCER's National Registry. To learn more, go to **www.nccer.org** or contact us at **1.888.622.3720**. Our website has information on the latest product releases and training, as well as online versions of our *Cornerstone* magazine and Pearson's product catalog.

Your feedback is welcome. You may email your comments to **curriculum@nccer.org**, send general comments and inquiries to **info@nccer.org**, or fill in the User Update form at the back of this module.

This information is general in nature and intended for training purposes only. Actual performance of activities described in this manual requires compliance with all applicable operating, service, maintenance, and safety procedures under the direction of qualified personnel. References in this manual to patented or proprietary devices do not constitute a recommendation of their use.

Objectives

When you have completed this module, you will be able to do the following:

1. Describe how to troubleshoot various HVAC system accessories.
   a. Describe how to approach the troubleshooting process.
   b. Describe how to troubleshoot humidifiers.
   c. Describe how to troubleshoot electronic air cleaners.
   d. Describe how to troubleshoot UV lighting devices.
2. Describe how to troubleshoot accessories related to the introduction of outside air.
   a. Describe how to troubleshoot economizers.
   b. Describe how to troubleshoot recovery ventilators.

Performance Task

Under the supervision of your instructor, you should be able to do the following:

1. Using the correct tools and circuit diagrams, isolate and correct malfunctions in selected accessories.

Trade Terms

Free cooling
Mechanical cooling
Water hammer

Industry-Recognized Credentials

If you are training through an NCCER-accredited sponsor, you may be eligible for credentials from NCCER's Registry. The ID number for this module is 03312-13. Note that this module may have been used in other NCCER curricula and may apply to other level completions. Contact NCCER's Registry at 888.622.3720 or go to **www.nccer.org** for more information.

Contents

Topics to be presented in this module include:

Figures and Tables

1.0.0 ACCESSORY TROUBLESHOOTING

Objectives

Describe how to troubleshoot various HVAC system accessories.

   a. Describe how to approach the troubleshooting process.
   b. Describe how to troubleshoot humidifiers.
   c. Describe how to troubleshoot electronic air cleaners.
   d. Describe how to troubleshoot UV lighting devices.

Performance Task 1

Using the correct tools and circuit diagrams, isolate and correct malfunctions in selected accessories.

Trade Terms

Water hammer: A banging noise in water pipes that occurs when the water supply solenoid valve on a humidifier abruptly shuts off water to the humidifier.

Troubleshooting the accessories and optional equipment used with HVAC systems is done in the same way as troubleshooting basic heating, cooling, or air distribution equipment. It involves close examination of the equipment to detect telltale signs of what may be wrong. Following standard procedures makes troubleshooting accessories easier than a trial-and-error method.

By using the manufacturer's wiring diagrams, specifications, and troubleshooting aids, the cause of almost any problem can be quickly isolated. The HVAC technician must be able to listen to the owner's complaint, observe the system in operation, then identify and pursue the most probable cause of the malfunction. Accurate diagnosis and repair of HVAC accessories also requires the proper selection and safe use of test equipment and instruments.

Many kinds of accessories can be used with HVAC equipment. In addition, accessories used with residential equipment are often different from those used with commercial equipment.

This module will cover troubleshooting of the common accessories and optional equipment introduced in the *HVAC Level Two* module, *Air Quality Equipment*. They include the following:

- Humidifiers
- Electronic air cleaners
- Economizers
- Heat recovery ventilators

1.1.0 Troubleshooting Approach

For the purpose of this module, problems are divided into two categories: electrical and mechanical. This does not mean that all malfunctions fit easily into these two categories. Obviously, there are problems in which one affects the other. Whether troubleshooting an electrical or mechanical problem, using a logical, structured approach to troubleshooting is essential. The technician must know how to diagnose specific control devices and components used with accessories. A good set of service tools and the ability to read electrical wiring diagrams are a must. Do not guess at the cause of a problem. Also, review the material presented in the *HVAC Level Three* module, *Control Circuit and Motor Troubleshooting*. It describes the elements necessary to perform a systematic approach to control troubleshooting. Many of the problems related to accessories are control-related. Even when this is not the case, some testing and manipulating the controls is often required to pinpoint the real problem area. The module also discusses the different kinds of troubleshooting aids provided by equipment manufacturers, both in their equipment and in their service literature.

1.1.1 Electrical Troubleshooting

The electrical troubleshooting information covered in this module is limited to those electrical devices specific to certain accessories. Troubleshooting most of the electrical problems and components in HVAC equipment is performed in the same way, regardless of the type of equipment. Troubleshooting also requires the use of various common tools and test instruments (*Figure 1*).

For quick reference, troubleshooting topics that are relevant to accessories are listed here:

- Isolating to a faulty circuit first via the process of elimination.
- Isolating to a faulty component.
- Single-phase and three-phase input voltage measurements.
- Fuse and circuit breaker checks.

Figure 1 Technician troubleshooting with a multimeter.

- Resistive and inductive load checks.
- Switch and relay/contactor checks.
- Control transformer checks.

When troubleshooting an accessory that uses an electronic control, do not arbitrarily replace the board. First make sure the module or board has failed. In some equipment, this means using a built-in diagnostic system to indicate that a board has failed. If the equipment does not have a built-in diagnostic capability, make sure to isolate the board as the cause of a problem by using the troubleshooting information in the product literature. Often this means checking that the correct input signals are available to the control. If the correct inputs are available, and the control is providing the correct output signals, the problem is probably not with the control board, but elsewhere in the system.

1.1.2 Troubleshooting Aids

The first action, when troubleshooting an accessory, is to consult the product literature. This literature usually describes the mechanical and electrical operation of the equipment, including sequence of operation. Voltage or signal inputs and outputs related to the operation of the accessory are also defined.

Service literature almost always includes troubleshooting aids such as wiring diagrams and troubleshooting charts. These aids will help to isolate an equipment problem to the specific failed component.

1.1.3 Sequence of Operation

Knowing the sequence of operation of any accessory is absolutely necessary to successfully troubleshoot a problem. While each manufacturer may have very specific sequences for their equipment, most follow a common sequence, regardless of the brand. For example, in gas forced-air furnaces, the burner ignition sequence is nearly universal: on a call-for-heat, the inducer energizes. When inducer airflow is proven, the igniter warms up, the gas valve opens, flame is proved, and the blower operates. Similarly, common sequences occur with accessories.

The purpose of a humidifier used with a forced-air furnace is to introduce moisture into the structure through the air distribution system. Without the blower operating, no moisture will be distributed. For that reason, many humidifiers have their operation tied directly to blower operation. If the blower is not running, the humidifier will not operate. Electronic air cleaners (EACs) also require air movement through them for proper operation. If airflow (blower operation) is missing from the sequence of operation of these accessories, they will not operate or will function poorly. With either of these accessories, there may be some component that proves airflow to allow normal operation.

The sequence of operation can be presented in several ways in the product literature. One way is a list of steps in a normally occurring sequence. Step one is a certain event that must occur before step two (another event) can occur. Sometimes the operating sequence is presented graphically, using a flow chart or a spreadsheet to show events and their relationship to other events over time. *Figure 2* shows an operating sequence arranged as the steps occur. *Figure 3* is a graphical representation of an operating sequence. Both sequences of operation are for steam humidifiers.

1.2.0 Troubleshooting Humidifiers

A humidifier is a device used to add humidity. All humidifiers introduce water vapor into a building's conditioned air at a predetermined rate; therefore, they must be connected to a water supply. Some humidifiers also require a drain connection. Humidifiers must be carefully installed and mounted as directed in the manufacturer's installation instructions. Some are motor-driven while others are not. They are controlled by a humidistat in almost all applications.

Most humidifier problems can be prevented with periodic maintenance. Evaporative humidifiers leave behind minerals contained in the water that has evaporated. These minerals are left on the media and in the water reservoir. The float, float valve orifice, and valve seat for most humidifiers need periodic cleaning to remove mineral

OPERATION	UNIT ACTIVITY
Power is provided or restored to the unit	• HPC-450i controller is energized. • LEDs three times; On LED remains lit. • All sensors activated.
Water level sensor determines the steam chamber is empty *Unit is prepared for a humidification cycle*	• Following short delay, the fill valve opens and the Fill LED illuminates in unison. • Water flows into the steam chamber for up to 14 minutes, depending upon chamber size. • Water level sensor determines chamber is full. The HPC-450i de-energizes the fill valve and related LED.
On a call for humidification by the control device	• A DC signal is completed at the control, applying DC power to terminal H1 on humidifier. The signal will range from 1 to 10 VDC based on the ΔRH between the new setpoint and the measured space RH. • The fan relay will energize in 5-7 seconds. • The HPC-450i will monitor for closed circuits in the fan proving switch and the high limit switch. • If circuits remain closed, the STEAM LED will illuminate and remain on as long as the demand for humidity is present. • The STEAM LED will de-energize if the monitored circuits open or the DC demand signal falls below 1 VDC. • After a 20-second delay, the contactor will close to energize the electric elements. • In roughly 9 minutes, the water in the steam chamber will boil and steam will be generated. Steam enters the dispersion hose and enters the manifold. **WARNING:** Steam chamber and hose will be very hot!
As humidification continues, the water level in the steam chamber will fall	• Fill sequence as noted above begins.
The demand for humidity ceases	• Steam LED cycles off. • The electric elements are now de-energized. • The fan continues to run for 4 minutes, and then the fan relay is de-energized by the HPC-450i.

03312-13_F02.EPS

Figure 2 Operating sequence.

deposits. If so equipped, strainers should also be cleaned. In addition, algae, bacteria, and virus growth can cause problems. Algaecide can be used to help neutralize algae growth.

Most manufacturers recommend that the water reservoir be drained and humidifier components cleaned about every two months. In hard water areas, more frequent cleaning may be necessary.

POWER ON	DRAINING CYCLE			FILLING CYCLE	BOILING CYCLE		
1	2	3	4	5	6	7	8
⏻							⏻
					�֍		✖
						♒	
		♨		♨			♨
	♒		♒				
Up to 5 sec	Up to 3 min	Up to 4 min	Up to 2 min	Up to 2 min	Up to 5 sec	Up to 6 min	Up to 20 sec
	Up to 9 minutes total for drain cycle				1 steam cycle		

03312-13_F03.EPS

Figure 3 Graphical presentation of an operating sequence.

Figures 4, 5, and *6* show typical control circuits for three different humidifiers.

Figure 4 is a fairly simple control circuit. The furnace blower motor and a 24V transformer are connected in parallel and are simultaneously energized when the furnace fan control closes. Note that this transformer is dedicated to the humidifier only; a separate transformer serves the primary equipment. A humidistat in the low-voltage circuit is in series with a 24V motor in the humidifier. In this control circuit scheme, the humidifier will operate only if the blower motor and transformer are energized and the humidistat is calling for moisture. Sometimes a 24V solenoid will be connected in parallel with the 24V motor, allowing water to flow to the humidifier.

Figure 5 is a variation of the circuit shown in *Figure 4*. In *Figure 5*, the humidifier motor and water solenoid valve are powered by 120VAC supplied through a normally open set of relay (R) contacts. In this circuit, the humidistat contacts close to power the coil of the relay. When energized, the relay contacts provide a circuit through which the humidifier fan and solenoid valve are energized.

Figure 6 is the control circuit for an infrared humidifier. Unlike other humidifiers, an infrared humidifier does not depend on heat supplied by the furnace to evaporate water. Instead, infrared heaters heat a pan of water and evaporate it in much the same way that sunlight evaporates a puddle of rainwater. The infrared heaters are powered by line voltage and energized by a contactor that is controlled by a timer. Devices in the control circuit are connected in series. For example, the humidistat may be calling for moisture, but if the sail switch is not sensing airflow in the duct and/or the high-limit aquastat is open, the humidifier will not operate.

Troubleshooting all three of these different circuits involves checking for the presence of power at key locations; testing circuits for open and closed switches, either by continuity or voltage testing; testing the electrical integrity of various components; and checking the mechanical integrity of various load components.

Common problems encountered with humidifiers follow:

- *Humidifier not running/not providing moisture* – This is usually caused by an electrical control circuit problem. As applicable, check the overload protection, humidistat, and control circuit components. Also check to see if the humidifier motor has failed. Electrical failures such as a tripped overload or failed motor can result from the rotating media being bound as a result of a mineral buildup or mechanical failure. A solenoid valve that is stuck closed or has an open coil will not supply water to the humidi-

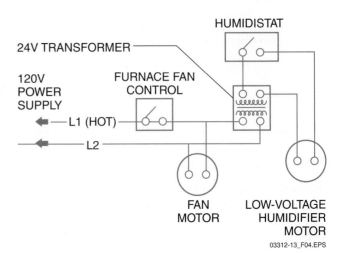

03312-13_F04.EPS

Figure 4 Control circuit schematic.

Figure 5 Schematic of a typical fan-powered humidifier circuit.

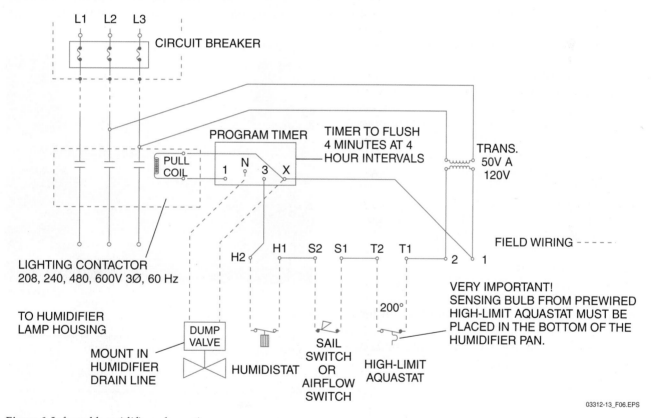

Figure 6 Infrared humidifier schematic.

Servicing Accessories

When servicing accessories, it is often necessary to consult the service literature for the main HVAC unit with which the accessory is being used. Also, consult the service literature for some individual components, such as the humidistat or thermostat, used in the accessory control system.

fier and no moisture will be added to the structure, even if the humidifier motor is running.

- *Water overflow* – Check for a dirty, misadjusted, or faulty float valve. If the humidifier is equipped with a solenoid valve, check for dirt or scale lodged between the valve plunger and seat that may cause the valve to stick open, even when de-energized. The valve may need cleaning, adjusting, or replacement. Also, check for water backing up the drain line as a result of the line being plugged or restricted.

> **CAUTION**
>
> Water overflowing from a humidifier can corrode metal parts, including furnace heat exchangers, and lead to premature failure of the part. Water can also damage electrical and electronic components.

- *Noise* – If the unit is equipped with a solenoid valve, the clicking noise it makes when it opens and closes can be amplified by sheet metal ductwork. To help reduce this problem, check to make sure the unit is snug against the plenum. The solenoid valve is powered by alternating current that operates at a frequency of 60Hz. That frequency can sometimes be heard as a humming sound that can also be amplified by the ductwork. One manufacturer supplies a rectifier kit that can be installed on the terminals of the solenoid. The rectifier converts AC to DC to help eliminate the hum. High pressure in the humidifier's water supply line may cause water hammer when the solenoid opens and closes. It may be necessary to place a length of high-pressure hose before the valve, or install a pressure-reducing device in the supply line to reduce this problem.
- *Low or high levels of humidity* – The recommended level of relative humidity (RH) that should be maintained in a building during the winter is dependent on several factors. Condensation on windows and inside exterior walls are signs of a relative humidity value that is too high. Results of low RH can include the following:

  – Dry, itchy skin
  – Static electricity shocks
  – Clothing static cling
  – Sinus problems
  – Chilly feeling
  – Sickly pets and plants
  – Loose furniture joints

When the humidity is too high or too low, first check the humidistat setpoint (*Figure 7*). Electromechanical humidistats like the one shown in *Figure 7* have been used for many years. They con-

tain a material that expands and contracts relative to the moisture level in the air. That material is connected to a switch in the humidistat, which opens and closes depending on the moisture levels in the room and the setting of the humidistat. Because these parts are electromechanical, their calibration can drift over time.

To check calibration, compare the humidistat dial setting to the relative humidity level in the conditioned space, as measured with a sling or digital psychrometer. First, turn the humidistat dial to its lowest setting, and then slowly increase the setting until a light click is heard. The relative humidity setting indicated by the dial should be close to the relative humidity measured with the sling psychrometer. If the humidistat is out of calibration, adjust it, if possible. *Figure 8* lists typical recommended levels of indoor relative humidity that should be maintained at various outdoor winter temperatures.

Figure 9 shows a digital humidistat that electronically senses moisture levels in the indoor air.

03312-13_F07.EPS

Figure 7 Humidistat.

AT OUTDOOR TEMPERATURE (°F)	RECOMMENDED (SAFE) INDOOR RH (%)
−20	15
−10	20
0	25
10	30
20	35
30	40

BASED ON AN INDOOR TEMPERATURE OF 72°F

03312-13_F08.EPS

Figure 8 Recommended indoor relative humidity in the winter.

Electronic humidistats often will use an optional thermistor to sense outdoor air dry-bulb temperature. With outdoor temperature known, the control's software can automatically maintain indoor relative humidity at a level that avoids indoor condensation on windows. Electronic humidistats offer several advantages over older electro-mechanical types, including accuracy and more reliable operation.

Modern room thermostats can contain a built-in humidistat that measures and controls humidity electronically. One manufacturer refers to this control as a Thermidistat™. The desired humidity level is entered into the control by pressing a button and selecting the desired humidity level on a digital display. The control then operates the humidifier to maintain the desired humidity level during heating operation as with a conventional humidistat. If an outdoor temperature sensor is used with the combination thermostat/humidistat, the control automatically adjusts the indoor humidity setpoint according to outdoor temperature.

This type of humidistat can also control the cooling system to dehumidify. Normal air conditioning operation removes humidity from the air. However, there may be times when humidity is high, but the temperature in the space is not high enough to initiate a call for cooling. In these cases, the control calls for compressor operation. Typically this arrangement is used with a variable-speed blower and an electronic control in the air handler. When a call for dehumidification is present without a call for cooling, the blower operates at a much lower speed to maximize moisture removal. The software in the control and in the air handler allows dehumidification without over-cooling the space. To enable this feature, the room control and/or the air handler must be properly configured according to the manufacturer's instructions.

Some reasons for low humidity not directly related to the humidifier itself are as follows:

- The water supply feed valve/saddle supply valve is closed or not fully open.
- Excessive air infiltration from permanently opened windows, exhaust fans, or open fireplace dampers can cause low humidity.
- An oversized furnace can cause low humidity because burner cycles are too short for the humidifier to do its job. Remember that many humidifiers can only add moisture to the air when the blower is also functioning. With this condition, it may be necessary to set the blower for continuous operation so that the humidifier can run as needed. However, it is important to note that the capacity of this type of humidifier (evaporative) is substantially reduced when only room air is being used, as opposed to using heated air for evaporation. The capacity of steam humidifiers would be unaffected.

Excessive humidity may be caused by continuous operation of the humidifier as a result of a failed humidistat or a defective or sticking contact on the humidifier relay. Either low or high humidity levels can also be caused by using a humidifier with the wrong capacity. Humidifiers are typically rated in gallons of water per day. Humidifier selection depends on the volume of the building or area in square feet (ft^2). It also depends on the building's air tightness. *Figure 10* shows a typical graph used for the selection of residential humid-

03312-13_F09.EPS

Figure 9 Digital humidistat.

Humidifier Selection

To reduce mineral deposit buildup or algae growth, select a humidifier design that minimizes these problems. For example, many bypass and fan-powered humidifiers lack a water reservoir. Water flows through the humidifier only when needed; excess water is drained away. Eliminating the reservoir reduces algae growth, while draining the excess water helps flush out mineral deposits.

ifiers. Similar graphs and/or charts are available for commercial and industrial humidifiers.

The terms *loose*, *average*, and *tight* are defined as follows:

- *Loose* – The building has little insulation, no vapor barriers, and no storm doors or windows. In homes, it can also mean an undampered fireplace. The air exchange rate is about 1.5 changes per hour.
- *Average* – The building is insulated, has vapor barriers, and has loose storm doors or windows. In homes, it can also mean a dampered fireplace. The air exchange rate is about 1.0 change per hour.
- *Tight* – The building is well insulated, has vapor barriers, and tight storm doors or windows. In homes, it can also mean a dampered fireplace. The air exchange rate is about 0.5 change per hour.

1.3.0 Troubleshooting Electronic Air Cleaners

Electronic air cleaners (EACs) (*Figure 11*) remove airborne particles and odors from the air circulated to the conditioned space. They can be stand-alone units or may be mounted in the return-air side of a duct system. EACs create an electrostatic field that produces a positive or negative charge on the particles being carried along in the air stream. The EAC then removes the charged particles from the air stream by drawing them to oppositely charged collector plates.

EACs have a high-voltage, solid-state power supply. The high voltage produced by the power supply, which can range from 6,000 to 10,000 volts DC, is used to electrically charge (ionize) all particles in the air that pass through the filter. After the particles are ionized, the high voltage is then used to attract, collect, and hold the particles in the filter. *Figure 12* shows typical EAC operation.

Figure 10 Humidifier capacity chart.

GALLONS PER DAY HUMIDIFICATION REQUIRED

03312-13_F10.EPS

03312-13_F11.EPS

Figure 11 Electronic air cleaner (EAC).

As shown, the filter portion of an EAC consists of a prefilter, ionizer section, collector section, and charcoal filter section. As the air enters the EAC, larger particles are trapped by the prefilter section. Smaller particles pass through the prefilter to the ionizing section.

This section consists of a fine tungsten wire grid (ionizer grid) connected to the high-voltage DC power supply. An electrostatic ionizing field created by the high voltage on the wire grid gives particles a positive or negative charge as they pass through the grid. These charged particles are then drawn into the collector section.

The collector section consists of a series of equally spaced, parallel collector plates connected to the high-voltage DC power supply; even- and odd-numbered plates are at a positive and negative DC voltage, respectively. As ionized particles flow between the plates, they are attracted to and held on oppositely charged collector plates. The air, cleaned of pollutants, then passes through the charcoal filter section where odors are absorbed. From there, air passes to the conditioned space. Pollutants remain held in the collector section until the filter is cleaned.

Figure 13 shows the schematic diagram for a typical EAC. In the unit shown, the high-voltage power supply is made from discrete components such as diodes, capacitors, and resistors. In newer

8,000VDC
IONIZING VOLTAGE

HIGH-VOLTAGE
DC POWER
SUPPLY

3,000VDC
COLLECTOR VOLTAGE

LARGE PARTICLES
ARE SCREENED BY
PREFILTER

CHARGED PARTICLES
ARE TRAPPED IN
COLLECTOR SECTION

SMALLER PARTICLES
RECEIVE ELECTRIC
CHARGE IN IONIZER
SECTION

CHARCOAL FILTER
ABSORBS ODORS

03312-13_F12.EPS

Figure 12 Electronic air cleaner operation.

Bypass Humidifiers

A bypass humidifier is normally installed on the supply duct with the bypass duct connecting to the return. The humidifier can operate just as effectively if installed on the return duct with the bypass connecting to the supply plenum. This makes bypass humidifiers more versatile than other types of humidifiers, which must always be installed on the supply plenum.

HUMIDIFIER   SOLENOID

BYPASS DAMPER   BYPASS DUCT   RETURN AIR
(CLOSED IN   DUCT
SUMMER)

03312-13_SA01.EPS

modules, the power supply may be a solid-state circuit that is fully enclosed in a non-repairable, sealed assembly.

The EAC is normally electrically connected to the furnace fan circuit and operates only when the furnace blower motor is energized. On some units, this control is through contacts of a sail

Figure 13 Electronic air cleaner schematic.

① INTERLOCK SWITCH.

② SHORTING BAR.

③ AIRFLOW SWITCH DISABLE JUMPER.

④ OPTIONAL W8600F AIR CLEANER MONITOR. DRIVER
   BOARD AND CABLE PROVIDED WITH W8600F.

⑤ NEON LIGHT.

03312-13_F13.EPS

switch installed in the air stream. Others may use contacts of the fan control relay.

Some EACs are electrically interlocked so that the unit shuts off automatically when the service doors are opened to gain access to the unit.

EACs do not clean air effectively when dirty. Electronic cells and protective screens are generally cleaned every two to three months. The electronic cells of most residential models can be safely placed in a dishwasher. Large commercial

Humidifiers and Hot Water

Some bypass and fan-powered humidifier installers tap into a hot water line instead of a cold water line to supply water to the humidifier. However, some manufacturers prohibit this practice. Hotter water vaporizes much more quickly in the humidifier, but the energy used to heat the water increases its cost. If hot water is used to supply a humidifier, the flow rate must be carefully controlled to prevent waste.

models may be equipped with built-in washing apparatus and automated washing controls.

Some manufacturers have designed their equipment so that the power indicator flashes when the EAC needs cleaning. If cleaned regularly, many EAC problems can be avoided. However, some conditions can cause an EAC to become dirty and overloaded in an abnormally short time. These conditions include the following:

• New carpeting has been installed.
• Untreated concrete floors exist.
• Dusty construction work is being performed in the area.
• Extreme amounts of pollen enter the building during the spring.

EAC operation is sensitive to airflow. When operated at airflow rates above those recommended by the manufacturer, they can become quite inefficient. When the airflow is reduced below the manufacturer's recommended minimum, enough ozone can be generated to cause an annoying odor. If the EAC is operating and is clean, but is not cleaning the air, or if the smell of ozone is present, check to make sure that the system airflow matches the EAC specifications.

The recommended procedure for troubleshooting problems in an EAC is to first perform a visual inspection to identify any mechanical damage that might be the cause of the problem. For electrical problems, use the troubleshooting aids in the manufacturer's service literature to fault-isolate the unit. *Figure 14* shows a typical manufacturer's troubleshooting flow chart. Note that the chart calls for a reduced ionizer voltage (7,500 to 8,500VDC) when the cells are in place, compared to the value with the cells removed. This is due to the voltage drop that occurs when a load is imposed on the high-voltage transformer. This is also true of the collector voltage, at roughly the same percentage of reduction.

> **CAUTION**
>
> If measuring high voltage while troubleshooting an electronic air cleaner, use a voltmeter that is designed and rated for measuring the high voltages encountered in the unit.

1.4.0 Troubleshooting Ultraviolet Lamps

Ultraviolet (UV) lamps (*Figure 15*) are used for microbe control on evaporative coils, drain pans, and duct insulation. They are also used to kill airborne pathogens and may be used in conjunction with an electronic air cleaner. Two different methods are used for air purification. One highly effective method constantly bathes a fixed object such as an evaporator coil with UV light. The other method purifies the air with UV lamps positioned in a moving airstream. Since microbes in a moving airstream can move past the lamps fairly fast, their exposure to UV rays is limited, making this method of air purification less effective.

Saddle Valves

To obtain water for a humidifier, a nearby water line is often tapped using a saddle valve. This valve is clamped around a copper water line and contains a piece of rubber that acts as a gasket to prevent water leakage. The water line is tapped, or penetrated, by turning in the handle of the valve. This causes a steel point to penetrate the wall of the copper tube. When the tube is pierced, the steel point seats itself against the copper to shut off or allow water flow. One manufacturer states that adequate humidifier water flow can be obtained by opening the valve no more than one turn. Some local codes do not allow the use of saddle valves. They do have a tendency to leak over time, as the rubber seal dries out and hardens.

03312-13_SA02.EPS

Ionizer Voltage (Y to Ground)

    No Cells – 8,500VDC Minimum
    With Cells – 7,500 to 8,500VDC

Collector Voltage (R to Ground)
    No Cells – 3,500VDC Minimum
    With Cells – 3,000VDC Minimum

Secondary Voltage Without Cells – 3,200 to 4,100VAC

03312-13_F14.EPS

Figure 14 Electronic air cleaner manufacturer's troubleshooting flow chart.

03312-13_F15.EPS

Figure 15 Ultraviolet lamp.

WARNING!

The light from ultraviolet lamps can damage eyes. When servicing these lamps, follow all manufacturer safety warnings and make sure the lamp is switched off and disconnected from the power supply.

There is not much to troubleshooting most ultraviolet lamps. If a lamp is not working, check its power supply. Check the lamp itself for physical damage and make sure no debris is in the lamp socket.

Ultraviolet lamps must be replaced after a certain number of hours of operation. Consult the manufacturer's instructions for information about the life span of the lamps and at what intervals they should be changed.

WARNING!

Ultraviolet lamps contain a small quantity of mercury, a hazardous substance. If a lamp breaks during handling, use caution when cleaning up the debris and dispose according to local regulations.

Connecting Accessory Devices

Many HVAC equipment manufacturers pre-wire their products with sets of terminals to which accessory devices such as humidifiers and air cleaner may be connected. This eliminates wiring errors and field modification of their product to accommodate the accessory device. The manufacturer of this furnace has supplied line-voltage terminals to which an electronic air cleaner can be connected and 24V terminals for connecting a humidifier.

Additional Resources

Refrigeration and Air Conditioning, An Introduction to HVAC/R, Fourth Edition. Larry Jeffus. Air Conditioning and Refrigeration Institute. Prentice Hall.

1.0.0 Section Review

1. A humidifier that works with a forced-air furnace system will operate if the blower is not running.

   a. True
   b. False

2. In the control circuit of an infrared humidifier, devices are typically connected _____.

   a. in parallel with the humidistat
   b. next to the infrared lamps
   c. in series with the humidistat
   d. in parallel with the blower motor

3. What does the presence of an ozone odor during electronic air cleaner operation indicate?

   a. Low airflow
   b. Normal airflow
   c. High airflow
   d. Low voltage

4. Why are microbes in a moving airstream less likely to be killed by UV light?

   a. They are moving too fast.
   b. They are immune to UV light.
   c. The light is not concentrated.
   d. The air acts as an insulator.

SECTION TWO

2.0.0 OUTSIDE AIR ACCESSORY TROUBLESHOOTING

Objective

Describe how to troubleshoot accessories related to the introduction of outside air.
  a. Describe how to troubleshoot economizers.
  b. Describe how to troubleshoot recovery ventilators.

Performance Task 1

Using the correct tools and circuit diagrams, isolate and correct malfunctions in selected accessories.

Trade Terms

Free cooling: A mode of economizer operation. During cooling operation, the system compressor is turned off and the indoor fan is used to bring outside air into the building to provide cooling.

Mechanical cooling: The cooling provided in the conventional manner by the compressor and mechanical refrigeration system.

Before HVAC equipment was readily available, staying comfortable was a much different task. If it became too warm in a building or the air seemed a little stale, the solution was simple: open a window and let in cool, fresh air. The result was not often effective; the biggest drawback was no control over the quality and temperature of the air entering the building. In addition, dust and pollen could easily penetrate many window screens. Modern HVAC equipment also has the ability to bring in outside air, but the difference is that equipment is able to control the quantity, temperature, and quality of the air that enters. Two devices commonly used for this purpose are economizers and recovery ventilators.

2.1.0 Economizer Operation

An economizer (*Figure 16*) is an accessory for commercial units that reduce operating cost during cooling by using outdoor air for cooling (**free cooling**) and, whenever possible, reducing compressor (**mechanical cooling**) operation. It does

this by controlling the amount of outside air that is brought into a conditioned space. *Figure 17* shows a basic economizer system. It consists of a damper actuator assembly and related economizer control module.

Control signals applied to the economizer control module come from three sources: the thermostat located in the conditioned space, an enthalpy (total heat) sensor located in the outdoor air duct, and the discharge air sensor located on the discharge side of the system evaporator coil.

The setpoint controls the changeover from mechanical to free cooling. The outdoor air enthalpy sensor monitors the air dry-bulb temperature and humidity, allowing the use of outdoor air at higher temperatures for free cooling, as long as the humidity is low. The mixed, or discharge, air sensor monitors the average air temperature on the leaving side of the indoor coil and compares this to the predetermined setpoint. It also monitors the temperature of the return air coming from the conditioned space.

There are many types of economizers. A simplified schematic for an economizer is shown in *Figure 18*. The operation of this unit, as briefly described here, is similar to that of many economizers found in the field.

On a call for cooling by the space thermostat, the system operates as follows:

When the enthalpy of the outdoor air is below the setpoint, the outdoor air damper is driven open and the return air damper is driven closed to maintain a temperature selected by the setpoint of the mixed/discharge air sensor. Typically, this is between 50°F and 56°F.

If the enthalpy of the outdoor air is above the setpoint, the outdoor air damper closes to its preset minimum position. A call for cooling from the

Figure 16 Rooftop unit with an economizer installed.

OUTSIDE AIR INTAKE

RELIEF AIR DAMPERS

03312-13_F16.EPS

MIXED AIR AND MINIMUM POSITION ADJUSTMENT CONTROLS UNDER COVER

DAMPER ACTUATOR

DAMPER MOTOR

ECONOMIZER CONTROL MODULE

SPACE THERMOSTAT

ENTHALPY SENSOR

OUTDOOR AIR

DISCHARGE AIR SENSOR

INDOOR COIL

ECONOMIZER CONTROL MODULE

DAMPER ACTUATOR

ENTHALPY SENSOR

RETURN AIR

03312-13_F17.EPS

Figure 17 Basic economizer system.

space thermostat turns on mechanical cooling. During the unoccupied period, the damper actuator spring returns the outdoor damper to the fully closed position. A minimum position for outside air is typically not required when the building is unoccupied. Closing the outside air damper during this period saves energy.

With a two-stage cooling thermostat, the first stage will always be satisfied by economizer operation, assuming outdoor temperature conditions are suitable. The mechanical cooling can be operated by the second stage of a cooling thermostat.

Some economizer systems use a second enthalpy sensor located in the return duct. This control method is called differential enthalpy. The economizer control selects the air (outside or re-turn) with the lower enthalpy for cooling. If the outdoor air has lower enthalpy than the return air, then the outdoor air damper will be opened to bring in outdoor air for free cooling.

2.1.1 Economizer Types

When troubleshooting economizers, three basic types will be encountered: dry-bulb economizers, enthalpy economizers, and integrated economizers. The dry-bulb economizer (*Figure 19*) uses an outdoor air temperature sensor (O_{AT}) to activate a control relay whenever the outdoor dry-bulb temperature falls below a user-specified value (typically 55°F). The control relay (CR) locks out the compressor. Space cooling is accomplished by

Figure 18 Simplified schematic of an economizer.

modulating the amount of outdoor air entering the return side of the supply air fan. As the outdoor air temperature falls below 55°F, more return air may be mixed with outdoor air to maintain a 55°F supply air temperature. This is the simplest and least costly type of economizer control.

The enthalpy economizer (*Figure 20*) uses an outdoor air temperature sensor (O_{AT}) wired in series with an enthalpy sensor (hOA) and a control relay (CR). When the outdoor air temperature is below 60°F and the outdoor enthalpy is below 24.7 Btu/lb (both common setpoints), the control relay activates to prevent compressor operation.

Economizers

In addition to energy savings and reduced long-term maintenance costs, economizers provide the natural advantage of good building ventilation. While in the free-cooling mode, the system is using anywhere from 50 to 100 percent outside air. This level typically exceeds the code-required ventilation rates for most building applications. Economizers also provide the option of ventilating a building at the needed rate based on occupancy. This is normally achieved by using CO_2 sensors mounted in rooms that may have high occupancy rates; conference rooms, assembly halls, restaurants, training rooms, and meeting centers are examples. These rooms are often used on a sporadic scheduling basis and therefore less than fully occupied much of the time. By interlocking a CO_2 sensor with the economizer, the ventilation rate can be controlled when the system is either in a design cooling or design heating mode, when 100 percent outside air is not desirable. The CO_2 sensor only increases the rate of ventilation to match the human occupancy load, thus maximizing both energy savings and human comfort with superior ventilation. This control sequence is called demand control ventilation (DCV) and is becoming an expected feature in medium-to-heavy-commercial applications.

Figure 19 Dry-bulb economizer.

Zone cooling is accomplished by modulating the amount of outdoor air entering the return side of the supply air fan. An enthalpy sensor is able to sense the total heat in the air (sensible heat and latent heat). The outdoor air temperature sensor used with dry-bulb economizers can only detect sensible heat. An enthalpy economizer will prevent the entry of overly-humid outdoor air into a structure, even if the dry-bulb temperature of the air is fairly low.

The integrated economizer (*Figure 21*) uses two sensors to measure enthalpy: a temperature sensor and humidity sensor. An outdoor air temperature sensor (O_{AT}), outdoor air humidity sensor (hOA), return air temperature sensor (R_{AT}), and return air humidity sensor (hOA) are wired to the

controls of the central unit. The controller calculates outdoor air and return air enthalpy values. Below a pre-determined outdoor air temperature (55°F in this example), the economizer provides cooling with the compressor off, just like a dry-bulb economizer.

For outdoor air temperatures above 55°F, when the outdoor enthalpy is less than the return air enthalpy, 100 percent outdoor air is used to assist compression cooling. When the outdoor air enthalpy is above the return air enthalpy, outdoor air dampers are set to the minimum ventilation position and the compressor provides the needed cooling capacity.

2.1.2 Troubleshooting Economizers

Like many HVAC products, economizers operate basically the same regardless of the manufacturer. For that reason, basic troubleshooting steps are the same. However, before troubleshooting any economizer, always consult the specific manufacturer's service literature before proceeding. Although the functional concepts may remain the same, the controls used can differ significantly.

Economizers depend on dampers to control and mix air. Devices which control the dampers are electronic modules and electronic sensors that provide inputs to the modules. The electronic controls and their sensors tend to be very reliable and are not as prone to problems or failure as the elec-

Figure 20 Enthalpy economizer.

Figure 21 Integrated economizer.

Economizer Control Modules

Older economizer control modules used conventional relay circuitry, while most new ones are solid-state electronic modules.

tromechanical components in the system, such as the dampers and their drive motors (actuators). When first approaching a system equipped with an economizer, it is important to check for normal operation. If any problems are found, proceed to find the root cause.

The following procedure should be used to confirm normal operation:

- *Check sensors* – Temperature sensors used with modern economizers tend to be electronic and very reliable. However, their placement and connections to the control module can affect economizer operation. Check that all sensors are placed in locations according to the manufacturer's instructions. Check the sensors for cut or pinched leads, or loose and/or corroded connections.
- *Check outdoor air damper operation* – The outdoor air damper can be seen under the rain hood of the economizer, or from the return air plenum. Under certain occupancy conditions, the damper may be fully closed. Shut off power to the rooftop unit at the disconnect. With power removed, the outdoor air damper will slowly close. Restoring power should return the damper to its minimum position. With the damper in its minimum position, vary the potentiometer on the control module that controls the minimum position. If determined that varying the potentiometer changes damper position, return the potentiometer to its original position. It is also important to ensure that the mechanical operation of the damper and linkages is smooth and does not bind.
- *Check economizer operation* – Disconnect the room thermostat from the economizer and use a jumper wire to simulate a first-stage call-for-cooling. With a two-stage cooling thermostat, the first stage will always be satisfied by economizer operation, assuming outdoor temperature conditions are suitable. If outdoor temperature is low enough, the outdoor air damper and return air damper will modulate and the blower in the rooftop unit will mix the air supplied to the conditioned space. A low outdoor air temperature can be simulated by cooling the outdoor air temperature sensor with ice in a plastic bag for testing purposes.

The dampers and damper motors must function properly for an economizer to be effective. Leaky or inoperative dampers waste energy. To troubleshoot dampers (*Figure 22*), make the following checks:

- Check that all voltages and electrical signals used to actuate the damper are within specifi-

Figure 22 Motorized rectangular damper.

cations. Ensure that all sensors are positioned correctly for proper sensing.
- Observe damper motors and actuators through an operating cycle to check for defects or binding. Make sure that all mounting bolts are securely fastened.
- Check that the linkage from the actuators is adjusted to ensure that the damper blades fully open or close within the stroke or travel of the actuator arm.
- Check that all blades close tightly in the closed position. If necessary, adjustments should be made to damper linkage(s) to close any partially open blades. Remember that dampers will not always be commanded to close fully, especially in an occupied mode of operation. Do not make the error of adjusting dampers to a fully closed position when the controls are not calling for this setting.
- Make sure that all damaged blades are replaced. Dirt, soot, and lint should be removed, especially around moving parts.
- Check the blade edge and side seals. Replace where necessary.
- Check the pins, straps, and bushings/bearings for wear, rust, or corrosion and replace any as required.
- Inspect the condition of any caulking used to make damper frames tight to the unit.

2.2.0 Recovery Ventilators

Energy recovery ventilators (ERVs) and heat recovery ventilators (HRVs) draw fresh outdoor air into the home. At the same time, they expel stale indoor air to the outside (*Figure 23*). Expelled air carries odors and other unwanted contaminants with it.

While removing stale air from the home, recovery ventilators also recover much of the heat and/or cooling energy from the outgoing indoor air and transfer it to the fresh incoming air.

(A)

ENERGY RECOVERY
VENTILATOR

FRESH
OUTDOOR AIR

EXHAUST AIR

(B)

03312-13_F23.EPS

Figure 23 Energy recovery ventilator.

ERV and HRV units are similar in construction and in operation. The heart of both units is a special heat exchanger. In the ERV, this heat exchanger may be referred to as an energy recovery core, since it transfers both sensible and latent heat between the fresh incoming air and the stale air, allowing it to remove moisture from the air during the summer cooling season. In the HRV, the heat exchanger operates in a similar manner, except that it is designed to transfer only sensible heat between the entering and leaving air streams.

ERVs provide more efficient year-round operation than HRVs. For this reason, they are used in most areas of the United States and some parts of Canada. HRVs are used to supply fresh air, quickly remove excessive moisture, and recover heat energy during the heating season. They are more popular in areas with colder climates and extended heating seasons.

An ERV/HRV can be installed as a standalone unit with its own ductwork. However, many are installed as part of a forced-air heating/cooling system. This can influence the load imposed on the HVAC system blower. Also, for the ventilator

Combustion Air

A recovery ventilator does more than just refresh indoor air. In a tightly built structure with a fossil fuel furnace, the system can be designed so that the ventilator brings fresh combustion air into the structure.

to operate properly in these cases, the system capacity must be able to handle the additional heating and/or cooling load from the introduction of outside air. Although the heat exchange process is significant, these units are not capable of total energy transfer. As a result, air that is colder or warmer than the desired temperature indoors is almost always being introduced.

The ERV/HRV can sometimes be controlled independently by a simple wall-mounted control. A blower interlock relay incorporated in the furnace/fan coil control circuit ties the operation of the ERV/HRV unit to the furnace or fan coil blower.

When the ERV/HRV is in operation, the interlock relay is energized, causing the furnace or fan coil blower to operate. To maintain a good transfer of energy and proper airflow, the air system must be properly balanced and the ERV/HRV components clean of dust and dirt. The system must be balanced so intake and exhaust airflows are about equal. Typically, the ventilator is set up to bring in slightly more air than it exhausts. This results in a slightly positive pressure in the structure. If the structure is at a negative pressure, outdoor air is drawn into the space uncontrolled around windows and doors. Balancing is done with a balancing kit at installation and any time an airflow problem is suspected.

The ERV/HRV air filters, screens, and recovery core must be cleaned to remove dust and dirt. These tasks should be performed at intervals recommended by the ventilator manufacturer. Typically, the recovery core of an ERV should be cleaned about every three months. The core of an HRV should be cleaned about once each year at the end of the heating season.

> **NOTE**
>
> Some cores are made of cardboard and cannot be washed with water. Follow the manufacturer's instructions for the specific model being serviced carefully.

Troubleshooting an inoperative or malfunctioning ERV/HRV involves isolating the problem to one of its three main parts: the wall control,

electronic control board, and fan motors. Normally, the wall control is either functional or defective, which can be determined easily enough with a VOM. In many cases, it is simply one or two switches. To troubleshoot the electronic control board, the wall control must be electrically connected before the unit will operate properly. Any configuration jumpers located on the board must match the configuration specified in the service literature. The easiest way to check their operation is to use the wall control to initiate low-speed and high-speed operation.

Troubleshooting electrical problems in a ventilator must always be done in accordance with the detailed troubleshooting procedures and diagrams provided in the service literature supplied with the unit.

Typical recovery ventilator problems and possible causes are as follows:

- *Unit not operating* – Check power supply. Check that operating controls are properly set.
- *Inadequate airflow* – Check outside air intake and connecting duct for blockage. Ensure that any dampers are open and no duct air leaks or disconnected ducts are present. Check that the heat exchanger is not dirty or frozen.
- *Frozen coil* – Some frost on the heat exchanger is normal. Check the defrost circuits. Check system airflow balance.
- *Excess humidity* – Adjust dehumidifier setpoint. Remove sources of excess humidity. Check system airflow balance.
- *Inadequate humidity* – Adjust dehumidifier setpoint. Check/adjust recovery ventilator blower speed.
- *Water in unit* – Check for plugged drain pan/drain lines. Check if unit is level.
- *Noisy operation* – Check fans and fan motors. Check recovery core for cleanliness.

Figure 24 is an example of troubleshooting information provided by one HRV manufacturer.

Some form of air filtration is usually a part of an ERV/HRV unit as well. The filter must be serviced periodically. If the unit is operated without a filter, costly and possibly permanent damage can occur to the heat exchanger of the unit.

TROUBLESHOOTING YOUR HRV SYSTEM		
SYMPTOM	**CAUSE**	**SOLUTION**
Poor air flows	• 1/4" (6 mm) mesh on the outside hoods is plugged • filters plugged • core obstructed • house grilles closed or blocked • dampers are closed if installed • poor power supply at site • ductwork is restricting HRV • improper speed control setting • HRV airflow improperly balanced	• clean exterior hoods or vents • remove and clean filter • remove and clean core • check and open grilles • open and adjust dampers • have electrician check supply voltage at house • check duct installation • increase the speed of the HRV • have contractor balance HRV
Supply air feels cold	• poor location of supply grilles, the airflow may irritate the occupant • outdoor temperature extremely cold	• locate the grilles high on the walls or under the baseboards, install ceiling mounted diffuser or grilles so as not to directly spill the supply air on the occupant (eg, over a sofa) • turn down the HRV supply speed. A small duct heater (1 kw) could be used to temper the supply air • placement of furniture or closed doors is restricting the movement of air in the home • if supply air is ducted into furnace return, the furnace fan may need to run continuously to distribute ventilation air comfortably
Dehumidistat is not operating	• outdoor temperature is above 15°C (59°F) • improper low voltage connection • external low voltage is shorted out by a staple or nail • check dehumidistat setting; it may be on OFF	• dehumidistat is functioning normally • check that the correct terminals have been used • check external wiring for a short • set the dehumidistat at the desired setting
Humidity levels are too high. Condensation is appearing on the windows	• dehumidistat is set too high • HRV is undersized to handle a hot tub, indoor pool, etc. • lifestyle of the occupants • moisture coming into the home from an unvented or unheated crawl space • moisture is remaining in the washroom and kitchen areas • condensation seems to form in the spring and fall • HRV is set at too low a speed	• set dehumidistat lower • cover pools, hot tubs when they are not in use • avoid hanging clothes to dry, storing wood and venting clothes dryer inside. Heating wood may have to be moved outside • vent crawl space and place a vapor barrier on the floor of the crawl space • ducts from the washroom should be sized to remove moist air as effectively as possible, use of a bathroom fan for short periods will remove additional moisture • on humid days, as the seasons change, some condensation may appear, but the home's air quality will remain high with some HRV use • increase speed of the HRV
Humidity levels are too low	• dehumidistat control is set too low • blower speed of HRV is too high • lifestyle of occupants • HRV air flows may be improperly balanced	• set dehumidistat higher • decrease HRV blower speed • humidity may have to be added through the use of humidifiers • have a contractor balance HRV airflows
HRV and / or ducts frosting up	• HRV air flows are improperly balanced • malfunction of the HRV defrost system	• Note: minimal frost build-up is expected on cores before unit initiates defrost cycle functions • have HVAC contractor balance the HRV • ensure damper defrost is operating during self-test
Condensation or ice build-up in insulated duct to the outside	• incomplete vapor barrier around insulated duct • a hole or tear in outer duct covering	• tape and seal all joints • tape any holes or tears made in the outer duct covering • ensure that the vapor barrier is completely sealed
Water in the bottom of the HRV	• drain pans plugged • improper connection of HRV's drain lines • HRV is not level • drain lines are obstructed • HRV heat exchange core is not properly installed	• ensure O-ring on drain nozzle sits properly • look for kinks in line • check water drain connections • make sure water drains properly from pan

03312-13_F24.EPS

Figure 24 Recovery ventilator troubleshooting chart.

Make-Up Air Devices

Recovery ventilators refresh the indoor air and can also replace and supply air used for combustion in tightly-built structures. Combustion air can be supplied to a fuel-burning appliance in other ways without the expense of installing a recovery ventilator. Make-up air devices simply supply combustion air without the complexity found in an energy-recovery device. The simplest make-up air device is an air duct that brings in outside air to the vicinity of the furnace or boiler. A screen in the intake stops debris from entering the duct. One manufacturer offers a device that installs on the return duct of a forced air furnace. When the blower runs, the negative pressure created in the return duct draws in outside air through a duct connected to the device. A barometric damper in the device opens under negative pressure and helps keep air from entering the return duct when the blower is not running. Several other manufacturers offer motorized make-up air controls that provide air to the vicinity of a furnace or boiler. Combustion air is only supplied when there is a call-for-heat signal from the room thermostat. On a call-for-heat, the blower in the make-up air control starts, drawing in outdoor air through a duct. When combustion airflow is proven by a flow-sensing switch, the make-up air control allows the normal burner ignition sequence to occur. Some motorized make-up air devices temper the cold incoming air by mixing it with indoor air. In cold climates, the duct that brings in outside air must be insulated to prevent condensation and/or frost from forming on the duct.

03312-13_SA04.EPS

Additional Resources

Refrigeration and Air Conditioning, An Introduction to HVAC/R, Fourth Edition. Larry Jeffus. Air Conditioning and Refrigeration Institute. Prentice Hall.

2.0.0 Section Review

1. What about an enthalpy economizer prevents overly humid air from entering a structure?
   a. It contains a de-humidifier.
   b. It can sense latent heat in the air.
   c. It contains tightly fitted dampers.
   d. It contains multiple enthalpy sensors.

2. The recovery core of a heat recovery ventilator (HRV) recovers _____.
   a. all indoor heat
   b. latent heat only
   c. sensible and latent heat
   d. sensible heat only

SUMMARY

To effectively troubleshoot HVAC accessories, the HVAC technician listens to a customer's complaint, performs an independent analysis of a problem, then initiates and performs a systematic, step-by-step troubleshooting method that results in the correction of the problem. The technician must understand the purpose and principles of operation of each component in the accessory being serviced. Also, the relationship between the operation of the accessory and the main HVAC equipment with which it is used must also be understood. The technician must be able to tell whether a given device is functioning properly and be able to recognize the symptoms arising from the improper operation of any part of the equipment. On the basis of symptoms revealed through analysis of a system's operation, the technician should use the troubleshooting aids provided by the system manufacturer to identify and repair the cause of the problem.

1. Some humidifiers are controlled by a(n) _____.
   a. check valve
   b. economizer
   c. ventilator
   d. humidistat

2. If a humidifier is not running, the problem is usually caused by a(n) _____.
   a. faulty control circuit
   b. poor water quality
   c. high RH condition
   d. oversized furnace

3. When a humidifier experiences water overflow, the problem is often caused by a(n) _____.
   a. faulty control circuit
   b. dirty float valve
   c. high RH condition
   d. open solenoid coil

4. Signs of high relative humidity include _____.
   a. dry, itchy skin
   b. loose furniture joints
   c. static electricity shocks
   d. condensation on windows

5. One possible cause of a low humidity complaint could be the _____.
   a. float valve is set too high
   b. water supply feed valve closed
   c. furnace blower running in high speed
   d. condensing unit is unable to start

6. A tight building is characterized by the _____.
   a. lack of a vapor barrier
   b. presence of a fireplace
   c. lack of insulation
   d. presence of tight windows

7. In an electronic air cleaner, the output of the high-voltage power supply is applied to _____.
   a. the ionizer section
   b. the collector section
   c. both the collector and ionizer
   d. the prefilter

8. When an electronic air cleaner is operated at an airflow above that recommended by the manufacturer, the filter's cleaning ability is _____.
   a. slightly better
   b. worse
   c. the same
   d. doubled

9. The enthalpy sensor(s) used with an economizer control _____.
   a. room temperature enthalpy only
   b. outdoor air humidity enthalpy only
   c. changeover from free cooling to mechanical cooling
   d. both room temperature and humidity enthalpy

10. An economizer uses a mixed/discharge air sensor to control the open or closed position of _____.
    a. both the outdoor and return dampers during free cooling
    b. the outdoor damper during free cooling
    c. the outdoor damper during mechanical cooling
    d. both the outdoor and return dampers during both free and mechanical cooling

11. In an economizer that uses differential enthalpy sensing, what type of air does the economizer control module select for cooling?
    a. only outside
    b. only return
    c. either outside or return
    d. mixed/discharge

12. How many air temperature sensors does a dry-bulb economizer use?
    a. Multiple
    b. One
    c. Two
    d. Three

13. When troubleshooting a problem suspected to be caused by a malfunctioning economizer damper, one of the first things to do is _____.

    a. check the control voltage applied to the compressor
    b. check an operating damper for defects and binding
    c. jumper the Y-C contacts on the room thermostat
    d. replace the damper with a known good one

14. A device that uses a heat exchanger to transfer heat from warmed indoor air to ventilation air entering a building is a(n) _____.

    a. economizer
    b. enthalpy control
    c. recovery ventilator
    d. mixed air thermostat

15. Energy recovery ventilators are more likely to be found _____.

    a. only in cold climates
    b. throughout the United States
    c. installed with heat pumps
    d. installed with hydronic systems

Trade Terms Introduced in This Module

Free cooling: A mode of economizer operation. During cooling operation, the system compressor is turned off and the indoor fan is used to bring outside air into the building to provide cooling.

Mechanical cooling: The cooling provided in the conventional manner by the compressor and mechanical refrigeration system.

Water hammer: A banging noise in water pipes that occurs when the water supply solenoid valve on a humidifier abruptly shuts off water to the humidifier.

Additional Resources

This module presents thorough resources for task training. The following resource material is suggested for further study.

Refrigeration and Air Conditioning, An Introduction to HVAC/R, Fourth Edition. Larry Jeffus. Air Conditioning and Refrigeration Institute. Prentice Hall.

Figure Credits

Section Review Answer Key

Answer	Section Reference	Objective
Section One		
1. b	1.1.3	1a
2. c	1.2.0	1b
3. a	1.3.0	1c
4. a	1.4.0	1d
Section Two		
1. b	2.1.1	2a
2. d	2.2.0	2b

NCCER CURRICULA — USER UPDATE

NCCER makes every effort to keep its textbooks up-to-date and free of technical errors. We appreciate your help in this process. If you find an error, a typographical mistake, or an inaccuracy in NCCER's curricula, please fill out this form (or a photocopy), or complete the online form at **www.nccer.org/olf**. Be sure to include the exact module ID number, page number, a detailed description, and your recommended correction. Your input will be brought to the attention of the Authoring Team. Thank you for your assistance.

Instructors – If you have an idea for improving this textbook, or have found that additional materials were necessary to teach this module effectively, please let us know so that we may present your suggestions to the Authoring Team.

NCCER Product Development and Revision
13614 Progress Blvd., Alachua, FL 32615

Email: curriculum@nccer.org
Online: www.nccer.org/olf

❏ Trainee Guide   ❏ Lesson Plans   ❏ Exam   ❏ PowerPoints   Other _____

Craft / Level: _____ Copyright Date: _____

Module ID Number / Title: _____

Section Number(s): _____

Description: _____

_____

_____

_____

_____

Recommended Correction: _____

_____

_____

_____

_____

Your Name: _____

Address: _____

_____

Email: _____ Phone: _____

Zoning, Ductless, and Variable Refrigerant Flow Systems

OVERVIEW

Although VAV systems provide excellent zoned comfort in large commercial applications, they are not practical or cost-effective for smaller applications. A zoning system that uses a single air source but distributes the air in a highly controlled manner is one approach that makes a standard constant-volume system more effective. Ductless systems may be single zone or multi-zone, offering small capacity heating and cooling to one or more areas. Variable refrigerant volume systems take this approach a step further with increased capacity. This module explores the operation and application of these innovative systems.

Module Eight

Trainees with successful module completions may be eligible for credentialing through NCCER's National Registry. To learn more, go to **www.nccer.org** or contact us at **1.888.622.3720**. Our website has information on the latest product releases and training, as well as online versions of our *Cornerstone* magazine and Pearson's product catalog.

Your feedback is welcome. You may email your comments to **curriculum@nccer.org**, send general comments and inquiries to **info@nccer.org**, or fill in the User Update form at the back of this module.

This information is general in nature and intended for training purposes only. Actual performance of activities described in this manual requires compliance with all applicable operating, service, maintenance, and safety procedures under the direction of qualified personnel. References in this manual to patented or proprietary devices do not constitute a recommendation of their use.

Objectives

When you have completed this module, you will be able to do the following:

1. Identify common zoning systems and describe the basic approach to troubleshooting.
   a. Describe common zoning system components.
   b. Describe the sequence of operation for common zoning systems.
   c. Describe the basic approach to troubleshooting common zoning systems.
   d. Describe common VVT system zoning components.
   e. Describe the operating characteristics of VVT control systems.
2. Identify ductless and variable refrigerant flow systems and describe the basic approach to troubleshooting.
   a. Identify and describe the operation of ductless split-system equipment.
   b. Describe the installation and troubleshooting of typical ductless systems.
   c. Identify and describe the operation of variable refrigerant flow systems.
   d. Describe the installation and troubleshooting of variable refrigerant flow systems.

Performance Tasks

Under the supervision of your instructor, you should be able to do the following:

1. Troubleshoot and repair a typical ductless or variable refrigerant flow system.
2. Program the controller for a ductless or variable refrigerant flow system.

Trade Terms

Dump zone
Polling
Pressure-dependent operation

Pressure-independent operation
Refnet joints

Industry-Recognized Credentials

If you are training through an NCCER-accredited sponsor, you may be eligible for credentials from NCCER's Registry. The ID number for this module is 03315-13. Note that this module may have been used in other NCCER curricula and may apply to other level completions. Contact NCCER's Registry at 888.622.3720 or go to **www.nccer.org** for more information.

Contents

Topics to be presented in this module include:

Figures and Tables

1.0.0 COMMON ZONING SYSTEMS

Objectives

Identify common zoning systems and describe the basic approach to troubleshooting.

  a. Describe common zoning system components.
  b. Describe the sequence of operation for common zoning systems.
  c. Describe the basic approach to troubleshooting common zoning systems.
  d. Describe common VVT system zoning components.
  e. Describe the operating characteristics of VVT control systems.

Trade Terms

Dump zone: Area used as a point to bypass unneeded supply air from a zoning system. The dump zone is typically in the conditioned space, but not an area where maintaining the space temperature at a particular setpoint is required. A dump zone generally has no supply air grilles, but a return grille is often installed to relieve room pressure and return mixed air back to the system.

Polling: A network communications arrangement where a VVT system seeks information and needs from individual zones, determines what demands have priority, and then responds accordingly. Essentially, members of a polling system are voters.

Pressure-dependent operation: A VVT-system approach where dampers are positioned based on demand only, or operate in two positions. The result is that the design volume of air may or may not be flowing through the damper at any time. The volume of air flow through each damper is dependent on the duct pressure; the air volume fluctuates as the pressure fluctuates. A minimum damper position is set, but air volume is not a controlling factor, and so the minimum damper position is a fixed point.

Pressure-independent operation: A VVT-system approach where dampers are positioned by demand and by the actual volume of air flowing through the damper. To achieve a specific amount of airflow, the damper positions itself as necessary. A maximum and minimum airflow setpoint is effective, and the damper will modulate as necessary to remain within these boundaries.

Traditional heating and cooling systems maintain the desired conditions throughout a home or small business by answering to a single thermostat. The system has no idea of the conditions in every room or area; it responds only to the conditions at the thermostat location. This often results in areas that are too hot or too cold, even with a properly balanced system. Throughout a typical day and night, the load in a given area can change dramatically as the sun moves across the sky, or as internal loads cycle on and off. In addition, the system can respond to a single programmable schedule of operation, using a programmable thermostat. A zoning system allows a single heating and cooling system to respond to two or more thermostats, providing conditioned air only to those zones that are presently in need of heating or cooling. This enables a single system to accommodate the needs of the structure in a more precise way, and saves energy. For example, the bedrooms of a home may be considered one zone, while the living area is considered as another zone. If the bedrooms are unoccupied all day, the temperature in that zone(s) can be set up or set back. The system can then focus only on providing conditioned air to the living area during the day. At night, the living area can then be set up or set back, while the system focuses on providing comfort in the bedroom areas. However, the system also possesses the ability to heat or cool all the zones at the same time.

This section will explore basic zoning systems and how to test and troubleshoot them when problems arise.

1.1.0 Zoning System Components

Zone control provides a number of separately controlled spaces or zones within a building where different heating or cooling temperatures can be controlled at the same time. Each of these areas may be referred to as a control zone. This provides a way to overcome variations in cooling or heating loads that occur in different areas of the building at different times. In addition, it provides occupants with a means to provided different temperatures in various areas, and at different times of the day.

A typical zoned system (*Figure 1*) consists of an electronic control panel that is connected to several zone thermostats and motorized dampers. Each thermostat reports its needs, if any, to the control panel in much the same way that all thermostats operate. They are normally powered by a 24VAC control circuit, and contacts open or close in the thermostats when heating or cooling is needed. The zone damper, which is installed in the supply duct to each zone, controls the flow of conditioned air to the zone. The control panel is also connected to the control circuits of the building heating/cooling equipment.

1.1.1 Zone Thermostats

Most zone thermostats (*Figure 2*) look like any other thermostat. They usually have buttons or a touch screen that allow the user to change the setpoint(s) and the mode of operation. Programmable models allow the user to set schedules for setpoint changes for individual zones. For most zoning systems, the type of thermostat used must be compatible with the HVAC equipment. For example, if a heat pump is being used, the zone thermostats must be heat pump models. However, this is not true for every system.

In some older simple systems, the thermostats in individual zones do not have a means to change the mode of operation. In these systems, one thermostat is designated as the master (*Figure 3*). Only the master thermostat provides mode switching capability. The system will operate in the heating or cooling mode, or can be set for auto-changeover if the thermostat is equipped with this function. All other zone thermostats, such as the one

also shown in *Figure 3*, are programmed with setpoints, but are permanently set in an auto-changeover mode. If a zone calls for heating, heating can only be provided if the master thermostat is set to allow that mode. Otherwise, there is no system response.

03315-13_F02.EPS

Figure 2 Programmable zone thermostat.

03315-13_F01.EPS

Figure 1 Zoned system schematic.

SYSTEM AND FAN SWITCHES

MASTER THERMOSTAT

NO SYSTEM AND FAN SWITCHES

ZONE THERMOSTAT

03315-13_F03.EPS

Figure 3 Master thermostat and matching zone thermostat.

1.1.2 Zone Dampers

Zone dampers are rectangular or round (*Figure 4*). Both are available in a wide array of sizes. Most zone dampers are two-position models. They are usually powered to their closed position, and are equipped with springs that return the damper to its open position when the motor is de-energized. This may seem backwards, as one might expect the damper to be powered open when heating or cooling is called for by the thermostat. However, by maintaining power on the damper when it is closed and using a spring to open it when it is de-energized, the damper becomes fail-safe. If the motor or its power source fails, the damper will open and remain in this position. This ensures that a zone can still be conditioned in exceptionally cold or hot weather, in spite of the failure.

Some dampers can modulate their position, stopping at any point between open and closed. Rather than being equipped with a spring, they are powered by a motor in both directions. This is a function of the motor rather than the damper. Either type of motor can generally be attached to the same damper assembly.

Even two-position dampers can be adjusted so that they do not fully close when powered in that direction. This is referred to as a minimum position setting. Preventing a zone damper from fully closing ensures that the zone will always receive a small amount of air for ventilation any time the system is running. It can also reduce the amount of air that must be bypassed when some zones are satisfied.

RECTANGULAR ZONE DAMPER

ROUND ZONE DAMPER

03315-13_F04.EPS

Figure 4 Zone dampers.

Although zone dampers are most often placed in trunk and/or branch ducts, they can also be attached to a single diffuser (*Figure 5*). This setup can be used to help improve the conditions in a single room or area that is consistently overcooled or over-heated by a constant volume system, without installing an entire zoning system. The assembly is equipped with a temperature sensor that determines if the system is in the heating or cooling mode by the supply duct temperature, and then reacts accordingly. For example, if the cooling system is running and the wall thermostat falls below the setpoint, the damper will close to stop the airflow to this one diffuser. If it later senses that the heat is running, based on the duct temperature, it will open and allow the cold zone to warm up.

1.1.3 Bypass Dampers

Airflow in a heating and cooling system is important. With a furnace, a certain amount of airflow must be maintained across the heat exchanger to keep the furnace from overheating. In cooling systems, a certain amount of air flow must be maintained to keep the coil from freezing, and to ensure there is enough load to vaporize the refrigerant in the coil and prevent liquid floodback. This is especially true of systems using a fixed restrictor. For heat pumps operating in the heating mode, a significant reduction in airflow will cause the head pressure to increase and shut the system down. The problem of airflow across coils and/or heat exchangers is a significant concern in zoning systems.

Since one or more zone dampers may be closed at any time while a system is in operation, the total amount of airflow into the conditioned space changes. There must be a method of ensuring that the airflow through the furnace or air handling unit remains nearly constant, without overheating or overcooling satisfied zones. Small fluctuations may be acceptable, especially if a thermostatic expansion valve is used as a metering device. Furnaces generally have at least a 30°F range of acceptable temperatures in their temperature rise. But large reductions in airflow cannot be tolerated.

In typical zoning systems, this function is provided by the bypass damper (*Figure 6*). The bypass damper is the most important damper in a zoning system. As zone dampers close, the airflow through the air handling unit or furnace is reduced. To maintain a consistent volume of flow through it, air must bypass the closed zone dampers and go elsewhere. Since the number of closed zone dampers is unpredictable moment to moment, the amount of air to be bypassed also changes. While zone dampers are often two-position models (fully open or fully closed), bypass dampers modulate their position. This means that they can stop at any point between full open and full closed, to satisfy the needs of the system.

MOTORIZED, MODULATING
BYPASS DAMPER

BAROMETRIC
BYPASS DAMPER

03315-13_F06.EPS

Figure 6 Bypass dampers.

03315-13_F05.EPS

Figure 5 Diffuser/zone damper assembly.

Two types of bypass dampers are in common use: barometric dampers and motorized dampers. Both types are shown in *Figure 6*. The barometric damper is sometimes chosen as a cost-cutting measure, because it does not require electrical connections and control wiring. It is not motor-driven. The barometric damper's balancing mechanism is adjusted to open the damper when the system pressure reaches a predetermined point. Note that, due to the action of the barometric damper, it cannot be installed in a vertical position. The other type is a motorized, modulating damper. This damper is also positioned based on system pressure, but does so using a motor.

There are two primary ways to apply a bypass damper, as shown in *Figure 7*. Air can be bypassed to a **dump zone**, or it can be bypassed directly from the supply air duct into the return air duct. In either case, the amount of air moving across the coil or heat exchanger remains consistent. A dump zone is an area located within a zoned building that does not require precise temperature control, such as a basement, large foyer, or hallway. In a commercial building, especially one that uses the ceiling plenum as a return air plenum, the air can be dumped into the ceiling area. The farther away from the air handling unit it is dumped, the better the system functions. As the zone thermostats in the controlled areas are satisfied and the related zone dampers close, any excess system airflow is routed through the bypass damper into the dump zone. The bypass damper acts as a pressure regulator to maintain the design pressure in the supply trunk for the operating zones.

One benefit of using a dump zone bypass is that the heating or cooling capacity supplied by the bypassed air will pick up part of the building load. Another advantage is that the return air temperature to the furnace or coil does not change dramatically when the dump zone opens. The air that is dumped continues to mix with room air before returning to the unit. A return air grille is normally located in a dump zone area to avoid pressure buildup in the space. The disadvantage is that the area where the air is discharged is not controlled, and the temperature may become too hot or too cold. Some installations using a dump zone may waste conditioned air, if it is an area where the occupants do not care about the temperature. If a barometric damper is used as a bypass, it is best used with a dump zone. When a direct bypass to the return duct is installed, barometric dampers do not perform as well.

The direct-return bypass uses a damper installed in a bypass duct near the furnace or fan coil. Since the air discharged by the damper is by-

passed directly to the return inlet of the furnace or air handler, it has no chance to pick up any part of the building heating or cooling load. This is a disadvantage because when the bypass damper is wide open, the air temperature returning to the equipment will significantly increase during heating and decrease during cooling. Bypassed air that is too hot can cause the furnace to shut off on high limit. Air that is too cold can cause the air conditioning coil to freeze up. To avoid excessive temperature swings in the heating or cooling equipment, a direct-return bypass is usually sized to handle only about 25 percent of the total system air volume when the bypass damper is fully

DUMP ZONE BYPASS IN BASEMENT

DIRECT RETURN BYPASS

03315-13_F07.EPS

Figure 7 Dump zone and direct return bypass zones.

open. In addition, protection is provided by using a duct temperature sensor. The sensor monitors the temperature and will cycle the equipment off when it reaches preset low and high setpoints. As a result, the equipment may cycle on and off far more often when only a small part of the system requires heating or cooling.

In some areas of the country, the use of bypass dampers is being reconsidered. When a great deal of air is being bypassed consistently, energy efficiency suffers. One manufacturer has come up with a new approach that eliminates the use of a single bypass damper. In the process, it also eliminates the problem of hot or cold air being dumped directly into the return duct of the systems. A BZD Barometric Zone Damper™ (*Figure 8*) is a new type of zone damper that combines a motorized damper blade with barometric operation. Although the damper blade is motorized, half of the blade area can also swing open barometrically. A small weight is provided for adjustment. When a zone calls for heating or cooling, the entire blade moves to the open position. When it is closed, and pressure begins to build throughout the duct system, the barometric blade of each closed zone damper can open partially. This relieves the some of the pressure in the duct through each satisfied zone. Although the satisfied zones are receiving unwanted air, it should not be enough to significantly raise or lower the room temperature.

03315-13_F08.EPS

Figure 8 BZD Barometric Zone Damper™.

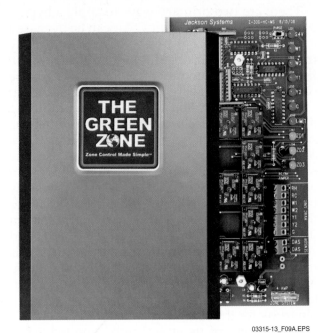

03315-13_F09A.EPS

Figure 9 Zoning system control panel. (1 of 2)

1.1.4 Zoning Control Panels

The zoning control panel (*Figure 9*) provides overall system management. To provide that function, it serves as a centralized point of connection for thermostats, dampers, sensors, and the HVAC equipment to be controlled. Although zoning control panels can vary significantly in layout, most provide the same basic type of connections shown in *Figure 9*.

Zoning control panels must be selected based on the application requirements. All manufacturers offer panels that can control some finite number of zones. For example, a two-zone panel can handle only two zones and may be appropriate for a simple two-story. Another panel may handle six zones. Panels that can accommodate as many as 20 zones are also available. The number of

HVAC Equipment for Zoned Systems

A single-speed compressor and a fixed-speed indoor blower deliver a fixed amount of capacity, regardless of the number of zones calling for conditioned air. If all the zones are calling for air, the system works properly. If fewer than all zones call for air, the excess system capacity must be handled in some way, normally with a bypass arrangement.

Two-speed or modulating-capacity compressors, two-stage or modulating-capacity gas furnaces, and variable-speed blower motors can respond much better to a reduced number of zones calling for conditioned air. When fewer than all zones call for conditioned air, the system responds by operating at a lower compressor capacity, a lower heating capacity, and/or a lower fan speed. The result is better indoor comfort, increased energy efficiency, and a quieter system.

PURGE SWITCH
(WHEN SET TO ON, FAN WILL CONTINUE TO RUN AND AIR WILL BE PURGED TO LAST ZONE CALLING FOR 1 MINUTE AFTER CALL IS SATISFIED)

JACKSON SYSTEMS, LLC Z-301-HC-MS

ZONE 1 THERMOSTAT

ZONE 1 DAMPER

ZONE 2 THERMOSTAT

ZONE 2 DAMPER

ZONE 3 THERMOSTAT

ZONE 3 DAMPER

24 VOLTS

ADJUSTABLE HIGH LIMIT

SWITCH		LIMIT
1	2	
OFF	OFF	130
ON	OFF	140
OFF	ON	150
ON	ON	160

24V — 24 VOLT POWER LED
W1 — FIRST STAGE HEAT LED
W2 — SECOND STAGE HEAT LED
Y1 — FIRST STAGE COOL LED
Y2 — SECOND STAGE COOL LED
G — FAN LED
LIMIT — HIGH / LOW LIMIT LED
ZD1 — ZONE 1 DAMPER LED
ZD2 — ZONE 2 DAMPER LED
ZD3 — ZONE 3 DAMPER LED
RH/RC JUMPER

HVAC UNIT TERMINALS

DISCHARGE AIR SENSOR

ZONE SYSTEM 4 AMP

HVAC EQUIPMENT 4 AMP

4 AMP FUSE PROTECTS PRINTED CIRCUIT BOARD

4 AMP FUSE PROTECTS THE RELAY CONTACTS

03315-13_F09B.EPS

Figure 9 Zoning system control panel. (2 of 2)

zones that the designer anticipates using is generally the first concern in panel selection. Note however, that panels capable of handling some fixed number of zones are not required to make use of every zone output. A six-zone panel, for example, can function just fine with only five zones in use.

The type of HVAC equipment to be controlled is the next consideration. Standard, single-stage heating and cooling equipment is fairly simple. Multi-stage systems may require a different panel. Note that the panel shown in *Figure 9* is capable of controlling two stages of cooling (through the Y1 and Y2 terminals) and two stages of heating (through the W1 and W2 terminals). If a heat pump or dual-fuel system is in use, this must also be factored into the selection. The control panel shown here can be switched between heat pump or conventional fossil-fuel furnace service. Not every panel can properly control every type of system. Today, there are many unique equipment capabilities to consider, such as variable-speed blowers, variable-speed compressors, and humidity control. Most HVAC equipment manufacturers have their own zoning system products, some of which are specifically designed to work with their more advanced products. Using the proper control panel for a given type of equipment allows the user to take advantage of their advanced system features.

The control panel must also be selected to coordinate with the type of zone and bypass dampers that will be used. Zone dampers that are powered closed and spring open when de-energized (two-positions) are often referred to as two-wire dampers. Only two wires are required to operate them. Zone dampers that modulate their position require three wires, and may be referred to as three-wire dampers. The panel shown in *Figure 9* can accommodate three zones with three-wire dampers. Although some panels my offer the flexibility of working with either type of damper, many do not. This is a cost-saving measure, rather than a lack of technology. While programmable panels can be built that are capable of working with most every conceivable design and equipment, it may not be a cost-effective product for many installers and users.

One other very significant option available in zoning systems is wireless communication. Damper motors are normally wired back to the zone panel, which provides their source of power. However, wireless systems are now available.

The zone dampers must still be powered, but that can be accomplished using a 120VAC/24VAC control transformer located much closer, or even adjacent to, the damper locations. For an attic or crawl-space system, a single transformer can be installed to power all of the dampers. The control panel itself may be a significant distance away. The thermostats are also wireless for such a system. The panel shown in *Figure 10* has no terminals for thermostats or zone dampers. The system temperature sensors are connected to the panel and do not communicate wirelessly in this case.

Some zoning panels, such as the one shown in *Figure 10*, require the installer to program a variety of operating parameters. However, the majority of today's systems use a series of small switches on the panel for configuration instead. Most of these switches are two-position, although some may be three-position. Choosing a specific position for each switch configures the panel as needed for the application. One switch, for example, may allow the installer to configure the panel for use with a heat pump system or a gas furnace. Another may allow the installer to choose whether the heat pump reversing valve is energized in the heating mode or the cooling mode.

1.1.5 Additional Devices

Most systems, especially those that are equipped with a bypass damper, require some additional controls. One such control is the device that provides the information needed to position the bypass damper. A static pressure sensor (*Figure 11*) monitors the pressure in the duct and generates a signal that is used to position a motorized bypass damper. Barometric dampers are manually adjusted and require no further control.

Unique Zoning Damper

Zoning dampers are available to meet most every possible installation situation. The model shown here replaces the take-off at the main trunk. As a result, it saves on the cost of buying a take-off, and the installation labor for the damper itself is no more than that of the take-off.

03315-13_SA01.EPS

ZONE DAMPER
STATUS LIGHTS

HVAC EQUIPMENT
STATUS LIGHTS

SYSTEM STATUS
LIGHTS

PROGRAMMING
MODULE

FUSES

HVAC EQUIPMENT
INTERFACE

SENSOR INPUTS

POWER IN TO PANEL;
POWER OUT TO
DAMPERS

03315-13_F10.EPS

Figure 10 Wireless zone control panel.

03315-13_F11.EPS

Figure 11 Duct static pressure sensor.

High- and low-limit switches or sensors are also required components for systems with bypass dampers. They may also be needed in two-zone systems without a bypass damper for maximum

protection. The high-limit is used to protect the heating unit, and typically monitors the discharge air temperature of the system. It opens when a preset limit is reached. However, this does not lockout the equipment. It simply cycles the heating unit off until the temperature is reduced to an acceptable level. A low-limit protects the cooling unit from frozen coils and possible liquid refrigerant floodback. If the discharge air temperature falls below a preset level, the switch opens and the compressor is cycled off until the air temperature rises significantly. Since most systems are equipped with short-cycle protection, the compressor will generally be off for at least five minutes. Instead of monitoring air temperatures, a low-limit switch may be placed on the suction line of the compressor instead. As noted earlier, different manufacturers have different approaches to solving many potential problems.

Instead of switches, a single sensor can provide a discharge temperature indication to the control

Zone Dampers for Existing Systems

Existing systems can often be retrofitted for zoning. Installing common zone damper assemblies in these systems can be somewhat labor-intensive. A type of damper has been developed specifically for this purpose. The damper is installed directly into an existing round duct. Since the assembly consists only of a blade and a motor, they save on both material and labor costs.

03315-13_SA02.EPS

panel. The high- and low-limit switching is then accomplished by the panel. The logic of the panel determines the proper course of action when temperature limits are reached. Some systems may also use an outdoor air sensor. The sensor can provide an outdoor temperature indication that the panel can use for such tasks as controlling the supplementary heat of a heat pump system. If the temperature outdoors is above a preset balance point, the supplemental heat will remain locked out (except in the defrost mode) until the temperature falls below the setpoint.

The interconnecting wiring for most zone system components is done using standard 18-gauge thermostat wire. This is possible since the current loads are quite low. Damper motors draw very little current (10VA or roughly 0.4 amps, for example), and they typically represent the highest electrical loads in the zoning system.

1.2.0 Zoning System Operation

There are many variations in zoned systems, but the principles of operation are basically the same. Each zone is controlled by its own thermostat which, through the control panel, operates the zone damper. When all zones are satisfied and the HVAC equipment is idle, the zone dampers are all in their open position. This usually means that the damper motor is de-energized and the spring-return mechanism has opened the damper. Most dampers provide a means of visually determining the damper position (*Figure 12*). In addition to lights, the damper blade shaft may also have a visible line on it, showing the position of the blade.

The user may wish to simply circulate air, and can do so by placing the Fan switch of any zone thermostat into the On position. The system fan will start, again as instructed through the zone control panel. In some systems, all the zone dampers will simply remain in their present, open position. In others, only the zones with the Fan switch set to On will remain open. The other dampers will close.

When a zone thermostat calls for heating (or cooling), the control panel starts the heating (or cooling) equipment, and closes the dampers to all zones that are satisfied and not calling for conditioned air. When all of the zone thermostats are again satisfied, the control panel zone relays are de-energized. This causes all the zone dampers to return to the open position as selected by a switch provided on the control panel for each zone. Indoor fan operation may be allowed to continue for a preset period of time in order to provide for continuous air circulation in any or all of the

03315-13_F12.EPS

Figure 12 Zone damper in the open position.

zones. In addition, it makes use of remaining boiling refrigerant in the evaporator coil, or heat that remains in the heat exchanger of a furnace. This is often referred to as a purge cycle, but it should not be confused with the purge cycle of fossil-fuel furnace or burner.

Some zoned system control panels provide for automatic changeover of the system from the heating mode to the cooling mode. Other zoned systems must be manually switched between the two operating modes. Older systems especially may use a master thermostat. This particular zone is then configured as Zone 1. There, the thermostat is switched to the desired Heat, Off, or Cool mode. An Auto mode may also be provided. The operation of the fan, either On or Auto, is also controlled from the master thermostat. When a master thermostat is used, the zone thermostats cannot control the mode of operation.

There are a number of variations on this basic operating sequence. Some of those variations are discussed in the following sections. It is very important to always read the manufacturer's literature for each system before servicing or troubleshooting. Never assume that one system works the same way as another. Even minor differences in the sequence of operation can lead a technician to the wrong conclusion, wasting both time and money.

1.2.1 Ventilation Modes

In most zoning systems, each thermostat is equipped with both fan and system mode switches. Any of the zone thermostats can start the system fan by turning the fan switch from Auto to On. The zone damper for any thermostat set to On will remain open and allow ventilation air to enter. Any zone thermostat with the fan switch remaining in Auto will power the zone damper closed during the ventilation mode. If any of the zones call for cooling or heating, the system will then respond normally to the call by closing dampers for zones that do not need conditioned air. In other words, even with the fan switch set to On for ventilation, the damper will close if unwanted heated or cooled air is being provided. Once the heating or cooling operation stops, the damper will reopen to restore ventilation airflow.

1.2.2 Heating and Cooling

There are a number of variations related to heating and cooling. Sometimes, one zone will request cooling while another requests heating. Generally, the zone calling first takes priority. The zone panel will start the appropriate equipment to satisfy the needs of that zone first. Once that zone is satisfied, the system will switch to the opposite mode and satisfy the other zone. Before switching however, there will usually be a purge period where the heating or cooling equipment switches off and the fan runs for several more minutes.

Some systems also have programming that allows for conditioning to be shared. For example, assume a call for heating and cooling from different zones is made at nearly the same time, and that the cooling request was made first. The cooling equipment will start, and the damper will close on all systems that do not need cooling. Without the ability to share operating time, this status could continue indefinitely; the zone needing cooling could continue to dominate until it is satisfied. With programming that allows sharing, the dominating call will only be served for a fixed period of time; twenty minutes is common. At that point, the system will stop cooling, purge itself, and then start the heating equipment to answer the heating call. If the call for heat is satisfied within the time period, the system will switch back to cooling again. If not, it will switch back to cooling again when the same time limit is reached.

Some systems have both minimum on and off times for the cooling equipment. Short-cycling HVAC equipment increases wear and tear on the compressor especially. Compressors experience more electrical stress during startup than any other time. Cooling systems often have a 3 to 5 minute minimum off time before they are restarted. Note that technicians must pay attention to the setup of such systems. Thermostats often have time delays built in for this purpose. Zone panels may also have such a timer, and delay-on-break short cycle timers may be installed in the condensing unit. Depending on how some of these systems are wired and programmed, these time periods may be compounded. Minimum on times may be shorter than minimum off times. At

Additional Heat

In many homes, there is one small room or area that always seems to be a bit too cold or hot, while the rest of the home is comfortable. The Hot Pod™ is one way such problems can be solved. Equipped with a small fan and connected to its own wall thermostat, the fan can help boost the flow of air into a single room or branch duct. It is also fitted with an electric heating element. When the thermostat calls for additional heat, both the integral fan and the electric heater are energized to warm the area.

03315-13_SA03.EPS

Note that it is not connected to a zoning system in any way; it is a stand-alone device. In fact, using it with a zone system could create some problems. If the zone damper for the area the Hot Pod™ serves is closed, then its source of air is cut off. Although there would be little harm in the cooling mode, the electric heater would overheat without airflow across it, and would then cycle off and on through its limit switch. Although it is protected, there would certainly be no benefit from its operation.

least two minutes of on time after a start is common.

1.2.3 Multistage Systems

Ideally, a zone's need for heating or cooling is taken care of by the first stage of heating or cooling over a reasonable period of time. This is the energy-efficient point of having a multistage system. However, zone control panels must have some sort of programming that determines when the second stages of heating or cooling are called upon. This eliminates the need to use multistage thermostats in each zone.

A common method is based on the amount of time that a call for heating or calling remains. The timing period begins when the zone thermostat first calls for cooling or heating. After a preset period of time, if the zone thermostat has not been satisfied, the zone panel will start the second stage of cooling or heating. The time period is adjustable on most systems.

More sophisticated systems can allow for each zone to have a different time period allowed before the second stages of cooling or heating are called upon. For example, a living area zone could be set to allow 5 minutes run time before a second stage of capacity can start. A bedroom area may then be set to allow 11 minutes of run time before a second stage can start. Note that these are not suggestions, but examples only. Each individual installation and the habits of the occupants must be considered. These more sophisticated systems may also allow for a minimum number of zones to be calling for conditioned air before a second stage can start. This intelligent feature allows the installer to prevent the second stage of heating or cooling to start when only one or two zones in a large system are calling. In a well-designed system, all stages of heating or cooling would not normally be required to satisfy the needs of a single zone. There are occasional exceptions however.

1.2.4 High- and Low-Limit Operation

Many two-zone systems do not require bypass dampers or high- and low-limits for equipment protection. However, as the number of zones increases and become smaller, both bypass dampers and temperature limit devices are usually required. Remember though, that there is now an alternative to the traditional bypass damper. The barometric zone damper discussed earlier can also be used.

High- and low-limit switches or sensors cycle the cooling and heating equipment to prevent

overheating, coil freezing, and liquid refrigerant floodback. The primary idea is to prevent equipment safety controls, such as a furnace high-limit switch or a condensing unit low-pressure safety switch, from activating and shutting the equipment down completely. Some of these equipment safety devices require a manual reset. In addition, they are generally designed to activate near the performance limits of the equipment. When applying a zoning system, there is no need or desire to push the equipment performance to this level. By using high- and low-limit switches or sensors, the zoning system can cycle off the equipment before equipment safety devices are activated.

When a bypass damper is partially or fully open, excessively cool or warm air can be circulated through the heat exchanger or coil. Monitoring the discharge air temperature allows the system to cycle the heating or cooling equipment off until the air temperatures return to a safe level. If the equipment is cycled off due to a high- or low- temperature condition, the rest of the zoning system continues to operate normally. The zone dampers of zones that are calling for conditioned air remain open and the blower continues to circulate air. Once the air temperature changes sufficiently, the equipment will be cycled back on. However, the start time will be subject to other limits, such as the minimum off time programmed into the panel. In other words, both time and air temperature likely play a part in how long the equipment remains off.

Low-limits are sometimes fixed and cannot be changed by the installer or user. A common setpoint is 45°F. Other panels may allow an adjustment within a small range, as shown in *Figure 13*. The high-limit is usually adjustable within a range, such as 110°F to 160°F.

03315-13_F13.EPS

Figure 13 High- and low-limit setpoint adjustment.

1.3.0 Zoning System Troubleshooting

Troubleshooting a zoning system is not difficult as long as the technician understands basic control troubleshooting. However, like thermostatic expansion valves, zoning controls are often suspected or blamed for a problem before other, easily understood components. This is usually due to a lack of understanding of what the electronic controls are supposed to do. To successfully troubleshoot any zoning system, it is essential that the technician study the literature for that specific system. Without a clear sequence of operation, troubleshooting cannot be accomplished accurately.

The troubleshooting process for a non-zoned system often starts as a no heat or no cooling service call. In these cases, technicians begin the process by determining if the system has power applied to it. If the equipment is powered, the process moves to determining if a control voltage is provided. When line voltage power and control power is present, the next step is to determine if a call for cooling or heating is coming from the thermostat.

With a zoning system, this initial troubleshooting process should be the same. The existence of a zoning system should not change that process dramatically. When there is a lack of cooling or heating throughout the facility, start with these simple tests to verify power. If the heating or cooling equipment is operating, technicians should then verify that it is operating correctly before proceeding to the zoning system controls. Check the refrigerant charge, air filter condition, and other common causes for comfort complaints. However, before making these tests, it is best to do so with all zone thermostats calling for cooling or heating. This ensures that the maximum amount of load available is being applied to the equipment.

The real change in the troubleshooting process begins when a call for the cooling or heating equipment to operate is being checked. The zoning panel typically takes the place of the single wall thermostat. Control power to energize a furnace, blower, and/or cooling unit comes from the zoning panel connections instead of directly from a thermostat.

1.3.1 Zone Panels

As noted above, troubleshooting of the zone system typically begins when the heating or cooling equipment is not running at all, and the absence of power has already been eliminated. Of course, if the complaint is only related to a single zone, then the troubleshooting process begins directly at the zoning system. This type of complaint will be discussed later in this section.

After reviewing the literature for a zoning system, use a voltmeter and check to ensure the panel is being powered. Zone panels are usually powered by a 24VAC source. The voltage applied should generally be within 10 percent, but check the manufacturer's literature to be sure of the

Wireless Equipment That Stands Alone

Wireless zoning systems are becoming more popular, due to the significant labor savings in their installation. In addition to complete systems, there are also wireless dampers and thermostats that can operate independently and handle problem areas that overheat or overcool. They do not interface with a zoning control panel. The thermostat shown is battery-operated. It communicates with the damper through an RF wireless data link. The damper motor must be powered by 24VAC, but the thermostat operates on batteries and there is no wiring between the two devices. The damper also has a sensor to monitor the supply air temperature. This determines whether the thermostat operates in a heating or cooling mode. For example, if the sensor indicates that there is 58°F air in the supply duct, the system is obviously in the cooling mode. If the thermostat becomes satisfied and requires no more cooling, it will close the damper. The opposite action takes place when the sensor indicates the main system is in the heating mode.

03315-13_SA04.EPS

range. Most panels have LED lights to indicate a variety of conditions exist. If no lights at all are illuminated, this may indicate a lack of power as well. A panel power lamp is very common (*Figure 14*).

Once it is certain the panel is powered, it is a good practice to check any available programming next. There are many different panels in use, each with different capabilities and characteristics. If the programming or setup is incorrect for the application, problems will result. However, this step also provides information to the technician. During the rest of the process, it is very good to know precisely what the time and temperature settings are for various functions. Use the manufacturer's literature to review panel programming carefully. Also check the position of any switches located on the panel (*Figure 15*). Most panels are configured using switches of this type, rather than being programmable. For example, if a panel is set-up to require at least two zones to be calling for cooling to start the equipment, then having a single zone requesting cooling will not result in equipment operation. A technician can easily assume that the panel is defective, since there will be no output to the blower or condensing unit when the single zone calls for operation. In reality, the panel is doing exactly what it was programmed to do. Short-cycle timer operation on the board can also fool an impatient technician.

Assuming a sufficient number of zones are calling for cooling, the equipment outputs at the panel can be checked in the same way as any thermostat. R is the leg of control power being broken through the panel relays; it is the source of power that will feed back to the equipment via the W, Y, and G wires (as well as other wires, if required). If an output for cooling is expected, check for voltage at the panel terminal board. Jumper wires can also be used at the panel outputs to check the integrity of controls downstream. For example, placing a jumper wire from R/RC to Y at the panel should start the condensing unit. If it does, all wiring and controls downstream are functioning. If outputs to the HVAC equipment are not present at the panel, and it has been confirmed that other conditions are correct, the panel is likely defective.

Remember that temperature sensors can also prevent heating and cooling equipment from operating. If a failed sensor indicates to the panel that the supply air temperature is too high in the heating mode, or too cool in the cooling mode, outputs to the equipment will be prevented. Before condemning a panel that does not provide outputs when all other factors are correct, first check the accuracy of the temperature sensor input. Testing and troubleshooting thermistors was covered in a previous module.

Since most panels have a number of LEDs to indicate the status of the system and various conditions (*Figure 16*), pay attention to them while troubleshooting and observing system operation. They can be a valuable aid. As shown in *Figure 16*, some panels have lights that indicate that the high- or low limit programming is active. If these indicators are lit, but the discharge air tempera-

PANEL POWER
INDICATOR LIGHT

03315-13_F14.EPS

Figure 14 Panel power indicator light.

03315-13_F15.EPS

Figure 15 Panel configuration switches.

Figure 16 Zoning system status lights.

ture is normal, then either the sensor or the panel is likely at fault.

1.3.2 Zone Thermostats and Dampers

Except in wireless systems, zone thermostats are wired back to the zoning control panel (*Figure 17*). The zone thermostats can be tested for proper outputs like any other thermostat. From the panel location, a technician can test to see if thermostat circuits are open or closed by measuring voltage at the appropriate terminals. If a thermostat is suspected of being faulty, it is always a good idea to confirm it at the thermostat before replacing it. This eliminates the possibility that there is a problem in the interconnecting wiring.

Also remember that individual thermostats may be programmable. Sometimes a problem can be as simple as a user not realizing that a program has been invoked, changing the setpoint at some time of day and causing discomfort. A program can even create issues during the troubleshooting process. If a setpoint is scheduled to change during the hours the technician is present, it may occur during the troubleshooting process and create confusion.

A good way to start the process of checking thermostats and dampers is to begin with set-

Figure 17 Zone thermostat terminal blocks.

ting all zones to call for either heating or cooling. During this period, depending on the conditions, technicians may wish to cycle the power to the heating and cooling equipment off. This will prevent the equipment from running during a test of zone thermostats and dampers. With all zones calling, evaluate the system airflow and visually confirm that every zone damper is open as it should be. The bypass damper should be completely closed when all the zone dampers are open. This should be confirmed visually as well. Remember that barometric bypass dampers are controlled only by weight and balance. If they are not fully closed, make the necessary adjustments.

Check That Panel

Looking at the thermostat brand is not an accurate way to determine the brand of zoning system that has been installed. Unlike many older systems, many of today's zoning systems can function with any common thermostat used for the zones. Therefore, a Honeywell thermostat may be found operating with a Jackson Systems zone panel. Always check the zone panel to confirm the manufacturer. Securing the installation and troubleshooting data for the panel is most important when troubleshooting zone controls.

Once proper operation has been verified with all zones calling, begin adjusting one zone thermostat at a time. Set it to a low or high enough setting to satisfy the thermostat. Confirm that this damper closes fully. When possible, observe the damper shaft or blade as it travels. More often than not, dampers fail to open or close properly due to binding or stress placed on the housing during or after the installation. Often, there will be some distinguishable increase in airflow from the other zones. The bypass damper may or may not move slightly. This depends primarily on the size of the zone compared to the total system and the number of zones. Also remember that systems using a variable-speed blower motor will automatically reduce the system air flow; opening the bypass damper is likely unnecessary for these systems. In fact, they may have no bypass damper at all.

Continue this process for each of the connected zones. Although it may be tempting to control the entire process from the zoning control panel, doing so eliminates too many variables that need to be checked. Control each zone from its thermostat, and observe the damper motor as it travels whenever possible.

When zone dampers do not respond as expected, especially when commanded to close, begin by checking to ensure that an output is being sent from the zoning control panel. If not, check to ensure that the thermostat is signaling the panel appropriately. However, remember that two-wire damper motors use a spring to open the damper. Therefore, a two-wire damper motor with no input power to close the damper should always open on its own. Again, binding or stress on the damper blade may be the culprit. Three-wire dampers (*Figure 18*) must be powered to both the open and closed positions. Therefore, these dampers will not move at all unless powered in one direction or the other.

In *Figure 18*, note the markings on the terminal board. The COM wire is the common wire from the 24VAC power source at the panel. The center terminal, marked PO, provides the second leg of power to open the damper. The PC terminal powers the damper closed. To determine if the damper is being commanded to move in either direction, the technician need only check for 24VAC between COM and either of the other two terminals. If 24VAC is present, that command is being sent to the damper. The PO and PC terminals should never be energized at the same time. That would likely indicate a serious problem with the panel itself, or possibly wiring damage that is allowing the tow conductors to make contact.

Also in *Figure 18*, note the words Manual Gear Release above the terminal board. Many dampers have a means of disengaging the motor from the damper blade. This allows the technician to position the damper manually. If a zoning panel fails, a technician can manually disengage and open all the dampers. This allows the system to operate as a non-zoned system until repairs can be made.

When in doubt about the operation of a system or an individual component, make direct contact with the manufacturer. Many companies provide technical assistance by phone for challenging problems, and prefer to do so while the technician is on site. They can then ask for specific troubleshooting steps to be taken. However, relying on the manufacturer should not be substituted for knowledge and self-study of the system.

1.4.0 VVT System Components

Variable volume and temperature (VVT) systems have a lot in common with standard zoning systems. However, they are more suited to the commercial environment. The components of a VVT system also communicate in a more sophisticated way. Although there are many similarities, there are also a number of ways that the two types of systems differ.

It is very important to note that VVT system have changed in many ways through the years. Each generation offers different capabilities and may communicate in different ways. The task of programming most VVT systems is as simple as following a checklist and understanding the de-

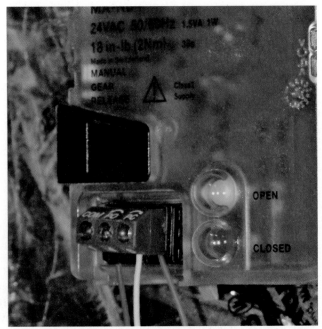

03315-13_F18.EPS

Figure 18 Three-wire zone damper terminal block.

tails of the design to make good decisions. However, even though the task itself is fairly consistent from one product or model to another, terminology and features can differ dramatically. Whenever working with a VVT system, it is crucial to obtain and review the technical data for that specific system and its specific components.

Figure 19 shows the layout of a typical VVT system. VVT systems are best used in commercial buildings where they control systems up to 25 tons. They are used on larger systems than most common zoning systems, and usually have more zones as a result. The arrangement shown here allows a single constant-volume packaged unit to meet individual zone cooling and heating needs. Each room is a control zone with its own communicating zone damper controller and room sensor.

Many of the components shown in *Figure 19* will look familiar, as many are also a part of common residential zoning systems. Zone dampers are in place, and there is a wall-mounted sensor in each zone. In a common zoning system, this would be a thermostat that switches 24VAC. In a VVT system, the zone thermostat or sensor communicates its needs and status in a completely different way. Another notable difference is found in the interconnecting wiring. In the VVT system,

the zone dampers are communicating with each other through a bus. Further, they are communicating with a device in the unit and with the bypass damper controller.

1.4.1 Bypass Damper

Because the central HVAC system is usually a constant-volume system, a bypass damper must be provided. As zone dampers throttle, the bypass damper is opened to keep the airflow through the unit consistent. They are controlled based on the static pressure of the supply air duct. The damper bypasses supply air directly to the return duct or to a ceiling return plenum.

This same concept applies to common zoning systems. However, there is often one distinct advantage to applying the system in a commercial building rather than a residence. One of the consistent complications of any zoning system is how to handle the bypass air. Bypassing it directly into the return duct works, but it is not a perfect solution and the equipment must be cycled off sometimes for its own protection. A dump zone also works, but is highly wasteful.

In many commercial buildings, the plenum area above a suspended ceiling often serves as a

1. RETURN AIR GRILLE
2. SUPPLY DIFFUSER
3. SUPPLY DUCT
4. ZONE DAMPER
5. ZONE DAMPER CONTROLLER
6. MASTER ZONE CONTROLLER
7. ZONE CONTROLLER
8. COMMUNICATIONS BUS
9. ROOFTOP UNIT CONTROLLER
10. BYPASS DAMPER & CONTROLLER

03315-13_F19.EPS

Figure 19 Variable volume/temperature system.

return air plenum. This eliminates the need to install an elaborate return air system; the unit simply draws its return air from the plenum through a single duct opening. One or more ceiling tiles are replaced with grilles to allow room air to easily migrate into the return air plenum.

This plenum approach can work in favor of a common zoning or VVT system, but such a plenum is extremely rare in residential construction. The bypass air can be dumped into the ceiling plenum, preferably a considerable distance away from the unit return air duct. This concept is present in *Figure 19*. As it naturally travels back toward the return duct, there is plenty of time for the cooled or heated bypass air to mix with air in the plenum before reaching the unit. This helps to stabilize the return air temperature and prevent it from reaching extremes.

In addition, this conditioned air is still being used in a positive way to condition the building, rather than being wasted into a dump zone. Allowing bypassed air to flow into a remote part of a ceiling return plenum offers the best performance in both VVT and zoned systems.

1.4.2 VVT Zone Dampers

Zone dampers (*Figure 20*), also referred to as VVT terminals, are available for VVT in both round and rectangular types. The most economical installation cost is achieved when the terminal size and shape match those of the branch duct in which it is installed, and when the number of dampers is minimized. However, it is important to properly size a damper for the volume of air to be passed. It should not be assumed that an existing branch duct size is correct, and then choosing the zone damper based only on the duct size.

Round air terminals are made in several sizes, typically ranging from 160 to 1,125 cfm (0.4 to 2.8 tons). They have dampers that range from 6 to 16 inches in diameter in order to match typical branch duct sizes. Round dampers come from the factory without any insulation. Insulation and vapor barriers are applied externally during installation. Round dampers are normally installed on jobs where the duct pressure at the terminal inlet is 1.0 in wg or less.

Rectangular air terminals are typically made in sizes ranging from 8" × 10" to 8" × 24", where 8" represents the height of the terminal. Rectangular air terminals are available from the factory with an insulated duct sleeve designed for new duct systems. The sheet metal sleeve has internal insulation that complies with the *National Fire Protection Association (NFPA) 90A* fire protection code. The internal damper is made of multiple opposing blades and is powered by a high-torque actuator. The actuator delivers sufficient torque to handle higher velocity, higher static inlet pressures, and more turbulence than can be handled by round dampers. The actuators for VVT systems are significantly more powerful than those used in standard residential zoning systems. This is necessary due to the higher pressures, velocities, and air volumes found in some commercial duct systems.

Normally in a VVT system, when a zone needs heat and the central unit is in the cooling mode, the zone needing heat must wait. However, VVT system terminals may be equipped with an auxiliary heat source. This is usually electric heat, but even a hydronic heating coil can be used. If a rectangular air terminal is used, the hot water or electric duct heater is mounted directly on the discharge of the terminal. The auxiliary heat

Figure 20 A VVT terminal, awaiting installation.

source is used to provide heat to a needy zone while other zones need cooling.

1.4.3 Zone Controllers

In older VVT systems, there is one primary zone controller (also referred to as a linkage coordinator, master zone controller, or monitor thermostat) and a number of secondary, or zone, controllers. The primary zone controller sends time of day, system mode of operation, and outdoor temperature information to each of the zone controllers. It also monitors the status of each of the zone controllers on its communications bus and selects the mode of operation for the central packaged unit.

While providing these functions in most installations, it is also responsible for managing its own zone damper, like the rest of the zones. When a sufficient number of zones call for heat, the master controller switches the central packaged unit into heat. When all heating zones have been satisfied, it switches the central packaged unit back into cooling. In this manner, the system provides cooling and heating to a building that has some zones needing cooling while others need heating.

Note however, that the system cannot provide simultaneous heating and cooling. In this respect, VVT systems suffer the same limitation as a standard zoning system. However, as noted earlier, some terminals can be fitted with auxiliary heat. Although they help maintain comfort, auxiliary heaters are generally energy-wasteful devices, since the main unit is operating in the cooling mode. This duplicates energy consumption as they work against each other.

The monitor thermostat is where the vast majority of the system programming is accomplished. It can be equipped with or without a time clock. The zone thermostats do not have their own time clock. Instead, they share this information from the monitor thermostat via the communication bus. The monitor thermostat also has system switches, while the zone models do not. Time-of-day scheduling can often be accomplished either at the zone thermostat or at the monitor. Each zone can operate on its own schedule if desired.

Newer VVT control systems have done away with the concept of monitor and zone thermostats with independent programming. The monitor thermostat has been replaced by computer software that interfaces with zone sensors, zone dampers, and the HVAC equipment through a network. Programming for each zone and the system as a whole is done at the PC keyboard. Each zone damper control module stores the information pertaining to its zone once programmed through the PC and network. It can then operate on its own regardless of whether the controlling PC is connected or not.

1.4.4 Relay Packs and Unit Controllers

Since VVT controls systems communicate differently, a method of communicating with the HVAC equipment is required. For most systems using a monitor thermostat, a relay pack installed in the equipment provides this service. Signals from the monitor thermostat are accepted, and the relays open and close in response to start and stop equipment operations.

Newer systems that are driven through computer software use a digital controller inside the equipment to provide this function. Like newer zone damper controllers, they also store their own programming information.

1.5.0 VVT System Operation

As noted earlier, VVT systems communicate differently than common zoning systems. Most zoning systems operate by closing and opening 24VAC circuits. The components respond to the addition of control power, or the lack of it. Each zone damper opens or closes; it does not provide or consume any information. VVT system components communicate important information and share actual values with each other, working as a team at all times.

1.5.1 Communication

Figure 21 shows the wiring arrangement for a zone controller on a system equipped with a monitor thermostat. The zone controller electronics are housed in the front portion of the device, which is then connected to its wiring subbase by a ribbon

---

Monitor-Only Usage

When a VVT system is first installed in a new building, comfort conditions may need to be provided before the complete control system has been installed and set up. Monitor thermostats typically have a monitor-only mode. In this mode, it does not operate its own zone damper. Instead, it merely controls the HVAC equipment. A monitor thermostat can used this way to operate a unit temporarily while the rest of the system is under construction.

---

cable. At the top of the diagram, the red, black, and green wires provide the path for communications between the devices. This is known as a bus. Signals are transferred through the red and black wires along the bus from each device, and all other devices share the information. Shielded wire is used for communication to prevent electrical noise generated by high-voltage and/or high-current load wiring nearby. Each device in the VVT control system is connected to the bus the same way.

Figure 22 shows a VVT system control communication schematic. The key to VVT system operation and control is the use of zone controllers that have the ability to communicate with each other through the bus. As shown, each zone has its own damper controller mounted on a VVT air terminal. A zone controller monitors temperature and allows the zone occupant to modify his own cooling and heating setpoints.

This type of VVT system uses a **polling** approach to determine the needs of the individual zones and to gather information from other system devices. In a polling network (*Figure 23*), the monitor thermostat requests, or polls, information from each of the other controllers, one at a time, and in a certain order. In addition, information from other accessory devices, such as an outdoor air temperature sensor, will also be polled. The polling process is executed in the same order continuously.

In such a network, the response time may be slow due to the time required to complete the polling sequence. This is particularly true when considering alarms. For example, consider that controller number 4 is being polled in a network consisting of 20 devices. If controller 3 goes into

03315-13_F21.EPS

Figure 21 Zone controller wiring.

The Popularity of VVT

You are likely to find the VVT type of system on many commercial buildings. It has gained widespread acceptance and offers building owners comfort at an affordable price. In addition, 80 percent of all the buildings in the United States are 20 tons or smaller in size—the type of building targeted by this system.

alarm at this moment, the alarm cannot be communicated until the polling device has communicated with the other controllers and returns to polling controller 3 again.

The single monitor thermostat in the system does the polling. It polls the other zone controllers and devices at regular intervals. The monitor thermostat then chooses the appropriate mode of operation and communicates that to the HVAC unit. Polling is a continuous process that occurs as long as the system is powered.

1.5.2 Pressure-Dependent vs. Pressure-Independent Operation

Most simple VVT systems operate in a pressure-dependent mode. This is a simpler approach and requires fewer and less-sophisticated controls. In a **pressure-dependent** system, the actual volume of airflow through a given zone damper is not monitored. The damper is positioned between the open and closed position based on the demand for the zone. *Figure 24* shows an example of this. The damper position is changed incrementally as the demand increases or decreases. Usually, when the setpoint and room temperature are more than 1.5°F apart, the damper is fully opened. However, this is generally a changeable value. A pressure-

dependent system has no idea how much air is flowing through a given damper, and has no means of control over the volume. The actual volume will depend on the pressure present in the duct at any given moment.

Pressure-independent systems are more sophisticated. In these systems, the airflow through a zone damper is monitored using a velocity sensor. Based on the measured velocity and the area of the duct in which the damper is installed, the system calculates the actual air volume flowing through. Programmed values for the minimum and maximum airflow in a zone are then used to determine damper position. Basic operation re-

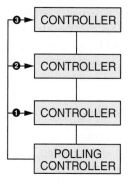

03315-13_F23.EPS

Figure 23 Polling network.

1. PACKAGED AIR CONDITIONER OR SPLIT SYSTEM
2. MONITOR THERMOSTAT WITH TIME CLOCK
3. ZONE CONTROLLER
4. ZONE DAMPER WITH ACTUATOR
5. BYPASS DAMPER WITH ACTUATOR AND STATIC PRESSURE SENSOR
6. BYPASS CONTROLLER

03315-13_F22.EPS

Figure 22 VVT control communication schematic.

mains the same: more demand for a given mode of operation (heating or cooling) results in the damper opening further. However, with the air volume constantly being monitored, the damper will only open until the maximum airflow setpoint is reached. Regardless of how high the demand may be, it will not open further. The airflow calculation is also used to set the damper at its minimum position for ventilation. In a pressure-dependent system, the only accurate way of determining the minimum position for the damper is to measure the airflow manually, and set the minimum position of the damper accordingly. The minimum position of the damper will always be the same, regardless of pressure changes in the duct. In a pressure-independent system, the actual minimum damper position will be different as the duct pressure changes. In a manner of speaking, pressure-independent systems are self-balancing.

1.5.3 Occupied Cooling

When one or more zones become occupied, the monitor thermostat determines that the building is occupied and may tell the packaged unit indoor fan to start. In many VVT systems used in the commercial market, the indoor fan runs continuously when the building is occupied. At the same time, the HVAC system's outside air dampers open to their minimum ventilation position. Running the fan continuously is the only way to ensure that each area gets the required amount of fresh air, regardless of whether heating or cooling is needed. Newer systems can also respond to the input of a CO_2 sensor and position the outside air dampers to meet the demand for fresh air, based on actual conditions. Dehumidification modes are yet another feature in advanced systems.

Another related fact is that zone damper assemblies are rarely programmed to go to a completely closed position, especially during occupied hours. Even when a zone is not calling for heating or cooling, fresh air needs to be provided in a commercial setting. Therefore, each zone damper is generally set to close down to a minimum position setpoint. This leaves the damper open slightly. Ventilation air can then enter the space during all occupied hours.

Each zone controller has its own setpoints. There is usually a separate setpoint for heating and cooling. The zone controller works with the monitor thermostat to determine how far away the actual zone temperature is from the setpoint. This is called the demand. The monitor thermostat regularly polls the individual zone controllers and some calculate the total average cooling demand and total average heating demand for the group. These demands may be displayed at the monitor thermostat or other user interface. The average demand is weighted based on the design airflow to be delivered to a zone. An 800 cfm zone will count twice as much in the demand average calculation as a 400 cfm zone.

Demand may also be based solely on individual zone votes. Each zone, regardless of size, gets one

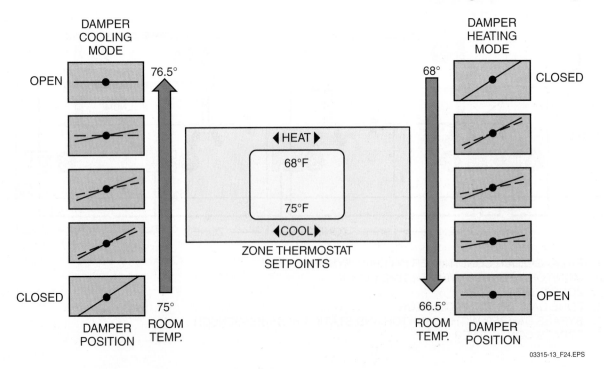

Figure 24 Pressure-dependent zone damper modulation.

03315-13_F24.EPS

vote. Zones vote for heating or cooling, and the majority wins. Cooling is usually the priority. If the average cooling demand exceeds the average heating demand and is greater than the minimum level of demand required for the cooling mode, the system enters the cooling mode of operation and communicates that information to each zone controller. More than one vote, and sometimes as many as three or four votes, is typically required to get the system to operate. The minimum level of demand required is a programmable feature of the system. This is true for systems that calculate weighted demand and for those that are simply counting individual votes. It prevents the system from operating a large HVAC unit and system to satisfy a small demand. This often results in significant energy use, since a great deal of air would need to be bypassed to serve a single small zone demand.

A mode timer is also started when the cooling mode begins. VVT systems can be programmed to operate in the cooling mode for a maximum specified period of time before switching to the heat mode to satisfy rising heating demand from other zones. This prevents a few zones from totally dominating the system. In addition, most systems have minimum operating times as well. Once the cooling mode starts, it will run for some preset period of time at least before shutting down or switching to the heating mode.

Since this is the occupied mode, the outside air damper should already be open to provide ventilation air when cooling begins. Compressor stages are added as necessary. If the system is equipped with an economizer, it may be activated as the first stage of cooling if the outdoor conditions are favorable. With cool air being supplied, each zone controller modulates its individual air terminal damper as needed to maintain the zone cooling setpoint. The farther the room temperature is from the setpoint, the more the damper will open. As the temperature closes in on the setpoint, the damper will begin to modulate toward the closed position.

As the air terminals throttle airflow, the supply duct pressure rises and the airflow through the central packaged unit decreases. A bypass controller monitors the rise in duct pressure and opens the bypass damper accordingly. This allows supply air to be recirculated back to the return side of the central unit and maintains a constant airflow across the DX cooling coil.

Like common zoning systems, temperature sensors are used to monitor the discharge air temperature. Minimum and maximum values are programmed, and the heating and cooling equipment is cycled off as necessary to prevent problems.

If the average cooling demand continues to rise, the central unit controller adds cooling capacity stages as necessary. Capacity stages are then cycled on and off to satisfy the average demand setpoint for cooling. The cooling mode ends when the average cooling demand has dropped below the field-configured minimum average cooling demand, a greater average heating demand is encountered, or a heating demand has existed for a specified period of time while the unit was in the cooling mode.

1.5.4 Occupied Heating

Should any zone require heating while the central unit is in cooling, the zone controller closes that air terminal's damper to the minimum position and sends the zone heating demand to the monitor thermostat or linkage coordinator. Since the damper must remain at a minimum position for ventilation and is not fully closed, some continued cooling can and does occur. This drives the space temperature even further from the setpoint, increasing the demand. If the average demand for heating throughout the system becomes greater than the average demand for cooling, and the cooling mode minimum timer has elapsed, the air source controller switches the central unit from cooling to heating. It then communicates that the heating system mode is active to the zone controllers. The mode timer is also restarted.

Those zones needing heat modulate their air terminal dampers open to meet the zone heating setpoint. All other air terminal dampers are set to their minimum ventilation position, even if any zone starts to need cooling. The monitor thermostat or linkage coordinator continues to poll the zones and accessory devices and responds when necessary.

If the average heating demand continues to rise, the central unit controller adds heating stages (if available) as necessary. The stages are then cycled on and off to satisfy the average demand setpoint for heating. The heating mode ends when the average heating demand has dropped below the field-configured minimum average heating demand, when a greater average cooling demand is encountered, or cooling demands have gone unanswered for a preset period of time.

During the heating mode, the bypass controller continues to function and position the bypass damper, ensuring a constant air volume flows across the heat exchanger and/or coil.

1.5.5 Unoccupied Period

When the monitor thermostat determines that all zones are unoccupied, it requests the central unit to stop and all zone dampers to be returned to their minimum positions. The central unit compressors and supply air fan stop, and the outside air damper closes. If the zone controllers are equipped with setback temperatures, the central unit operates just as it did during the occupied period to maintain the new setpoints, except that the outside air dampers remain closed.

1.5.6 Simultaneous Cooling and Heating

When the central unit is in cooling, simultaneous heating at the zone level can be provided if zone air terminals are equipped with a heat source. If the system is so configured, a zone heater can be activated by its controller. The zone damper will remain at it minimum open position, since ventilation rates must still be maintained. The zone heater, whether it be electric or hydronic, reheats the supply air to warm the zone. When the central unit is in heating, however, simultaneous cooling cannot be provided by a VVT system.

Normal operation of the VVT system does not involve the use of zone heaters and comes close to satisfying a building's simultaneous need for cooling and heating by time-sharing central equipment heating and cooling modes. In fact, many VVT systems are not equipped with zone heaters at all. Their installation drives up the installation cost significantly, due to the added equipment and electrical wiring costs. The simultaneous need for cooling and heating in small buildings tends to be seasonal, occurring in the spring and fall when the building loads are significantly less than the peak design loads. However, internal loads are also a factor. Because VVT uses a central unit designed for the peak cooling and heating loads, the unit can quickly switch from one mode to another and satisfy the building's smaller part-load conditions.

VVT, with its time-cycling of capacity, works well in building sizes up to 25 tons. They are best suited for buildings with zones that have fairly equal loads. If one or two of many zones have an intense amount of heat gain compared to the others, comfort problems will usually result. These unique situations often need to be addressed with unique solutions, such as a dedicated ductless system.

Ductless Hydronic Systems

Ductless split systems are also available that are hydronic. The indoor units look much the same, but are equipped with water coils instead of refrigerant coils. Water is generally easier to manage and move through a building in piping systems than refrigerant. The piping systems themselves are not as tedious to install either. Heating, however, is generally not available. They can be added to buildings that have a separate heating system effectively. As a result, they are probably more popular in the north, where hydronic or steam heating systems are more likely to be in place already. A few heat pump chillers have been developed, providing both chilled and hot water. However, they haven't become popular.

Additional Resources

ANSI/ACCA Manual Zr – 2012, Residential Zoning. Air Conditioning Contractors of America.

1.0.0 Section Review

1. In a system with a master thermostat, _____.

   a. the system mode switches on the master and the zone thermostat must be set to the same position
   b. the master thermostat alone determines the mode
   c. each individual zone can determine the operating mode of the system
   d. an auto-changeover feature cannot be used

2. In most modern systems, the thermostat for any zone can start the indoor fan for ventilation.

   a. True
   b. False

3. When a user with a zoning system complains that they have no cooling at all, the best way to begin is by _____.

   a. checking the individual zone dampers
   b. testing the individual zone thermostats
   c. checking for power at the HVAC equipment
   d. checking the zoning panel programming

4. An older VVT system will have _____.

   a. only one primary controller
   b. two primary controllers
   c. three primary controllers
   d. at least four primary controllers

5. In which VVT system operating mode would the outside air damper most likely remain closed?

   a. Occupied cooling
   b. Occupied heating
   c. Ventilation
   d. Unoccupied

2.0.0 DUCTLESS AND VARIABLE REFRIGERANT FLOW (VRF) SYSTEMS

Objectives

Identify ductless and variable refrigerant flow systems and describe the basic approach to troubleshooting.

a. Identify and describe the operation of ductless split-system equipment.
b. Describe the installation and troubleshooting of typical ductless systems.
c. Identify and describe the operation of variable refrigerant flow systems.
d. Describe the installation and troubleshooting of variable refrigerant flow systems.

Performance Tasks

Troubleshoot and repair a typical ductless or variable refrigerant flow system.

Program the controller for a ductless system.

Trade Terms

Refnet joints: Engineered joints used as connection points in a VRF-system refrigerant network. Refnet joints connect the main refrigerant lines to branch runs or to indoor units.

Ductless split systems have been extremely popular in Europe and Asia for many years, dominating the housing industry in many places. Ductless systems have even been used extensively overseas in high-rise residential buildings. Their popularity has greatly increased in the United States, after a rather slow start. Variable refrigerant flow (VRF) systems are a newer introduction. Although they share some characteristics with ductless systems, they represent a significantly different approach to zoning with equipment rather than zoning through air distribution system control. VRF systems are also not necessarily ductless.

2.1.0 Ductless Split Systems

In some heating and cooling applications, it is difficult or impractical to install a ducted air distribution system. Examples include a home heated with a hydronic system, or a historic building that does not have the space to accommodate ducts without undesirable structural modifications. Other unique and challenging applications include highway tollbooths, guard shacks, stand-alone ATMs outside of banks, fast-food restaurant drive-through windows, and add-on sunrooms. The options for cooling in these applications include the use of ductless split systems. *Figure 25* shows one common example of a ductless split system installation. Ductless split systems are available in both straight-cooling and heat pump configurations.

As the name implies, these are split systems that use a ductless air handling unit. The narrow profile and horizontal discharge of air from the air handler make it ideal for installations where space is limited. The air handlers used in ductless split systems can be mounted on the floor, high on a wall, and either on or in the ceiling. Although ductless systems, also referred to as mini-splits, are normally installed without any duct at all, some manufacturers do offer indoor units that can be concealed. They can usually accommodate a small amount of duct routed to one or two grilles. However, the blowers used cannot overcome a substantial amount of external static pressure, so the amount of ductwork must be limited.

The indoor and outdoor components of the system are connected together by refrigerant tubing and interconnecting wiring. The system is typically controlled with a handheld remote control that is similar to a TV control (*Figure 26*). Some systems also allow the use of an optional wall-mounted room thermostat that is hardwired.

2.1.1 Ductless Split-System Condensing Units

The outdoor unit (*Figure 27*) contains the compressor, condenser coil, condenser fan, and controls like any other split-system condenser. However, other aspects of their construction differ considerably. They are generally designed as rectangular slabs that resemble a large suitcase. The condenser coil is mounted horizontally in the cabinet and air is either blown or drawn through the coil and discharged horizontally. Higher capacity units may have the appearance of two units stacked on top of each other, with two condenser fans. Units are typically powered by single-phase 115V or 230V power. Power for the indoor fan coil is typically

WEATHERPROOF
FUSED DISCONNECT†
PER *NEC*®

2 WIRES (+) GROUND

AIRFLOW

AIRFLOW

OUTDOOR
UNIT

INSULATED
SUCTION
LINE

INSULATED
LIQUID LINE

CONTROL
CONNECTIONS WIRES†
(5) HEAT PUMP
(4) COOLING ONLY

2 WIRES
(+) GROUND

AIRFLOW

24V CONTROL POWER
PROVIDED INSIDE UNIT

FUSED DISCONNECT†

CONDENSATE DRAINLINE†

WIRELESS
CONTROLLER*

TO OPEN SIGHT DRAIN
(DO NOT TRAP)

LEGEND

NEC® — *National Electrical Code*®

Piping

24V

*Standard
†Field supplied

NOTES:
1. All piping must follow standard refrigerant piping techniques.
2. All wiring must comply with the applicable local and national codes.
3. Wiring and piping shown are general points-of-connection guides only and are not intended for a specific installation.
4. Insulate condensate drain if installed in a conditioned space.
5. If local code permits, power connections may be run from the outdoor unit to the indoor unit as shown on schematic drawings.

03315-13_F25.EPS

Figure 25 Common ductless split-system installation.

supplied through the outdoor unit. The metering device is often located in the outdoor unit and not

Figure 26 Handheld remote control for a ductless system.

inside the fan coil as it is in traditional systems. An electronic expansion valve is usually required for best performance. The result is that the tube that carries refrigerant to the indoor fan coil in the cooling mode does not contain subcooled liquid refrigerant, as is the case with a traditional split system. Instead, it carries a low-pressure, low-temperature liquid/gas refrigerant mix. For that reason, both of the refrigerant tubes must be insulated to prevent capacity loss and condensation on the lines. Electronic controls are widely used in ductless split systems.

Many ductless systems use variable speed compressors powered by inverters that change the provided AC power to DC. This allows them to provide the precise amount of capacity needed. Since many ductless systems are capable of supporting more than one indoor unit, the required capacity can vary widely as indoor units start and stop in response to their individual temperature controls. Variable-speed compressors also provide a major portion of the energy savings ductless systems offer.

As mentioned earlier, the indoor unit(s) is typically powered from the outdoor unit. The interconnecting wiring between the two, therefore, includes both line voltage power as well as control inputs.

03315-13_F27.EPS

Figure 27 Ductless split-system condensing unit.

2.1.2 Ductless Split System Air Handlers

The indoor unit contains the evaporator coil, indoor blower, and a control board. As mentioned earlier, the metering device is typically in the outdoor unit. A washable air filter is also included. The blower wheel is the centrifugal type, small in diameter but elongated. The refrigerant piping and wiring can enter through the back of the unit, where it remains unseen on the inside. It can also enter from the side and be exposed on the inside wall of the room where appearances are not important.

The variety of different air handlers available for ductless split systems greatly enhances their versatility. Air handler types include the following:

- High sidewall
- Ceiling mount
- In-ceiling cassette
- Floor mount

High-sidewall air handlers (*Figure 28*) are usually rectangular and shallow in depth. By being mounted high on a wall, the fan in the air handler can more effectively move the air into and through the room. The forward-inclined centrifugal blower in this type of air handler is small in diameter and almost as long as the air handler is wide. The moving motorized louvers on the air outlet help distribute the air more evenly throughout the room if desired. If the air handler is mounted on an outside wall, the condensate can be drained through the wall by gravity. If mounted on an inside wall, an accessory condensate pump can be used for condensate disposal (*Figure 29*). Some condensate pumps fit inside the unit cabinet, while others must be mounted exposed on the wall. These small pumps cannot

---

Figure 28 High-sidewall air handler.

raise the condensate very high, so drainage must be carefully considered in every installation.

High-sidewall air handlers are typically attached to the wall on a mounting plate. The mounting plate also acts as a template for drilling holes through the wall for the refrigerant piping, wiring, and condensate drain. Control of the air handler and outdoor unit is done by way of a wireless handheld remote. Some systems offer the option of a traditional hard-wired room controller. Regardless of which type is used, room temperature sensing is done at the indoor unit return air inlet.

Ceiling-mounted air handlers, often called under-ceiling air handlers, have many of the same characteristics as high-sidewall air handlers. The main difference is that they are suspended beneath a ceiling. They are often located close to a wall so that refrigerant piping, wiring, and condensate piping can easily be routed to the unit without being exposed. If located away from an outside wall, an accessory pump must be used to dispose of condensate. Control of the unit is the same as that of a high sidewall unit.

In-ceiling cassettes (*Figure 30*) are usually mounted in drop ceilings, leaving the face of the unit flush with the ceiling. All that may protrude slightly below the ceiling line is the return air intake (located in the center) and the supply air distribution louvers (located around the perimeter). When installed, the unit has the appearance of a concentric ceiling diffuser. The blower draws air in through the center of the diffuser and distributes it out through the four supply louvers. Refrigerant lines, wiring, and condensate lines are all located above the drop ceiling (*Figure 31*). A condensate pump is often required in this application. Control is similar to that of other ductless split-system air handlers.

In-ceiling cassettes are different from the other types of ductless split-system air handlers in that a limited length and size of ductwork can sometimes be attached to the air handler to supply an adjoining space. A knockout, visible on the side of the cassette, is provided for attaching the duct. If the ducted air is supplied to an adjoining room, air return must be accomplished using an undercut door or other method. Some cassettes also

(A)

(B)

03315-13_F29.EPS

Figure 29 Ductless system condensate pumps.

have provisions for the introduction of outside air for ventilation. Note the use of a power ventilation accessory fan to overcome the friction loss of the interconnecting duct.

On floor-mounted or console air handlers (*Figure 32*), air louvers direct airflow into the room. They are typically mounted against a wall for a through-the-wall connection to the outdoor unit. Condensate disposal and control options are similar to those of the other types of air handlers.

2.1.3 Multiple Ductless Split Systems

Ductless split systems are also available that allow several air handlers to be attached to a single outdoor unit (*Figure 33*). The different areas where the air handlers are located become, in effect, control zones. Each zone has independent temperature control. The outdoor unit has multiple service valves or other connection points where the refrigerant lines supplying each fan coil are connected. The larger the capacity of the condensing unit, the more fan coils it can likely supply with refrigerant. Again, the metering device(s), typically an electronic expansion valve, is located in the outdoor unit.

Due to the zoned nature of multiple configurations, there will be times when the full capacity of the outdoor unit will not be required. For example, if only one zone of a three-zone system requires cooling, the compressor will have excess capacity. Different manufacturers deal with compressor capacity control differently. Methods used include dual compressors, two-stage compressors, and variable-speed compressors.

Increased versatility can be realized by using different types of air handlers in different zones. For example, in a four-zone system, the first zone may use a ceiling-mounted air handler, the second and third zones may use high-sidewall air handlers, and the fourth zone may use an in-

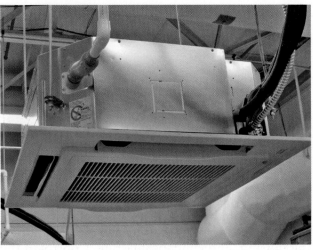

03315-13_F31.EPS

Figure 31 Piping and wiring.

ceiling cassette air handler. Each air handler has its own separate set of refrigerant lines that are routed to the condensing unit.

(A)

(B)

03315-13_F30.EPS

Figure 30 In-ceiling cassette air handler.

Figure 32 Floor-mounted (universal) air handler.

03315-13_F32.EPS

2.2.0 Ductless Split System Installation and Troubleshooting

Although some generic installation issues will be discussed here, it is important to always check the manufacturer's installation manual for the exact model being installed. The installation instructions vary from one manufacturer to another and from one model to another. The same is true for servicing ductless units; the proper information must be acquired to successfully complete troubleshooting and repair tasks.

2.2.1 Installing Ductless Systems

Ductless split systems are installed similar to conventional split system air conditioners and heat pumps. Outdoor units should be installed to al-low proper airflow and access for service. One difference to consider here is that the condenser air is horizontally discharged. The condenser air from standard condensing units is generally discharged vertically. The horizontal discharge enables these units to be installed underneath a structure such as a deck.

When locating the outdoor unit, installers must ensure that the condenser air intake and discharge areas are not obstructed. Like all units, they should not be located in areas where flammable gases are present. The potential for snowfall and the prevailing wind direction must also be considered. Due to the shape, size, and weight of ductless condensing units, it is always recommended that they be bolted securely to their foundation. Otherwise, this style of unit is far more likely to be blown over by strong winds or quickly removed by vandals. Before securing the unit in its final location, ensure that all clearances required per the installation manual have been addressed. Also note that some heat pump units may require a short section of drain line be attached to the pan of the outdoor unit. This is best done before the unit is secured.

In many countries, ductless condensing units can be seen mounted on the outside wall of residential high-rise buildings, from top to bottom (*Figure 34*). This has not been a popular approach in the United States, where centralized systems are the norm. Accessing the equipment for service is certainly not without some level of risk. However, thousands of buildings like this in Europe and Asia are heated and cooled by ductless systems. Since the condenser air can be drawn in

Figure 33 Direct expansion multiple air handler hookup.

03315-13_F33.EPS

Figure 34 Ductless condensing units serving a residential building in Shanghai, China.

through the sides as well as the back, the unit can be placed close to a wall. This approach can also be used at a residence where heavy snowfall is expected (*Figure 35*).

Although several different types of connections may be encountered, the refrigerant piping is most often connected to the indoor and outdoor units by flare fittings. It is common for the installation guidelines to provide a specific torque to be applied to the two different flare nut sizes. The installation instructions will also provide other refrigerant piping guidelines, such as the following:

• The proper line size
• Maximum length of the lines
• Maximum height of the refrigerant piping
• The amount of refrigerant to be added for each foot of refrigerant piping over a certain length

Both refrigerant lines are usually insulated, as both can produce condensation and suffer from heat transfer to the surrounding environment. Following the installation, the refrigerant lines must be properly leak-tested and evacuated before the refrigerant contained in the outdoor unit is released into the system.

Note that filter-driers are prohibited by some manufacturers. Therefore, any refrigerant piping work must be done using best practices for cleanliness and joint quality. A nitrogen purge is required for all brazing work on the system. Triple-evacuations are a must to ensure all moisture has been removed. One reason for this policy is that many ductless and related systems from Asian manufacturers use polyvinylether (PVE) oil instead of polyolester (POE) oil. Both oils are very hygroscopic, taking in water at every opportunity. POE oils however, tend to create molecular bonds with water that makes it impractical to remove it through an evacuation process. PVE oils do not have this characteristic. There are some additional benefits to PVE oils, but the fact that water can indeed be removed through evacuation is one of the most significant to technicians.

The installation procedure for the indoor unit depends on the chosen style. The typical wall-mounted indoor unit simply hangs from a metal

Figure 35 Wall-mounted ductless system condensing unit.

Figure 36 Ductless indoor unit mounting plate.

WALL SLEEVE

Figure 37 Wall opening for piping and wiring.

plate mounted securely to the wall (*Figure 36*). Note the torpedo level used across the two tabs to level the mounting plate. The unit must be level to avoid condensate drainage problems, and to present a neat appearance. Per the manufacturer's instructions, a hole is bored through the wall to accommodate the piping and wiring. Piping and wiring can also be routed within the wall itself. The hole through the wall is bored at a slight angle down towards the outside (*Figure 37*). This prevents rain from entering the opening. In addition to a structural sleeve that is often required, such as a piece of SCH 40 or SCH 80 pipe, a plastic sleeve is often used to line the opening.

The refrigerant piping and drain connections are not usually made at the unit itself. It would be an awkward chore to connect the piping while someone holds the unit in the air. The indoor unit is typically equipped with flexible copper refrigerant lines and a plastic or copper drain line that are long enough to pass through the wall. The connections are then made outdoors with the indoor unit already installed on the mounting plate. The installation instructions for most all units include various options for the connections to approach the unit from different directions, such as right-side, right-back, or right-bottom. The drain line is not generally trapped on ductless units.

The wiring from the outdoor unit both powers the indoor unit and provides communication between the two pieces. All applicable codes for wiring must be followed. Marked terminal blocks are provided on both units to make the wiring connections simple. For a neat appearance, the refrigerant piping and wiring can be concealed in special line set covers (*Figure 38*). A variety of fittings helps to accommodate most any situation. The material from some manufacturers can also be painted to blend in with the environment.

Figure 38 Line set cover.

Most ductless split systems do not use a 24VAC control circuit like that found in conventional split systems. Instead, DC communication signals between the outdoor unit and indoor unit(s) control all functions of the system. Such electronic controls are widely used in ductless split systems. These controls may have to be set up or configured during the installation to ensure that the outdoor unit is compatible with its various air handlers.

Configuring the unit may include placing jumpers on the PC board across specific pins, or setting a series of DIP switches. Since each manufacturer has its own electronic controls, it is important to read the manufacturer's installation instructions carefully during the initial setup.

The remote control (*Figure 39*) must be set up prior to startup. They are battery-operated, usually using two or three AAA batteries. Once power has been applied to the outdoor unit, the remote control is programmed for the present day of the week and time of day. The system controlled by this remote has five modes of operation:

- *Heat* – The unit operates in the heating mode, as a heat pump. Note however, that most do not have supplementary heat. Therefore, when they enter the defrost mode, cold air is blown into the space. They also may lack sufficient heating capacity at very cold outdoor temperatures.
- *Cool* – The unit operates in its normal cooling mode.
- *Dry* – When placed in the Dry mode, the ability to manually control the temperature setpoint and the airflow is locked out. The unit attempts to dehumidify by slowing the fan speed signif-

icantly, while attempting to also keep the room temperature the same.
- *Auto* – The unit switches automatically between Cool and Heat as needed to maintain the desired settings.
- *Fan* – This mode simply provides air movement without conditioning.

There are a number of other settings that are user-controlled. Note however, that the terminology and features vary between companies and products. Always follow the manufacturer's operating guidelines for the specific unit. Moving from the top of *Figure 39* clockwise, they are as follows:

- The Home Leave feature allows the user to press a single button to force the system to respond to settings preferred when the space is vacant. Both a temperature and an airflow rate can be preset.
- The Vertical/Horizontal Louver or Swing buttons allow the user to set the position of the discharge blades on the supply air outlet. This affects the direction of the airflow.
- The Intelligent Eye or Sensor button controls an energy-saving feature. The sensor can de-

STANDARD REMOTE CONTROLLER FOR ALL RA WALLMOUNT FAN COILS

03315-13_F39.EPS

Figure 39 Remote control button layout.

tect human movement. After a period of time passes without movement being sensed, the unit will change the setpoint to save energy. This particular unit changes the temperature up or down 3.6°F (2°C) up or down, depending on the operating mode. The angle of the sensing eye on the face of the unit can be adjusted to look in the appropriate direction for movement.

- The Timer On (and Timer Off) buttons, along with the Select button, allow the user to set times of day for the system to start and stop.
- The Outdoor Quiet button affects the outdoor unit operation. At night or other times when the sound of the outdoor unit may be a nuisance, the button can be used to change the speed and frequency of the condenser fan. This feature is not available in Powerful mode.
- The Powerful setting disregards any previous settings for airflow control and maximizes the cooling capacity for 20 minutes.

Some systems may offer full programming with different settings based on time of day. During start-up, all the functions and settings of the remote should be checked and proper operation confirmed. The remote control described here has a Trial mode that is available to the operator when the system is first started. In Trial mode, the system operates on its own for roughly 30 minutes, performing various operational tests. Any problems are reported through fault codes displayed on the remote or at the unit control board outdoors.

Training the homeowner in remote control use is also an essential step. Note that most of the button labels are non-technical in nature and user-friendly. The Quiet button for example, slows the condenser fan motor. Average homeowners may not know what a condenser fan motor is, but they certainly understand what it means for the system to operate more quietly.

2.2.2 Troubleshooting Ductless Systems

Ductless systems have proven to be very reliable. However, eventually troubleshooting and repair will likely be required. Troubleshooting the refrigerant circuit is not significantly different from common split systems. However, it is important

to remember that the metering device is usually located in the outdoor unit. The service valves located at the outdoor unit provide the only refrigerant circuit access. Since the refrigerant lines are virtually inaccessible at the indoor unit, all pressure and temperature readings are taken at the condensing unit. Procedures such as refrigerant charging should be done according to the manufacturer's instructions using commonly accepted service practices.

The controls on ductless systems are significantly different from common systems. Electronic controls often have built-in diagnostics in the form of fault code light-emitting diodes (LEDs). The LEDs flash in a particular sequence or for varying periods of time to indicate the fault that was sensed. Labels on the equipment and the service literature can help decipher any fault codes. Manufacturers also provide hotlines for technicians to call, as understanding the fault codes is essential to the troubleshooting process.

The remote control on some systems can also report faults with the system. An astute user can view the trouble codes and report them to the service agency when the service call is placed. *Figure 40* shows a typical list of fault codes for one specific system. As noted at the top of the figure, the fault codes can be viewed by repeatedly pressing a specific key.

Ductless system controls can be extremely complex. It is essential that technicians acquire the correct troubleshooting information for the specific unit being serviced. It is virtually impossible to troubleshoot these systems without the correct information. The literature usually provides troubleshooting charts, such as the one shown in *Figure 41*, to help technicians start the process. As shown here, the chart guides the technician to a particular page of the manual for additional information. A chart showing all of the possible error codes (*Figure 42*) also guides technicians to a specific page that provides additional instructions to resolve the problem.

Once an error code is defined and a referenced page number is accessed, a flow chart is often found to guide the technician through determining what action must be taken. For example, in *Figure 42*, note the error code EA, which refers to a four-way valve abnormality. The code description

Ductless System Sophistication

Here's a clue to the level of sophistication that can be found in ductless systems. The service manual for a relatively standard indoor and outdoor matched combination, portions of which were used here in the training guide, is 414 pages in length.

2. Press the TIMER CANCEL button repeatedly until a continuous beep is produced.

• The code indication changes as shown below, and notifies with a long beep.

	CODE	MEANING
SYSTEM	00	NORMAL
	U0	REFRIGERANT SHORTAGE
	U2	DROP VOLTAGE OR MAIN CIRCUIT OVERVOLTAGE
	U4	FAILURE OF TRANSMISSION (BETWEEN INDOOR UNIT AND OUTDOOR UNIT)
INDOOR UNIT	A1	INDOOR PCB DEFECTIVENESS
	A5	HIGH PRESSURE CONTROL OR FREEZE-UP PROTECTOR
	A6	FAN MOTOR FAULT
	C4	FAULTY HEAT EXCHANGER TEMPERATURE SENSOR
	C9	FAULTY SUCTION AIR TEMPERATURE SENSOR
OUTDOOR UNIT	EA	COOLING-HEATING SWITCHING ERROR
	E5	OL STARTED
	E6	FAULTY COMPRESSOR START UP
	E7	DC FAN MOTOR FAULT
	E8	OPERATION HALT DUE TO DETECTION OF INPUT OVER CURRENT
	F3	HIGH TEMPERATURE DISCHARGE PIPE CONTROL
	F6	HIGH PRESSURE CONTROL (IN COOLING)
	H6	OPERATION HALT DUE TO FAULTY POSITION DETECTION SENSOR
	H8	CT ABNORMALITY
	H9	FAULTY SUCTION AIR TEMPERATURE SENSOR
	J3	FAULTY DISCHARGE PIPE TEMPERATURE SENSOR
	J6	FAULTY HEAT EXCHANGER TEMPERATURE SENSOR
	L4	HIGH TEMPERATURE AT INVERTER CIRCUIT HEATSINK
	L5	OUTPUT OVERCURRENT
	P4	FAULTY INVERTER CIRCUIT HEATSINK TEMPERATURE SENSOR

NOTE

1. A short beep and two consecutive beeps indicate non-corresponding codes.
2. To cancel the code display, hold the TIMER CANCEL button down for 5 seconds. The code display also cancels itself if the button is not pressed for 1 minute.

03315-13_F40.EPS

Figure 40 Ductless system fault codes.

then refers the technician to page 119 of the service manual. A deeper description of the problem is found there, as shown in *Figure 43*. Next, a flow chart to follow to resolve the problem is shown in *Figure 44*. Due to the complexity of the system controls, these troubleshooting aids are indispensable.

2.3.0 Variable Refrigerant Flow Systems

Variable refrigerant flow (VRF) systems represent a complete different approach to zoning than ductless systems. Although they share some similarities with ductless systems, a larger variety of ducted indoor units is available. In fact, ducted indoor units represent roughly half of the equipment connected to VRF systems worldwide.

Like ductless systems, a single VRF outdoor unit can support a number of indoor units. With ductless systems, each indoor unit has an independent set of refrigerant lines that must be connected directly to the outdoor unit. A VRF system uses only one set of refrigerant lines and distributes the refrigerant to each indoor unit through the same circuit (*Figure 45*).

VRF systems control the flow of refrigerant to each connected indoor unit as necessary. Although the systems do provide for a great deal of communication between the indoor and outdoor unit, this communication has little to do with control. In fact, the outdoor and outdoor units are hardly aware of the others existence from a functional point of view. The primary chore of the outdoor unit is to produce a varying amount of refrigerant flow at a preset subcooling value. Daikin, the industry leader in VRF systems and the original developer of the technology, uses a subcooling value of 9°F. Rather than using information gathered from the indoor units, the condensing unit modulates the refrigerant flow volume to maintain the 9°F subcooling setpoint. Instead of using the generic term variable refrigerant flow, Daikin has trademarked the term *variable refrigerant volume (VRV™)*. The indoor units are equipped with electronic expansion valves (EEVs) and are designed to maintain a 9°F superheat value when they are in operation.

Symptom	Check Item	Details of Measure	Reference Page
The unit does not operate.	Check the power supply.	Check if the rated voltage is supplied.	—
	Check the type of the indoor unit.	Check if the indoor unit type is compatible with the outdoor unit.	—
	Check the outdoor temperature.	Heating operation cannot be used when the outdoor temperature is 24°C (75.2°F) or higher, and cooling operation cannot be used when the outdoor temperature is below 10°C (50°F).	—
	Diagnose with remote controller indication.	—	94
	Check the remote controller addresses.	Check if address settings for the remote controller and indoor unit are correct.	388
Operation sometimes stops.	Check the power supply.	A power failure of 2 to 10 cycles stops air conditioner operation. (Operation lamp OFF)	—
	Check the outdoor temperature.	Heating operation cannot be used when the outdoor temperature is 24°C (75.2°F) or higher, and cooling operation cannot be used when the outdoor temperature is below 10°C (50°F).	—
	Diagnose with remote controller indication.	—	94
The unit operates but does not cool, or does not heat.	Check for wiring and piping errors in the connection between the indoor unit and outdoor unit.	—	—
	Check for thermistor detection errors.	Check if the thermistor is mounted securely.	—
	Check for faulty operation of the electronic expansion valve.	Set the unit to cooling operation, and check the temperature of the liquid pipe to see if the electronic expansion valve works.	—
	Diagnose with remote controller indication.	—	94
	Diagnose by service port pressure and operating current.	Check for refrigerant shortage.	104
Operating noise and vibrations	Check the output voltage of the power module.	—	153
	Check the power module.	—	—
	Check the installation condition.	Check if the required spaces for installation (specified in the installation manual) are provided.	—

03315-13_F41.EPS

Figure 41 Ductless system troubleshooting chart.

As individual indoor units start and stop, the refrigerant flow in the circuit changes. This change in the flow rate results in a change in the subcooling value of refrigerant leaving the condenser. For example, assume two indoor units at 9,000 Btuh each are in steady operation. A third unit at 12,000 Btuh capacity is then brought online to cool a third area. As the indoor unit EEV opens to admit refrigerant, the subcooling value at the condensing unit drops. This is because the refrigerant is now moving through the condenser more rapidly, spending less time exposed to condenser airflow. The condensing unit responds by increasing the speed of the compressor(s) to catch up, settling at a speed that again results in a 9°F subcooling value.

VRF systems are available in straight cooling, heat pump, and heat recovery models. Like conventional heat pumps, VRF heat pumps can provide both heating and cooling, but not at the same time. Heat recovery units however, also provide both heating and cooling, but it can be done si-

Metering Devices

Ductless systems place the system metering device, which is usually an EEV, in the outdoor unit. VRF systems use EEVs exclusively, but those responsible for metering refrigerant for the indoor units are located there, not in the outdoor unit. However, EEVs may also be found in the outdoor unit to control the flow of refrigerant within the unit itself for various purposes.

	Error Codes	Description		Reference Page
System	00	Normal		—
	U0★	Refrigerant shortage		104
	U2	Low-voltage detection or over-voltage detection		107
	U4	Signal transmission error (between indoor unit and outdoor unit)		109
	UA	Unspecified voltage (between indoor unit and outdoor unit)		112
Indoor Unit	A1	Indoor unit PCB abnormality		95
	A5	Freeze-up protection control or heating peak-cut control		97
	A6	Fan motor or related abnormality	DC motor (FTXS series)	99
			AC motor (FDXS series)	101
	C4	Indoor heat exchanger thermistor or related abnormality		103
	C9	Room temperature thermistor or related abnormality		103
Outdoor Unit	E1	Outdoor unit PCB abnormality		113
	E5★	OL activation (compressor overload)		115
	E6★	Compressor lock		116
	E7★	DC fan lock		117
	E8	Input overcurrent detection		118
	EA	Four-Way valve abnormality		119
	F3	Discharge pipe temperature control		121
	F6	High pressure control in cooling		123
	H0	Compressor system sensor abnormality		125
	H6	Position sensor abnormality		128
	H8	DC voltage / current sensor abnormality (09/12 class only)		131
		CT or related abnormality (24/30/36 class only)		132
	H9	Outdoor temperature thermistor or related abnormality		134
	J3 ★	Discharge pipe thermistor or related abnormality		134
	J6	Outdoor heat exchanger thermistor or related abnormality		134
	L3	Electrical box temperature rise		136
	L4	Radiation fin temperature rise		138
	L5★	Output overcurrent detection		140
	P4	Radiation fin thermistor or related abnormality		134
	U7	Signal transmission error on outdoor unit PCB (24/30/36 class only)		111

★: Displayed only when system-down occurs.

03315-13_F42.EPS

Figure 42 Fault code guide.

multaneously. With heat recovery VRF systems, while one or more indoor units are operating in the cooling mode, other units can operate in the heating mode. This is done by routing a hot gas line to each indoor unit, along with the liquid and suction lines. Heat that is absorbed by the system from units in the cooling mode is then transferred to areas that require heating. This feature is very unique among heating and cooling systems, other than VAV systems.

4.16 Four-Way Valve Abnormality

Remote Controller Display	EA
Method of Malfunction Detection	The room temperature thermistor and the indoor heat exchanger thermistor are checked if they function within their normal ranges in each opeartion mode.
Malfunction Decision Conditions	A following condition continues over 1 ~ 10 minutes after operating for 5 ~ 10 minutes. ■ Cooling / Dry (room thermistor temp. – indoor heat exchanger temp.) < −5°C (−9°F) ■ Heating (indoor heat exchanger temp. – room thermistor temp.) < −5°C (−9°F) ■ If the error repeats, the system is shut down. ■ Reset condition: Continuous run for about 60 minutes without any other error
Supposed Causes	■ Disconnection of fourway valve coil ■ Defective fourway valve, coil, or harness ■ Defective outdoor unit PCB ■ Defective thermistor ■ Refrigerant shortage ■ Water mixed in refrigerant ■ Defective stop valve

03315-13_F43.EPS

Figure 43 Problem description.

2.3.1 VRF Equipment

VRF condensing units (*Figure 46*) are typically available up to 12 tons in size. Units with 8 tons of capacity or more usually have multiple compressors. When two compressors are used, one may be a fixed-speed model, while the other is a variable speed model powered by an inverter. Both compressors may be variable speed in other models or brands. The inverter controls the frequency of power to the compressor(s), thereby controlling the speed. Compressors using this technology can typically reduce their capacity to roughly 15 percent of maximum, and operate at virtually any point between 15 and 100 percent. Other benefits in the use of variable frequency drive compressors include the following:

- Very low starting currents
- Soft, quiet starts that do not cause lights to flicker due to the starting current
- Systems pressures that build slowly, reducing stress
- Fewer start and stop cycles, which are stressful to common compressors
- Much improved ability to dehumidify without overcooling

Single units can also be manifolded together as a modular system. The units share a common refrigerant circuit and can provide up to roughly 30 tons of capacity (three 10-ton units) this way. Even a single condensing unit can be connected to an astounding 62 separate indoor units (varies by manufacturer).

In a system where the refrigerant flow rate is constantly changing, the return of refrigeration oil to the compressor has to be a concern. If the refrigerant lines were sized to maintain a proper velocity for oil flow at the lowest possible flow rate, they would be far too small to support the flow needed at maximum capacity. Therefore, VRF systems incorporate oil separators at the outdoor unit to minimize the amount of oil in circulation. In addition, some manufacturers apply what may be referred to as an oil recovery mode. After a predetermined period of operation, the system positions all the EEVs to maximize the flow rate, and then maximizes the speed of the compressor(s). This increases the refrigerant velocity to its maximum point throughout the refrigerant circuit, giving any oil remaining in the circuit an opportunity to return to the compressor. Oil recovery cycles typically last less than 10 minutes.

Check No.01
Refer to P.142

Check No.13
Refer to P.145

Check No.14
Refer to P.145

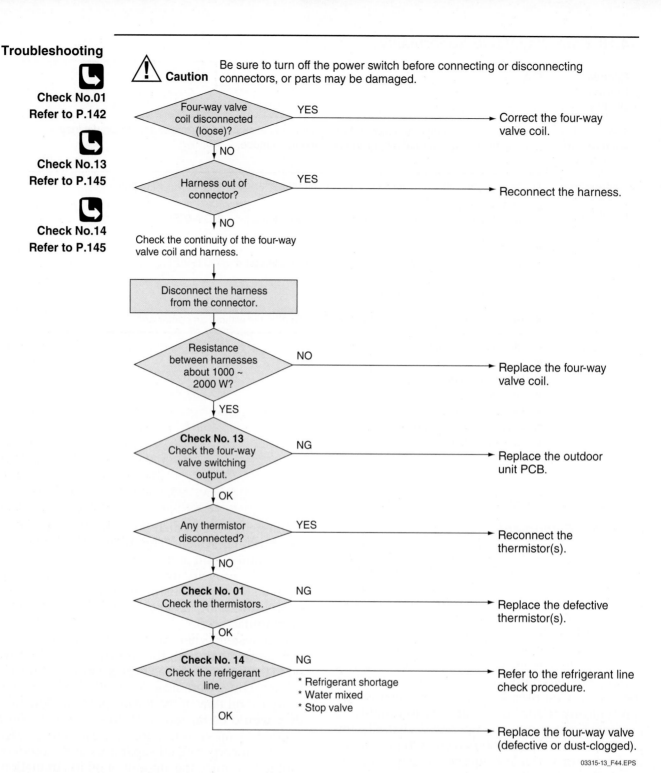

⚠ **Caution** Be sure to turn off the power switch before connecting or disconnecting connectors, or parts may be damaged.

Figure 44 Troubleshooting flow chart.

VRF condensing units typically air-cooled, but water-cooled units are also available. Like commercial water-cooled heat pumps systems, a means of heat rejection such as a cooling tower or dry cooler is required. In addition, a boiler or similar source of heat is required. The two parts of the system work to maintain the loop water temperature in the appropriate range. A water temperature range of 67°F to 95°F is one example.

VRF indoor units are available in a wide variety of configurations. Models that are ductless and look the same as other ductless units are quite popular. However, VRF systems have a much broader selection of indoor equipment that includes a number of ducted units as well. Units that are virtually the same as common fan coil units used with standard split systems are available, as are units designed to be concealed in small spaces

Figure 45 VRF system refrigerant flow schematic.

Figure 46 VRF condensing unit.

(*Figure 47*). Some units can be fitted with electric heat if necessary. However, these units have more controls and devices installed than common systems in their size range. Thermistors are required to provide temperature information and calculate superheat values at each individual unit. Electronic devices are then needed to translate this

Figure 47 VRF fan coil unit designed for concealment.

information and position the EEV correctly. Most ducted models can accommodate outside air connections as well

VRF system indoor and outdoor units are in constant communication with each other. A daisy-chain configuration, much like that used in VVT zoning systems, is used to transmit information. However, as mentioned earlier, this transfer of information has little to do with system control. The primary focus is to collect all possible operating information and provide it to operators and service personnel to evaluate system operation. For example, by connecting a laptop to the system, a technician can see every operating parameter that is being monitored by the system. If a specific indoor unit is having problems, he or she can see the readings from each thermistor, the calculated superheat, the precise position of the EEV, and information related to the blower. Again, the indoor and outdoor units share a great deal of information, but their function is primarily based on information collected within their own cabinet.

The indoor units of VRF systems can be controlled several different ways. One way is through a wireless remote, much like those used with ductless systems. A simple wired controller is also available. However, even the simple control devices for VRF systems, depending on the manufacturer, can control up to 16 indoor units from a single device. The wired control shown in *Figure 48* is known by Daikin as the Navigation Remote Controller™. This controller also provides advanced scheduling in addition to controlling up to 16 units.

2.4.0 VRF System Installation and Troubleshooting

VRF systems have some unique refrigerant piping requirements, compared to common systems. Failing to follow the specific installation guidelines for the piping will undoubtedly result in functional problems. The interconnecting communications wiring must also be done with great care and attention to the details provided by the system manufacturer. Due to the level of sophis-

Figure 48 Navigation Remote Controller™.

tication in the controls, troubleshooting can often be done with very few tools in hand; the system can provide most of the needed information to make an informed repair decision.

2.4.1 VRF System Installation

VRF system indoor and outdoor units are installed with the same considerations as most other systems. The unique installation requirements are found in the refrigerant piping and control wiring. *Figure 49* provides an overview of the refrigerant piping and control wiring for a typical VRF system.

All of the indoor units are served by the same refrigerant circuit. This requires careful management of the refrigerant flow and great attention to detail in the installation process. Special fittings and piping assemblies have been engineered for VRF systems and must be utilized. These special piping accessories are available through the product manufacturer.

All refrigerant lines, including the hot gas line used for systems capable of heat recovery and simultaneous heating and cooling, must be insulated individually. The insulated lines are then generally wrapped together as a group with a special tape, with communication wiring also included in the bundle.

Refrigerant filter-driers should never be used. Although it is important in every installation, it is even more important with VRF systems that nitrogen be circulated through the piping at all times when brazing. Since these systems have many small openings throughout the system, solid contaminants must be avoided. It is also important that traps in the refrigerant line be avoided. Remember that the refrigerant flow rate can become quite slow, which hinders oil from returning to the compressor. Traps will further hinder oil return.

Engineered branch line connections and headers (*Figure 50*) are copper and must be used to make primary connections. These components are often referred to as **refnet** (refrigerant net-

NOTE: ONLY ONE OF TWO REFRIGERANT LINES SHOWN.

Figure 49 VRF system piping and wiring overview.

work) **joints**. The refnet joints include a special distributor to help ensure the balanced distribution of refrigerant in both directions. Note the stepped-down construction of the tubing for the connections. The components have been designed so that a single style of refnet joint can be used with multiple pipe sizes. The installer simply uses the tube size required for the installation, cutting off any portion of the connection that is smaller than the desired size. This reduces the required inventory of installation parts. Manufacturers also provide preformed insulation that fits these special piping components properly. Field-insulating these parts properly would be challenging at best.

The refnet joints must be installed in accordance with an important set of guidelines. In addition, the system designer must lay out and size the piping according to the manufacturer, and choose the location of the joints wisely. Common installation considerations for refnet joints and headers include the following:

• The refnet joint can be oriented looking straight up, straight down, or laid horizontally with the

main and branch lines side-by-side and level with each other. It should never be installed in a manner that places the main line and branch line above or below each other.

• Headers must always be installed so that the branch lines leaving the manifold are horizontal, never looking up or down. Note, as shown in *Figure 50*, that the liquid and suction headers are constructed differently. Both headers are shown properly oriented for installation in the figure. Unused ports on a header must be closed by brazing or soldering.

• A distance of 20" is generally required after an elbow or similar fitting before a refnet joint is placed. The next refnet joint in the main line must then be at least 40" downstream. Refer to *Figure 51*.

Manufacturers also provide specific guidelines related to the size and length of each portion of the refrigerant network. It is essential that the guidelines from the specific equipment manufacturer be followed. The requirements will differ from one brand to another, and possibly from one model series to another. Always follow the manufacturer's instructions for the specific system being installed. In most cases, the installing technician is not responsible for sizing or designing the layout of the refrigerant piping. But it is important for the installer to understand the guidelines that support the design. The piping layout may also include expansion loops or flexible sections of refrigerant line in key locations that may be subjected to significant levels of expansion and contraction. Otherwise, the piping may be damaged by these powerful forces.

Most VRF systems will provide maximum and/or minimum values for the following portions of the refrigerant network:

• The maximum distance from the first refnet joint to the farthest indoor unit.
• The maximum distance from the outdoor unit to the nearest refnet joint.
• The maximum distance from a refnet joint to the indoor unit it serves.
• The minimum distance between refnet joints.
• The maximum difference in length between the longest and shortest run connecting refnet joints to an indoor unit.

The refrigerant piping between condensing units that are manifolded together on a single circuit also requires great attention to detail. If multiple units are connected improperly, oil return problems will likely result, as well as functional difficulties.

03315-13_F50.EPS

Figure 50 Refnet joints and headers.

For systems that provide heat recovery, which allows the system to provide simultaneous heating and cooling, an additional device is required. A branch selector box (*Figure 52*) or similar apparatus is required to manage the flow of refrigerant to units that are in different modes of operation. A heat recovery system has three refrigerant lines, including the necessary hot gas line. However, indoor units have only two line connections—one for liquid and one for vapor. The branch selector box manages the refrigerant flow as necessary between the outdoor and indoor units. To do so, it must also be made a part of the communications network. Branch selector boxes often support a single indoor unit. However, models are available that support up to eight indoor units. This style can take the place of refnet headers in the refrigerant piping, taking the three refrigerant lines from the outdoor systems and distributing it to multiple indoor units from a single enclosure.

The communication and control wiring between the indoor units, outdoor unit(s), and wired controllers is also critical to proper system operation. Incorrect wiring methods will result in lost communications. Terminals are provided on each unit for the interconnecting wiring. *Figure 53* shows the terminal designations for one line of equipment. Note that the communication wiring terminals are designated as transmission terminals by this manufacturer. The terminals designated Forced Off are used to interlock the equipment with emergency shutdown devices, such as a fire panel or even the float switch of a condensate pump. On the outdoor unit, the terminals marked Q1 and Q2 are used when multiple condensing units are connected together on the same system. The outdoor units would communicate through this circuit. The P1 and P2 terminals

are used to connect a wired controller. As the wiring travels between one unit and another, the polarity must remain the same; the same wire must be connected to all F1 terminals throughout the system, for example. Remember that these termi-

03315-13_F52.EPS

Figure 52 Branch selector boxes.

03315-13_F51.EPS

Figure 51 Refnet joint positioning.

(A) INDOOR UNIT TERMINALS

(B) BRANCH SELECTOR BOX TERMINALS

(C) OUTDOOR UNIT TERMINALS

03315-13_F53.EPS

Figure 53 VRF system communication wiring terminals.

nal designations can differ from manufacturer to manufacturer.

The wire typically used for communications is two-conductor #18 AWG stranded, non-shielded wire. #16 AWG may be required if the length is excessive. Note that common thermostat wire used for 24VAC field control wiring has a solid conductor and is not suitable for this application. Three conductors may be required for some applications. This type of wire rarely needs to be in conduit, so it is often routed through ceiling cavities by following other runs of conduit, piping, or building structures. However, the communications wiring should never be strapped along the outside of conduit carrying AC-powered wiring to avoid interference. Shielded wire can be used but is not required. If it is used, the shield must be grounded at every component to which the conductors are connected. The wire should not be spliced between components; a continuous piece of wire should be routed from component to component.

As mentioned earlier, communication wiring is routed in a daisy-chain configuration (*Figure 54*). Star connection wiring configurations, for example, should not be used. All terminations and connections occur at the equipment. There should be no connections or junctions in the wiring elsewhere. Although other strategies may work fine for distributing power, they do not allow for proper communication signals in this type of system.

Once the installation of a system is complete, the system must be charged with refrigerant. Most standard HVAC units of 5 tons capacity or less come with a sufficient amount of refrigerant charge to handle a specific length of piping. In charging VRF systems, the entire piping system, all the way back to the condensing unit must be considered. The required weight of refrigerant is calculated based on the length of the liquid lines and system components, and then weighed into the system. Each foot of liquid line of a given size can hold a specific volume of refrigerant. Therefore, the installed length of each different liquid line size must be calculated separately. Each specific indoor unit model and piping component requires a certain amount

Special Tool Kits

Since ductless and VRF system manufacturers specify torque values for each size of flare nut on their systems, special tool kits have been developed for these installations. Tool manufacturers and equipment manufacturers alike offer kits specifically to support the installation of these systems. They typically include a refrigerant gauge manifold with hoses; flaring tools; gauges to check the quality of a formed flare; a torque wrench with various sizes of open-end attachments to fit the flare nuts; metric and/ or imperial hex wrenches; and various brass adapters and valve core tools. Since these systems, primarily from Asian manufacturers, use PVE oils, it is preferred that a refrigerant gauge manifold set be used exclusively on their equipment to prevent cross-contamination of oil charges.

03315-13_SA05.EPS

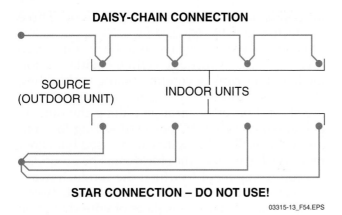

DAISY-CHAIN CONNECTION

SOURCE
(OUTDOOR UNIT)   INDOOR UNITS

STAR CONNECTION – DO NOT USE!

03315-13_F54.EPS

Figure 54 Daisy-chain wiring.

of refrigerant to be added. Adding refrigerant randomly to these systems rarely results in proper performance. Most manufacturers suggest that the system be charged exclusively by weight. When the charge becomes incorrect due to leaks, the remaining charge is recovered and the system is recharged from scratch by weight. This significant task is yet another reason why the quality of piping and joints must be at its highest for VRF systems.

2.4.2 VRF System Troubleshooting

Proper training is essential to successfully troubleshooting VRF systems. Although they may seem complex in many ways, the sophisticated controls and communications can actually do most of the troubleshooting for you. With standard systems, technicians often reach for the refrigeration gauge manifold first, in every troubleshooting situation. However, with VRF systems, troubleshooting should never begin with connecting a set of gauges. Troubleshooting some common problems in VRF systems can begin with the symptoms at hand. Charts and tables that provide guidance based on symptoms (*Figure 55*) are often offered by the manufacturers. In some cases, the solution offered may seem overly simplified, but many problems are in fact caused by simple errors or a misunderstanding of how the system operates.

For more complex problems or those that do not have a simple solution (such as moving the curtains out of the way of the unit discharge), troubleshooting often begins at the remote control. Troubleshooting is where the real value of the communications network shines through. In many VRF systems, the operating parameters of each piece of equipment can viewed from a single point, regardless of how large or expansive the system may be.

Each manufacturer likely has a different way of accessing the needed information from a remote control. This is done by using menus to access test functions where error codes related to the problem will be displayed, or by using a unique sequence of key-presses to access the desired information. In some cases, error codes may be displayed on the user's main display. Since many settings are not understood by users but can lead to serious problems if they are casually or accidently changed, unique methods of accessing service and maintenance menus provide a layer of protection.

The first menus encountered can provide operating and maintenance information for one or more units on the system (*Figure 56*). All communicating components in the system have a unique address, allowing the technician to access the information for a specific unit. Of course, problems that affect the entire system are likely related to the outdoor unit or the refrigeration cycle. When only one area experiences a performance issue, the indoor unit serving that area is likely where the problem lies. In most systems though, the information from a specific piece of equipment can be accessed from any single remote. Once a unit number or address is selected, the information for that unit is displayed. As shown here, even the operating hours for each piece of equipment can be viewed.

Some distinction is usually made between problems that cause system shutdown and those that merely represent a reason for concern. The latter are usually referred to as warnings. In this

Thermostat? What thermostat?

Throughout this section, the word thermostat has seldom been used. With standard heating and cooling systems, a wall thermostat is typically in control of the system. However, with ductless and VRF systems, the device used to control the systems is referred to as a remote control. We tend to think of remote controls as portable devices, like the ones used to control TVs. With these systems, though, even wall-mounted devices are referred to as remote controls, or simply remotes. The term is actually more accurate, as the remote control for ductless and VRF systems does not monitor room temperature. Room temperature is usually monitored by a thermistor at the return air inlet of the indoor unit(s). The remote is used to change settings and acquire information, but not to stop and start the unit in response to room temperature.

46   NCCER – *HVAC Level Three*   03315-13

	Symptom		Supposed Cause	Countermeasure
1	The system does not start operation at all.		Blowout of fuse(s)	Turn Off the power supply and then replace the fuse(s).
			Cutout of breaker(s)	• If the knob of any breaker is in its OFF position, turn ON the power supply. • If the knob of any circuit breaker is in its tripped position, do not turn ON the power supply.
			Power failure	After the power failure is reset, restart the system.
2	The system starts operation but makes an immediate stop.		Blocked air inlet or outlet of indoor or outdoor unit	Remove obstacle(s).
			Clogged air filter(s)	Clean the air filter(s).
3	The system does not cool or heat air well.		Blocked air inlet or outlet of indoor or outdoor unit	Remove obstacle(s).
			Clogged air filter(s)	Clean the air filter(s).
			Enclosed outdoor unit(s)	Remove the enclosure.
			Improper set temperature	Set the temperature to a proper degree.
			Airflow rate set to **LOW**	Set it to a proper airflow rate.
			Improper direction of air diffusion	Set it to a proper direction.
			Open window(s) or door(s)	Shut it tightly.
		[In cooling]	Direct sunlight received	Hang curtains or shades on windows.
		[In cooling]	Too many persons staying in a room	
		[In cooling]	Too many heat sources (e.g. OA equipment) located in a room	
4	The system does not operate.	The system stops and immediately restarts operation.	If the **OPERATION** lamp on the remote controller turns ON, the system will be normal. These symptoms indicate that the system is controlled so as not to put unreasonable loads on the system.	Normal operation. The system will automatically start operation after a lapse of five minutes.
		Pressing the **TEMP ADJUST** button immediately resets the system.		
		The remote controller displays **UNDER CENTRALIZED CONTROL**, which blinks for a period of several seconds when the **OPERATION** button is depressed.	The system is controlled with centralized controller. Blinking display indicates that the system cannot be operated using the remote controller.	Operate the system using the **COOL/HEAT** centralized remote controller.
		The system stops immediately after turning **ON** the power supply.	The system is in preparation mode of micro computer operation.	Wait for a period of approximately one minute.
5	The system ~~controller displays~~		The system stops due to an ~~interruption in communication caused by electrical~~	Remove causes of electrical noises. If these causes are removed, the system will automatically restart

03315-13_F55.EPS

Figure 55 Partial symptom-based troubleshooting guide.

Maintenance Menu	Item 2	Remarks		
1. Model Name	1. Unit No.	Select the Unit No. you want to check.		
	2. Indoor unit			
	3. Outdoor unit			
2. Operating Hours	1. Unit No.	Select the Unit No. you want to check.		
	2. Indoor unit operating time	All of these are displayed in hours.		
	3. Indoor fan operation			
	4. Indoor unit energized time			
	5. Outdoor operating time			
	6. Outdoor fan 1 operation			
	7. Outdoor fan 2 operation			
	8. Outdoor comp. 1 operation			
	9. Outdoor comp. 2 operation			
3. Indoor Unit Status	1. Unit No.	Select the Unit No. you want to check.		
	2. FAN	Fan tap		
	3. FLAP	Swing, fixed		
	4. Speed	Fan speed (rpm)		
	5. EV	Degree that electronic expansion valve is open (pls)		
	6. MP	Drain pump ON/OFF		
	7. EH	Electric heater ON/OFF		
	8. Hu	Humidifier ON/OFF		
	9. TBF	Anti-freezing control ON/OFF		
	10. FLOAT			
	11. T1/T2			
	12. Unit No.	Select the Unit No. you want to check.		
		SkyAir		VRV
	13. Th1	Suction air thermistor		Suction air thermistor
	14. Th2	Heat exchanger thermistor		Heat exchanger liquid pipe thermistor
	15. Th3	—		Heat exchanger gas pipe thermistor
	16. Th4	Discharge air thermistor		Discharge air thermistor
	17. Th5	—		—
	18. Th6	—		—
4. Outdoor Unit Status	1. Unit No.	Select the Unit No. you want to check.		
	2. FAN step	Fan tap		
	3. COMP	Compressor power supply frequency (Hz)		
	4. EV1	Degree that electronic expansion valve is open (pls)		
	5. SV1	Solenoid valve ON/OFF		
		SkyAir		VRV
	6. Th1	Outdoor air thermistor		—
	7. Th2	Heat exchanger thermistor		—
	8. Th3	Discharge pipe thermistor		—
	9. Th4	Heat exchanger deicer thermistor		—
	10. Th5	Heat exchanger gas pipe thermistor		—
	11. Th5	Liquid pipe thermistor		—
5. Forced Defrost (SkyAir only)	1. Forced defrost ON	Enables the forced defrost operation.		
	2. Forced defrost OFF	Disables the forced defrost operation.		

03315-13_F56.EPS

Figure 56 Partial remote control maintenance menu.

case, a light on the remote and/or indoor unit may remain on, but a warning message is displayed on the remote. For problems that cause system or unit shutdown, this same light may blink rapidly and an error code is displayed on the remote. The responses to problems will be different from system to system.

Once the correct menu is accessed, one or more error codes will be displayed. Each individual code should be carefully considered. In some cases, one problem causes a number of other problems that may also result in an error code. For example, on the error code chart shown in *Figure 57*, there are several error codes that are related to refrigerant condensing section. Error code E3 indicates the activation of a high pressure switch; code E7 is related to a malfunction of the condenser fan motor; and code H3 is related to the failure of the high-pressure switch. Only one problem may truly exist, but the problem can possibly cause more than one error code to be displayed.

Once the error code is determined, the technician is guided to specific pages where more information about the problem is found. There, an explanation of the actual circumstance that caused the error code to be generated is usually found. In addition, a flow chart to guide the technician's troubleshooting activities is provided.

Our previous example of error code E7 will be used to demonstrate the use of the troubleshooting guide. Code E7 refers to a malfunction of the outdoor unit fan motor. The technician is directed to page 138 of the service manual to proceed further. One page 138, an explanation of the error code is provided (*Figure 58*). The text explains that it occurs as the result of detecting abnormal currents from the main inverter or the fan motor inverter. It may also be caused by detecting a lower-than-expected fan rpm. Further, it lets the technician know that detecting this condition four times will cause unit shutdown. Otherwise, only a warning message is displayed on the remote and unit operation continues. Possible causes include a fan motor failure; a defect in the wiring or connections to the motor; or a foreign object interfering with the rotation of the motor. The final note indicates that the warning message will go away if normal operation is detected for a full five minutes.

A flow chart is provided on the next page to further guide the technician (*Figure 59*). Specific steps to follow are shown, with a resolution for each step based on the result. In some cases, the technician may be guided to another page for a procedure. For example, the final step in the flow chart shown in *Figure 59,* tagged as Check 6, directs the technician to page 215 for guidance on checking the fan motor connector. The procedure from that page is shown in *Figure 60*. The technician then uses a meter to measure the resistance from terminal Vcc to the other three wires, as well as checking the resistance from the Ground wire to the other three terminals. Returning to the flow chart, the technician reads the statement, answers yes or no based on the test results, and completes the repair as prescribed. For testing thermistors, charts are provided that indicate the proper resistance reading for a given temperature. Charts are also provided for pressure sensors to check their accuracy (*Figure 61*). Note that the high pressure and low pressure sensors have different voltage characteristics, as shown on the chart.

There are even more powerful ways to view and troubleshoot many VRF systems. Some manufacturers have testing equipment that serves as an interface between the system and a laptop computer. The device shown in *Figure 62* is known as the Service Checker. It allows the technician to communicate with the entire system, using a laptop instead of the very limited displays and keys offered by remote controls.

Troubleshooting VRF systems can be very challenging for any technician. Some of the difficulty may arise from the quality of the documentation following its translation. Proper training in the product line in use is essential to success with VRF systems. However, it is also important to note that properly installed VRF systems have a high degree of reliability, and many problems are simpler than they may initially appear. Many problems stem from a misunderstanding of their operation by the user, as well as by untrained technicians who attempt to troubleshoot these systems using a conventional approach.

	Malfunction code	Operation lamp	Malfunction contents	Page Referred
Indoor Unit	A0	l	Error of External Protection Device	109
	A1	l	PCB Defect	110
	A3	l	Malfunction of Drain Level Control System (S1L)	111
	A6	l	Fan Motor (M1F) Lock, Overload	113 115 116
	A7	k	Malfunction of Swing Flap Motor (M1S)	119
	A9	l	Malfunction of Moving Part of Electronic Expansion Valve (Y1E)	121
	AF	k	Drain Level above Limit	123
	AJ	l	Malfunction of Capacity Determination Device	124
	C4	l	Malfunction of Thermistor (R2T) for Heat Exchanger	125
	C5	l	Malfunction of Thermistor (R3T) for Gas Pipes	126
	C9	l	Malfunction of Thermistor (R1T) for Suction Air	127
	CJ	k	Malfunction of Thermostat Sensor in Remote Controller	128
Outdoor Unit	E1	l	PCB Defect	129
	E3	l	Actuation of High Pressure Switch	130
	E4	l	Actuation of Low Pressure Sensor	133
	E5	l	Inverter Compressor Motor Lock	135
	E6	l	STD Compressor Motor Overcurrent/Lock	—
	E7	l	Malfunction of Outdoor Unit Fan Motor	138
	E9	l	Malfunction of Moving Part of Electronic Expansion Valve (Y1E, Y2E)	141
	F3	l	Abnormal Discharge Pipe Temperature	143
	F6	l	Refrigerant Overcharged	145
	H3	k	Failure of High Pressure Switch	—
	H4	l	Actuation of Low Pressure Switch	—
	H7	l	Abnormal Outdoor Fan Motor Signal	—
	H9	l	Malfunction of Thermistor (R1T) for Outdoor Air	146
	J2	l	Current Sensor Malfunction	—
	J3	l	Malfunction of Discharge Pipe Thermistor (R2T)	147
	J5	l	Malfunction of Thermistor (R3T, R5T) for Suction Pipe	148
	J6	l	Malfunction of Thermistor (R4T) for Outdoor Unit Heat Exchanger	149
	J7	l	Malfunction of Liquid Pipe Thermistor (R7T)	150
	J9	l	Malfunction of Subcooling Heat Exchanger Gas Pipe Thermistor (R6T)	151
	JA	l	Malfunction of High Pressure Sensor	152
	JC	l	Malfunction of Low Pressure Sensor	154
	L0	l	Inverter System Error	—
	L1	l	Malfunction of PCB	157
	L4	l	Malfunction of Inverter Radiating Fin Temperature Rise	159
	L5	l	Inverter Compressor Abnormal	162
	L8	l	Inverter Current Abnormal	164
			Start up Error	166
				—

03315-13_F57.EPS

Figure 57 Example error code chart.

3.17 "E7" Malfunction of Outdoor Unit Fan Motor

Remote Controller Display	E7
Applicable Models	RXYMQ36 · 48PVJU
Method of Malfunction Detection	Detect a malfunction based on the current value in the inverter PCB (as for motor 2, current value in the fan PCB). Detect a malfunction for the fan motor circuit based on the number of rotation detected by hole IC during the fan motor operation.
Malfunction Decision Conditions	■ Overcurrent is detected for inverter PCB (A2P) or fan inverter PCB (A3P) (System down is caused by 4 times of detection.) ■ In the condition of fan motor rotation, the number of rotation is below the fixed number for more than 6 seconds. (System down is caused by 4 times of detection.)
Supposed Causes	■ Failure of fan motor ■ Defect or connection error of the connectors/ harness between the fan motor and PCB ■ The fan cannot rotate due to any foreign substances entangled. ■ Clear condition: Continue normal operation for 5 minutes

03315-13_F58.EPS

Figure 58 Error code E7 explained.

Caution Be sure to turn off the power switch before connecting or disconnecting connector, or parts may be damaged.

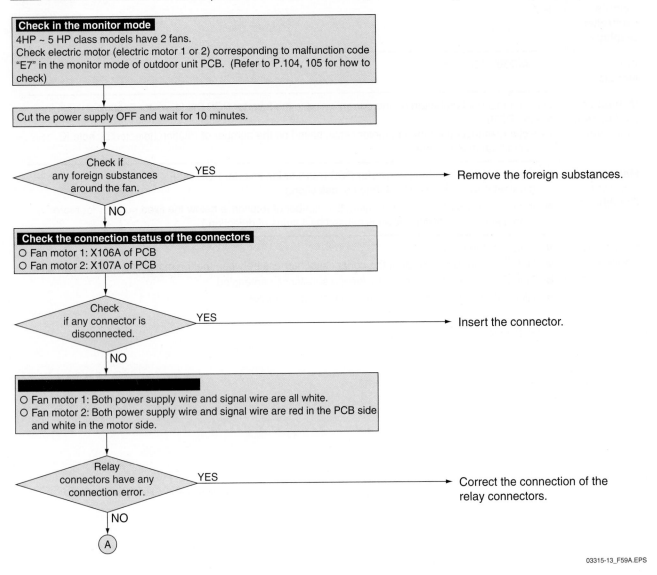

Check in the monitor mode
4HP ~ 5 HP class models have 2 fans.
Check electric motor (electric motor 1 or 2) corresponding to malfunction code "E7" in the monitor mode of outdoor unit PCB. (Refer to P.104, 105 for how to check)

Cut the power supply OFF and wait for 10 minutes.

Check if any foreign substances around the fan. — YES → Remove the foreign substances.

NO

Check the connection status of the connectors
O Fan motor 1: X106A of PCB
O Fan motor 2: X107A of PCB

Check if any connector is disconnected. — YES → Insert the connector.

NO

O Fan motor 1: Both power supply wire and signal wire are all white.
O Fan motor 2: Both power supply wire and signal wire are red in the PCB side and white in the motor side.

Relay connectors have any connection error. — YES → Correct the connection of the relay connectors.

NO

(A)

03315-13_F59A.EPS

Figure 59 Flow chart to troubleshoot the condenser fan motor. (1 of 2)

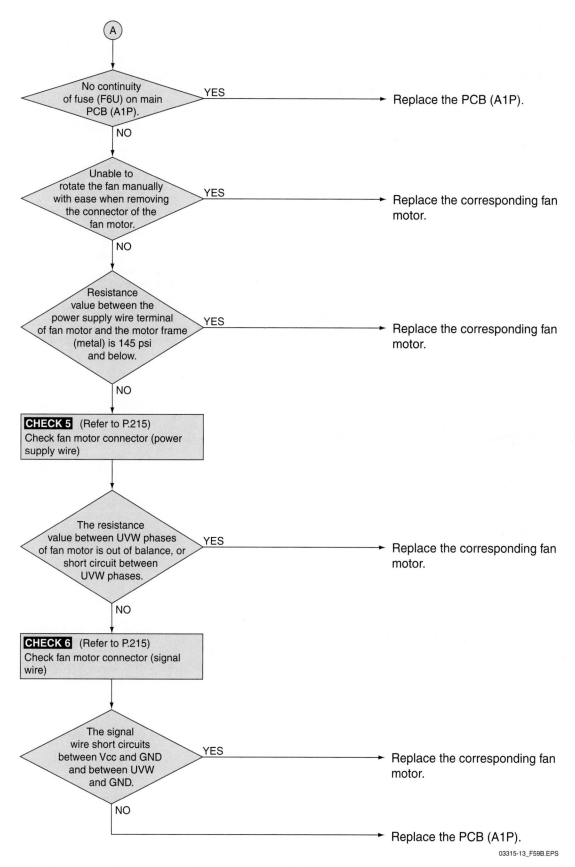

Figure 59 Flow chart to troubleshoot the condenser fan motor. (2 of 2)

03315-13_F59B.EPS

(1) Turn off the power supply. (Signal wire)
(2) Measure the resistance between Vcc and each phase of U,V,W, and GND and each phase at the motor side connectors (five-core wire) to check that the values are balanced within the range of ± 20 %, while connector or relay connector is disconnected.
Furthermore, to use a multi-meter for measurement, connect the probe of negative pole to Vcc and that of positive pole to GND.

03315-13_F60.EPS

Figure 60 The Check 6 troubleshooting procedure.

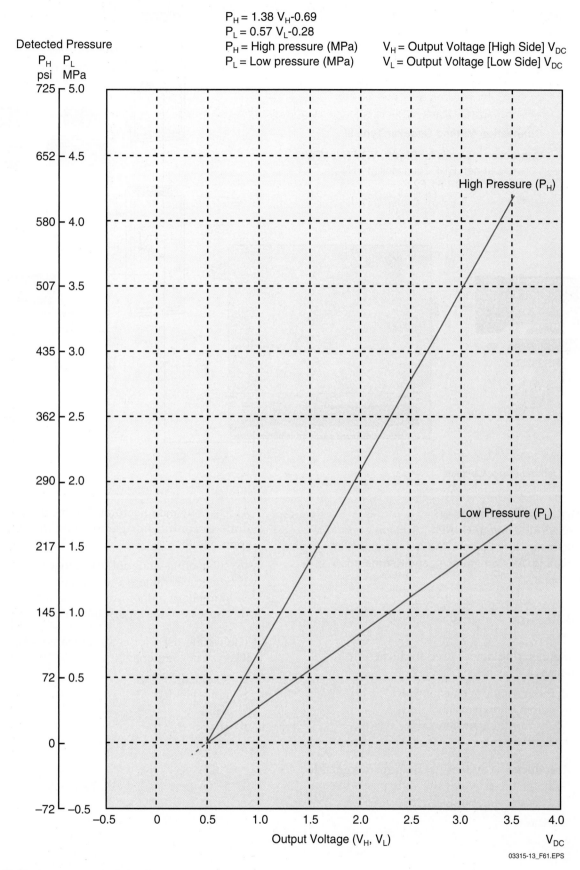

$P_H = 1.38 V_H - 0.69$
$P_L = 0.57 V_L - 0.28$
P_H = High pressure (MPa) V_H = Output Voltage [High Side] V_{DC}
P_L = Low pressure (MPa) V_L = Output Voltage [Low Side] V_{DC}

Figure 61 Pressure sensor test chart.

Connection Method Checker Type III

Connection methods for SkyAir, VRV-S & VRV:
- "DIII-net" F1-F2 connection VRV
- OUTDOOR F1-F2

● F1-F2 IN/D
● F1-F2 OUT/D

RS232C

03315-13_F62.EPS

Figure 62 Daikin Service Checker.

Additional Resources

ASHRAE Handbook – HVAC Systems and Equipment. 2012 Edition. The American Society of Heating, Air Conditioning, and Refrigeration Engineers.

2.0.0 Section Review

1. The metering device used in ductless split systems typically a(n) _____.

   a. thermostatic expansion valve
   b. automatic expansion valve
   c. electronic expansion valve
   d. capillary tube

2. Most ductless condensing units are designed to discharge the air vertically, as opposed to standard condensing units.

   a. True
   b. False

3. At least one manufacturer designs the outdoor units to control refrigerant flow in direct response to the _____.

   a. subcooling value at the outdoor unit
   b. superheat value at the outdoor unit
   c. number of indoor units in operation
   d. capacity of the largest indoor unit

4. The wiring used for communications between the system components *must* be _____.

   a. shielded
   b. non-shielded
   c. solid
   d. stranded

Summary

Zoning, ductless, and VRF systems offer different approaches to making indoor comfort more personal and accurate. Although they are generally thought of as being used for small applications, this is not necessarily the case. Light and even heavy commercial systems of most any size can employ zoning to increase their effectiveness. VRF systems can also provide significant capacity, typically up to 30 tons per system. However, as the capacity requirements increase, VAV systems become more cost-effective and practical.

Zoning systems use motorized dampers and related controls to manage system airflow with more precision, and provide users with multiple thermostats to control the conditions in a group of smaller areas. Ductless systems offer the same basic user flexibility, but do so using multiple sources of airflow from small capacity, compact indoor units. VRF systems use refrigerant management techniques and sophisticated control and communication circuits to serve multiple air sources from a single refrigeration circuit. In all three cases, both heating and cooling can be provided.

Troubleshooting these systems requires excellent electrical skills and the ability to read and understand technical information and guidance. Without the proper information on the specific system in use, problem-solving is virtually impossible. However, with the proper literature in hand, a well-trained technician can set himself or herself apart from other technicians. Developing the advanced skills needed to absorb and understand these systems and effectively install or repair them is crucial to technicians who seek to place themselves at the top of their trade.

1. When a heat pump is used in a zoning system, all zone thermostats must be heat pump-capable models.

   a. True
   b. False

2. When a single zone damper is installed in a constant-volume system to control the temperature in one space, it is aware of the system operating mode through _____.

   a. communication with the main control system
   b. a direct connection to the system thermostat
   c. a temperature sensor placed in the supply air duct
   d. a temperature sensor placed in the return air duct

3. To avoid air pressure build-up in a dump zone, _____.

   a. a return air grille can be installed there
   b. an extra zone damper is used to control pressure
   c. a pressure transmitter in the space controls bypass damper position
   d. most of the zone dampers must remain open at all times

4. Motorized zone dampers that modulate their position may be referred to as _____.

   a. barometric dampers
   b. bypass dampers
   c. two-wire dampers
   d. three-wire dampers

5. The status of the cooling control wiring between a zone control panel and the condensing unit can be quickly checked from the panel by placing a jumper across _____.

   a. R/RC and G
   b. R/RC and W
   c. R/RC and Y
   d. R/RC and O

6. In some VVT systems, the individual zone thermostats receive time of day and system operating information from _____.

   a. a monitor thermostat
   b. their own internal programs
   c. the HVAC equipment
   d. no source; the information is not necessary

7. Pressure-dependent VVT systems _____.

   a. do *not* have modulating zone dampers
   b. will cease to function if the duct pressure is more than 0.5 in wg
   c. monitor the actual volume of air flowing through a zone damper
   d. do *not* monitor the actual volume of air flowing through the zone damper

8. Some overcooling or overheating of zones can occur in commercial VVT systems because _____.

   a. VVT system controls tend to respond slowly to changing conditions
   b. the zone dampers are set to a minimum open position to provide required ventilation
   c. VVT zone dampers are known to leak consistently
   d. the bypass damper cannot be allowed fully open

9. In a typical ductless system operating in the cooling mode, _____.

   a. both refrigerant lines are at a low pressure
   b. both refrigerant lines are at a high pressure
   c. the vapor line is always at a high pressure
   d. the liquid line is always at a high pressure

10. In-ceiling cassettes used with ductless systems differ from other ductless indoor units in that _____.

    a. the metering device is in the indoor unit
    b. they can often accommodate some supply ductwork
    c. they cannot be combined with the other types in a system
    d. they use propeller fans instead of centrifugal blowers

11. Which of the following types of oil are more commonly found in ductless systems from Asian manufacturers?

    a. Mineral oil
    b. POE oil
    c. PVE oil
    d. PAG oil

12. On a ductless system remote control, what function would a Quiet setting provide?

    a. It reduces the speed and capacity of the system compressor.
    b. It causes the indoor air supply to be directed away from occupants.
    c. It slows the blower speed of the indoor unit.
    d. It slows the condenser fan speed of the outdoor unit.

13. Which of the following wiring configurations is required for VRF system communication?

    a. Homerun
    b. Star
    c. Synchronous
    d. Daisy-chain

14. Which of the following statements about VRF system installation is correct?

    a. Refnet joints can only be installed in specific positions.
    b. Stranded or solid wire can be used for communication between components.
    c. Refrigerant charging is best done using the subcooling method.
    d. Only the vapor line needs to be insulated.

15. When multiple error codes are displayed on a VRF system remote, it indicates that more than one device has failed.

    a. True
    b. False

Trade Terms Introduced in This Module

Dump zone: Area used as a point to bypass unneeded supply air from a zoning system. The dump zone is typically in the conditioned space, but not an area where maintaining the space temperature at a particular setpoint is required. A dump zone generally has no supply air grilles, but a return grille is often installed to relieve room pressure and return mixed air back to the system.

Polling: A network communications arrangement where a VVT system seeks information and needs from individual zones, determines what demands have priority, and then responds accordingly. Essentially, members of a polling system are voters.

Pressure-dependent operation: A VVT-system approach where dampers are positioned based on demand only, or operate in two positions. The result is that the design volume of air may or may not be flowing through the damper at any time. The volume of air flow through each damper is dependent on the duct pressure; it fluctuates as the pressure fluctuates. A minimum damper position is set, but air volume is not a controlling factor, and so the minimum damper position is a fixed point.

Pressure-independent operation: A VVT-system approach where dampers are positioned by demand and by the actual volume of air flowing through the damper. To achieve a specific amount of airflow, the damper positions itself as necessary. A maximum and minimum airflow setpoint is effective, and the damper will modulate as necessary to remain within these boundaries.

Refnet joints: Engineered joints used as connection points in a VRF-system refrigerant network. Refnet joints connect the main refrigerant lines to branch runs or to indoor units.

Additional Resources

This module presents thorough resources for task training. The following resource material is suggested for further study.

ASHRAE Handbook – HVAC Systems and Equipment. 2012 Edition. The American Society of Heating, Air Conditioning, and Refrigeration Engineers.

ANSI/ACCA Manual Zr – 2012, Residential Zoning. Air Conditioning Contractors of America.

Figure Credits

Topaz Publications, Inc., Module opener, Figures 3, 12, 13, 15, 17, 18, 20, 31, 34

Jackson Systems, LLC - Controls Done Right™, Figures 2, 4–6, 8–10, 14, 16, SA02–SA04

Courtesy of ZONEFIRST, SA01, Figure 11

Carrier Corporation, Figures 25, 30B

Daikin Industries, Ltd., Figures 26–28, 33, 36, 39–44, 46–48, 51, 52 (photo), 53, 55–62

Courtesy of Sauermann N.A. Corp., Figure 29A

Courtesy of DiversiTech, Figures 29B, 37 (photo), 38

Mitsubishi Electric and Electronics USA, Inc., Figure 30A

EMI - an ECR International Brand, Figure 32

Courtesy of Michael Chasse, the quad in **Quadomated.com**, Figure 35

Courtesy of CPS Products, SA05

Section Review Answer Key

Answer	Section Reference	Objective
Section One		
1. b	1.1.1	1a
2. a	1.2.1	1b
3. c	1.3.0; 1.3.1	1c
4. a	1.4.3	1d
5. d	1.5.5	1e
Section Two		
1. c	2.1.1	2a
2. b	2.2.1	2b
3. a	2.3.0	2c
4. d	2.4.1	2d

NCCER CURRICULA — USER UPDATE

NCCER makes every effort to keep its textbooks up-to-date and free of technical errors. We appreciate your help in this process. If you find an error, a typographical mistake, or an inaccuracy in NCCER's curricula, please fill out this form (or a photocopy), or complete the online form at **www.nccer.org/olf**. Be sure to include the exact module ID number, page number, a detailed description, and your recommended correction. Your input will be brought to the attention of the Authoring Team. Thank you for your assistance.

Instructors – If you have an idea for improving this textbook, or have found that additional materials were necessary to teach this module effectively, please let us know so that we may present your suggestions to the Authoring Team.

NCCER Product Development and Revision

13614 Progress Blvd., Alachua, FL 32615

Email: curriculum@nccer.org
Online: www.nccer.org/olf

❏ Trainee Guide   ❏ Lesson Plans   ❏ Exam   ❏ PowerPoints   Other _____

Craft / Level: _____   Copyright Date: _____

Module ID Number / Title: _____

Section Number(s): _____

Description: _____

_____

_____

_____

Recommended Correction: _____

_____

_____

_____

Your Name: _____

Address: _____

_____

Email: _____   Phone: _____

03305-13

Commercial Hydronic Systems

OVERVIEW

In commercial and industrial hydronic systems, water is used as the heat exchange medium for both heating and cooling. Chillers provide the cold water for cooling, typically using different types of compressors. Hot water for heating is provided by various types of boilers. Commercial hydronic systems have unique servicing and maintenance requirements, including maintaining proper chemical balance of the water to prevent fouling and similar problems. The layout of water distribution systems varies with the size and type of building.

Module Nine

Trainees with successful module completions may be eligible for credentialing through NCCER's National Registry. To learn more, go to **www.nccer.org** or contact us at **1.888.622.3720**. Our website has information on the latest product releases and training, as well as online versions of our *Cornerstone* magazine and Pearson's product catalog.

Your feedback is welcome. You may email your comments to **curriculum@nccer.org**, send general comments and inquiries to **info@nccer.org**, or fill in the User Update form at the back of this module.

This information is general in nature and intended for training purposes only. Actual performance of activities described in this manual requires compliance with all applicable operating, service, maintenance, and safety procedures under the direction of qualified personnel. References in this manual to patented or proprietary devices do not constitute a recommendation of their use.

Objectives

When you have completed this module, you will be able to do the following:

1. Describe basic concepts related to water as a substance and its movement.
   a. Describe the basic properties of water and water pressure.
   b. Explain how pressure drop, head pressure, and static pressure are related to hydronics.
2. Describe various commercial hot-water heating system components and subsystems.
   a. Identify various types of hot-water boilers and their common controls.
   b. Describe the construction of common centrifugal pumps.
   c. Identify various types of valves and other commercial hydronic system components.
   d. Identify common commercial piping systems and their characteristics.
   e. Describe how a typical hydronic piping system is balanced.
3. Explain the basic concepts of chilled water systems and the related components.
   a. Explain the basic concepts of chilled-water cooling systems.
   b. Identify various types of chillers and their common controls.
   c. Identify various types of cooling towers and evaporative condensers.

Performance Tasks

Under the supervision of your instructor, you should be able to do the following:

1. Identify the major components of commercial hot-water heating and chilled-water cooling hydronic systems.
2. Identify the types of common piping configurations used with commercial hot-water and chilled-water hydronic systems.

Trade Terms

Boiler horsepower (Bohp)
Firetube boiler
Flash economizer
Lithium bromide
Secondary coolant

Superheated steam
Thermal economizer
Venturi tube
Watertube boiler

Industry-Recognized Credentials

If you are training through an NCCER-accredited sponsor, you may be eligible for credentials from NCCER's Registry. The ID number for this module is 03305-13. Note that this module may have been used in other NCCER curricula and may apply to other level completions. Contact NCCER's Registry at 888.622.3720 or go to **www.nccer.org** for more information.

Contents ————————————

Topics to be presented in this module include:

Figures and Tables

SECTION ONE

1.0.0 WATER AND MOVEMENT

Objective

Describe basic concepts related to water as a substance and its movement.

   a. Describe the basic properties of water and water pressure.

   b. Explain how pressure drop, head pressure, and static pressure are related to hydronics.

Before addressing more complex hydronic subjects, a review of basic water terminology and concepts will prove helpful. Many problems with hydronic system operation can be identified or resolved if the technician has a thorough understanding of these basic concepts.

1.1.0 Properties of Water

Before studying how commercial hydronic systems work, it is important to understand some water concepts and know the meaning of common terms. Water is a chemical compound of two elements: oxygen and hydrogen. It can exist as ice, water, or steam (solid, liquid, or vapor) due to changes in temperature. These three forms of water are referred to as different states. At the atmospheric pressure found at sea level, it freezes at 32°F (0°C) and boils at 212°F (100°C).

Water changes weight with a change in temperature; that is, the higher the temperature of water, the less it weighs. This change in weight is due to the expansion and reduction in water volume, as the molecules become closer or farther apart. For example, one cubic foot of water weighs 62.41 pounds at 32°F and 59.82 pounds at 212°F, a difference of roughly 2.5 pounds.

Different water weights resulting from different temperature levels can cause natural gravity circulation of water in a closed system. This natural circulation is referred to as thermal circulation. Water itself is non-corrosive to most materials used in hydronic systems. The oxygen that makes up the water is basically inert. However, a variable amount of air, which consists primarily of nitrogen and oxygen, is also dissolved in water. Although nitrogen is also relatively inert, the free, dissolved oxygen will cause corrosion and rusting. For example, iron oxide, known as common rust, is formed from iron and oxygen.

For this reason, only minimal amounts of fresh water should be admitted to the system after the initial fill, since it contains some free, dissolved oxygen. This minimizes the amount of oxygen available. Corrosion stops when the available free oxygen has been consumed. Open systems (meaning they are open to the atmosphere at one or more points in the system) allow the water to absorb additional oxygen continuously. These systems are not used for hydronic heating and cooling systems that circulate throughout a building. However, open systems are still used for water-cooled condensers and similar applications.

Each time that fresh water is admitted in a closed system, following a leak repair for example, additional oxygen is admitted and fresh corrosion begins until all free oxygen is consumed. If the admission of fresh water continues, components in the system will suffer additional corrosion and possible failure.

Because most systems use automatic valves to admit fresh water, hydronic systems should be inspected regularly for leaks. A small leak can result in the constant and unmonitored addition of fresh, oxygenated water. On larger systems, water meters may be installed to monitor the volume of fresh water being added over a given period of time. Checking the meter provides a way to see if a significant amount of water is being added.

As water is heated in a boiler, air dissolved in the water is liberated in the form of bubbles. Eventually, most of the dissolved air is released into the system. That air must be removed from the system for it to function properly.

Because water contains minerals, scale can form in the piping. Scale reduces both heat transfer and water flow, another reason to minimize the admittance of fresh water. Special chemicals are often added to systems to reduce scaling. Such chemicals work to reduce free oxygen by using the oxygen to produce harmless materials and preventing the oxygen from bonding with iron to form rust. Always check with the boiler manufacturer before adding any water treatment chemical to boiler water. The wrong chemical may damage the boiler and void any warranties. The subject of water treatment is covered in depth in another module.

1.1.1 Water Pressure

Figure 1 provides the general formula for calculating the pressure exerted by water in pounds per square foot at various depths, as well as the formula to calculate the same information in pounds per square inch. You can use the formula shown, or simply divide your resulting answer in pounds per square foot units by 144 to obtain pounds per square inch. One cubic foot of water at sea level weighs 62.4 pounds, the basis for the calculations in *Figure 1*.

1.2.0 Movement of Water Through Piping Systems

Factors such as the resistance to water flow within the piping and the pressure available to force water past that resistance affect water movement in the system.

1.2.1 Pressure Drop

Pressure drop is defined as the difference in pressure between two points in a hydronic system. It is an important factor in all hydronic systems. In *Figure 2*, there is a 5 pounds per square inch (psi) drop in pressure between two points when water is flowing at a rate of 5 gallons per minute (gpm). Because of the length and volume of flow required in many commercial or industrial applications, as opposed to smaller residential systems,

PRESSURE DROP EQUALS 5 PSI

03305-13_F02.EPS

Figure 2 Pressure drop.

pressure drop in piping must be fully understood and given careful consideration. Where there is no flow or water movement, no pressure drop is encountered.

When flow is present, friction is encountered and increases in direct proportion to the velocity of flow. The change in pressure drop resulting from a change in flow in gallons per minute can be calculated as follows:

$$(\text{Final gpm} \div \text{initial gpm})^2 \times \text{initial pressure drop} = \text{final pressure drop}$$

Hydronic system pumps must provide sufficient power to overcome the effects of pressure drop encountered in the piping system and all components. Because piping systems differ, the pressure drop for piping must be calculated for each system. The manufacturer's specifications for other hydronic system components, such as coils and heat exchangers, provide pressure drop data for various flow rates. This information may be expressed in a variety of units, so the conversions below may be helpful:

- 1 pound per square inch = 2.31 feet of water
- 1 foot of water = 0.433 pounds per square inch
- 1 foot of water = 12,000 milli-inches
- 1 inch of water = 1,000 milli-inches

1.2.2 Head Pressure

Head pressure is another measure of pressure, often expressed in feet of water. A head pressure value is, therefore, often referred to as feet of head. This term is most often used to describe the capacity of a circulating pump, indicating the height of a column of water that the pump can lift, without consideration of pressure drop from friction losses.

Since a column of water 1 foot in height produces a pressure of 0.0433 psi, simple calculations prove that a column of water 2.31 feet in height results in a pressure of 1 psi. *Figure 3* provides examples of these conversion formulas.

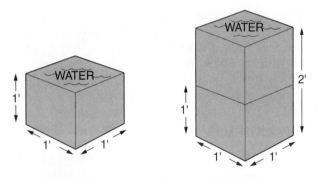

FORMULA FOR COMPUTING PRESSURE IN POUNDS PER SQUARE FOOT (psf)

WHERE: P = PRESSURE PER SQUARE FOOT
D = DEPTH OF WATER IN FEET

P = 62.4 × D	P = 62.4 × D
P = 62.4 × 1	P = 62.4 × 2
P = 62.4 psf	P = 124.8 psf

FORMULA FOR COMPUTING PRESSURE IN POUNDS PER SQUARE INCH (psi)

WHERE: P = PRESSURE PER SQUARE INCH
D = DEPTH OF WATER IN FEET

$P = \dfrac{62.4}{144} \times D$	$P = \dfrac{62.4}{144} \times D$
$P = \dfrac{62.4}{144} \times 1$	$P = \dfrac{62.4}{144} \times 2$
P = 0.433 × 1	P = 0.433 × 2
P = 0.433 psi	P = 0.866 psi

03305-13_F01.EPS

Figure 1 Water pressure concepts.

1.2.3 Static Pressure

Static pressure (*Figure 4*) is created by the weight of the water in the system. We know that a column of water that is 1' high produces a pressure of 0.43 psi. Therefore, static pressure is equal to 0.43 psi for each foot of height above the system gauge. Consider a simple water system to be an upright loop of water confined in a pipe. As such, the static pressure in one of the vertical pipes of the loop is identical to the static pressure at the same level in the opposite vertical pipe. Static pressure levels in a system decrease as we move from the lowest point in the system toward the highest point in the system. At the highest point, the static pressure is 0 psig. Remember that this measured value must be stated in gauge-pressure units, since the weight of the atmosphere is also pushing down on the water column.

For example, if the highest point holding water in a hydronic system is 30' above the hot-water boiler, the ascending static pressure levels within the system are as follows:

30' (highest point in system) = 0 psig
20' (10' below highest point) = 4.3 psig (10 × 0.43)
10' (20' below highest point) = 8.6 psig (20 × 0.43)
0' (30' below highest point) = 12.9 psig (30 × 0.43)

> **Think About It**
> # Head Pressure Relationships
>
> The water level in a water system reservoir is maintained at 150' above its use level. What is the pressure resulting from this head indicated by a gauge installed at the system use level?
>
> A gauge installed at the use level of a water system reads 64.5 psi. What is the water level (head pressure) in the water system reservoir that causes this reading?

FORMULA FOR CONVERTING HEAD PRESSURE OF WATER IN FEET TO PRESSURE IN PSI

WHERE: P = PRESSURE (PSI)
H = HEAD PRESSURE (FEET OF WATER)

P = H × 0.43

FORMULA FOR CONVERTING PRESSURE IN PSI TO HEAD PRESSURE OF WATER IN FEET

WHERE: H = HEAD PRESSURE (FEET OF WATER)
P = PRESSURE (PSI)

H = 2.31 × P

03305-13_F03.EPS

Figure 3 Relationship of head pressure in feet to pounds per square inch.

03305-13_F04.EPS

Figure 4 Static pressure.

Additional Resources

Air-Cooled Chillers, TDP-622, Latest Edition. Syracuse, NY: Carrier Corporation.

Refrigeration and Air Conditioning, An Introduction to HVAC/R, Latest Edition. Larry Jeffus. Upper Saddle River, NJ: Prentice Hall.

Water-Cooled Chillers, TDP-623, Latest Edition. Syracuse NY: Carrier Corporation.

1.0.0 Section Review

1. What substance in water causes it to be corrosive to hydronic system components?

    a. Dissolved oxygen
    b. Suspended particles
    c. Dissolved nitrogen
    d. Oxides of iron

2. No pressure drop in water piping will occur if _____.

    a. there are no elbows or fittings in the pipe
    b. there is no flow of water in the pipe
    c. the pipe interior is coated with Teflon®
    d. the water is chemically treated

SECTION TWO

2.0.0 COMMERCIAL HOT-WATER HEATING SYSTEM COMPONENTS

Objective

Describe various commercial hot-water heating system components and subsystems.
- a. Identify various types of hot-water boilers and their common controls.
- b. Describe the construction of common centrifugal pumps.
- c. Identify various types of valves and other commercial hydronic system components.
- d. Identify common commercial piping systems and their characteristics.
- e. Describe how a typical hydronic piping system is balanced.

Performance Tasks 1 and 2

Identify the major components of commercial hot-water heating and chilled-water cooling hydronic systems.

Identify the types of common piping configurations used with commercial hot-water and chilled-water hydronic systems.

Trade Terms

Boiler horsepower (Bohp): One Bohp is equal to 9.803 kilowatts. Also, it is the heat necessary to vaporize 34.5 pounds of water per hour at 212°F. This is equal to a heat output of 33,475 Btus per hour or about 140 square feet of steam radiation.

Firetube boiler: A boiler with flue gases contained inside the tubes and heated water on the outside.

Superheated steam: Steam that has been heated above its saturation temperature.

Venturi tube: A short tube with flaring ends and a constricted throat. It is used for measuring flow velocity by measurement of the throat pressure, which decreases as the velocity increases.

Watertube boiler: A boiler where the heated water is contained inside the tubes, with the flue gases on the outside.

A number of hydronic devices and components previously examined apply to commercial and industrial systems. Those that are used in the same configuration and style in commercial systems will be briefly reviewed, while those that differ in application in larger systems are discussed.

Primary commercial hot-water system components are the following:

- Hot-water boilers
- Operating and safety controls associated with commercial boilers
- Expansion and compression tanks
- Air control devices
- Circulating pumps
- Valves

2.1.0 Commercial Hot-Water Boilers

A boiler heats water using different fuels or electricity as the heat source. Geothermal heat pumps are sometimes used to produce hot water for comfort heating. Some boilers can be fired with more than one fuel. This is done by burner conversion or by using dual-fuel burners. Gaseous fuels include natural gas, manufactured gas, and liquefied petroleum (propane and butane). Oil fuels include both lightweight and heavy oils. Solid fuels include coal and wood.

Commercial and industrial markets are dominated by fuel oil and natural gas, due to their availability and their high heat content. Natural gas is enjoying resurgence in use as new and plentiful domestic sources are being discovered and exploited. Natural gas is delivered to a given facility via pipeline, eliminating dependence upon ground transportation and requiring no on-site storage.

Although rare, large electric boilers are also used in some applications. They are often used by electric power utilities in company-owned facilities as a means of proving their commitment to the electric power they provide.

Most boiler combustion systems have a prepurge and/or purge sequence to make sure that there is not a combustible mixture already in the combustion chamber that could cause an explosion during startup. The methods used to vent boilers are similar to those used with furnaces.

The construction and operation of hot-water and steam boilers are similar, with two exceptions. The operating and safety controls used with hot-water boilers are different from those used with steam boilers. Also, hot-water boilers are entirely filled with water, while steam boilers are not.

Low-temperature boilers are the most widely used type of boiler. Low-pressure hot-water boilers can be built to have working pressures of up to 160 psi. However, they are typically designed for a 30-psi maximum working pressure. They are generally operated below that maximum pressure level, with 12 to 15 psi being common. Low-pressure hot-water boilers are generally limited to a maximum operating temperature of 250°F. Above this temperature, even water under low pressure will begin to boil and change to steam, creating a dangerous condition. In most cases, low-pressure boilers operate below 200°F.

Medium- and high-pressure hot-water boilers are built to operate at pressures above 160 psig and temperatures above 250°F. Keeping the pressure higher raises the boiling point of the water. In extremely large systems, such high pressures and temperatures allow the water to be circulated great distances (from a central heating plant to other buildings, for example) without losing all heating capacity before arriving at the intended destination. Such systems can also be extremely dangerous.

Very large hydronic heating systems often use high-temperature, high-pressure water, especially in central plant applications for college campuses and military bases. Escaping water, at such extreme temperatures (over 450°F in some cases) and under high pressure, instantly flashes to steam when even a small fracture or tiny leak develops in piping or components. Quite often, evaporation of the superheated steam occurs so rapidly, no steam plume is seen. This is due to its sudden exposure to atmospheric pressures at which water boils at 212°F.

> **WARNING!**
>
> Serious physical harm can result by simply walking near or moving a hand across the path of escaping steam from a high-pressure system. Low-pressure steam can also produce serious physical injury. Use extreme caution when servicing or in the presence of such systems.

Most boilers in larger commercial applications are made of cast iron or steel. Small commercial facilities are more likely to use boilers of copper-finned tube construction. This design continues to increase in popularity because of heat transfer efficiency, reduced installation cost, and application flexibility.

2.1.1 Copper-Finned Tube Boilers

Boilers using the copper-finned tube design offer some distinct advantages in commercial applications. Because traditional steel and cast-iron units generally have a significant volume of hot water stored on board, they are subject to greater heat losses when heating needs are low. Also, they can require more space for installation and service.

With the copper-finned tube design, heat transfer is greatly enhanced, providing for rapid recovery. These also eliminate the need for hot water to be stored within the boiler itself. Such boilers are gas-fired, generally using natural gas or propane.

Atmospheric models, such as the one shown in *Figure 5*, often offer modulating or staged-heating capacity to further enhance operating efficiency. When compared to traditional cast iron or steel alternatives, they are also very lightweight, making for easier installation and/or replacement.

One other unique feature these boilers can offer is outdoor installation. The unit pictured in *Figure 5* does not have a top connection for flue venting and can be installed outdoors without any additional protection.

Boilers installed outdoors are subject to damage from freezing in the event of failure of the burner or pumping system. For outdoor installations, an antifreeze solution is added to protect the boiler and water circuit from freeze damage. The addition of antifreeze impacts the heat transfer and viscosity of the water. Both issues must be considered in the pump selection process. Generally, to achieve the desired overall heat transfer ef-

---

Clean Natural Gas

With today's concerns about global warming, people are looking at natural gas as the best fuel to burn to reduce carbon emissions. Natural gas produces 30 percent less carbon emissions than oil and 45 percent less carbon emissions than coal. Plus, natural gas does not produce ash particles like coal and oil. Due to recent advances in drilling technology, the United States now has about a 100-year supply of this very clean fuel. Natural gas-fired boilers are very popular in the HVAC industry.

Combustion in a boiler occurs by combining the fuel with oxygen and igniting the mixture. The methods used for combustion in gas-fired and oil-fired boilers are basically the same as those used with furnaces. Fuel ignition can be achieved by use of a standing gas pilot, electric spark-type intermittent igniter, or hot surface igniter. The presence of a pilot or flame is proven by a protective device before the main gas or oil valve is allowed to open.

Figure 5 Commercial copper-finned tube boiler cutaway.

fect, system flow rates will need to be increased when antifreeze is used. Galvanized piping is not recommended when glycol-based antifreeze is used. The zinc in the coating may react with the glycol to form sludge in the piping system.

> CAUTION
>
> Proper water flow in gpm through a copper finned-tube boiler is critical to the life of the boiler. Consult the manufacturer's instructions and engineering guides to determine the correct pipe size and pump.

Other copper-finned boiler models feature induced-draft combustion, allowing for additional installation flexibility and higher levels of efficiency. *Figure 6* provides an example of an induced-draft model. The induced-draft design allows the combustion chamber to be sealed; air for the combustion process can be ducted in from outdoors.

Figure 6 Boiler with sealed combustion chamber.

Although very high heating capacities are available in copper-finned tube units, multiple units are often selected and piped in parallel to further enhance efficiency by matching capacity to the heating load. Using multiple boilers to achieve desired maximum capacity also provides redundancy when one or more units are in need of repair or maintenance.

2.1.2 Cast-Iron Boilers

Cast-iron boilers are formed by assembling together individual cast-iron heat exchanger sections (*Figure 7*). Each section is basically a separate boiler. The number of sections used determines the size of the boiler and its energy rating. Cast-iron boiler capacities range from those required for small residences up to large commercial systems of 13,000 MBh (one MBh is equal to 1,000 Btus).

In a cast-iron boiler, the water circulates inside the cast sections with the flue gases on the outside of the sections (*Figure 8*). The cast sections are usually mounted vertically, but they can be mounted horizontally. In both arrangements, the heating surface is large relative to the volume of water. This allows the water to heat up quickly.

The boiler depicted in *Figure 8* uses modular construction. This allows different combinations of boiler sections, bases, and flue collectors to be assembled to match heating requirements. Cast-iron boilers used for residential and smaller commercial jobs usually are supplied completely assembled (packaged). A packaged boiler is one

Figure 7 Cast-iron sections in a packaged boiler.

DURING BOILER DESIGN, SECTIONS ARE ADDED AS NEEDED TO MEET HEAT LOAD REQUIREMENTS

HOT WATER

BOILER NOZZLE

CAST-IRON BOILER SECTIONS

BOILER WATER

HOT GASES

ACCESS DOOR

FIREBRICK OR OTHER FIRE-RETARDANT MATERIAL

BURNER

COMBUSTION CHAMBER

REINFORCED CONCRETE PAD

03305-13_F08.EPS

Figure 8 Cast-iron packaged boiler operation.

that includes the burner, boiler, controls, and auxiliary equipment. Larger boilers can be packaged units, or they can be assembled on the job site.

The terms *dry-base*, *wet-leg*, or *wet-base* may be used by manufacturers when describing cast-iron and other boilers. These terms refer to the location of the boiler's water-filled sections relative to the combustion chamber (firebox). They are defined as follows:

- *Dry-base* – Firebox is located beneath the water vessels.
- *Wet-leg* – Firebox top and sides are enclosed by water vessels.
- *Wet-base* –Firebox is surrounded by water vessels, except for necessary openings.

2.1.3 *Steel Firetube and Watertube Boilers*

Steel boilers are fabricated into one assembly of a given size and rating. The heat exchanger surface is usually formed by vertical, horizontal, or slanted tubes (*Figure 9*). Two types of steel boilers are the firetube boiler and the watertube boiler. In the firetube boiler, flue gases are contained inside the tubes, with the heated water on the outside. In the watertube boiler, the heated water is contained inside the tubes, with the flue gases on the outside.

Smaller capacity steel boilers are usually vertical firetube units. Medium- and large-capacity steel boilers normally use horizontal or slant-mounted tubes. They can be either firetube or watertube boilers. Steel boilers range in size from small residential units of 150,000 Btuh to large systems of 23,500 MBh and above.

A common steel firetube boiler is the Scotch Marine boiler (*Figure 10*). As with most boilers, Scotch Marine boilers are made to produce either hot water or steam. They are used mainly for heating and industrial applications. As shown, this boiler has a low profile, allowing it to be installed in areas where headroom is limited. Steel boilers, including the Scotch Marine boiler, are usually built as packaged units.

2.1.4 *Steel Vertical Tubeless Boilers*

Steel vertical tubeless boilers are another type of boiler used to produce hot water or steam (*Figure 11*). In this boiler, the water vessel surrounds the combustion chamber. A top-mounted, fuel-fired burner sends a spinning (cyclonic) flame down the length of the furnace. Flame retarders in the furnace retain the flame for maximum heat transfer.

As the flame swirls downward, the water spirals upward in the water vessel. Hot gases from

FIRETUBE BOILER

BOILER WATER

FIRETUBES

ACCESS PLATE

COMBUSTION CHAMBER

BURNER

FINISHED FLOOR

STEAM OR HOT WATER

BREECHING

WATER LEVEL FOR STEAM OPERATION

HOT GASES

ACCESS PLATE

SUPPORT

FIREBRICK OR OTHER FIRE-RESISTANT MATERIAL

WATER DRUM

BOILER WATER

WATER TUBES

ACCESS PLATE

WATER WALL FURNACE EXTENSION

BURNER

COMBUSTION CHAMBER

BREECHING

FIREBRICK

FIREBRICK BAFFLE

WATERTUBE BOILER

03305-13_F09.EPS

Figure 9 Firetube and watertube boilers.

03305-13_F10.EPS

Figure 10 Scotch Marine boiler.

the flame are carried up the outside of the water vessel in a secondary flue passage convection area. Full-length convection fins welded on the outside of the water vessel transfer the remaining heat to the water.

2.1.5 Electric Boilers

Electric boilers use immersion resistance heater elements to heat the water in the boiler (*Figure 12*). Heating results from electrical current flowing through the heater elements. These elements are totally immersed in the water within the boiler pressure vessel. The heater elements are uniformly stretched, helical resistance elements made from stainless steel. They are installed in flanges mounted at the top of the boiler assembly. This makes them easily removable for inspection and maintenance.

FLUID OUTLET

POWER BURNER

SECONDARY FLUE PASSAGE

SPINNING CYCLONIC FLAME

RISING HEAT

FLUID INLET

FLUID CHAMBER

COMBUSTION CHAMBER

FLAME RETARDER

UPWARD SPIRALING FLUID

CONVECTION FINS

03305-13_F11.EPS

Figure 11 Vertical tubeless boiler.

In smaller boilers, heater elements are often mounted horizontally. Capacities of resistance-type electric boilers can range from those required for small residences and light commercial use up to a maximum capacity of about two megawatts, or 200 **boiler horsepower (Bohp).** (One Bohp equals 9.803 kilowatts.) Boiler operation is controlled by an electric/electronic controller. This controller activates and deactivates the individual heater elements in response to load demands. It also allows for incremental loading of the electric service to reduce line fluctuations and power surges during startup.

Another type of electric boiler, called an electrode boiler, relies on the electrical resistance of

> **NOTE**
> Do not confuse boiler horsepower with brake horsepower, which is abbreviated as bhp.

water for creating heat. In this boiler, an electric current passes through the water between several solid metal electrodes that are immersed in the water. Current flow through the electrical resistance of the water causes the water to be heated.

The capacity of an electrode boiler is controlled by regulating the amount of water that comes in contact with the electrodes. If there is no contact, there is no current flow, and therefore no heating of the water. Partial contact will cause some heat to be generated. Maximum heat is generated when there is total contact between the electrodes and the water. Hot-water electrode boilers are available with capacities up to 20 megawatts (about 2,000 Bohp).

2.1.6 Boiler Operating/Safety Controls

A variety of controls are used to manage and maintain safe conditions on hot-water boilers. Most of these devices, such as the pressure/temperature gauge, pressure (safety) relief valve, and thermal/electronic probe safety and operating controls, have been discussed in prior modules. If you are unfamiliar with these devices, review them before proceeding.

One device not previously presented is the float-operated, low-water cutoff control (*Figure 13*) used to protect boilers from damage caused by operating with too low or no water. This control consists of a float that actuates a related electrical switch. The switch contacts are connected in the boiler burner control circuit.

FLUID OUTLET

ELECTRIC ELEMENT

SPIRALING FLUID

HEATER VESSEL

FLUID INLET

03305-13_F12.EPS

Figure 12 Electric hot-water boiler.

EXPANSION TANKS

SUPPLY MAIN

RELIEF VALVE

CITY WATER SUPPLY

FLOAT-OPERATED, LOW-WATER FUEL CUTOFF CONTROL

FILL VALVE

STEAM BOILER

RETURN MAIN

BURNER CONTROL

PUMP

03305-13_F13.EPS

Figure 13 Float-operated, low-water cutoff control.

If the water in the boiler drops below the minimum safe level, the float drops and the switch contacts open, turning off fuel supply to the burners. Operating levels for the control are factory-set and nonadjustable. Some low-water fuel cutoff controls have an opening through which a tool can be inserted and positioned to simulate a low-water condition. Other controls have this feature built in. *Figure 13* shows a typical installation using a float-operated, low-water fuel cutoff control. However, several locations and piping arrangements can be used. Because there is no normal water line to be maintained in a hot-water boiler, any location of the control above the lowest permissible water level is satisfactory.

Copper-finned tube style units transfer heat effectively with no significant volume of water contained on board. Because of this, they are subject to damage very quickly if the water flow is stopped while combustion continues. Since there is no water-filled vessel or space other than the tubes, a low-water float-type device is of no value for protection. For this reason, copper-finned tube units should be equipped with a water flow switch (*Figure 14*). These simple units have a paddle placed into the water piping. They are adjusted so that a flow volume insufficient for safe boiler operation will not push the paddle forward, allowing the integral switch to remain open. The switch is usually wired directly in series with the combustion controls to shut the unit down when water flow is too low or non-existent. Flow switches are also used in chilled water systems.

Because of the large amount of gas used to fuel large commercial burners, these boilers are equipped with special flame safeguard controls.

2.2.0 Centrifugal Pumps

A pump is a mechanical device designed to increase the energy of a fluid so that it can be moved from one location to another. Pumps are generally driven by electric motors and are used to pump a variety of liquids.

All pumps operate by the general principle that fluid is drawn in at one end (the suction side) and

Indirect Electric Boilers

Although most electric boilers directly heat the water flowing through a building for heating purposes, others heat this water stream indirectly. By maintaining a reasonable pressure inside the boiler, the water inside can be heated well above the 212°F boiling point at sea level (325°F, for example). The building heating water loop is then routed through a heat exchanger in or near the boiler tank and is heated by the surrounding water inside. The water for building heat never makes direct contact with the boiler tank or boiler water.

Figure 14 Flow switch.

03305-13_F14.EPS

forced out at the other end (the discharge side) at an increased velocity.

Pumps are classified as either positive-displacement or centrifugal. Positive-displacement pumps are often used to pump high-viscosity fluids (thick fluids that flow slowly). Positive-displacement pumps are generally high-pressure pumps.

Centrifugal pumps are used to pump liquids that have low viscosities, such as water or thin fluids. They are generally used in applications that do not require extreme pressure differentials. They are by far the most common type used in commercial hydronic systems. To accommodate a variety of design situations and installations, they are available in a number of different configurations.

2.2.1 Centrifugal Pump Construction

A centrifugal pump operates by increasing the velocity of a liquid. Fluid entering the pump is rotated by an impeller. This rotation increases centrifugal force within a stationary casing. The force raises the pressure and causes the fluid to be discharged at high speed and at a higher pressure. It is the difference in pressure between the inlet and outlet of the pump that causes the water to circulate through the piping system. *Figure 15* shows a centrifugal pump.

The parts of centrifugal pumps may vary in size and shape depending on the manufacturer, but will have the same functions. Centrifugal pump parts, shown in *Figure 16*, include the following primary parts:

- *Pump casing* – The part surrounding the shaft, bearings, packing gland, and impeller. Pump casings can be of split or solid design.

- *Suction port* – Where fluid enters the pump.
- *Discharge port* – Where fluid is discharged into the piping system.
- *Pump shaft* – A bearing-supported part that holds and turns the impeller when the shaft is coupled to a motor.
- *Bearings* – Parts that support the shaft and impeller in the casing.
- *Impeller* – A rotating part that increases the speed of the fluid. There are many different types of impellers used for different purposes.
- *Impeller vanes or blades* – Parts of the impeller that direct flow of fluid within the pump.
- *Impeller shrouds* – Parts that enclose the blades and keep the flow of fluid in the impeller area.
- *Wearing rings* – Replaceable rings used in some pumps to allow some fluid leakage between the impeller and the casing in the suction area. The leakage makes a hydraulic seal and helps the pump operate more efficiently.
- *Packing gland* – Contains a packing gland nut that exerts force upon the packing to control fluid leakage around the shaft.
- *Mechanical seal* – Seals the fluid flow in the pump. In some pumps, they are used instead of packing.

A variety of impeller styles can be chosen, depending upon the application and operating characteristics desired. *Figure 17* provides examples of some different styles. The most common designs used in commercial hydronic comfort systems are the single-suction closed impeller and the double-suction impeller. Some impellers, such as the single-suction closed style, can be precisely machined to a given diameter to provide very specific flow performance to meet unique project design specifications.

Cavitation can occur in a water-circulating pump when the pressure at the inlet of the impeller falls below the vapor pressure of the water, causing the water to vaporize and form bubbles. These bubbles are carried through the pump impeller inlet to an area of higher pressure where they implode (burst inward) with extreme force that can cause the impeller vane tips or inlet to be damaged due to pitting or erosion.

The problem of cavitation can be eliminated by maintaining a minimum suction pressure at the pump inlet to overcome the pump's internal losses. This minimum suction pressure is called the net positive suction head required (NPSHR). Some symptoms that indicate cavitation is occurring in a pump include the following:

- Snapping and crackling noises at the pump inlet

Figure 15 Centrifugal pump.

- Pump vibration
- Reduced or no water flow
- A drop in pressure

The type of centrifugal pump used depends upon the requirements of the system. Some systems require that large amounts of fluids be pumped, others require high pressures. Two types of centrifugal pumps are double-suction and multistage.

2.2.2 *Double-Suction Centrifugal Pumps*

A double-suction centrifugal pump is used to pump large volumes of fluid. Fluid is drawn in through openings on both sides of the impeller and passes out through a single discharge opening. *Figure 18* shows a double-suction centrifugal pump with a horizontally split case.

DISCHARGE PORT

IMPELLER VANES

WEAR RING

PACKING GLAND

PUMP SHAFT

BEARINGS

PUMP CASING

IMPELLER SHROUDS

WEAR RING

SUCTION PORT

IMPELLER

SUCTION COVER

03305-13_F16.EPS

Figure 16 Centrifugal pump components.

2.2.3 Multistage Centrifugal Pumps

Multistage centrifugal pumps (*Figure 19*) contain two or more stages and can be either single- or double-suction pumps. Fluid is discharged from one stage to the next through passages in the pump casing. Each stage has an impeller, used to increase the velocity of the fluid being pumped until the desired pressure is reached.

2.3.0 Valves and Components

Common valves and their uses, as well as some specialty valves, have been discussed in earlier modules. Larger commercial and industrial systems generally use these same valves along with additional types for proper and flexible system operation.

2.3.1 Multi-Purpose Valves

In larger systems, a multi-purpose valve is commonly installed in the discharge side of the circulating pump (*Figure 20*). They are also referred to as triple-duty valves. This single valve functions as a shutoff valve, a check valve, and a balancing valve. It has a calibrated stem to return the valve to the set position after the valve is used as a shutoff valve. When in the set position, it acts as a balancing valve.

The valve is tapped to install two readout valves that enable it to be used as a flow meter. When balancing a system, the pressure drop across the valve can be measured using a differential pressure gauge connected to these readout valves. The measured pressure drop can then be used with a conversion chart to find the corresponding total flow rate of water passing through the valve.

2.3.2 Balancing Valves and Flow Meters

Balancing valves (*Figure 21*) and flow meters provide a means of balancing a water system. A balancing valve can function as both a precision balancing device as well as an isolation valve.

As a balancing valve, it is adjusted to a predetermined setting that relates to the degree of closure needed to achieve the desired water flow rate at a given pressure. The initial setting is usually

Pump Selection

The websites for many major pump manufacturers have an online software program that can be used to help select a circulating pump for a particular application. User inputs to this program typically include the required flow rate, total pump head, speed of pump operation, and the type or model of pump to use.

OPEN IMPELLER   SEMI-OPEN IMPELLER

SINGLE-SUCTION CLOSED IMPELLER

DOUBLE-SUCTION
IMPELLER

MIXED-FLOW
IMPELLER

03305-13_F17.EPS

Figure 17 Impeller types.

03305-13_F18.EPS

Figure 18 Double-suction centrifugal pump.

03305-13_F19.EPS

Figure 19 Multistage pump.

determined by the system designer, and is based on a specific design pressure. The installer sets it accordingly. When set to the closed position, the valve is fully closed and serves as an isolation valve.

Most balancing valves are equipped with two ports that enable them to be connected to a flow meter. When a flow meter, also known as a differential pressure gauge, is connected to these ports, the pressure drop across the balancing valve can be measured. By using a conversion chart or calculator, the measured pressure drop can be converted to the corresponding total flow rate of water through the valve. Some balancing valves give direct readouts of flow, but they are quite rare.

Flow measurements made to balance a water system are done by measuring the pressure drops at specific locations throughout the system. Venturi tubes and orifice plate devices installed in the system can be used for this purpose (*Figure 22*).

These devices insert a specific fixed-area restriction in the path of water flow, thereby creating a predictable pressure drop. The flow rate through the device can be determined by accurate measurement of this pressure drop. Measure-

PRESSURE READOUT VALVES

OPEN
0
10
20
30
40
CLOSED

CIRCUIT SETTER

BALANCE VALVE

ADJUSTMENT
KNOB

ADJUSTMENT

DIRECT
PRESSURE
READOUT

4
3
2
1

CALIBRATED
NAMEPLATE

READOUT
VALVES/PORTS

**STRAIGHT-THROUGH
MODEL**

READOUT
VALVES/PORTS

ANGLE MODEL

03305-13_F20.EPS

Figure 20 Multi-purpose valve.

03305-13_F21.EPS

Figure 21 Balancing valve.

ment of the pressure drop is typically made using a manometer or differential pressure gauge. A differential pressure gauge used for this purpose typically reads in feet of head. For proper operation, the manufacturer's installation instructions must be followed.

Because of its streamlined shape, a venturi tube is more accurate than an orifice plate. Water passing through the reduced area of the venturi throat increases in velocity, creating a pressure differential between the inlet and the throat areas. After passing through the throat, the flow area is gradually increased, which decreases the velocity and allows pressure recovery. The orifice plate may be installed between a single pair of flanges that have pressure differential taps built in by the manufacturer.

2.3.3 Butterfly Valves

Butterfly valves are popular in large piping systems as a simpler means of controlling flow, when compared to globe or gate valves. A butterfly valve (*Figure 23*) has a round disc that fits tightly in its mating seat and rotates 90 degrees in one direction to open and allow fluid to pass through the valve. The butterfly valve can be operated quickly by turning the hand wheel or hand lever one-quarter of a turn, or 90 degrees, to open or close the valve. These valves can be used completely open, completely closed, or partially open for non-critical throttling operations. Butterfly valves are typically used in low- to medium-pressure and low- to medium-temperature applications. They generally weigh less than other types of valves because of their narrow body design.

When the butterfly valve is equipped with a hand lever, the position of the lever indicates whether the valve is open, closed, or partially open. When the lever is perpendicular to the flow line through the valve, the valve is open. Butterfly valves that are 12 inches in diameter and larger are usually equipped with a hand wheel or gear-operated actuator because of the large amount of fluid flowing through the valve and the great amount of pressure pushing against the seat when the valve is fully closed.

Butterfly valves have an arrow stamped on the side, indicating the direction of flow through the valve. They must be installed in the proper flow direction, or the seat will not seal and the valve will leak.

There are three common types of butterfly valves in use are wafer-type, wafer lug, and two-flange.

The wafer-type butterfly valve (*Figure 24*) is designed for quick installation between two flanges. No gasket is needed between the valve and the flanges because the valve seat is lapped over the edge of the body to make contact with the valve faces. Bolt holes are provided in larger wafer valves to help line up the valve with the flanges.

Wafer lug valves (*Figure 25*) are the same as wafer valves except that they have bolt lugs completely around the valve body. Like the wafer valve, no gasket is needed between the valve

Variable-Speed Centrifugal Pumps

GOING GREEN

Energy-saving variable-speed centrifugal pumps are becoming more common in hydronic systems, especially in zoned systems where loads are constantly changing. An electronic control senses temperature conditions in the hydronic system and its zones. The control then signals the pump to increase or decrease its speed. This, in turn, delivers a correct volume of water at the correct temperature to satisfy the load.

03305-13_SA01.EPS

VENTURI TUBE

PRESSURE TAPS

THROAT

VENTURI TUBE

READOUT VALVES

CALIBRATION CHART

ORIFICE PLATE

ORIFICE PLATE

ORIFICE PLATE

03305-13_F22.EPS

Figure 22 Venturi tube and orifice plate flow metering device.

and the flanges because the valve seat is lapped over the edges of the body to make contact with the valve faces. The lugs are normally drilled to match ANSI 150-pound steel drilling templates. The lugs on some wafer lug valves are drilled and tapped so that when the valve is closed, downstream piping can be dismantled for cleaning or maintenance while the upstream piping is left intact. When tapped wafer lug valves are used for pipe end applications, only one pipe flange is necessary. During installation, both wafer-type and wafer-lug valves must be in the open position prior to tightening the flange-mating hardware in order to prevent damage to the valve seat.

The body of a two-flange butterfly valve is made with a flange on each end (*Figure 26*). The valve seat is not lapped over the flange ends of the valve, so gaskets are required between the flanged body and the mating flanges. The two-flange valve is made with either flat-faced flanges or raised-face flanges, and the mating flanges must match the valve flanges.

2.3.4 Expansion and Compression Tanks

Commercial hot water systems also incorporate expansion or compression tanks (*Figure 27*), as is the case with smaller systems. The tanks will often be larger to accommodate the greater volume of water in the system, since the water will require more room for expansion as it is heated.

If a pressurized expansion tank fails to hold the pre-charge of air, you may suspect a leaking air valve or a leaking bladder/diaphragm. Most bladder-type tanks used commercially are larger than those used in smaller applications. The bladder is accessible for replacement through a large flange. With most diaphragm tanks (now used less often than bladder tanks), the diaphragm is not accessible and the tank must be replaced.

2.3.5 Air Management

Air management in commercial systems is extremely important to ensure proper water circulation and prevent noise. The same types of air management devices used in smaller applications are generally employed in commercial systems.

Figure 23 Butterfly valve.

Figure 25 Wafer lug butterfly valve.

Figure 24 Wafer butterfly valve.

Figure 26 Two-flange butterfly valve.

However, because of the greater volume of water and the potential for much larger volumes of air to collect and create problems with circulation, large commercial systems are often equipped with an air separator (*Figure 28*). This type of air separator is also used in chilled-water systems. They are highly efficient at collecting entrained air from water and enable large systems to operate more effectively.

The unit shown is available in sizes large enough to accommodate 36" pipe. It reduces the

Figure 27 Pressurized expansion tanks.

water velocity to a speed such that the buoyancy of the air bubbles causes them to rise, preventing them from being swept along by the water stream. At the same time, it imparts a centrifugal spin on the water flow. This causes heavier water to move outward and lighter air to move inward. The air is collected there and discharged at the top of the tank.

Air management in hydronic systems is accomplished through one of two strategies, air control or air elimination. In air control (*Figure 29*), a compression tank is mounted on the air separator discharge and air is discharged to it. The air and water are in contact inside the tank, but all air is eliminated from the moving loop of water. Heated water expands into the compression tank, pushing against the air cushion. When air elimination (*Figure 30*) is used, a bladder-type tank is used on the suction side of the pump, providing the area for heated water to expand into. All air is vented to atmosphere through a high-capacity air vent on top of the separator.

2.4.0 Common Commercial Piping Systems

Water piping systems provide routing and distribution of hot water, chilled water, or both. Piping systems used in residential and light commercial

03305-13_F28.EPS

Figure 28 Commercial air separator.

03305-13_F29.EPS

Figure 29 Air control component arrangement.

applications tend to be simple. Those piping systems were covered in an earlier module. Large commercial and industrial piping systems tend to be more complex because the heating and cooling loads in a large building can vary widely. Some areas of a large building might need cool-

Waterlogged Expansion Tanks

You should suspect a waterlogged expansion tank if the boiler relief valve discharges whenever the boiler water is being heated. With no room for the expanding water to go, the system pressure can rise very quickly. If this happens, check the expansion tank and any related air-control components for leaks.

Figure 30 Air elimination component arrangement.

ing, while at the same time another area would require heat.

Although there are some two- and three-pipe systems remaining in older buildings, they are rare. For this reason, this module will focus on conventional systems in common commercial use today.

2.4.1 Four-Pipe Systems

The four-pipe system is a dual-temperature piping system. Because of its design simplicity, room control, and operational economy, it is commonly used in large commercial applications. It is constructed of two separate two-pipe circuits that use separate supply and return lines to carry hot and chilled water to the terminals. One supply and return circuit carries the hot water, while the other supply and return circuit carries the chilled water. *Figure 31* shows a simplified four-pipe system. The chilled and hot water streams never mix together in the four-pipe system.

In this system, three-way valves are often used at the input and output of a coil. The system also uses terminals with separate chilled- and hot-water coils. This allows heating and cooling to potentially occur at the same time, thus providing for humidification and reheating of the circulated air. However, this operating mode is rare, since it uses a great deal of energy.

Two-way valves can also be used at the input to each of the coils. However, when two-way valves close, all flow ceases in the portion of the circuit. When many or all of the valves close, this may cause problems for the pump. To overcome this problem, the pump must either be cycled on and off as demand changes. A better way is to modulate the speed of the pump. Three-way valves typically maintain the same amount of flow at all times. Regardless of whether the water is flowing through a coil or around it, the flow rate remains steady.

The hot- and chilled-water piping systems are separate, so pressure regulation is less critical in the four-pipe system than in older two- and three-pipe systems. Also, much higher hot-water supply temperatures are possible, especially when separate heating coils are used in the terminals. This allows a reduction in the size of the piping for

Figure 31 Four-pipe water system with one-coil and two-coil terminals.

that circuit. Another advantage of the four-pipe system is that various types of heating terminals, such as baseboard and finned-tube units, can be connected to the system in spaces that do not need cooling.

2.4.2 Primary-Secondary Piping Systems

Primary-secondary hot-water or chilled-water systems are used in applications that have peak and varying loads. They supply variable-flow, constant-temperature chilled/hot water using methods that provide for energy conservation. *Figure 32* shows a simplified diagram for a primary-secondary chilled-water system. For a hot-water system, the chillers would be replaced by boilers.

As shown, a primary chiller loop for this system consists of two 500-ton chillers and one 250-ton chiller, each with a related circulating pump. Each of the primary pumps is sized to handle only the low head and steady gpm for the related chiller. The building load is divided into three secondary loops, each with a secondary circulating pump, terminal devices, and control valves.

Again, each of the pumps in the secondary loop is sized to handle the flow and pressure requirements for this specific loop. In this system, primary loop pumps circulate the water through the primary loop; the secondary loop pumps move water through the secondary loops. The transfer of chilled water between the primary loop and each of the secondary loops is made by a short section of pipe (the common pipe) to both the primary and secondary circuits.

During periods of peak load, all three chillers in the primary loop operate to produce 45°F water for application to the secondary loops. The pump in each of the secondary loops is turned on and the maximum amount of flow is in demand.

During periods of reduced load, only the 250-ton chiller and its pump may need to operate to supply 45°F water to the terminals. With variable-speed pumps used in the secondary loops, each of the secondary pumps may be operating at different speeds to satisfy the individual loop flow requirements. Different combinations of chillers and pumps are used in the primary and secondary loops to satisfy the cooling load as it changes.

2.5.0 Water System Balancing

The basic concepts regarding water balance, as well as the calculation of needed flow rates and pumping requirements, were presented in a previous module. Here, the hydronic system balancing procedures will be discussed. In a large building, the procedure can be a lengthy and complex task. Hydronic system balancing is most often done by firms that specialize in the proce-

03305-13_F32.EPS

Figure 32 Simplified primary-secondary chilled-water system.

dure, especially when a new building or system is first commissioned. There are several reasons for this. One reason is that balancing can be a complex task in large facilities. The job often requires skills that many technicians do not possess. In addition, especially when a new building is being commissioned, owners and mechanical engineering firms prefer to have the analysis of system performance provided by a company other than the installing contractor.

Testing, balancing, and adjusting hydronic heating or cooling systems can best be done by following a systematic procedure. The entire process should be carefully planned and performed. Properly scheduling all activities, including organization, procurement of test instruments and equipment, and execution of the balancing procedure, is a must. Coordinate air balancing with water balancing, because many systems function differently on a seasonal basis and temperature performance is a big factor when testing and balancing.

Before attempting to balance a water system, the boiler and/or chiller and related components should be operating in accordance with the manufacturer's instructions. Normally, the system startup procedures include making sure that the chiller is correctly charged with refrigerant and that the boiler and chiller water systems are correctly filled with treated water. They also include ensuring that all chiller/boiler-related safety and operating controls have been tested, adjusted, and set for proper operation.

2.5.1 Pre-Balance Checks

Before beginning the balancing procedure for a selected water system, complete the following pre-balancing procedure:

Step 1 Obtain all technical data, including original plans and specifications or as-built plans pertaining to the hydronic equipment. This includes:

- Pump curves and performance data

- Pressure, flow-temperature characteristics, and other performance data for boilers, chillers, heat exchangers, and coils
- Motor nameplate data and ratings
- Performance characteristics of balancing and flow-meter devices

Step 2 Locate and study the following diagrams:

- Temperature control and control valve characteristics
- Special piping circuits

Step 3 Locate and record the ratings and settings of pressure-reducing valves, pressure-relief valves, and temperature-relief valves.

Step 4 Locate the operating and maintenance instructions for important equipment.

Step 5 Prepare a schematic layout of each system, if necessary. Identify connection points of primary and secondary piping circuits, manual valves, automatic valves, flow meters, air vents, coils, and expansion tanks. Note any changes in the design layout that were made during construction. Number all water-flow balance points and metering points for rapid identification and reporting.

Step 6 Study the water system piping arrangement to determine the steps and sequence required for balancing and the instruments needed for the task. Note system features that would contribute to an unbalanced condition, such as dirty fins on terminals. Determine which circuits must be balanced by temperature rather than fluid flow because of the absence of flow-regulating devices. Note the location of strainers and other piping configurations that may contribute to the collection of dirt and sludge.

Air Separation Science

The solubility of air in water at different temperatures and pressures is based on a law of physics known as Henry's law. This law describes how both the pressure and temperature of water affects solubility. As pressure increases, water is able to keep more air in suspension. Conversely, as pressure decreases, air separates more readily. Higher temperature also decreases water's ability to hold air in suspension. As temperature falls, more air can be held in suspension. The result is that air is separated more readily from water at high temperature and lower pressure. This also means that chilled water systems, which must operate at higher pressures because of the needs of the water distribution system, represent the most challenging air management condition.

Step 7 Obtain all necessary instruments and calibration data. If more than one instrument of a similar type is used, calibrate each one precisely.

Step 8 Make a complete visual inspection of the system and make sure that all air distribution equipment and controls are functioning properly. Do the same for hydronic equipment and controls.

2.5.2 Water System Balancing Procedure

Proper water balancing is essential. A general procedure that can be used for testing, balancing, and adjusting a water system follows:

Step 1 Make sure that all preliminary visual checks and service and maintenance tasks have been performed.

> **WARNING!**
>
> Watch out for rotating, pressurized, or hot components. Follow all safety instructions labeled on the equipment and given in the manufacturer's service manual. Air balancing requires that equipment and systems be energized and operational; use extreme caution when working in the vicinity of energized load and circuit controls.

Step 2 Check all electrical wiring in power and control circuits.

Step 3 Check that all manual valves are open or properly preset.

Step 4 Verify that all automatic valves are in the fully open position.

Step 5 Look for broken thermometers and gauges on the equipment.

Step 6 Read and record the system pressure at the pump with the pump de-energized.

Step 7 Operate the system, bleed the air from the piping, and allow flow conditions to become stabilized.

Step 8 Record the operating voltage and amperage draw of all electrical motors.

Step 9 Record the pump speed (rpm).

Step 10 Slowly close the balancing cock in the pump discharge line with the pump running and record the shutoff discharge and suction pressures at the pump gauge connections.

Step 11 Use the shutoff head to determine the pump operating curve (impeller size) and compare it with the design curve data. Check the pump differential at the pump. Use one gauge to read the differential pressure. If two gauges are used, the gauge readings must be corrected to the centerline elevation of the pump. If there is a discrepancy in the pump output compared to design specifications, either correct the velocity head or report the information to the proper authority.

Step 12 Record the pump discharge pressure by opening the discharge balancing cock slowly to the fully open position. Record the suction pressure and total head in feet at the same time.

Step 13 Using the total head and the pump curve established in Step 11, read the water flow. If the total head is greater than design specifications, the water flow will be less than design. If the total head is less than design, water flow will be more than the selected rating. If the water flow is more than the selected rating, the pump discharge pressure should be increased by partially closing the balancing cock until water flow is approximately 110 percent of design specifications.

Step 14 Start any secondary pumps and readjust the balancing cock in the primary circuit pump discharge if necessary. Record the new adjusted readings.

Step 15 Take a complete set of pressure-drop readings through all equipment such as chillers, boilers, hot-water coils, and chilled-water coils before adjusting any balancing cocks at the equipment. Compare the readings with design specifications. Determine which units are above/below design requirements.

Step 16 Make a preliminary adjustment in the balancing device on all units with high water flow. Adjust them to about 10 percent above the design flow rate.

Step 17 Perform another complete set of pressure and amperage draw readings in the pump system. If the pump total flow has fallen below design, open the balancing cock and the pump discharge to raise the pump output, bringing the flow rate within 105 percent to 110 percent of design.

Step 18 Decrease the flow a second time on all units that are still receiving more than their design flow rate. This will also serve to increase water flow through units that are not presently receiving sufficient water.

Step 19 Keep repeating the procedure (adjusting unit flow devices then pump output) until all units are within ±10 percent of the design water flow.

Step 20 Make a final check of the water flow rates at the pump and equipment and of the pump motor amperage draw and record the findings.

Step 21 After all the balancing is completed, mark all balancing cocks, gauges, and thermometers at the final settings or range of operation for future reference.

Step 22 Check all water flow safety shutdown controls for proper action.

Step 23 Prepare and submit all necessary reports.

These are the basic steps that can be used for balancing regular water (hydronic) systems. However, special types of hydronic distribution systems, such as condenser water and cooling tower systems, steam and hot-water boilers, and heat exchangers will require additional procedures by certified professionals.

Typical System Parameters

The following system parameters are typical for chilled-water systems using R-22 refrigerant chillers. All values are based on 67°F wet-bulb air entering the evaporator coil.

Air-Cooled Chiller Operation

Outside ambient air temperatures (°F)	75	85	95
System head pressure (psig)	213	246	280
System suction pressure (psig)	63	66	69

Water-Cooled Chiller Operation

Outside dry-bulb air temperatures (°F)		75	85	95
Outside wet-bulb air temperatures (°F)		65	73	80
Tower basin temperature (°F)	73	81	88	
System head pressure (psig)	170	198	213	
System suction pressure (psig)	63	66	69	

Chilled-Water Temperatures
Supply water leaving chiller: 42°F to 48°F, typically 45°F
Temperature differential (return water – leaving water): 5°F to 10°F, typically 8°F
Freezestat setting (in leaving water): 35°F to 38°F
Leaving chilled-water control thermostat setting: 44°F to 48°F
Return chilled-water control thermostat setting: 42°F to 45°F

Condenser Water Temperatures
Temperature differential (condenser out – condenser in): typically 5°F to 10°F.

Note
A range of less than 5°F through the condenser with normal head pressure usually indicates a dirty condenser. A range greater than 10°F usually indicates restricted water flow.

The following system parameters are typical for a hot-water heating system:

- Boiler water temperatures (cold weather) – Leaving water 180°F, return water 160°F
- Boiler water temperatures (mild weather) – Leaving water 160°F, return water 140°F
- Boiler water temperatures (minimum setting) – Leaving water 140°F, return water 120°F

Additional Resources

Air-Cooled Chillers, TDP-622, Latest Edition. Syracuse, NY: Carrier Corporation.

Refrigeration and Air Conditioning, An Introduction to HVAC/R, Latest Edition. Larry Jeffus. Upper Saddle River, NJ: Prentice Hall.

Water-Cooled Chillers, TDP-623, Latest Edition. Syracuse NY: Carrier Corporation.

2.0.0 Section Review

1. The maximum working pressure of a low-pressure hot-water boiler is _____.

   a. 250 psi
   b. 160 psi
   c. 30 psi
   d. 10 psi

2. A centrifugal pump operates by increasing the velocity of a(n) _____.

   a. impeller
   b. piston
   c. fluid
   d. refrigerant

3. A butterfly valve is fully open when the handle is rotated _____.

   a. two full turns
   b. one full turn
   c. one-half turn
   d. one-quarter turn

4. In a four-pipe water system, what type of valve is used on each single-coil terminal?

   a. Butterfly valve
   b. Three-way valve
   c. Venturi tube
   d. Two-way valve

5. Which information is *not* needed when performing a pre-balance check?

   a. Chiller refrigerant type
   b. Motor nameplate data
   c. Pump curves
   d. Boiler performance data

3.0.0 CHILLED-WATER SYSTEMS

Objective

Explain the basic concepts of chilled water systems and the related components.

    a. Explain the basic concepts of chilled-water cooling systems.

    b. Identify various types of chillers and their common controls.

    c. Identify various types of cooling towers and evaporative condensers.

Trade Terms

Flash economizer: An energy-conservation process used in chillers that accomplishes intercooling of the refrigerant between the condenser and evaporator using a flash chamber or similar device. Flash gas of coolant produced in the flash chamber flows directly to the second stage of the chiller compressor, bypassing the evaporator and the first compressor stage.

Lithium bromide: A salt solution commonly used in absorption chillers.

Secondary coolant: Any liquid, such as water, that is cooled by a system refrigerant and then used to absorb heat from the conditioned space without a change in state. Also known as indirect coolant.

Thermal economizer: An energy-conservation process used in chillers in which warm condensed refrigerant is brought into contact with the coldest (inlet) water tubes in a condenser or heat exchanger. This causes the condensed refrigerant to subcool to a temperature below the condensing temperature.

A chilled-water system eliminates the need to circulate refrigerant through a building to provide comfort cooling. Instead, a piece of equipment called a chiller (*Figure 33*) uses refrigerant to cool the water. The chilled water is circulated throughout the building and back to the chiller where the process is repeated. This uses basic water piping throughout the building rather than refrigerant piping, which is more expensive and represents a unique challenge to install in large buildings.

After chilled water is delivered to the building, it may be circulated through a large cooling coil in a central air handler to cool a large area, or it can be delivered to the conditioned space and circulated through a small cooling coil housed in a terminal to cool the room or zone.

3.1.0 Basic Chilled-Water System Concepts

This section focuses on chilled-water systems that operate above 40°F and are used for comfort cooling. Other types of chilled-water systems may be used to provide comfort air conditioning or support an industrial process. These systems cool a brine or other type of water and antifreeze solution, operating at water temperatures below 32°F.

Chilled-water systems are used to cool buildings that have either a separate heating system or no heating system at all. They can also be used as part of a dual-temperature system. *Figure 34* shows a simplified diagram of a basic chilled-water system. It consists of a chiller, cooling terminal, chilled-water pump, and related chilled-water piping. The chiller unit shown is a mechanical refrigeration system. Some chilled-water systems use absorption-type chillers to produce cooling.

The chiller referred to in *Figure 34* is based on the same refrigeration cycle as all other common cooling systems. The chiller evaporator (cooler) is a water-to-refrigerant heat exchanger through which both chilled water and system refrigerant flow. The chilled water is cooled by the vaporization of the liquid refrigerant.

For comfort cooling applications, the chilled water temperature supplied to the building is generally around 45°F, with the returning water temperature expected to be around 10°F warmer.

03305-13_F33.EPS

Figure 33 Packaged air-cooled chiller.

Of course, the return water temperature can vary with the load. The chilled water produced acts as a **secondary coolant**, which is any liquid cooled by a system refrigerant, then used to transfer heat from the conditioned space without a change in state.

The refrigerant vapor from the chiller evaporator is drawn into the chiller compressor. There, the low-temperature, low-pressure refrigerant vapor is converted into a high-temperature, high-pressure vapor so that it can be condensed into a liquid in the condenser. The system shown in the example has an air-cooled condenser.

In some systems, the condenser may be a separate unit and not built into the chiller itself. The air-cooled condenser in this type of chiller has a series of tubing coils through which the refrigerant flows. As cooler outside air is forced across the condenser tubing, the hot refrigerant vapor condenses into a liquid, giving up its heat to the outside air. The high-temperature, high-pressure liquid refrigerant then flows to the evaporator through the expansion device.

The expansion device regulates the flow of refrigerant into the evaporator and decreases its pressure and temperature. This converts the high-temperature, high-pressure liquid refrigerant into the low-temperature, low-pressure refrigerant needed to absorb heat in the evaporator.

The water cooled by the chiller is circulated through the piping system by the chilled-water pump. When there is a call for cooling from the building controls, the water is recirculated through the chiller evaporator and cooling terminal.

In a typical system, the terminal is an air handler containing chilled water coils and a fan. Cooling of the room air takes place by blowing the warm room air over the chilled water coils. This transfers the heat from the warmer room air through the coil tubing to the chilled water.

An isometric view of the system shown in *Figure 34* is shown in *Figure 35*. Note that the chiller unit used in this system is a packaged air-cooled reciprocating chiller.

3.2.0 Chillers

Chillers fall into two classes: mechanical and non-mechanical. Mechanical chillers are by far the most common and use various types of compressors to provide cooling. Compressor operation lowers the pressure and temperature of the refrigerant in the chiller evaporator, allowing the refrigerant to boil and absorb heat from the chilled water. Also, the pressure and temperature of the refrigerant in the chiller condenser is raised, allowing the refrigerant to condense and give up its heat to either air or water. The compressor also maintains a pressure difference between the high-pressure and low-pressure sides of the chiller system. This pressure difference causes the refrigerant to flow throughout the refrigerant circuit.

Non-mechanical chillers do not have a compressor. Instead, they use a chemical cycle to provide cooling. This type of chiller is called an absorption chiller.

Three common types of mechanical chillers are reciprocating or scroll compressor chillers, centrifugal chillers, and screw chillers.

Mechanical water chillers can be assembled from components in the field or they may be assembled at the factory into a packaged unit. This section will focus on packaged chillers. For our discussion, a packaged chiller is one that includes the compressor, condenser, evaporator, expansion device, related controls, and auxiliary equipment. The system water pumps are often not a part of the package. This is because they are not part of the refrigerant circuit, and every application has different pumping requirements. The basic refrig-

03305-13_F34.EPS

Figure 34 Basic chilled-water system.

Figure 35 Chilled-water system with a packaged air-cooled chiller.

eration components used in mechanical chillers operate the same way as similar components in other, more basic, cooling systems.

3.2.1 Reciprocating and Scroll Chillers

Reciprocating or scroll chillers (*Figure 36*) use one or more compressors to provide the necessary cooling capacity. When multiple compressors are used, the compressors are generally staged by temperature controls. As load increases and the water temperature begins to increase from the setpoint, more compressors are started. In some cases, two compressors may be connected on a single refrigeration circuit.

Although hermetic reciprocating and scroll compressors dominate the small- to medium-sized chiller market today, models using one or more semi-hermetic compressors remain in use and are still available from some manufacturers.

Figure 36 Packaged air-cooled scroll chiller.

The chiller condenser can be air cooled, water cooled, or evaporative. Water-cooled condensers may be tube-in-tube, shell-and-tube, or shell-and-coil.

The expansion device controls refrigerant flow from the condenser to the evaporator to maintain enough suction superheat to prevent liquid refrigerant from flooding the compressor. HCFC-22 refrigerant was widely used in new reciprocating water chillers. Today's models use non-chlorine refrigerants such HFC-410A. Reciprocating and scroll chillers have cooling capacity sizes ranging from about 5 tons to as much as 225 tons. However, other types of chillers typically dominate the market over 100 tons. Capacity control is generally accomplished using on-off cycling of the compressor(s), cylinder or scroll unloading, or hot gas bypass for applications requiring extreme accuracy in water temperature.

3.2.2 Centrifugal Chillers

The centrifugal chiller (*Figure 37*) uses a centrifugal compressor consisting of one or more stages. Both open and hermetic compressors can be used. Open compressors can be driven by gas or steam turbines or engines, or electric motors. Centrifugal compressors use a high-speed impeller with many blades that rotate in a spiral-shaped housing. The impeller is driven at high speed inside the compressor housing. Refrigerant vapor is fed into the housing at the center of the impeller. The impeller throws the incoming vapor outward in a circular path from between the blades and into the compressor housing. This action, called centrifugal force, creates pressure on the high-velocity gas and forces it out the discharge port.

The condenser in a centrifugal chiller is usually water-cooled, with the refrigerant condensing on

03305-13_F37.EPS

Figure 37 Centrifugal chiller.

the outside of the tubes. Air-cooled condensers can also be used, but their use greatly increases the unit's power consumption and they are very rarely seen. The evaporator (cooler) can be a direct-expansion type in smaller units, but is usually a flooded type in the larger units.

Centrifugal chillers have cooling capacities ranging from about 200 tons to 8,000 tons. Capacity control is normally accomplished using adjustable compressor inlet guide vanes or compressor speed control.

HCFC-123 and HFC-134a are common refrigerants used in centrifugal chillers. HCFC-123 is a low-pressure refrigerant, while HFC-134a is a high-pressure refrigerant. A variety of low-pressure CFC refrigerants such as CFC-11 were used in the past, but most have either been replaced or modified for use with new, environmentally friendly refrigerants. In light commercial applications, some chillers using CFC refrigerants do remain in service. As all chlorine-bearing refrigerants will eventually be phased out, those based on HCFC-123 will be replaced or modified.

Subcooling or intercooling devices are often used in a centrifugal chiller to conserve energy. Subcooling is a process by which the condensed refrigerant is subcooled in either a subcooling section of the water-cooled condenser or in a separate heat exchanger. This is done by bringing the warm, condensed liquid refrigerant into contact with the coldest (inlet) water tubes in the condenser or in a heat exchanger. This causes the condensed refrigerant to subcool further, increasing the refrigeration cycle efficiency and lowering power consumption. Devices that accomplish subcooling in this way are referred to as **thermal economizers**.

Intercooling is a process in which the condensed refrigerant leaving the condenser is passed through a flash (surge) chamber before reaching the evaporator. The pressure in this chamber falls between the condenser and evaporator pressures. Because its pressure is lower than the condenser pressure, some of the liquid refrigerant flashes into a vapor. This cools the remaining liquid. The flash gas, having absorbed heat, is drawn off and returned directly to the inlet of the compressor second stage, where it is mixed with discharge gas already compressed by the first-stage impeller. Since the flash gas must pass through only one stage of compression to reach condenser pressure, there is a savings in power.

The liquid refrigerant remaining in the flash chamber is metered through a valve and flows into the evaporator. Because pressure in the evaporator is lower than the flash chamber pressure, some of the liquid flashes and cools the remainder of the refrigerant to evaporator temperature. Devices that accomplish intercooling using a flash chamber or similar device are referred to as **flash economizers.**

Purge units are another type of device used on centrifugal chillers that operate with low-pressure refrigerants, such as HCFC-123. Chillers using these refrigerants operate at below-atmospheric pressures. Because of this, air and water tend to leak into and accumulate in the system, rather than the refrigerant leaking out. A purge unit must be used to separate and remove noncondensable gases and water vapor from the system.

Today's purge units are generally small refrigeration units mounted on the chiller assembly. They draw in refrigerant vapors from the chiller, and then chill the vapor below the point of condensation for the refrigerant itself. Any vapor that remains is considered a noncondensable gas other than refrigerant and is released to the atmosphere. Any moisture is captured by filter-driers. The condensed liquid refrigerant is then returned to the chiller. The purge units operate on high-pressure refrigerants, like most common cooling and refrigeration units.

Older purge systems that used CFC refrigerants were quite inefficient and tended to release a significant amount of refrigerant along with the air. Because the refrigerants were relatively inexpensive and environmental issues were not being considered, this issue presented little concern. The chiller was simply recharged periodically. To comply with new EPA guidelines, purge units have been redesigned to operate much more effectively.

3.2.3 Screw Chillers

Screw chillers (*Figure 38*) use an oil-injected hermetic or open-screw compressor. Screw compressors use a matched set of screw-shaped rotors, one male and one female, enclosed within a cylinder. The male rotor is driven by the compressor motor, and in turn drives the female rotor. Normally, the driven male rotor turns faster than the female ro-

Condenserless Packages

Some chillers are supplied without a condenser assembly attached, saving space in the mechanical room. Due to design issues, this option is generally found only with reciprocating, scroll, or screw chillers. The remote condenser can be air-cooled or water-cooled. As with packaged chillers that are equipped with a condenser on board, a standard direct-expansion refrigerant circuit is employed.

03305-13_F38.EPS

Figure 38 Air-cooled screw chiller.

tor because it has fewer lobes. Typically, the male has four lobes, while the female has six. As these rotors turn, they mesh with each other and compress the gas between them. The screw threads form boundaries that separate several compression chambers. Gas entering the compressor is moved through a series of progressively smaller compression stages until it exits at the compressor discharge in its fully compressed state.

Screw chillers can have a water-cooled shell-and-tube condenser, with the refrigerant condensing on the outside of the tubes. However, air-cooled condensers are also very popular. Water-cooled chillers are located indoors, in equipment rooms, while air-cooled chillers are located outside the building. When the chiller is used with a remotely located air-cooled or evaporative condenser, a liquid receiver may replace the condenser on the package assembly. A remote air-cooled condenser allows the chiller to be inside the building. This can be an important consideration, as screw chillers can be quite noisy. This is especially true when they operate at less than 100 percent of their capacity.

The evaporator can be either a direct-expansion or flooded type. When the direct-expansion evaporator is used, refrigerant usually flows through the tubes. With a flooded evaporator, refrigerant is normally outside the tubes. Because the screw compressor is oil-injected, an oil separator is also part of the package. A flash economizer, similar to the one described for the centrifugal chiller, can be used to improve chiller efficiency. In the screw compressor, flash gas produced in the economizer is injected into the compressor at an intermediate point of the compression cycle.

New packaged screw chillers use HFC-134a, HFC-407C, or HFC-410A refrigerant. Screw chillers have cooling capacity sizes ranging from about 40 to 850 tons. Capacity control is normally accomplished with a hydraulically operated slide valve.

3.2.4 Absorption Chillers

Absorption chillers use a chemical cycle to provide cooling. In the absorption cycle, evaporation and condensation of the refrigerant occur at different pressure levels. The cycles differ in that the absorption cycle uses a heat-operated generator to produce the pressure difference, while the mechanical refrigeration cycle uses a compressor. Both cycles require energy for operation, heat in the absorption cycle, and mechanical energy in the compression cycle.

The refrigerant (water) and absorption solution (lithium bromide) are the working fluids of the absorption cycle. Lithium bromide is a salt solution. The absorption cycle, or lithium bromide cycle, is based on two principles:

- Water can be made to boil in the temperature range of 40°F to 45°F by maintaining a low absolute pressure (high level of vacuum) of about 0.25 to 0.30 inches of mercury (in Hg) absolute. This means that water can actually be used like a refrigerant to absorb heat from the chilled water flowing through tubes in the chiller evaporator.
- Lithium bromide solution has the ability to absorb water vapor. This ability is best demonstrated when common table salt clogs up a salt shaker in humid weather, because the salt has absorbed moisture from the air.

> **WARNING!**
>
> Lithium bromide is potentially very harmful. It reacts with skin moisture to cause burns, damages the respiratory tract and lungs if inhaled, and is corrosive. Contact with the eyes can cause irritation, blurred vision, and possibly serious eye damage. When handling lithium bromide, always use appropriate personal protection equipment as recommended in the MSDS. Disposal of lithium bromide is controlled by local, state, and federal agencies. Check local, state, and Federal EPA regulations and comply with all disposal requirements.

GOING GREEN

Central Plant Facilities

Central plant facilities that provide hot and chilled water to large service areas have long been used for college campuses, airports, high-technology manufacturing centers, and some municipalities. Central plant technology is experiencing renewed interest from energy-conscious system designers. Designers of new buildings and retrofitted high-density complexes are leaning more toward this technology. Many of these facilities combine cogeneration capability that uses the heat byproducts of standalone electricity generation to drive the heating water system or power chillers. Typical cooling tonnages for these facilities begin in the 3,000-ton range and are seen consistently in the 8,000- to 10,000-ton range.

Absorption chillers are available in capacities ranging from 50 tons to as high as 1,800 tons. As a rule, they are used in applications where a significant amount of steam is readily available, often where refuse-burning facilities are nearby or on board sea-going vessels with large steam boilers in operation. They are relatively rare in commercial applications, and the operating cycle is quite complex and detailed. For this reason, an in-depth study of their operation is not included here.

3.2.5 Chiller Operating/Safety Controls

Almost all modern chillers have one or more control or interface panels used by the operator to control or monitor the operating status of the chiller (*Figure 39*). Operation of the chiller from these panels is under the control of a microprocessor programmed for the task. Status inputs are supplied to the microprocessor by remote sensors. The sensors may be thermistors for measuring temperature, transducers for measuring pressure, and contacts that supply or interrupt various signals.

Typically, chiller performance status is displayed on a various types of screens. Status messages can be generated whether or not the chiller is running. Operation of the chiller from the panel(s) is often done by entering commands using a computer-type keyboard in response to menu-driven displays. When troubleshooting, these same panels can be used to receive diagnostic messages or run tests. Many chillers are designed to interface easily with laptops or tablets.

Common chiller operations performed either automatically or manually by a typical microprocessor control center are the following:

- Startup and shutdown of the chiller
- Adjustment of chiller compressor capacity control to maintain the proper leaving chilled-water temperature
- Control of system operation per programmed occupied/unoccupied schedules
- Registering cooler, condenser, and lubricating system pressures and temperatures

03305-13_F39.EPS

Figure 39 Chiller operator interface.

- Generating alarm and alert messages for abnormal conditions and de-energizing the chiller when system safety dictates
- Recording the total machine operating hours

Because chilled-water systems represent a significant investment, use large amounts of energy, and are essential to the comfort of large numbers of occupants, they are often integrated into computer-controlled building automation systems. Through the use of these systems, even the smallest details of the chiller's current operation and operating history are readily available to building managers on site and other personnel in remote locations. When problems occur, service personnel can review detailed system information and the nature of the condition before arriving on site.

3.3.0 Cooling Towers/Evaporative Condensers

Condensers are used to remove heat from the chiller refrigerant system. They take in high-pressure, high-temperature refrigerant gas from the

Impellers in Series

Several impellers can be put in series (stages) to create greater pressure difference and to pump a sufficient volume of vapor. A compressor that uses one impeller is called a single-stage machine; one that uses two impellers is called a double-stage machine, and so on. When more than one stage is used, the discharge from the first stage is fed into the inlet of the next stage.

compressor and change it into high-temperature, high-pressure liquid refrigerant. They do this by removing the superheat and latent heat of vaporization from the refrigerant by transferring this heat to air, water, or both.

For the condenser to operate properly, the condensing medium of air or water must always be at a lower temperature than the refrigerant it is condensing. Water-cooled condensers reject their heat to circulating water, which then circulates through a cooling tower or evaporative condenser. Water-cooled condensers are normally used in conjunction with a cooling tower. This section will focus on the cooling towers and evaporative condensers typically used.

3.3.1 Cooling Towers

The water portion of a chiller's water-cooled condenser is connected by piping to a cooling tower that is located outdoors. In the cooling tower, the heat absorbed by the water in the condenser is rejected from the system into the atmosphere by evaporation. The cooled water is then returned from the cooling tower to the condenser for reuse.

Cooling towers pass outdoor air over the warm water to remove the heat from the water. There are many kinds of cooling towers used with water-cooled condensers. One type is a direct-contact, natural-draft tower (*Figure 40*). It is called a direct-contact tower because the heated water comes in direct contact with the air. These towers are mounted outdoors, usually on the roof, to make use of natural air currents and heights for the net positive suction head on the pump. They are made of a metal frame covering several layers or tiers of fiberglass-reinforced plastic decks or splash bars.

Because they use natural draft, no blowers are used to move air through the tower. Water is piped up from the condenser located in the building below to the top of the decks and is discharged in sprays over the decks. Spaces between the boards in the decks permit the water to drip or run from deck to deck, while being spread out and exposed to breezes that enter the tower from the open sides. The cooled water is collected in a sump at the bottom of the tower, and is pumped back to the condenser for reuse.

Natural-draft towers are becoming increasingly rare. They are generally much larger than newer designs. Packing more capacity into smaller packages is a common goal for most all HVAC/R equipment manufacturers.

Because cooling towers work partially on evaporation, any water lost in the evaporation process must be replaced in order to maintain correct system operation. This is done by a float-controlled makeup water system (*Figure 41*) that senses the water level in the sump and adds water as needed. Electronic water level sensing systems are also used.

A variation of the natural-draft tower is an indirect-contact tower, sometimes called a closed-circuit cooler (*Figure 42*). The indirect-contact tower has two separate water circuits. One circuit consists of cooling tubes through which the water from the chiller condenser is flowing. This circuit is closed to the air, thereby protecting the chiller condenser water from being contaminated by airborne dirt and impurities.

The second water circuit is open to the air and consists of a pump that delivers water from a sump at the bottom of the tower to the top of the tower. There it is discharged in sprays over the tower decking in the same way as in the direct-contact tower. In this tower, heat in the closed tubes is first transferred to the water, then from the water to outdoor air. Drift eliminators prevent moisture discharge from the cooling tower, reducing the amount of makeup water needed.

Centrifugal and other types of fans may be used to increase the volume of airflow through cooling towers and indirect-contact towers. When fans are used, the tower is called a forced-draft or induced-draft tower.

A forced-draft tower is one in which the fans push the air so that it flows up and over the wet decking surfaces. In an induced-draft tower, the fans pull the air rather than push it, so that it flows up and over the decking. Because they use fans, forced-draft and induced-draft towers tend to be small in comparison to natural-draft cooling towers. *Figure 43* shows an example of an indirect-contact, induced-draft cooling tower.

Figure 40 Direct-contact natural-draft cooling tower.

Figure 41 Cooling tower float-controlled makeup water system.

3.3.2 Evaporative Condensers

Like the water-cooled condenser and water tower combination, evaporative condensers first transfer heat to the water, then from the water to the

Figure 42 Indirect-contact natural-draft cooling tower.

outdoor air. The evaporative condenser combines the functions of a water-cooled condenser and cooling tower in one enclosure or package (*Figure 44*). The condenser water evaporates directly off the tubes of the condenser with each pound of water evaporated, removing about 1,000 Btus from the refrigerant flowing through the tubes.

Air enters the bottom of the unit and flows upward over the refrigerant-filled condensing tube

AIR OUTLET

FAN

WATER INLET

DRIFT ELIMINATORS

DECKING

DECKING

AIR IN

AIR IN

COILS

LOWER COLLECTION BASIN

COLD WATER OUTLET

HOT WATER OUTLET

PUMP

03305-13_F43.EPS

Figure 43 Indirect-contact induced-draft cooling tower.

bundle. At the same time, water is being sprayed over the tube bundle. Both the air and sprayed water absorb heat from the refrigerant inside. Water eliminators located above the water spray keep water entrained in the rising air from escaping the condenser. Water in the airstream strikes the eliminators and falls back into the condenser basin. Air is moved through the condenser assembly by one or more fans. Cooled by both air and water, the refrigerant in the coil condenses into a subcooled liquid at the output of the coil.

Relocating the refrigerant condenser to the base of the evaporative condenser generally requires a considerable amount of additional refrig-

Belly Heaters

Belly heaters are used on low-pressure systems to keep the pressure from falling and remaining below the atmospheric pressure during the off-cycle, allowing noncondensable substances to enter from the surrounding environment.

Absorption Chiller Heat Sources

Because heat is the primary energy source needed for absorption chillers, they are normally used where a source of steam or high-temperature water is readily available. The heat can also be from direct natural-gas combustion or even hot industrial gases.

Figure 44 Evaporative condenser.

03305-13_F44.EPS

erant piping. Because many water-cooled systems that use an evaporative condenser are quite large, this factor also significantly increases the refrigerant charge required and exposes the circuit to a greater potential for damage and leakage.

Refrigerant piping is usually more expensive, and its installation is more complex than water piping. For that reason, cooling towers are far more popular for most applications. Evaporative condensers are more likely to be found in refrigeration and food processing applications than in comfort cooling. They are often used in large refrigeration systems using anhydrous ammonia as the refrigerant.

Cooling Tower Water Treatment

Proper water treatment is essential to prevent and control biological growth and mineral deposits in cooling towers. Without proper treatment, a wide variety of biological growth will occur, sometimes with hazardous results to humans in the vicinity of the evaporating moisture. These same growths also cause a loss of capacity by impeding the heat transfer process.

Mineral deposits also impede heat transfer; as water is evaporated, minerals contained in the water supply remain behind. Although chemical treatment can assist in reducing these deposits, most cooling towers require bleed-off (a consistent minimal draining of the tower water) to dilute and flush away minerals in circulation. The chemical treatment approach should be designed with consideration for local conditions, the anticipated or identified types of biological growth, and the surrounding environment.

Cooling Towers

Cooling towers are usually designed to reduce water temperature by approximately 10°F based on the location of the system. A cooling tower can only cool to a point about 3°F to 7°F above the ambient wet-bulb temperature. Because they depend upon evaporation of water to the atmosphere for a cooling effect, they are more effective and possibly smaller in areas of the country where humidity levels are low.

03305-13_SA02.EPS

Additional Resources

Air-Cooled Chillers, TDP-622, Latest Edition. Syracuse, NY: Carrier Corporation.

Refrigeration and Air Conditioning, An Introduction to HVAC/R, Latest Edition. Larry Jeffus. Upper Saddle River, NJ: Prentice Hall.

Water-Cooled Chillers, TDP-623, Latest Edition. Syracuse NY: Carrier Corporation.

3.0.0 Section Review

1. Chillers used for comfort cooling operate above _____.

   a. 20°F
   b. 30°F
   c. 40°F
   d. 50°F

2. What does a mechanical chiller use to provide cooling?

   a. Compressors
   b. Centrifugal pump
   c. Air compressor
   d. Absorption compressor

3. If the cooling tower you are working on contains a refrigerant circuit, you have a(n) _____.

   a. natural-draft cooling tower
   b. evaporative condenser
   c. hybrid cooling tower
   d. direct-contact cooling tower

SUMMARY

Commercial hydronic systems use pipes to transport heated or cooled water from its source to where it is needed. The term hydronic refers to all water systems. An air conditioning system that circulates chilled water to cooling coils is known as a chilled-water system.

Hot-water heating systems are used mainly to provide comfort heating. They can be separate systems used to heat buildings that have separate or no cooling systems. They can also be used in dual-temperature systems. Commercial hot water systems are of the forced hot-water type, because mechanical pumping is generally necessary to overcome the friction loss and length of large piping systems.

A boiler is a heat exchanger that uses radiant heat and hot flue gases to generate hot water for heating and process loads. Low-pressure, hot-water boilers can have operating pressures up to 160 psi, but their operating temperatures must not exceed 250°F. The main differences between hot-water and steam boilers are in the operating and safety controls and accessories. Also, hot-water boilers are filled to capacity with water, while steam boilers are not.

Piping in hydronic systems provide for the routing and distribution of hot water, cold water, or both. In hydronic comfort systems, water balancing means the proper delivery of hot water, chilled water, or both, in the correct amounts to each of the areas being conditioned in a structure. The satisfactory distribution of conditioned water depends upon a well-designed piping system and properly chosen water system components.

Balancing the water system requires adjusting flow controls so that the right amount of hot or chilled water is circulated in the required spaces at the proper velocity to provide satisfactory heating and cooling.

Chilled-water systems eliminate the need to circulate refrigerant through refrigerant system components installed in a building to provide comfort cooling. Instead, a chiller uses refrigerant to cool the water. Then, the chilled water is circulated throughout the building and back to the chiller where the process is repeated.

Chillers are either the mechanical compression type or the absorption type. Mechanical compression chillers may use reciprocal, centrifugal, or screw-type compressors. The compressor produces the pressure differential needed to maintain the refrigeration cycle. Absorption chillers use a heat and chemical process involving water as a refrigerant, and lithium bromide as an absorbent to produce the pressure differential for the refrigeration cycle. Both types of chillers evaporate and condense the refrigerant liquid.

1. The difference in weight of a cubic foot of water as its temperature is raised from its freezing point to its boiling point _____.

   a. is roughly 2.5 pounds
   b. is roughly 14.7 pounds
   c. is roughly 62.4 pounds
   d. cannot be determined

2. Assume a hydronic piping system was originally designed for a flow rate of 50 gpm with an initial pressure drop of 6 psi. With an increase in flow rate to 100 gpm, the new pressure drop would be _____.

   a. 2 psi
   b. 12 psi
   c. 24 psi
   d. 72 psi

3. In any hydronic system, static pressure at the lowest point in the system will always be zero pounds per square inch (psi).

   a. True
   b. False

4. Low-pressure hot-water boilers are limited to a maximum operating temperature of _____.

   a. 180°F
   b. 212°F
   c. 250°F
   d. 500°F

5. One characteristic the copper-finned tube atmospheric boiler offers compared to other commercial types is _____.

   a. greater total heating capacity
   b. an extended life
   c. they cannot be installed outdoors
   d. they are very lightweight

6. Which boiler type allows the heating capacity to be increased by adding more sections?

   a. Firetube
   b. Watertube
   c. Cast-iron
   d. Vertical tubeless

7. In steel firetube boilers, _____.

   a. flue gases are contained in the tubes, with heated water on the outside
   b. heated water circulates through the tubes, with the flue gases on the outside
   c. cast-iron sections contain the flue gases
   d. the water and the flue gases both travel through separate tubes

8. Which type of boiler generally uses a water flow switch?

   a. Steel firetube
   b. Steel watertube
   c. Copper-finned tube
   d. Cast-iron

9. In some centrifugal pumps, the packing gland _____.

   a. reduces pump shaft friction
   b. attaches the impeller to the shaft
   c. stores lubricating oil
   d. helps control fluid leakage around the shaft

10. Multi-purpose valves are commonly installed _____.

    a. on the inlet side of the a pump
    b. on the discharge side of a pump
    c. at the inlet of each terminal device
    d. on the boiler makeup water line

11. Commercial air separators allow air to separate from the water stream by significantly reducing the water _____.

a. pressure
b. velocity
c. temperature
d. flow volume

12. The chilled water produced by a chiller is considered to be _____.

a. a secondary coolant
b. the primary coolant
c. a brine solution
d. more difficult to pump

13. The purpose of a purge unit on a low-pressure chiller is to _____.

a. provide additional subcooling of the refrigerant
b. automatically maintain the proper refrigerant charge
c. remove noncondensable materials from the refrigerant circuit
d. remove the remaining air from the chilled water

14. All indirect-contact cooling towers rely on natural air drafts for cooling.

a. True
b. False

15. Evaporative condensers differ from cooling towers in that _____.

a. they do not use water for cooling
b. cooling towers do not transfer heat through evaporation
c. the refrigerant condenser itself is contained inside the evaporative condenser
d. they cannot be used with most refrigerants

Trade Terms Introduced in This Module

Boiler horsepower (Bohp): One Bohp is equal to 9.803 kilowatts. Also, it is the heat necessary to vaporize 34.5 pounds of water per hour at 212°F. This is equal to a heat output of 33,475 Btus per hour or about 140 square feet of steam radiation.

Firetube boiler: A boiler in which the flue gases are contained inside the tubes, with the heated water on the outside.

Flash economizer: An energy-conservation process used in chillers that accomplishes intercooling of the refrigerant between the condenser and evaporator using a flash chamber or similar device. Flash gas of coolant produced in the flash chamber flows directly to the second stage of the chiller compressor, bypassing the evaporator and the first compressor stage.

Lithium bromide: A salt solution commonly used in absorption chillers.

Secondary coolant: Any liquid, such as water, that is cooled by a system refrigerant and then used to absorb heat from the conditioned space without a change in state. Also known as indirect coolant.

Superheated steam: Steam that has been heated above its saturation temperature.

Thermal economizer: An energy-conservation process used in chillers in which warm condensed refrigerant is brought into contact with the coldest (inlet) water tubes in a condenser or heat exchanger. This causes the condensed refrigerant to subcool to a temperature below the condensing temperature.

Venturi tube: A short tube with flaring ends and a constricted throat. It is used for measuring flow velocity by measurement of the throat pressure, which decreases as the velocity increases.

Watertube boiler: A boiler where the heated water is contained inside the tubes, with the flue gases on the outside.

Additional Resources

This module presents thorough resources for task training. The following resource material is suggested for further study.

Air-Cooled Chillers, TDP-622, Latest Edition. Syracuse, NY: Carrier Corporation.

Refrigeration and Air Conditioning, An Introduction to HVAC/R, Latest Edition. Larry Jeffus. Upper Saddle River, NJ: Prentice Hall.

Water-Cooled Chillers, TDP-623, Latest Edition. Syracuse NY: Carrier Corporation.

Figure Credits

Courtesy of Daikin McQuay, Module opener, Figures 36, 38

Raypak, Inc., Figures 5, 6

Utica Boilers, Figure 7

Burnham Hydronics, Figure 10

Topaz Publications, Inc., Figures 14, 20 (photo), 21 (photo), 37, SA02

Photo courtesy of Paragon Pump Co., Figure 18

Courtesy of Taco, Inc., SA01

Courtesy of Dwyer Instruments, Inc., Figure 23 (photo)

Xylem Applied Water Systems, Figure 27

Courtesy of Xylem Inc., Figures 28–30

McQuay International, Figure 33

Courtesy of John C. Schaub, Inc., Figure 39

EVAPCO, Inc., Figure 41

Section Review Answer Key

Answer	Section Reference	Objective
Section One		
1.a	1.1.0	1a
2.b	1.2.1	1b
Section Two		
1.b	2.1.0	2a
2.c	2.2.1	2b
3.d	2.3.3	2c
4.b	2.4.1	2d
5.a	2.5.1	2e
Section Three		
1.c	3.1.0	3a
2.a	3.2.0	3b
3.b	3.3.2	3c

NCCER CURRICULA — USER UPDATE

NCCER makes every effort to keep its textbooks up-to-date and free of technical errors. We appreciate your help in this process. If you find an error, a typographical mistake, or an inaccuracy in NCCER's curricula, please fill out this form (or a photocopy), or complete the online form at **www.nccer.org/olf**. Be sure to include the exact module ID number, page number, a detailed description, and your recommended correction. Your input will be brought to the attention of the Authoring Team. Thank you for your assistance.

Instructors – If you have an idea for improving this textbook, or have found that additional materials were necessary to teach this module effectively, please let us know so that we may present your suggestions to the Authoring Team.

NCCER Product Development and Revision

13614 Progress Blvd., Alachua, FL 32615

Email: curriculum@nccer.org
Online: www.nccer.org/olf

❏ Trainee Guide   ❏ Lesson Plans   ❏ Exam   ❏ PowerPoints   Other _____

Craft / Level: _____   Copyright Date: _____

Module ID Number / Title: _____

Section Number(s): _____

Description: _____

_____

_____

_____

Recommended Correction: _____

_____

_____

_____

Your Name: _____

Address: _____

_____

Email: _____   Phone: _____

03306-13

Steam Systems

OVERVIEW

Steam has a wide variety of uses in the commercial and industrial sectors. With the ability to carry large quantities of stored energy, steam can be applied in comfort heating, process heating, mechanical drive, pressure control, and a variety of other applications. The majority of steam's heat content is stored in its latent form, so it can be transported efficiently at low cost. It is generally a safer alternative to high-temperature water for carrying heat in large quantities.

Module Ten

Trainees with successful module completions may be eligible for credentialing through NCCER's National Registry. To learn more, go to **www.nccer.org** or contact us at **1.888.622.3720**. Our website has information on the latest product releases and training, as well as online versions of our *Cornerstone* magazine and Pearson's product catalog.

Your feedback is welcome. You may email your comments to **curriculum@nccer.org**, send general comments and inquiries to **info@nccer.org**, or fill in the User Update form at the back of this module.

This information is general in nature and intended for training purposes only. Actual performance of activities described in this manual requires compliance with all applicable operating, service, maintenance, and safety procedures under the direction of qualified personnel. References in this manual to patented or proprietary devices do not constitute a recommendation of their use.

Objectives

When you have completed this module, you will be able to do the following:

1. Describe the properties of water as they are related to steam systems.
   a. Describe the basic properties of water.
   b. Describe the pressure-temperature relationship of water.
2. Describe the basic steam cycle and the primary components related to its operation.
   a. Describe the steam cycle principles of operation.
   b. Identify steam boilers and their common accessories.
   c. Identify common steam heat exchangers and terminal devices.
3. Identify common steam system piping arrangements and valves.
   a. Identify common steam system piping arrangements.
   b. Identify common condensate return systems and describe waterside care.
   c. Explain how steam and condensate system piping is sized.
   d. Identify steam-based pressure-reducing and thermostatic control valves.
4. Identify, install, and maintain various types of steam traps.
   a. Identify various types of steam traps.
   b. Explain the basic concepts of installing steam traps.
   c. Explain how to maintain various steam traps.
   d. Explain how to troubleshoot various steam traps.

Performance Task

Under the supervision of your instructor, you should be able to do the following:

1. Perform two of the following:
   • Perform selected operating procedures on low-pressure steam boilers and systems.
   • Maintain selected steam traps.
   • Identify common piping configurations used with steam systems.

Trade Terms

Drip leg
Dry steam
Flash steam
Saturated steam

Trim
Turndown ratio
Water hammer

Industry-Recognized Credentials

If you are training through an NCCER-accredited sponsor, you may be eligible for credentials from NCCER's Registry. The ID number for this module is 03306-13. Note that this module may have been used in other NCCER curricula and may apply to other level completions. Contact NCCER's Registry at 888.622.3720 or go to **www.nccer.org** for more information.

Contents

Topics to be presented in this module include:

Figures and Tables

1.0.0 STEAM-RELATED FUNDAMENTALS OF WATER

Objectives

Describe the properties of water as they are related to steam systems.

a. Describe the basic properties of water.
b. Describe the pressure-temperature relationship of water.

Trade Terms

Dry steam: The steam that exists after all the water has been vaporized and is moving beyond its saturation temperature (being superheated).

Saturated steam: The steam produced and existing at the temperature that corresponds to the boiling temperature of water at the existing pressure (saturation).

Steam is the most widely used form of energy for industrial use in the world. Water itself is generally plentiful, as well as inert, making it a desirable compound to work with in many applications. The energy stored within water, once it is has been sufficiently heated to the point of vaporization and beyond, can be five to six times greater than liquid water can generally carry. Coupled with the fact that pumps are not required to move steam from the point of creation to the intended load, it easily becomes the best means of storing and moving large amounts of energy in a variety of applications.

1.1.0 Basic Properties of Water

Depending on its heat content and temperature, water can exist in three states: ice (solid), water (liquid), and water vapor (gas, such as steam). When water changes from one state to another, some in-

Did You Know?

Steam Energy

An astounding 80 percent of the energy needed to produce pulp and paper comes from steam, while roughly 35 percent of the total energy required for all product output from industry in the United States comes from this vapor phase of water.

teresting things occur. *Figure 1* shows this graphically for water at sea level. If heat is added to one pound of ice at 0°F, a thermometer will show a rise in temperature until the reading reaches 32°F. This is the point at which the ice starts changing into water. If heat is continually added, the thermometer reading remains fixed at 32°F instead of rising. It continues to read 32°F until the entire pound of ice is melted. The increase in temperature from 0°F to 32°F registered by the thermometer is sensible heat. Sensible heat is heat that can be sensed by a thermometer or by touch. The heat that was added to the ice and caused its change in state from a solid to a liquid, but did not register on the thermometer, is latent heat. Latent heat is the heat energy absorbed or rejected when a substance is changing state (solid to liquid, liquid to gas, or vice versa) without a change in the measured temperature.

If you continue to add heat to the pound of water, the thermometer will once again show an increase in temperature until the temperature reaches 212°F. At this point, the water starts boiling and changes state from water into steam (water vapor). As even more heat is added, the thermometer reading remains at 212°F until all the water has turned into steam. If you continue to add heat, the thermometer will once again register sensible heat.

As shown in *Figure 1*, it takes a great deal more heat to cause a change of state in water than is needed for a degree change in its temperature. It requires 144 Btus of latent heat to melt the pound of ice before the temperature began to rise. This is 144 times as much heat as is needed to raise the temperature of water one degree (specific heat). The change from water to steam requires an even greater amount of latent heat. It takes 970 Btus to change the water to steam. None of the latent heat consumed during the change of state of water registers on the thermometer. Once water reaches the threshold of changing states and begins the change process, the thermometer becomes stable and unchanged until the phase change is complete. This property of latent heat helps to explain why large amounts of energy can be stored and transported through the use of steam.

Figure 2 shows the common terms used in steam systems when defining the latent heat added or removed in changing water to and from the solid, liquid, or vapor states. It also shows the terms *subcooling* and *superheat* as they relate to sensible heat changes in water. Remember that the processes of subcooling and superheating are sensible processes. As a substance such as water is changing state, a latent process is at work and there is no change in temperature during this period.

GRAPH SHOWS
ONE POUND OF
ICE AT SEA LEVEL

SENSIBLE HEAT:
HEAT THAT CAUSES A TEMPERATURE CHANGE AND CAN BE SENSED BY A THERMOMETER OR BY TOUCH.

LATENT HEAT:
HEAT THAT PRODUCES A CHANGE IN STATE WITHOUT A CHANGE IN TEMPERATURE.

03306-13_F01.EPS

Figure 1 Changing states of water.

- *Latent heat of fusion* – The heat gained or lost in changing ice to water or water to ice.
- *Latent heat of vaporization* – The heat gained in changing water to steam at a given pressure. The temperature at which this occurs is known as the boiling point or the saturation temperature. As water absorbs latent heat and the change of state begins, steam can be seen rising from the water surface. This steam is commonly referred to as saturated steam. After all the water has been vaporized to steam at the saturation temperature and begins to become superheated, the steam is referred to as dry steam.
- *Latent heat of condensation* – The heat given up or removed from steam in changing back to water.
- *Superheat* – The measurable heat added to the steam after all the water has reached its boiling point and completely changed into steam. For example, the saturation temperature of water at atmospheric pressure is 212°F. If the resulting steam is superheated 10°F, the steam temperature is 222°F. Steam that has been heated above the saturation temperature is called superheated steam.
- *Subcooling* – The measurable heat that is removed after all the water has condensed from the vapor state to the liquid state. It is the reverse of superheat, and the subcooling process cannot begin until the change of state to liquid is complete. For example, the condensing temperature of water at atmospheric pressure is 212°F. If the water is subcooled 10°F, the temperature is cooled to 202°F.

Figures 1 and *2* show that most of the heat content of steam is stored as latent heat. It is this latent heat (970 Btu/lb at sea level) that makes steam so efficient as a carrier of heat. As it condenses, it will release this stored heat. Large quantities can be transmitted efficiently with little change in temperature. Remember that steam is pressure- and temperature-dependent. This means that the system can be controlled by varying either the steam pressure or temperature.

NCCER – *HVAC Level Three* 03306-13

1.2.0 Pressure-Temperature Relationship of Water

Steam tables illustrate the pressure-temperature relationships and other properties of saturated steam over a broad range of temperatures. *Figure 3* shows entries from such a table. One entry is for the properties of steam at sea level as just described. The other entries are given for the purpose of providing comparisons.

Columns A through C show the pressure-temperature relationships. Columns A and B show the corresponding gauge and absolute pressures. Absolute pressure is the pressure in pounds per square inch (psia) above a perfect vacuum. Gauge pressure (psig), which starts at zero psig, is pressure in pounds per square inch above atmospheric pressure. Gauge pressure plus 14.7 equals the related absolute pressure.

For each pressure, there is a corresponding temperature. Column C shows that the saturation temperature of steam at sea level (0 psig/14.7 psia) is 212°F. As steam cools down to these conditions, the latent process of condensation begins. If water is heated up to these conditions, the latent process of vaporization is initiated.

Column D shows the heat of saturated liquid. This is the amount of sensible heat required to raise the temperature of a pound of water from 32°F to the boiling point at the pressure and temperature shown. It is expressed in Btu/lb. In some tables, this value is called the enthalpy of the liquid (hf).

Column E shows the latent heat. This is the amount of heat required to change a pound of boiling water to steam, expressed in Btu/lb. This same amount of heat is released when a pound of

Metric Values

The changing states of water, as shown in *Figure 1*, are presented here using the equivalent metric values.

SENSIBLE HEAT:
HEAT THAT CAUSES A TEMPERATURE CHANGE AND CAN BE SENSED BY A THERMOMETER OR BY TOUCH.

LATENT HEAT:
HEAT THAT PRODUCES A CHANGE IN STATE WITHOUT A CHANGE IN TEMPERATURE.

03306-13_SA01.EPS

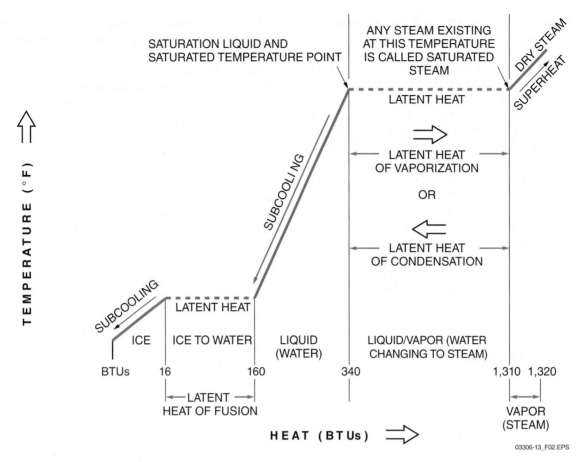

Figure 2 Change of state terminology.

steam is condensed back into a pound of water. This heat quantity is different for every pressure-temperature combination. In some tables, this value is called the enthalpy of the evaporation (hfg).

Column F shows the total heat of the steam. This is the sum of the sensible heat and the latent heat given in Columns D and E, expressed in Btu/lb. It is the total heat in the steam above 32°F.

In some tables, this value is called the enthalpy of the steam (hg).

Column G shows the specific volume of steam. This is the volume per unit of mass for the steam. It is the space that one pound of steam will occupy at the pressure-temperature shown. It is expressed in cubic feet per pound. Note that, as pressure and the total heat contained within the steam increases on the chart, the volume of the steam decreases.

You Know This...

The use of the terms *sensible* and *latent heat*, and the terms *superheat* and *subcooling*, have exactly the same meaning in a discussion of steam as they do in a discussion of refrigerants.

Properties of Saturated Steam Tables

Tables showing the various properties of saturated steam are often found in manufacturers' literature, as well as reference texts for steam systems. In addition to their use for determining pressure-temperature relationships, the various data about the other properties of steam given in such tables are commonly used in calculations made when designing and sizing components for a steam system.

Column A Gauge Pressure	Column B Absolute Pressure PSIA	Column C Steam Temp. °F	Column D Heat of Sat. Liquid btu/lb	Column E Latent Heat btu/lb	Column F Total Heat of Steam btu/lb	Column G Specific Volume cu ft/lb
Inches of Vacuum						
29.743	0.08854	32.00	0.00	1075.8	1075.8	3306.00
29.515	0.2	53.14	21.21	1063.8	1085.0	1526.00
27.886	1.0	101.74	69.70	1036.3	1106.0	333.60
19.742	5.0	162.24	130.13	1001.0	1131.1	73.52
9.562	10.0	193.21	161.17	982.1	1143.3	38.42
7.536	11.0	197.75	165.73	979.3	1145.0	35.14
5.490	12.0	201.96	169.96	976.6	1146.6	32.40
3.454	13.0	205.88	173.91	974.2	1148.1	30.06
1.418	14.0	209.56	177.61	971.9	1149.5	28.04
PSIG						
0.0	14.696	212.00	180.07	970.0	1150.4	26.80
1.3	16.0	216.32	184.42	967.6	1152.0	24.75
2.3	17.0	219.44	187.56	965.5	1153.1	23.39
5.3	20.0	227.96	196.16	960.1	1156.3	20.09
10.3	25.0	240.07	208.42	952.1	1160.6	16.30
15.3	30.0	250.33	218.82	945.3	1164.1	13.75
20.3	35.0	259.28	227.91	939.7	1167.1	11.90
25.3	40.0	267.25	236.03	933.7	1169.7	10.50
30.3	45.0	274.44	243.36	928.6	1172.0	9.40
40.3	55.0	287.07	256.30	919.6	1175.9	7.79
50.3		297.97	267.50	911.6	1179.1	6.66
60.3			277.43	904.5	1181.9	5.82
			286.39	897.8	1184.	

03306-13_F03.EPS

Figure 3 Example table entries for properties of saturated steam.

Additional Resources

Low Pressure Boilers, 3rd Edition. Fredrick M. Steingress, Daryl R. Walker. Orland Park, Illinois: American Technical Publishers.

High Pressure Boilers, 4th Edition. Fredrick M. Steingress, Harold J. Frost, Daryl R. Walker. Orland Park, Illinois: American Technical Publishers.

1.0.0 Section Review

1. Steam at a temperature of 222°F is referred to as _____.

    a. saturated
    b. superheated
    c. subcooled
    d. condensed

2. Steam tables provide a comparison of pressure-temperature relationships based on steam at sea level that is _____.

    a. saturated
    b. superheated
    c. subcooled
    d. vaporized

2.0.0 BASIC STEAM SYSTEM CYCLE AND OPERATIONAL COMPONENTS

Objectives

Describe the basic steam cycle and the primary components related to its operation.
  a. Describe the steam cycle principles of operation.
  b. Identify steam boilers and their common accessories.
  c. Identify common steam heat exchangers and terminal devices.

Performance Task 1

Under the supervision of the instructor, trainees must be capable of performing the following task:

Perform selected operating procedures on low-pressure steam boilers and systems.

Trade Terms

Trim: External controls and accessories attached to the boiler itself, such as sight glasses and water feeder controls.

The cycle for steam systems is essentially the same for all systems. But the type of equipment used with the steam system depends on the heat content and steam volume required. The heat content and volume of steam required usually dictates the steam pressure and temperature along with the type of boilers, their primary system equipment, and various types of ancillary equipment needed. Some of the types of boilers and their primary equipment along with some of ancillary equipment used with them are described in the following sections.

2.1.0 Steam Cycle Principles of Operation

Figure 4 shows the basic components of a simple steam cycle. They include the following:

- Boiler
- Control valve
- Heat exchanger
- Trap
- Deaerator and condensate receiver

- Condensate pump
- Vacuum breaker

The steam cycle begins at the boiler. In the boiler, water is vaporized to make steam. The initial source of energy used for this purpose is typically from the combustion of a fuel such as oil, gas, or coal. Once the water has reached the boiling point and vaporization begins, the pressure begins to rise. Remember that the water temperature required for boiling also increases due to this change in pressure. As shown in *Figure 3*, once the pressure reaches 1.3 psig, the boiling point increases to over 216°F and continues to increase as pressure rises. Steam is found at the highest temperature and pressure in the boiler itself. Steam flows through the supply piping and components as the result of natural movement from an area of higher pressure to an area of lower pressure, much like air through an air distribution system.

The control valve is an automatic, temperature-operated control device. Its purpose is to regulate or meter the flow of steam into the heat exchanger at a rate comparable to the load. This ensures that only the proper amount of steam needed to take full advantage of its stored latent heat is admitted. The steam system is designed so the control valve has the needed residual pressure to assure proper operation of the valve. Pressure drop across steam valves must be used to calculate the entering steam pressure to the final user of the steam.

The heat exchanger may be a steam coil in a system airstream, a steam-to-water heat exchanger, or any other type of terminal used with steam systems. Its purpose is to transfer the heat carried in the steam to the conditioned space or other

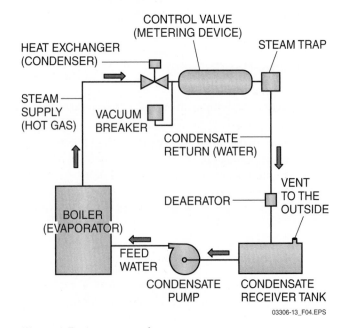

03306-13_F04.EPS

Figure 4 Basic steam cycle.

medium. The heat exchanger is like the condenser in a refrigeration system, where the steam (gas) is converted back into water (liquid). There, the latent heat and any sensible superheat gained in the boiler by vaporization is rejected to the cooler conditioned space, causing the steam to cool and condense back into water. The condensing process in the heat exchanger occurs at a lower pressure than in the boiler. The amount of pressure depends on the temperature of the medium being heated, the pressure losses in the supply piping, and the pressure drop across the control valve. It is this pressure difference that causes the steam to flow without a pump.

Steam traps are among the most important devices used in the steam cycle. The trap in the piping at the output of the heat exchanger holds the steam in the heat exchanger until it all condenses. Otherwise, the live steam would pass through the heat exchanger into the return line, wasting the useful heat released as a result of condensation. The trap will allow only liquid condensate and/or air, never steam, to flow away from the heat exchanger. There are a variety of styles and types suited to different applications. Traps also require the most attention and maintenance to ensure proper operation, because failure of a steam trap can create a number of problems.

Where elevations permit, steam systems can use a gravity return to route the flow of condensate water directly back to the boiler for re-evaporation into steam. However, most systems use gravity flow to route the condensate into a receiver tank, where it collects and is stored. When the boiler controls call for the water to replenish the amount boiled off, a condensate pump is turned on and produces the necessary pressure differential to pump the condensate (feed water) from the tank to the boiler. This method allows gravity to be used to the extent possible, with the condensate pump doing only the remaining work. Return of the condensate to the boiler for re-evaporation completes the steam cycle.

Steam systems that use a condensate receiver tank and pump in the condensate return line operate at atmospheric pressure, with the return opened to the atmosphere in one or more places. This has an advantage. When steam is applied to heat exchangers that have coils subjected to out-side airflow, or cold water such as in domestic hot-water heaters, the pressure in the heating coil can suddenly drop below atmospheric pressure (vacuum range). In extreme cases, this sudden drop in pressure can cause the tubes in the coil to collapse. This can be prevented by allowing air to enter the heat exchanger by way of a vacuum breaker device or through a thermostatic trap that senses the drop in pressure. A small amount of the air that enters the heat exchanger will be absorbed into the condensate, but most of it flows in the return line to the condensate receiver tank. From there it is discharged by vents back into the atmosphere.

2.2.0 Steam Boilers, Controls, and Accessories

Unlike hot-water boilers, steam boilers are not completely filled with water. A sufficient amount of space must be left at the top for steam to form. Steam boilers are generally categorized by their working pressure.

Many of the boiler controls and accessories used with steam systems are the same as those used with hot-water boilers and described earlier. Typically, the main difference is that these controls and accessories have different temperature and pressure ratings. For this reason, this section focuses on the controls and accessories that are unique to steam boilers and systems.

In a well-designed steam system, the boiler is equipped with several operating and safety controls needed to guarantee safe and proper operation of the boiler (*Figure 5*). Operating and safety controls normally used with all steam boilers include the following:

- Pressure gauge
- Water gauge glass/water column
- Low-water cutoff/water feeder
- High-pressure limit control
- Pressure-relief (safety) valve

2.2.1 Steam Boilers

Steam boilers are divided into three broad categories: those that produce low-pressure steam; those that produce medium-pressure steam; and

Piping Friction

The pressure of the steam leaving the boiler gradually falls as it flows through the supply piping because of the piping friction. The pressure differential at the boiler supply and return outlets allows steam to flow, just like a hydronic system. With steam however, no pump is required.

STEAM MAIN

STOP VALVE

SAFETY VALVE

OPERATING
AND
SAFETY
CONTROLS

STEAM
BOILER

CHECK VALVE

BLOWOFF VALVE

FROM
RECEIVER
TANK

HIGH-
PRESSURE
LIMIT

STEAM VALVE

PRESSURE
GAUGE

GAUGE GLASS

SNUBBER

SIPHON

OPERATING WATER LEVEL

WATER VALVE

LOW WATER
CUTOFF

DRAIN VALVE
(BLOWDOWN VALVE)

03306-13_F05.EPS

Figure 5 Typical cast-iron steam boiler controls and accessories.

those that produce high-pressure steam. Low-pressure steam boilers have working pressures of up to 15 psi (250°F). Medium-pressure boilers operate from 15 psi to 60 psi. High-pressure boilers operate at pressures from 60 psi and up. It is important to note that these ratings indicate the maximum working pressure of the boiler. For safety purposes, the actual operating pressure of the system would be somewhat lower than the maximum. Low-pressure steam boilers (*Figure 6*) are normally used in commercial applications, apartment houses, and single- or multiple-unit industrial facilities. High-pressure industrial boilers are used mainly to produce process steam for facilities such as college campuses, hospitals, or industrial plants.

COMMERCIAL GAS-FIRED
VERTICAL TUBELESS BOILER

INDUSTRIAL OIL-FIRED HORIZONTAL
FIRETUBE BOILER

03306-13_F06.EPS

Figure 6 Commercial and industrial steam boilers.

With the exceptions that follow, steam boilers operate in the same manner as hot-water boilers:

- Steam boilers are partially filled with water to allow room for steam creation and expansion.
- The pressure-temperature levels at which steam and hot-water boilers operate are different.
- The operating and safety controls are different.

Although larger boilers and higher-pressure boilers are generally designed specifically for steam applications, many low-pressure boilers are identical in construction to those used for hot water applications. For added safety, a number of such boilers are rated at a maximum working pressure of 30 psig for hot water, but have a reduced maximum rating of 15 psig for steam applications. Low-pressure steam boilers often differ from their hot water components only in the **trim**; that is, the external operating and safety controls or accessories attached to the boiler itself.

Overall, steam boiler designs vary more widely than hot-water designs. Typical designs include the following:

- Cast-iron, often built in matching sections for field assembly
- Steel firetube and watertube boilers
- Steel vertical tubeless boilers
- Electric boilers

Boilers used in residential (*Figure 7*) and some light commercial low-pressure applications are generally of cast-iron sectional or one-piece designs. Although many older units were quite inefficient, federal minimum efficiency standards, which took effect in 1992, require boilers to have a minimum Annual Fuel Utilization Efficiency (AFUE) rating of 80 percent. Boilers earning the EPA's ENERGY STAR designation must have a minimum AFUE rating of 85 percent. These boilers generally incorporate a number of energy-saving options, such as electronic ignition and sealed combustion systems using outside air for the combustion process.

2.2.2 Pressure Gauges

The pressure (steam) gauge shows the pressure within the boiler in pounds per square inch gauge (psig). This gauge is installed at the highest point of the boiler's steam space. A siphon assembly is generally installed between the boiler and the gauge. Two different types of siphons are used: pigtail siphons and gauge siphons. Both provide separation for the gauge or other control from the raw steam, which may cause damage to the gauge, as well as erroneous readings. *Figure 8* provides an example of a pigtail siphon. Pigtail

Using the Condensate Receiver and Pump

Another advantage of using the condensate receiver and pump is that they permit condensate return lines to be located below the water level in the boiler. This makes it possible to locate heat exchanger terminals at levels lower than the boiler. In large systems, more than one may be required.

GAS-FIRED STEAM BOILER

OIL-FIRED STEAM BOILER

03306-13_F07.EPS

Figure 7 Efficient residential cast-iron steam boilers.

siphons are always installed vertically and are filled with water before installing the gauge.

2.2.3 Water Level Indicators

The water gauge glass is used to check the water or steam level in the boiler (*Figure 9*). The steam level can be checked by opening the steam valve at the top of the water gauge. To check the water level, the water valve at the bottom of the water gauge is opened. Normally, the water gauge glass is connected directly to the boiler. A drain valve (blowdown valve) is usually connected to the bot-

03306-13_F08.EPS

Figure 8 Pigtail siphon.

tom of the water gauge. When opened, the valve is used to purge the gauge passages of trapped water and any buildup of sludge and sediment.

> **WARNING!**
>
> Blowdown or drain valves opened while a steam boiler is in operation can release water at a temperature above the atmospheric boiling point, causing flash steam. Use extreme caution when using the blowdown valve. The blowdown valve discharge must always be piped to an appropriate drain, and in a manner that will prevent hot water or flash steam from contacting personnel.

> **NOTE**
>
> The terms *bleed-off* and *blowdown* are often used interchangeably, but they apply to different systems. Bleed-off is generally used to define the process of bleeding water from a cooling tower or other hydronic loop to reduce the concentration of dissolved solids in the remaining water. Blowdown is associated with the process of removing dissolved solids and/or sludge from the surface or bottom of a steam boiler. Use of the proper term will ensure clear communication.

On larger boilers, the water gauge may be attached to a water column (*Figure 10*). It diverts water or steam from the boiler for application to the water gauge. This helps to stabilize water levels for more accurate water gauge readings.

Three globe valves on the water column provide a backup method for checking the boiler steam and water levels if the water gauge is damaged. These globe valves are often called gauge cocks or try cocks. Opening the top, middle, or bottom valve allows either steam, a mixture of steam and water, or water to escape. A blowdown valve connected to the bottom of the water col-

Figure 9 Water gauge glass.

Figure 10 Water column.

Figure 11 Water levels.

umn is used to clear the column of trapped water and any buildup of sludge and sediment.

In a steaming boiler, the gauge glass will generally indicate a lower water level than is accurate, as shown in *Figure 11*. This is because the water in the gauge glass is not subject to currents or agitation, it is cooler than the water in the boiler, and it does not contain steam bubbles.

2.2.4 Low- and High-Water Cutoff and Water Feeder Controls

Low-water cutoff controls provide valuable protection for steam boilers by interrupting power to the heat source when the water falls below the minimum safe level. Water feeders are used to control the flow of makeup water to the boiler. In systems that effectively collect and return steam condensate to the boiler for reuse, the water feeder uses the condensate reservoir and associated pumps as the primary water source. Fresh water, generally from a municipal water supply, becomes the backup source when condensate levels are insufficient. These boilers are generally classified as steam heating boilers. However, steam boilers are also used in many applications where little or no condensate is collected for reuse. In these cases, the water feeder may be quite active, using only the fresh water supply to make up for the lost steam. Boilers used in this manner are classified as steam-process boilers.

Typically, low-water cutoff control for steam boilers is done using the same methods as those for hot-water boilers, employing float-operated controls (*Figure 12*). However, steam boiler operation is adversely affected by a high water level as well. While a water boiler must be completely full of water, a steam boiler cannot operate in a flooded condition. As a result, both low- and high-water cutoff controls are needed for a steam boiler. Some steam boilers may be equipped with electronic probe-type controls (*Figure 13*) to provide these functions, much like hot-water boilers. A high-water cutoff is not mandatory in most cases. A low-water cutoff is required for safety purposes. A boiler firing with little or no water inside is a serious hazard.

Both the float-operated and electronic controls can be used to provide low-water and high-water cutoff duty. The control circuit switches have both normally open and normally closed contacts. The switch is then wired depending on the desired action. For example, a low-water float control would be wired normally open, and the contacts close when the water level is high enough for operation. Conversely, for high-water cutoff, the switch would be wired normally closed, and the contacts will open if the water level rises too high.

A marker on the float bowl of the float-operated control used with steam boilers shows the cutoff level for the device. The control is normally installed so that the marker on the float bowl is about ½ inch higher than the lowest visible point in the water gauge glass. For some float-operated controls, the operating level is established automatically by fittings on the device that permit its installation right in the gauge glass tappings of the boiler. If a low-water condition occurs, float-actuated switch contacts in the cutoff control open the electrical circuit to the boiler burner, preventing the burner from firing. Note that water will still be visible in the gauge glass. However, the boiler is shut down and cannot be operated until makeup water is manually added to the boiler. If this situation persists, it can cause a loss of heating until the water in the boiler is returned to a safe level. This problem is usually eliminated by installing a water feeder control between the boiler's supply water line input and its condensate return main line. This control functions to automatically maintain a constant water level in the boiler.

In some boiler installations, the water feeder control is a separate unit (*Figure 14*). There are many kinds of feeder controls. Typically, they use a solenoid-operated valve controlled by the related low-water cutoff control. The low-water cutoffs have a second switch that is actuated by the cutoff float. This switch is used to control the related solenoid-operated water valve (electric water feeder). This second switch operates at a higher water level than the cutoff level switch. If the boiler water drops to this first operating level, the float closes the switch contacts, completing the electrical circuit to the water feeder. This energizes the water feeder solenoid-operated valve, allowing makeup water to flow through the valve and into the boiler. When the boiler water level is restored, the float level actuates the switch to open the electrical circuit to the water feeder. This shuts off the supply of makeup water to the boiler.

In other boiler installations, the water feeder control is combined with the low-water cutoff control into one assembly. These controls are called feeder-cutoff combination controls. Usually, the water feeder part of the control is mechanically actuated by the float. It is inactive as long as the condensate water returning to the boiler keeps the water in the boiler at the normal operating level. If for some reason the boiler water drops below the normal operating level, the first thing the feeder cutoff combination control does is to add the amount of water required to keep the boiler in operation. It mechanically opens its feed valve, adds the small amount of makeup water necessary, and when the float reaches the normal water

Figure 12 Low-water cutoff.

Water Columns and ASME

Water gauges must be mounted on a boiler at a location that meets ASME code. ASME tests and approves gauges for mounting in specific locations. If the system fails and the gauge is not installed at the correct location, the installer may be liable for any damages that occur.

MINIMUM SAFE
WATER LEVEL

03306-13_F13.EPS

Figure 13 Electronic low-water cutoff.

level, it closes again. A marker on the float bowl casting shows the feeder closing level. Feeders are normally installed so that the marker on the float bowl is about 2 to 2½ inches below the normal water line.

If the feeder part of a combined feeder cutoff control cannot keep the water in the boiler at a safe level, the low-water fuel cutoff portion of the control will activate. When the water in the boiler drops below the minimum safe level, the float drops and opens the switch contacts, resulting in the fuel supply to the burners being turned off.

The selection of low-water cutoff and water feeder devices generally depends on design system pressure and desired mounting style. It is recommended that steam boilers over 400,000 Btuh input and/or operating above 15 psig be fit-

ted with two low water cutoff devices for redundancy. The first cutoff can be a type that resets itself automatically, while the second must be a manual-reset model. Water feeder and low water cutoff controls provide valuable protection and are essential for proper system operation. They should be disassembled, cleaned, calibrated, and reinstalled annually, at a minimum.

2.2.5 High-Pressure Limit and Other Controls

The high-pressure limit control (*Figure 15*), also known as a pressuretrol, is activated by a rise in boiler pressure. The high-pressure setpoint should always be lower than the setting of the pressure-relief valve, and a pressuretrol style that requires a manual reset should always be used for high-limit duty. Although a few models use snap switches, most use a mercury bulb much

03306-13_F14.EPS

Figure 14 Electric water feeder.

03306-13_F15.EPS

Figure 15 High-pressure limit control.

like older wall thermostats. Control circuit power is routed through the mercury bulb. Should the boiler pressure rise to the point where it exceeds the pressuretrol setpoint, the switch contacts open and break the burner control circuit. This shuts down the boiler.

Pressuretrols are also commonly used to provide on-off and sequential control of boilers. They can be adjusted in the field so that the boilers all start at high-fire (unison), or in sequence where the boilers start at some firing rate other than high-fire. The control is activated when the steam pressure drops below a preset level. This causes automatic feeding of an air/fuel mixture to the furnace, ignition of the burners, and opening of the main air damper(s). As the boiler steam pressure rises, it crosses fixed control points. At each control point, a predetermined mixture of fuel and airflow is selected and applied to the burners. When the boiler pressure reaches the operational pressure limit, the control shuts off the boiler.

2.2.6 Pressure-Relief Valve

A pressure-relief valve is used to protect the boiler and the system from steam pressures exceeding the pressure rating of the boiler or the system (*Figure 16*). It does not operate unless an overpressure condition exists. The low-pressure steam boiler is built for a maximum working pressure of 15 psi. The pressure-relief valve is set to open at the same pressure and discharge the steam into the atmosphere. More than one pressure-relief valve must be used on boilers having over 500 square feet of heating surface. This is a precaution to guard against the failure of a single valve. Safety relief valves should always be installed as directed by the manufacturer. They are usually installed in an upright position at the top of the boiler and at the highest point on the steam side of the boiler.

> **WARNING!**
>
> Unless stated otherwise, the pressure setting of a safety relief valve is not field-adjustable and must not be altered. Misadjustment may cause an explosion, which can result in death or personal injury. The relief valve must also be full-sized, without reductions on its outlet or inlet. A pressure-relief valve must be immediately replaced if it does not open when its pressure rating is exceeded, or if it leaks when the boiler is operating at normal pressures. Operation of the boiler with the relief valve leaking can cause a dangerous buildup of scale in the relief valve. This can cause the relief valve not to function should an over-pressurization condition occur.

2½ OR 3 INCH OUTLET PORT

SEAL

2 INCH INPUT PORT

03306-13_F16.EPS

Figure 16 Pressure-relief valve.

Pressure-relief valves should be checked per the manufacturer's instructions intermittently during operation and after long periods of inactivity.

2.3.0 Steam System Loads

Steam system loads can be manufacturing processes, steam-to-water heat exchangers used for heating potable water, or for supplying hot water used as a medium for space heating. They can also be loads comprised of various types of heating terminals that use steam directly as the medium for space heating.

2.3.1 Heat Exchangers/ Converters

A heat exchanger is used to transfer heat from one medium to another. A heat exchanger is used in steam systems mainly to generate hot water for hot-water heating systems, as a steam-to-water heat exchanger. *Figure 17* shows a horizontal shell-and-tube heat exchanger (converter) used in this manner. It consists of tubes encased in a shell. The tubes provide a closed water circuit that allows redirection of the water through the converter more than once. The hot water is circulated through the exchanger tubes, while the steam circulates in the shell that surrounds the tubes. There, the heat is transferred from the steam to the water. The result is that hot water at the output of the heat exchanger is available to heat all or part of a building.

Figure 18 shows a typical piping diagram for a heat exchanger used as a hot-water converter in a hot-water heating system. The modulating steam control valve shown is generally thermostatic,

Figure 17 Horizontal shell and tube heat exchanger.

03306-13_F17.EPS

regulating the flow of steam based on the leaving water temperature.

A steam-to-water heat exchanger can also be used to heat cold water to produce hot water at the output of the heat exchanger for domestic use. When used in this way, the heat exchanger is sometimes called an instantaneous heater.

For added safety against contamination, most heat exchangers used to heat potable water are of double-walled construction; that is, two walls separate the steam from the potable water. This is specifically required if conditioning chemicals have been added to the steam system.

2.3.2 Terminals

Many kinds of heating terminals can be used in steam-heating systems. Natural convection units include radiators, convectors, and baseboard convectors. Forced convection units include unit heaters (*Figure 19*), unit ventilators, cabinet heaters, induction units, fan coil units, and central air handling units.

With the exceptions covered here, the terminals used in steam-heating systems are built and operate in the same way as the terminals used in hot-water heating and dual-temperature systems. Fan coils, induction units, and unit ventilators can be used for both heating and cooling. Dual-coil units have a steam coil used for heating and a separate coil for cooling.

There are two types of steam heating coils, single-row steam coils (*Figure 20*) and steam distribution coils. It can be difficult to distinguish the two types of coils from the outside. Single-row coils are used for general heating applications where the following conditions exist:

- The coil will not be subjected to freezing temperatures
- The flow of steam will likely not be modulated
- Precise control of the air temperature across the entire face of the coil is not necessary

Steam distribution coils (*Figure 21*) are generally specified where one or more of these issues must be considered. These coils are built with perforated inner tubes that allow the steam to be distributed evenly throughout the coil. They are generally considered freeze-resistant. Many HVAC applications require this type of coil. Systems with outside air intakes that use steam to preheat the air are often exposed to temperatures below freezing. Most HVAC applications also require the flow of steam to be modulated,

Controlling Mercury Hazards

Mercury is a major environmental pollutant that is known to cause cancer. Every effort is being made to remove mercury from the environment and to regulate its disposal. The pressure switches in older boiler systems contain a large amount of mercury, and are therefore a target of this effort. As mercury switches reach the end of their useful lives, or as boilers are overhauled, the mercury switches are being replaced with other devices such as snap-action bimetal and electric switches. Disposal of switches and other devices containing mercury is carefully regulated. Most states have enacted laws governing the storage and disposal of mercury, so it is important to know the applicable regulations in your state.

STEAM MAIN

BY-PASS

MODULATING STEAM CONTROL VALVE

STRAINER

F & T TRAP

AIR VENT

DRAIN

HOT WATER OUT

HEATER EXCHANGER (CONVERTER)

COLD WATER IN

VACUUM BREAKER

AIR INPUT

F & T = FLOAT AND THERMOSTATIC
* GATE OR BALL VALVE
** GLOBE VALVE

PRIMARY TRAP

TO LOW-PRESSURE RETURN

03306-13_F18.EPS

Figure 18 Typical installation of a heat exchanger when used as a converter.

03306-13_F19.EPS

Figure 19 Unit heater.

03306-13_F20.EPS

Figure 20 Single-row steam coil.

making the steam distribution coil the best over-all choice. Single-row coils are used for general heating where no attempt is made to obtain equal air temperatures over the entire length of the coil. They are also used in airstreams where the temperature does not drop below 32°F. Steam distribution coils are made with perforated inner tubes that are used to distribute the steam along the full length of the inner surface of the tubes. They are

03306-13_F21.EPS

Figure 21 Steam distribution coil.

typically used where the steam must be throttled to obtain control. They are also used where even temperatures are required along the entire length of the tube, or where freezing temperatures are frequently encountered in the air moving over the coil.

Single-Coil Terminals

When a single-coil terminal is used, a steam-to-water heat exchanger must be used in the steam system to transfer the heat from the steam to a secondary hot-water source for use in the terminal.

Drainable Steam Coils

Draining the condensate from a steam coil helps prevent it from freezing. Some manufacturers offer drainable non-freeze coils for this reason. However, if the coil is to be out of service for an extended period, it should be blown out with low-pressure air. If it will be exposed to freezing temperatures, antifreeze may be added to the coil.

Additional Resources

Low Pressure Boilers, 3rd Edition. Fredrick M. Steingress, Daryl R. Walker. Orland Park, Illinois: American Technical Publishers.

High Pressure Boilers, 4th Edition. Fredrick M. Steingress, Harold J. Frost, Daryl R. Walker. Orland Park, Illinois: American Technical Publishers.

The Control of Boilers, 2nd Edition. Sam G. Dukelow. Research Triangle Park, North Carolina: International Society of Automation (ISA).

2.0.0 Section Review

1. The steam cycle begins at the _____.
   a. load
   b. control valve
   c. boiler
   d. condensate tank

2. Boilers with the EPA's ENERGY STAR logo must have a minimum AFUE rating of _____.
   a. 80 percent
   b. 85 percent
   c. 90 percent
   d. 95 percent

3. When used as a domestic hot water heater, a steam-to-water heat exchanger is sometimes called a(n) _____.
   a. flash steam heater
   b. instantaneous heater
   c. momentary demand heater
   d. water desuperheater

SECTION THREE

3.0.0 STEAM SYSTEM VALVES AND PIPING

Objectives

Identify common steam system piping arrangements and valves.

  a. Identify common steam system piping arrangements.

  b. Identify common condensate return systems and describe waterside care.

  c. Explain how steam and condensate system piping is sized.

  d. Identify steam-based pressure-reducing and thermostatic control valves.

Performance Task 1

Under the supervision of the instructor, trainees must be able to perform the following tasks:

Perform selected operating procedures on low-pressure steam boilers and systems.

Identify common piping configurations used with steam systems.

Trade Terms

Drip leg: A drain for condensate in a steam line placed at a low point or change of direction in the line and used with a steam trap.

Flash steam: Formed when hot condensate is released to a lower pressure and re-evaporated.

Turndown ratio: In steam systems, the ratio of downstream pressure to upstream pressure, usually applied to steam pressure-reducing valves. More than a 50 percent turndown ratio is generally considered a large pressure reduction.

Water hammer: A condition that occurs when hot steam comes into contact with cooled condensate, builds pressure, and pushes the water through the line at high speeds, slamming into valves and other devices. Water hammer also occurs in domestic water systems. If the cause is not corrected, water hammer can damage the system.

Various steam and auxiliary steam systems are used based on the size of the boiler system and the design steam temperature. A number of steam piping systems and valves, as well as auxiliary steam system piping that include condensate/feedwater piping and components, are covered in the following sections.

3.1.0 Steam System Piping

Steam piping systems are classified as either one-pipe or two-pipe systems. The type of system is determined by the piping method used to supply steam to and return condensate from the system terminals. Each of these piping systems is further classified by the method used to return the condensate (gravity or mechanical flow), and the directions of steam and condensate flow in the system.

3.1.1 One-Pipe Systems

In a one-pipe system, steam flows to and condensate is returned from the terminal unit through the same pipe. There is only one connection to each heating terminal unit, which functions both as the supply and the return. Normally, one-pipe systems use a gravity return, but they may use a condensate pump. This is done when there is insufficient height above the boiler water level to develop enough pressure to return the condensate directly to the boiler. One-pipe gravity systems do not use steam traps. Air vents are used at each terminal unit and at the ends of all supply mains to vent air so the system can fill with steam. If a one-pipe system uses a condensate pump, air vents are used at each terminal unit, but steam traps are used at the end of each supply main. One-pipe systems are used mostly in residential and small commercial systems. They are best applied in homes or small buildings with basements, where space and height exist for the needed drainage pitch. Though limited in potential applications, their advantage rests in their simplicity. Very few moving or functional components are required, increasing reliability and minimizing service on properly installed systems. However, they are more susceptible to water hammer because the condensate and steam frequently make contact and the condensate is not quickly removed from the system. This is especially true during system startup, when piping and components are relatively cold.

There are several configurations of one-pipe steam systems. *Figure 22* shows a basic one-pipe up-feed gravity system. The term *upfeed* means that the steam supply main is below the level of the heating terminals that they serve. Steam flow is then fed upward in the riser for input to the terminal units.

As shown in *Figure 22*, the steam supply main rises from the boiler to a high point, then

Figure 22 Basic one-pipe upfeed system.

is pitched downward from this point. The pitch should be no less than one inch down per 20 feet and should slope in the direction of the gravity flow of the condensate. The piping normally runs full size to the last terminal unit takeoff and is then reduced in size after it drops down below the boiler water line. When the return is above the boiler water line, it is called a dry return. In this case, it is only considered to be dry as it is not flooded or filled with water—it carries condensate, along with some steam and air. The section of return piping installed below the level of the water in the boiler is called the wet return. It is completely filled with water and does not carry steam or air. When the system is in operation, the steam and condensate flow in the same direction in the horizontal mains and in opposite directions in the branches and risers. Because the steam and condensate flow in the same direction in the horizontal mains, this type of system is also called a parallel-flow system.

A Hartford loop is a pressure-balancing loop that introduces full boiler pressure on the return side of the boiler. This pressure prevents reverse circulation or water leaving the boiler by the return line. The Hartford loop is actually two loops; the first is an equalizer line and the second is an extension from the steam main through the return and back to a connection at the equalizer loop.

3.1.2 Two-Pipe Systems

The two-pipe system uses separate pipes to supply the steam and return the condensate from the terminal units.

Thermostatic traps are used at the outlet of each terminal. They keep the steam in each terminal unit until it gives up its latent heat, then the trap opens to pass the condensate and permit more steam to enter the terminal. Float and thermostatic steam traps should be used at the end of each steam main, at the bottom of each riser, and along the horizontal mains. *Figure 23* shows a basic two-pipe gravity system.

Two-pipe systems can have either gravity or mechanical condensate returns. However, most systems use a mechanical return (*Figure 24*) provided either by a condensate pump or a vacuum pump. When steam systems increase in size, higher steam pressures are needed to obtain steam circulation, and some means other than gravity must be used to return the condensate to the boiler. A condensate pump and receiver tank are commonly used for this purpose. Normally, the condensate from the return piping system flows by gravity into a receiver tank. A condensate pump mounted near the receiver pumps the accumulated condensate into the boiler feedwater return piping. A typical condensate pump and receiver tank used for this purpose were covered earlier in this module. Note that in *Figure 24* air vents have been eliminated from the design. With a mechanical condensate return, air is released from the system through an open vent pipe on the receiver tank. This vent must open at a higher level than the boiler water level.

A vacuum-return system (*Figure 25*) is similar to a condensate return system, except that a vacuum pump is installed to provide a low vacuum in the return line to return the condensate to the boiler. A vacuum-return system is one that operates with a pressure in the condensate-return piping that is below atmospheric pressure. The steam

supply lines and terminal units are maintained at a positive pressure. The vacuum pump produces a vacuum in the system return lines by removing the air and vapor and venting it to the atmosphere in a related receiver tank.

It is important to note that the vacuum pump does not draw or pump the condensate itself. By quickly exhausting air from the system, the vacuum pump causes the steam to flow more rapidly, reducing warmup times, as well as noise from water hammer. The vacuum pump cycles as needed to maintain a predetermined vacuum level in the return lines. Operating the return line in a vacuum, rather than at or slightly above atmospheric pressure, increases the pressure differential in the overall system. As a result, steam pressure settings at the boiler can sometimes be lowered for significant fuel savings and improved efficiency. Much of this efficiency can easily be lost through air leaks. It is essential that return piping be leak-free to maximize the benefits associated with vacuum return and prevent excessive pump operation.

It is commonly understood that water boils, creating steam, at a lower temperature when exposed to pressures less than atmospheric. For example, at 15 in Hg vacuum, water boils at roughly 180°F. Some vacuum systems, known as variable vacuum systems, rely on this advantage by placing the supply side of the system, as well as the return piping, into a vacuum. This obviously requires greater capacity (expressed in cfm) from

Figure 23 Two-pipe gravity system.

Figure 24 Two-pipe system with mechanical condensate return.

the vacuum pump, and air tight piping systems become even more important. By reducing the boiling point of the water, even greater operating efficiencies can be obtained.

The vacuum pump controller, which determines the vacuum level that the pump must maintain, is generally designed to respond to outdoor temperatures. During mild weather, vacuum levels can be maintained at a lower value (18 to 25 in Hg), resulting in lower temperature steam and reduced fuel usage. As outdoor temperatures fall, creating the need for additional capacity, vacuum levels are maintained at a higher value (2 to 15 in Hg) to increase the temperature of the steam provided to the heating system. The controller is programmed to monitor outdoor temperatures and reset the vacuum level across the range of heating season outdoor temperatures expected for the specific geographic region.

Operation at lower pressures will increase the specific volume of the steam. Because of this increased volume, it takes less steam to fill the system, resulting in more efficient system operation.

3.2.0 Condensate Return/ Feedwater System Components

Condensate is the byproduct of a steam system. It forms in the distribution system because of the heat transfer from the steam to the area or substance being heated. Some condensation is also formed as a result of unavoidable radiation. Once the steam has condensed, giving up its latent heat, the hot condensate must be removed immediately and returned to the boiler. This is because the amount of heat in a pound of condensate is negligible when compared to the heat in a pound of steam.

The components that may be used in a condensate return/feedwater system are a condensate pump and receiver tank, vacuum pump, and de-aerating feedwater heater.

3.2.1 Condensate Pump and Receiver Tank

On larger steam-heating systems, a condensate pump and receiver are normally used to accumulate condensate and return it back to the boiler. There are many variations and types of equipment that can be used. Some systems may use a condensate receiver equipped with a single pump, while others may use a receiver with two pumps, often referred to as a duplex unit (*Figure 26*). Duplex units are often equipped with an alternator arrangement. In this arrangement, pumps are automatically switched on a timed basis, or when a pump failure occurs, to prevent the system from flooding and shutting down. Alternating their operation also balances pump wear.

Figure 25 Vacuum return system.

03306-13_F25.EPS

03306-13_F26.EPS

Figure 26 Duplex condensate receiver.

Figure 27 shows one type of condensate and receiver tank system. The receiver tank is an open reservoir into which the return piping discharges its condensate. This system uses a pump controller installed at the normal boiler water line to control the condensate (boiler feed) pump. Whenever the pump controller senses a drop in the normal boiler water level, it turns on the condensate pump. This causes water from the receiver tank to be pumped into the boiler as needed to replenish the boiler water level. A check valve in the condensate pump discharge line prevents the boiler water from backing up into the receiver when the condensate pump is turned off. A gate valve in the line allows the pump to be serviced without the need to drain the boiler.

Figure 27 Typical condensate pump and receiver system.

The receiver tank is also equipped with a float-actuated makeup water feeder control. If the level of condensate water in the receiver tank drops below the normal operating level, the control float drops and opens its valve to allow makeup water to enter the receiver tank from the city water supply. Additional boiler protection is provided by a combined feeder cutoff control. In the event of a power failure, condensate pump failure, or other problem, this control will first attempt to supply the boiler with sufficient water from the city water supply to prevent it from overheating. If the boiler water drops below a safe level, the control then acts to turn off the boiler burners.

3.2.2 *Vacuum Pumps*

Some systems use a vacuum pump in the condensate return line. Vacuum pump operation is similar to condensate pump operation in that the vacuum pump causes condensate from the condensate return line to flow into a receiver tank. Normally, a feed pump is then used to pump the condensate water from the receiver tank into the boiler. The vacuum unit and condensate receiver and pump are often combined into a single unit, such as the unit pictured in *Figure 28*.

At system startup, the condensate vacuum pump produces a vacuum in the system by removing the air and vapor from the system and venting it to the atmosphere. By quickly exhausting the air and condensate from the system, the vacuum pump causes the steam to circulate rapidly, reducing warm-up time and providing quieter system operation. In addition to causing condensate flow during system operation, the

Figure 28 Vacuum heating units.

Returning Condensate to the Boiler

In small systems, the condensate may be returned to the boiler by gravity. In most systems, however, pumping is required. Usually, the condensate flows by gravity to a collecting tank (receiver tank) from which it is pumped directly to the boiler or to a boiler feedwater system.

vacuum pump also turns on whenever needed to maintain the vacuum level in the system.

3.2.3 Deaerating Feedwater Heater

In some larger steam systems, the condensate water pumped out of the receiver tank does not get pumped into the boiler. Instead, it is pumped into a deaerating feedwater heater, also called a deaerator. This device uses system steam to preheat the water before it is supplied to the boiler in order to remove most of the dissolved oxygen. This is especially important on systems that lose 25 percent or more of the steam condensate from the circuit during normal operation, or are drained and refilled regularly. The reason is that fresh makeup water carries large quantities of dissolved oxygen. Deaeration is also beneficial in systems that operate at pressures above 75 psig.

3.2.4 Flash Tanks

When hot condensate under pressure is released to a lower pressure, part of it is re-evaporated and becomes **flash steam**. This normally happens when hot condensate is discharged into the condensate return line at a pressure lower than its saturation pressure. Some of the condensate flashes into steam and flows along with the liquid condensate through the return lines back to the boiler. This tends to cause an undesirable pressure increase in the condensate return lines. Flash tanks are used in medium-pressure and high-pressure systems to remove flash steam from the condensate lines by venting it either to the atmosphere or into a low-pressure steam main for reuse.

When flash steam is used as an energy source, high-pressure steam returns are usually piped to a flash tank (*Figure 29*). Flash tanks can be mounted vertically or horizontally. However, ver-

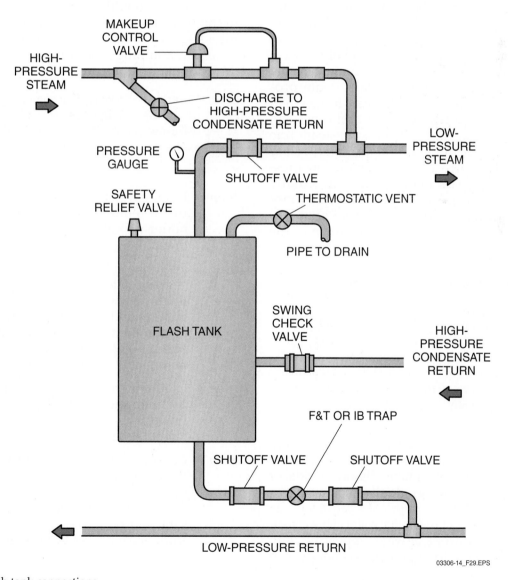

03306-14_F29.EPS

Figure 29 Flash tank connections.

tical mounting provides for better separation of steam and water.

When the condensate and any flash steam in the return line reach the tank, the high-pressure condensate is released into the lower pressure of the tank, causing some of it to flash into steam. This flash steam is discharged into the low-pressure steam main. The remaining condensate is pumped back to the boiler or discharged to a waste drain.

For proper flash tank operation, the condensate lines should pitch toward the tank. If more than one condensate line feeds the tank, a check valve should be installed in each line to prevent backflow. The top of the tank should have a thermostatic air vent to vent any air that accumulates in the tank. The bottom of the tank should have an inverted bucket or float and thermostatic-type steam trap.

The demand steam load on the flash tank should be greater than the amount of flash steam available from the tank. If it is not, the low-pressure system can be over-pressurized. A safety relief valve must be installed at the top of the flash tank to protect the low-pressure line from any over-pressurization. Since the flash steam produced in the flash tank should always be less than the amount of low-pressure steam that is needed, a makeup valve in the high-pressure steam line is used to supply the additional steam needed to maintain the correct pressure in the low-pressure line. This steam makeup valve is actually a pressure-reducing valve. It is much like those discussed earlier, but has an external pilot line to monitor the low pressure downstream of the valve itself.

3.2.5 Boiler Blowdown and Skimming

> **WARNING!**
> Boilers operate under high pressure and are potentially dangerous. When working with or around boilers, follow all applicable manufacturer's instructions, OSHA regulations, and job-site requirements relating to safety and the use of safety equipment.

During operation, makeup water is added to boilers to maintain the proper water level. As a result of adding this water, sediment, dissolved salts, or similar organic matter gradually accumulate in the boiler. Some of these impurities cause foam to form on the surface of the boiler water. This is usually indicated by drops of water appearing with the steam. If all the foreign matter is not purged

from the boiler on a regular basis, boiler operation will degrade. The process of purging the boiler is called blowdown. Every boiler is equipped with one or more valves for this purpose. Two kinds of blowdown valves may be found on a boiler: surface blowdown, also known as skimming valves, and/or bottom blowdown valves.

Surface blowdown valves are used to skim off impurities that cause foaming on the surface of the water inside the boiler. This foaming is caused by high surface tension due to a scum buildup from oil, grease, and/or sediment on the surface of the water. Foaming prevents steam bubbles from breaking through the water surface and hinders steam production. Surface blowdown valves are located on the boiler at the normal operating water level. Not all boilers are equipped with a surface blowdown valve. Use of the surface blowdown valve to skim off the impurities requires some experience.

Skimming should always be done according to the instructions given in the boiler manufacturer's service literature. The following represents common skimming procedures.

> **WARNING!**
> Never add cold water to a completely empty boiler when the burners are ignited. Doing so will cause serious damage and may create an explosion hazard.

Step 1 If not permanently connected to a drain, run a temporary connection from the boiler's skimming valve to a suitable drain.

Step 2 With the boiler empty and cool, slowly begin to add water. After a quantity of water has entered the boiler, turn on the burners and adjust the firing rate and/or burner controls so that the water being added is kept just below the boiling point. Boiling and turbulence must be avoided.

Step 3 Gradually raise the hot water level in the boiler to the point where the water just flows from the skimming valve, being careful not to raise it above this point.

Step 4 Continue to skim the boiler water until there is no trace of impurities. Water may be checked to make sure that it is free from oil by drawing off a sample. If the sample is reasonably free of oil, it will not froth when heated to the boiling point.

NOTE

Skimming will not clean the boiler of sediment that has accumulated at the bottom. After skimming, the boiler should be cleaned further by performing a blowdown procedure.

NOTE

Only attempt blowdown at a light load. Ideally, temporarily suspend the boiler's heating process to halt water turbulence and allow the solids to settle out.

Blowdown is another method used to purge the boiler of foreign matter. This method is used to drain off some of the water from the bottom of the boiler. The drained water is then replaced it with an equal amount of fresh, clean water. The bottom blowdown valve is used to drain off the sediment, scale, and other impurities that settle out of the water as it is heated and accumulate at the lowest point in the boiler. If the boiler is extremely dirty, blowdown may require that the boiler be completely drained and refilled with water one or more times. This process must be continued until the water discharged from the boiler runs clear. On smaller systems, the hot boiler water discharged through the bottom blowdown valve is drained directly into a nearby sewer. In larger systems, it is piped through a line to a tank where it is allowed to cool before going into the sewer. Codes require that wastewater from a steam or hot-water boiler not be hotter than 140°F. Boiler blowdown should always be performed as directed in the boiler manufacturer's service instructions. The following represents common blowdown procedures:

Step 1 Check the water level in the boiler.

Step 2 Partially open the bottom blowdown valve. Once the water starts draining, fully open the valve.

Step 3 Remain at the blowdown valve. Monitor the gauge glass during blowdown to make sure that the water level is not lowered to a dangerously low point.

Step 4 When the desired amount of water has been drained from the boiler, close the blowdown valve.

3.2.6 Boiler Water Treatment

The water used in a boiler must be kept clean for efficient operation. Never add dirty or rusty water to a boiler. Hard water may eventually interfere with the efficient operation of a boiler. For this reason, hard water should be chemically treated with water softeners before being used. Scale, corrosion, fouling, and foaming can all cause boiler problems. Scale deposits on heating surfaces increase boiler temperatures and lower operating efficiency. Corrosion can damage metal surfaces, resulting in metal fatigue and failure. Fouling clogs nozzles and pipes with solid materials, thereby restricting circulation and reducing the heat transfer efficiency. Foaming results in overheating and can cause water impurities to be carried along with the steam into the system.

There are various chemical and mechanical methods of water treatment used to help prevent problems caused by water impurities. At a minimum, proper water treatment includes blowdown, pH control, and the addition of chemicals to neutralize other contaminants. The specific methods and devices used for water treatment are covered in a separate module.

3.3.0 Steam and Condensate Pipe Sizing

Proper steam and condensate line sizing is essential for any steam system to operate efficiently and provide the needed capacity for each load. To size the piping system, information regarding the volume of steam needed, as well as the steam pressure in the system must be known.

As a general rule, it is best that steam boilers operate at a pressure near their design pressure. In some cases, especially in larger facilities with a significant number of loads, this may require that the steam system pressure be considerably higher than actually required at any one of the operating loads. Operating steam boilers at pressures significantly lower than the boiler is designed for reduces the volume of space in the boiler that normally is allowed for steam. This places the water level somewhat closer to the steam outlet, and the flow velocity of the steam to the outlet becomes faster due to the smaller area of confinement.

Using Flash Steam

The heat content of flash steam is the same as that of live steam at the same pressure. Use of the flash steam in a low-pressure steam main allows this heat to be used rather than wasted. It is commonly used for space heating, as well as heating or preheating of water, oil, and other liquids.

When operated in this manner, the likelihood of water droplet carryover is greatly increased.

There are additional advantages to controlling steam pressure higher than is needed downstream. The size of steam distribution mains becomes smaller to carry the same amount of energy, thus saving on installation costs. Due to smaller surface areas with smaller piping, heat loss is also reduced. Steam pressure can then be reduced using pressure-reducing valves near the inlet to each of the individual loads. These valves can be physically smaller, further reducing installation costs.

Appropriate pipe sizing is important to ensure a sufficient steam supply to the loads. Undersized piping causes high pressure drops and greater velocities, increasing noise and erosion of internal piping surfaces. It is also quite common for a new load to be added to a steam system long after the initial installation is complete. A piping system that is already undersized or marginal may prevent even a relatively small load from being added without re-piping, or an insufficient supply of steam to all loads could result. On the other hand, piping that is too large creates unnecessary heat losses and higher installation costs.

Two primary methods are used to size steam piping: sizing by desired velocity or sizing to maintain an acceptable pressure drop. Systems with shorter mains and branch lines are often sized by velocity. Longer piping runs may be sized first by velocity, then rechecked to ensure that acceptable pressure drops have not been exceeded. There are a number of factors that impact steam pipe sizing in various parts of the system.

3.3.1 Velocity Sizing

When sizing based on velocity, steam lines should be designed for speeds of 80 to 120 feet per second (fps), or 4,800 to 7,200 feet per minute (fpm). In very limited cases, where noise may not be an issue and pressure drops are not as important, velocities up to 200 fps or 12,000 fpm have been used. Again, these values are not recommended for the reasons stated earlier concerning undersized piping.

Figure 30 provides a steam velocity chart for pipe sizing. As mentioned earlier, the volume of steam to be carried, stated in lb/hr (pounds per hour) must be known. For new pipe sizing, the system design steam pressure must also be known. Note that this chart is for Schedule 80 pipe. The appropriate chart for the type of pipe being used in the installation must be used.

Example 1

A steam system is being designed that must carry a minimum of 1,500 lb/hr of steam through the steam main. Maximum design velocity must not exceed 100 fps, and the designated steam pressure will be 100 psig. Select an appropriate pipe size for the steam main.

Step 1 Enter the left side of the chart at 100 psig. Since the velocity cannot exceed 100 fps in this example, read across the chart horizontally at the line of values for 80 fps (120 fps is obviously too fast).

Step 2 Continue across the chart until you reach a value equal to, or greater than, 1,500 lb/hr. The closest value that is not less than our minimum of 1,500 lb/hr is 1,660 lb/hr.

Step 3 Read up the column to the top line of steam pipe sizes. Pipe size will be 2 inches.

When troubleshooting steam systems, the chart can also be used to check existing conditions. For example, if the system seems too noisy in operation, check the pipe size against the volume of steam being transported to see if velocities exceed the prescribed maximum of 120 fps for quiet operation. This can happen when new loads are added to an existing piping system, and steam pressures are increased to provide the necessary additional volume of steam through the existing piping.

3.3.2 Steam Line Pressure Drop

Steam lines are generally sized based on velocity first, then checked against the appropriate chart for pressure drop in the system. The allowable pressure drop varies widely, based on the system itself. Systems operating at 100 psig, with one or more loads that require a minimum of 80 psig at the inlet, certainly cannot be sized at a 25 psig allowable loss. Perhaps a more appropriate choice here would be a 5 to 10 psig loss. However, a system operating as low as 5 psig should probably be sized for a pressure drop no greater than 0.3 psig to maintain the needed pressure downstream. As a general rule, the higher the steam pressure, the greater the pressure drop that can likely be tolerated.

Figure 31 is one example of a steam pressure drop chart. Note that this chart is also designated for Schedule 80 pipe. Pressure drops for a given volume of flow will differ in other classes of pipe. Also note that the chart is based on a steam pressure of 100 psig. For pressures other than 100 psig, a correction factor must be applied to the value read from the chart. The correction factors for various pipe sizes are listed at the top of the

Pressure psi	Velocity ft/sec	½"	¾"	1"	1¼"	1½"	2"	2½"	3"	4"	5"	6"	8"	10"	12"
5	50	12	26	45	70	100	190	280	410	760	1250	1770	3100	5000	7100
	80	19	45	75	115	170	300	490	710	1250	1800	2700	5200	7600	11000
	120	29	60	110	175	245	460	700	1000	1800	2900	4000	7500	12000	16500
10	50	15	35	55	88	130	240	365	550	950	1500	2200	3770	6160	8500
	80	24	52	95	150	210	380	600	900	1500	2400	3300	5900	9700	13000
	120	35	72	135	210	330	590	850	1250	2200	3400	4800	9000	14400	20500
20	50	21	47	82	123	185	320	520	740	1340	1980	2900	5300	8000	11500
	80	32	70	120	190	260	520	810	1100	1900	3100	4500	8400	13200	18300
	120	50	105	190	300	440	840	1250	1720	3100	4850	6750	13000	19800	28000
30	50	26	56	100	160	230	420	650	950	1650	2600	3650	6500	10500	14500
	80	42	94	155	250	360	655	950	1460	2700	3900	5600	10700	16500	23500
	120	62	130	240	370	570	990	1550	2100	3950	6100	8700	16000	25000	35000
40	50	32	75	120	190	260	505	790	1100	1900	3100	4200	8200	12800	18000
	80	51	110	195	300	445	840	1250	1800	3120	4900	6800	13400	20300	28300
	120	75	160	290	460	660	1100	1900	2700	4700	7500	11000	19400	30500	42500
60	50	43	95	160	250	360	650	1000	1470	2700	3900	5700	10700	16500	24000
	80	65	140	250	400	600	1000	1650	2400	4400	6500	9400	17500	27200	38500
	120	102	240	410	610	950	1660	2600	3800	6500	10300	14700	26400	41000	58000
80	50	53	120	215	315	460	870	1300	1900	3200	5200	7000	13700	21200	29500
	80	85	190	320	500	730	1300	2100	3000	5000	8400	12200	21000	33800	47500
	120	130	290	500	750	1100	1900	3000	4200	7800	12000	17500	30600	51600	71700
100	50	63	130	240	360	570	980	1550	2100	4000	6100	8800	16300	26500	35500
	80	102	240	400	610	950	1660	2550	3700	6400	10200	14600	26000	41000	57300
	120	150	350	600	900	1370	2400	3700	5000	9100	15000	21600	38000	61500	86300
120	50	74	160	290	440	660	1100	1850	2600	4600	7000	10500	18600	29200	41000
	80	120	270	450	710	1030	1800	2800	4150	7200	11600	16500	29200	48000	73800
	120	175	400	680	1060	1520	2850	4300	6500	10700	17500	26000	44300	70200	97700
150	50	90	208	340	550	820	1380	2230	3220	5500	8800	12900	22000	35600	50000
	80	145	320	570	900	1250	2200	3400	4900	8500	14000	20000	35500	57500	79800
	120	215	450	850	1280	1890	3400	5300	7500	13400	20600	30000	55500	85500	120000
200	50	110	265	450	680	1020	1780	2800	4120	7100	11500	16300	28500	45300	64000
	80	180	410	700	1100	1560	2910	4400	6600	11000	18000	26600	46000	72300	100000
	120	250	600	1100	1630	2400	4350	6800	9400	16900	25900	37000	70600	109000	152000

03306-13_F30.EPS

Figure 30 Steam pipe sizing for steam velocity.

Blowdown Separators

Boilers may be equipped with blowdown separators. These devices help reduce corrosion and scaling by separating steam from water and removing dissolved solids. Blowdown separators also lower the temperature of the drain water, which may be necessary to comply with environmental regulations and municipal plumbing codes.

03306-13_SA02.EPS

chart. For example, if a steam system is to operate at 5 psig, then the value read from the chart at 100 psig must be multiplied by 5.2 to achieve the correct pressure drop value.

The values in the chart provide the pressure drop that exists in 100 feet of pipe. Your final answer must also be adjusted for the actual length of the pipe in your installation.

Example 2

Using the values and resulting pipe size in *Example 1*, calculate the pressure drop in 75 feet of pipe.

Step 1 Enter the chart at the bottom, locating the vertical line representing 1,500 lb/hr of steam flow.

Step 2 Follow the line vertically until it intersects with the diagonal line representing 2-inch pipe.

Step 3 Follow the horizontal line left from this point of intersection and read the resulting pressure drop for 100 feet of Schedule 80 pipe. The result would be 2.0 psig of pressure loss per 100 feet of pipe.

Step 4 To adjust for the actual length of pipe (75 feet in our example), multiply 2.0 psig/100 feet by 0.75. The resulting pressure drop will then total 1.5 psig in this particular steam line.

3.3.3 Condensate Line Sizing

The steam condensate can carry as much as 15 percent of the energy originally used in the process of creating steam. With ever-rising energy costs, it is imperative that steam systems capture and return the condensate back to the process to preserve this expended energy.

Condensate line sizing can be more difficult than sizing of the steam lines, due to the variety of differing conditions found in the system. A single set of recommendations does not apply. The lines carrying condensate from the load (such as a steam coil in ductwork for a comfort heating application) and into the trap represent a different set of conditions than those found in condensate lines transporting condensate away from a trap.

Condensate piping systems are classified based on the design pressures. Gravity returns describe those systems where the condensate line is at or very near atmospheric pressure, and the condensate generally returns to a vessel vented to the atmosphere. Low-pressure returns are those operating at a pressure below 15 psig, while medium-pressure returns are those operating between 15 and 100 psig. Systems that operate with the condensate line experiencing pressures above 100 psig are classified as high-pressure returns.

In all cases, the sizing of condensate lines is based on several primary factors:

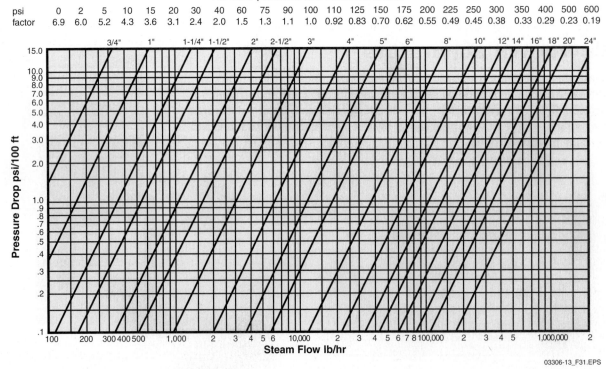

100 psig Saturated Steam
For other pressures use correction factors

psi	0	2	5	10	15	20	30	40	60	75	90	100	110	125	150	175	200	225	250	300	350	400	500	600
factor	6.9	6.0	5.2	4.3	3.6	3.1	2.4	2.0	1.5	1.3	1.1	1.0	0.92	0.83	0.70	0.62	0.55	0.49	0.45	0.38	0.33	0.29	0.23	0.19

Figure 31 Steam pipe sizing for pressure drop.

03306-13_F31.EPS

- *Pressure* – The difference in pressure from one end of the line to the other is certainly a factor. This pressure difference may cause some flash steam to be generated and may help promote flow in the line.
- *Condition of the condensate* – For a proper selection, the designer must know whether the condensate is primarily liquid, or whether a significant amount of flash steam can be expected.
- *Volume* – The anticipated volume of condensate flow is also an important factor. This condition changes dramatically between startup of a system and full, consistent operation. During startup, the temperature difference at the load and throughout the piping system, relative to the temperature of the steam supply, is greater. Far more condensate will be generated during this period in all locations, including steam mains, branch lines, and especially at the system loads.

Condensate must be drained away from load apparatus by drain line(s) and through trap(s) to prevent excessive water hammer and potential damage to piping and the internal components of the trap(s). As a general rule, there is no significant pressure drop in this line to consider, so flash steam resulting from changes in pressure is negligible or non-existent. There is also no pressure differential that would assist flow because condensate movement is by gravity. The condition and volume of the condensate will, however, change dramatically from startup to operation at full load. It is not uncommon for the volume of condensate to be three times greater during startup than at full load. The flow of noncondensables (air) through the system must be accommodated as well.

It cannot be assumed that the pipe outlet size of the load or apparatus is the proper size to promote good drainage. Once the steam trap has been selected based on design criteria, however, it is quite common for the trap inlet size to be the correct pipe size to use between the two points. In most cases, sizing the line for a volume twice that of the expected flow during a running load is sufficient to accommodate startup loads without being oversized for normal operating conditions.

Wherever possible, discharge lines from traps should incorporate gravity as the primary mover. A pressure drop does exist at the trap. Flash steam may result from this pressure difference, and will be more significant at steam pressures above 30 psig. As condensate is discharged from the trap, some small amounts of live steam may also make it past the trap. In the trap discharge line then, liquid and vapor are moving together. This is known as two-phase flow, much the same as the flow of liquid and vapor in sewer systems.

The mixture leaving the trap will exhibit more properties of steam than properties of liquid water, especially at steam pressures above 30 psig. Sizing is therefore evaluated much like steam lines. The same maximum velocities are sometimes used, although lower velocities generally work best to minimize water hammer and the creation of flash steam. Undersized drain lines from the trap, unable to handle the flow as needed, can dramatically slow down or flood the process.

As a result of heat loss through piping walls, condensate will also gather in steam mains, especially during system startup. Steam mains should be pitched, usually ½ inch down per 10 feet of run, to allow the condensate to gravity-drain toward a gathering point. Collection legs, also known as **drip legs**, are installed on the bottom of the main along the horizontal run at 150 to 300-foot intervals. The collection leg should be the same pipe diameter as the main itself. In cases where the pitch of the steam main causes it to be too low, a riser may be needed to route the line up higher. An additional collection leg should be installed at the base of all risers in stem mains. The inlet of a steam trap is then connected to each collection leg to drain the condensate away.

In most cases, it is necessary to drain condensate from more than one load into a common condensate return line. The line should never become completely filled, or flooded, with condensate. It should slope toward the vented receiver tank or flash tank, when one is included. In most cases, sizing for liquid-only (no steam, flash or otherwise) condensate drainage should be for velocities at or below 7 fps. The common line is generally sized based on the combined volume of flow used to size the individual trap discharge lines routed to the common line.

3.4.0 Valves

Many valves used to regulate and direct the flow of steam within a steam piping system are the same as those used in hot-water piping systems. These valves include gate valves, ball valves, globe valves, angle valves, and check valves. The focus of this section is on pressure-reducing valves and thermostatic valves used in steam systems.

3.4.1 Pressure-Reducing Valves

Pressure-reducing valves are used to reduce a higher steam pressure to a controlled lower pressure. Such a condition often exists when a process steam boiler supplies steam at a higher pressure than can be used in a related secondary steam-heating system. The valve output pressure level must be held constant, regardless of any pressure fluctuations in the input steam supply or variations in the downstream load. There are several kinds of pressure-reducing valves. One simple type of pressure reducing valve (*Figure 32*) controls the output pressure by achieving a balance between an internal spring pressure acting on a diaphragm and the pressure of the delivered steam. The spring pressure is usually adjusted manually to set the desired pressure downstream.

Larger and more complex pressure-reducing valves use external tubes to sense output pressure further downstream. Other valves are positioned using pneumatic or electric actuators for greater accuracy. This allows them to be interfaced with building or process automation systems using a variety of sensor inputs.

Figure 33 shows a typical single-stage pressure regulator. Notice that the pressure regulator in the figure can be bypassed by opening a globe valve. This allows steam to be supplied to the downstream load should the pressure regulator valve malfunction. Also, the gate valve installed on each side of the pressure-relief valve allows the pressure-reducing valve to be isolated for repair without shutting down the system.

Suitable drains and steam traps must be used to prevent accumulation of condensate on the input side of the pressure-reducing valve. The assembly, which consists of the pressure regulator and related valves and traps, is often called a pressure-reducing station.

Pressure reduction can be accomplished with one or more valves (stages). Multistage pressure reduction is commonly used where an intermediate pressure is required or where the total pressure reduction is large. A large pressure reduction, also referred to as the **turndown ratio**, is one in which the ratio of downstream pressure to upstream pressure is greater than 50 percent. *Figure 34* demonstrates the installation of pressure-reducing valves in series.

When there is a wide variation in the steam load, such as can occur between summer and winter load conditions, pressure-reducing valves are often installed in parallel. *Figure 35* shows a typical pressure-reducing station with valves installed in parallel. Normally, these valves have different operating pressures that provide for sequenced valve operation. A smaller regulator valve sized to handle about one-third of the maximum load is connected in parallel with a larger valve sized to handle the remaining load. The smaller regulator valve is set for a slightly higher pressure than the larger one. During times when the load is low, the smaller regulator handles the load. As the load increases, the reduced pressure causes the larger regulator to operate and provide any additional required flow.

There are several common factors that should be considered for a successful pressure-reducing valve installation:

- Follow the manufacturer's guidelines for proper sizing. Oversized valves tend to exceed the desired pressure output and experience greater wear, especially at light loads. Undersized valves are unable to pass a sufficient amount of steam to satisfy downstream needs.
- Pressure-reducing valves operate properly in one direction only. Make sure the valve is properly oriented in the line. Most have an arrow cast into the valve body to indicate the proper flow direction. Pressure-reducing valves also must be installed horizontally, and should always have a steam trap installed immediately upstream.

REGULATING SCREW AND LOCKNUT

SPRING CASE

SPRING

DIAPHRAGM

STEM

BODY

SEAT

BALL VALVE

BOTTOM SPRING

03306-13_F32.EPS

Figure 32 Pressure-reducing valve.

Figure 33 Single-stage pressure-reducing station.

03306-13_F33.EPS

Figure 34 Use of pressure-reducing valves connected in series.

03306-13_F34.EPS

- Pressure gauges should be installed near both the inlet and outlet. Valves that use a sensing tube for remote pressure input should also have a gauge installed at the same location as the sensing port.
- Dirt and other foreign matter are the most likely causes of valve failure. The use of strainers upstream of the valve is recommended. Strainers should be removed and cleaned soon after a new installation has been placed in operation.

3.4.2 *Thermostatic Valves*

Thermostatic valves control the flow of steam based on temperature inputs. They can be applied many different ways. Some possible applications include comfort heating radiation, steam water heaters, steam-jacketed vessels, various types of component molds, and tank heating coils. In the simplest applications, for example, a thermostatic valve can be used to control the volume of steam flow to a radiator or other type of steam terminal (*Figure 36*) to provide zone temperature control. A dial provides for operator setpoint adjustment. Much like pressure-reducing valves, many thermostatic valves are self-contained, with a liquid-filled temperature-sensing element attached directly to the valve such as the one in *Figure 37*. A number of thermostatic elements are available, with selection dependent upon the desired temperature range. As temperature at the sensing element increases, the liquid expands and applies

Figure 35 Use of pressure-reducing valves connected in parallel.

an opening force against the closing force of the valve spring. Setpoint adjustments change the spring tension, thereby changing the force required to open or close the valve.

Thermostatic valves share most of the same common installation considerations with pressure-reducing valves. Although a pressure gauge upstream of the valve is generally recommended, downstream pressure may not need to be monitored. Instead, thermometers should be placed in the controlled medium to provide a quick visual indication of the temperature.

Figure 37 Temperature-sensing element attached to a valve.

Figure 36 Typical individual thermostatic control.

Additional Resources

Low Pressure Boilers, 3rd Edition. Fredrick M. Steingress, Daryl R. Walker. Orland Park, Illinois: American Technical Publishers.

High Pressure Boilers, 4th Edition. Fredrick M. Steingress, Harold J. Frost, Daryl R. Walker. Orland Park, Illinois: American Technical Publishers.

Piping for High Pressure Boilers: The Installation and Inspection of High-Pressure Boiler Piping for Code Compliance with the ASME and National Board Code Requirements. Steve Kalmbach. Tucson, Arizona: Wheatmark®.

3.0.0 Section Review

1. A condensate system vacuum pump is used to pump condensate.

   a. True
   b. False

2. The load demand on the low-pressure steam line where a flash tank discharges its collected steam should always be _____.

   a. greater than the flash tank output
   b. the same as the flash tank output
   c. half as much as the flash tank outputs
   d. one-third as much as the flash tank output

3. To properly size the piping of a steam system, primary consideration must be given to _____.

   a. flash tank capacity
   b. condensate line size and lengths
   c. steam volume and pressure
   d. the number of steam traps to be used

4. The turndown ratio that normally requires the use of multiple pressure-reducing valves is a ratio of more than _____.

   a. 40 percent
   b. 50 percent
   c. 60 percent
   d. 70 percent

4.0.0 STEAM TRAPS AND STRAINERS

Objectives

Identify, install, and maintain various types of steam traps.

   a. Identify various types of steam traps.
   b. Explain the basic concepts of installing steam traps.
   c. Explain how to maintain various steam traps.
   d. Explain how to troubleshoot various steam traps.

Performance Task 1

Maintain selected steam traps.

The steam trap is one of the most important components used in steam systems. It is likely that they will demand a great deal of the total time invested in steam maintenance and repairs. Steam traps provide the point of separation between live steam and water in the system and must provide this service effectively for the system to function properly.

As steam is allowed to cool, it condenses to its liquid form. Most of this condensing process takes place at the load, such as a heat exchanger where steam gives up its stored energy to heat water (*Figure 38*). However, it can also take place in the distribution and piping system and collect in low spots. The condensate can then be picked up by steam as it passes, propelling it at a high rate of speed into bends and fittings. Known as water hammer, this phenomenon can cause serious damage and, at a minimum, create noise. For this reason, steam traps are not only used on the leaving side of loads such as heat exchangers and coils, but also at strategic low points throughout the piping system.

For steam to remain efficient, it must be dry and at the proper temperature and pressure. Essentially, a steam trap is a self-actuating drain valve that maintains effective heat transfer in the steam system by consistently purging it of collected condensate and air.

Air is always present during system startup and in the boiler feed water. In addition, the feed water may contain carbonates that dissolve and release carbon dioxide gas (CO_2). Air and carbon dioxide can cause problems in a steam line. Air is an excellent insulator and can reduce the efficiency of the steam at the terminal heat transfer surfaces. Carbon dioxide can dissolve in the condensate and form carbonic acid, which is extremely corrosive. Scale and debris are also transferred with the condensate to areas that are difficult to repair when clogged. Use of properly installed steam traps and strainers throughout the system can prevent these problems.

4.1.0 Steam Trap Types

There are three general categories of steam traps: mechanical, thermostatic, and thermodynamic. Among these many types, there are also various materials used, such as cast iron, stainless steel, forged steel, and cast steel. Steam traps are fitted to the pipe with screwed, flanged, socket-welded, or compression fittings. All types of steam traps automatically open an orifice, drain the condensate, then close before steam is lost.

4.1.1 Mechanical Steam Traps

Mechanical steam traps respond to the difference in density between steam and steam condensate. The trap opens to condensate and closes to steam. The typical mechanical trap is called an inverted bucket trap and has only two moving parts, the valve lever assembly and the bucket (*Figure 39*). The trap is normally installed in the drain line between the steam-heated unit and the condensate return header.

The cycle starts with the bucket down and the valve all the way open. The condensate fills the body and the bucket, causing the bucket to sink and open the discharge valve. Gases and air flow through first, then steam raises the bucket and closes the valve. The gas slowly bleeds out through a small vent in the top of the bucket. The valve remains open until enough steam has collected to float the bucket again. The inverted bucket trap operates intermittently, responding to the accumulation of condensate by opening, then reclosing to allow steam to pass.

Because the inverted bucket trap operates on a pressure differential, it must be primed before being put into service.

The inverted trap can handle high-pressure steam. It resists damage from pressure surges and water hammer, and tolerates freezing if the body is made of a ductile material. Its disadvantages are its limited air discharge capacity and a tendency to be noisy.

TRAP DRAINING CONDENSED WATER
FROM THE STEAM SUPPLY LINE

STEAM-TO-HOT WATER
HEAT EXCHANGER

CONDENSATE DRAIN LINES

TRAP SERVING THE
HEAT EXCHANGER

03306-13_F38.EPS

Figure 38 Installed steam traps.

4.1.2 Thermostatic Steam Traps

Thermostatic steam traps respond to temperature changes in the steam line. They open when the cooler condensate is present and close to the higher steam temperatures. The following descriptions, individually illustrated in *Figure 40*, represent four different types of thermostatic steam traps:

- *Liquid expansion thermostatic trap* – Remains closed until the condensate cools below 212°F, opening the thermostat downstream of the

LEVER

VENT

VALVE

BUCKET

STEAM AND
CONDENSATE IN

03306-13_F39.EPS

Figure 39 Inverted bucket trap.

valve. This discharges the cool condensate at a constant temperature.

- *Balanced pressure thermostatic trap* – Opens when condensate cools slightly below the steam saturation temperature at any pressure within the trap's range. Controlled by a liquid-filled bellows, the discharge valve closes when hot condensate vaporizes the liquid in the bellows. It opens when the condensate cools enough to lower the pressure inside the element. Balanced pressure traps are smaller than mechanical traps, open wide when cold, readily purge gases, and are unlikely to freeze.

- *Bimetal thermostatic trap* – Has bimetal strips that respond to cool condensate by bending to open the condensate valve. When steam hits the strips, they expand and close the valve. They require considerable cooling to reopen, and the pressure-compensating characteristics vary among the models. The closing force of some designs varies with the steam pressure, approximating the response of a balanced pressure trap. This trap has a large capacity, but responds slowly to process load changes. However, it vents gases well. It is not easily damaged by freezing, water hammer, or corrosion and it can handle the high temperatures of superheated steam.

- *Float and thermostatic (F and T) trap* – Has a float that rises when condensate enters and opens the main valve that lies below the water level. At the same time, another discharge valve at the top of the trap, operated by a thermostatic bellows, opens to release cooler gases. This trap design vents gases well and responds to wide and sudden pressure changes, discharging condensate continuously. In the failure mode, the main valve is closed but the air vent is usually open. Float thermostatic traps should always be protected from freezing because they usually contain water.

4.1.3 Thermodynamic Steam Traps

Thermodynamic steam traps use the heat energy in hot condensate and steam to control the opening and closing of the trap. They operate well at higher temperatures and can be installed in any position because their function does not depend on gravity.

The thermodynamic disc steam trap is the most widely used. In this unit, the only moving part is the disc (*Figure 41*). Flash steam pressure from hot condensate keeps the disc closed. Flash steam is steam that is formed when hot conden-

Figure 40 Thermostatic steam traps.

LIQUID EXPANSION

BALANCED PRESSURE

BIMETALLIC

FLOAT AND THERMOSTATIC

03306-13_F40.EPS

CAP CHAMBER — DISC

TRAP IS ACTUATED BY KINETIC FLOW

CONDENSATE DISCHARGE ORIFICE

03306-13_F41.EPS

Figure 41 Thermodynamic stainless-steel steam trap.

sate is released to a lower pressure and flashes back to steam as a result. When the flash steam above the disc is condensed by cooler condensate, the disc is pushed up and remains open until hot condensate flashes again to build up pressure in the cap chamber. At the same time, flashing condensate on the underside of the disc is discharged at a high velocity, lowering the pressure in this area and closing the disc before steam can escape.

The thermodynamic steam trap can withstand freezing, high pressure, superheating, and water hammer without damage. The audible clicking of the disc indicates when the discharge cycle occurs. These traps last longer if they are not over-sized.

4.1.4 Fixed-Orifice Steam Traps

As the name implies, the fixed-orifice trap depends on a single opening of a predetermined size to pass a calculated amount of condensate under specific pressure and load conditions.

In actual practice, it is rare that steam and condensate flow rates remain precise, rendering the fixed-orifice trap impractical under most conditions. They have a tendency to waterlog (hold back too much condensate) on startup when condensate loads are high, yet pass live steam once the piping system is hot if the load begins to drop. In cases where it can be successfully applied, the fixed-orifice steam trap has the distinct advantage of having no moving parts.

4.1.5 Strainers

In all systems, a strainer should be installed ahead of the steam trap. The scale and corrosion in any steam system has to be stopped before it enters the trap, or it will cause clogging and damage. Strainers need to be cleaned and inspected on a regular basis. Strainers are available in many styles and are sometimes incorporated into the steam trap. *Figure 42* shows three different types of strainers.

4.2.0 Installing Steam Traps

To mount a steam trap anywhere other than the correct position would render it useless. Two general rules for steam-trap mountings are that they should be lower than any line in the system, and that a strainer has to be upstream of the trap. Isolation drains, couplings, check valves, and drip leg placement are also important to each application. Their use depends on the needs and contents of the system. A drip leg is a drain for condensate in a steam line placed at a low point in the line and used with a steam trap.

Each steam trap manufacturer has resource books that detail how to install steam-trap systems. The following is a list of general guidelines for proper steam-trap installation:

- Provide a separate trap for each piece of equipment or apparatus. Short-circuiting may occur if more than one piece of equipment is connected to a single trap.
- When piping from the steam header or main to the load, always tap the steam supply off the top to prevent any collected condensate from draining into the load apparatus.
- Install a supply valve ahead of the steam trap and close to the steam header to allow maintenance or modifications to be performed without shutting down the steam header.
- Install a steam supply valve close to the equipment entrance to allow equipment maintenance work to be performed without shutting down the supply line.

SCREWED

BASKET

FLANGED Y

03306-23_F42.EPS

Figure 42 Strainers.

- Connect the condensate discharge line to the lowest point in the equipment to avoid water pockets and water hammer.
- Install a shutoff valve upstream of the condensate discharge piping to cut off discharge of condensate from equipment and allow service work to be performed.

- Install a strainer and strainer flush valve ahead of each trap to keep rust, dirt, and scale out of working parts and to allow blowdown removal of foreign material from the strainer basket.
- Provide unions on both sides of the trap for its removal or replacement.
- Install a test valve downstream of the trap to allow observation of discharge when testing.
- Install a check valve downstream or upstream of the trap to prevent condensate backflow during shutdown or in the event of unusual conditions.
- Install a downstream shutoff valve to cut off equipment condensate piping from the main condensate system for maintenance work.
- Do not install a bypass unless there is some urgent need for it. Bypasses are an additional expense to install and are frequently left open, resulting in the loss of steam and inefficient equipment operation. When a bypass is necessary, a globe valve should be used to allow manual balancing of condensate flow, as the steam process continues during maintenance of the primary trap.

Figure 43 shows a basic trap installation. A steam trap should have a provision for inspection without interrupting the process flow.

4.3.0 Maintaining Steam Traps

Most steam traps fail in the open position, which is difficult to pinpoint because it does not affect equipment operation. When the trap fails open, the steam flows freely through the system, keeping the heat transfer high but failing to hold back the condensate. Traps that fail in the closed position are easy to identify because the backup of condensate cools the system. Clogged strainers, which protect the trap from debris, also have the same effect.

An effective planned maintenance (PM) program includes scheduled checks of the entire system. Clogged strainers, leaking joints, leaking valve packing, or missing insulation are some of the items to check. When the internal parts of a steam trap wear out, the water seal deteriorates and steam flows through the valve assembly, further worsening the condition. Common causes of steam-trap failure include the following:

- Scale, rust, or corrosion buildup preventing the valve from closing
- Valve assembly wear
- Defective or damaged valve seat
- Physical damage from severe water hammer
- Foreign material lodged between seat and valve

STEAM HEADER

FLOW

DRIP LEG

TRAP ISOLATION VALVES

CHECK VALVE

TO CONDENSATION HEADER

GLOBE VALVE

SHUTOFF VALVE

STEAM TRAP

UNIONS OR FLANGES

BYPASS AROUND TRAP

DRAIN

TEST VALVE

STRAINER WITH BLOWOFF VALVE

03306-13_F43.EPS

Figure 43 Basic trap installation.

- Blocked, clogged, or damaged strainers
- Increased back pressure

Other failures are specific to the type of steam trap. The two primary indications of failure in a steam trap are allowing live steam to pass freely (will not close), or failure to allow condensate to pass through (will not open). The following sections explain the causes and solutions for failures on each type of steam trap.

4.3.1 Inverted Bucket Trap

If an inverted bucket trap blows steam, check for loss of the water seal. Isolate the trap, wait for the condensate to accumulate, and start the steam flow again. If this solves the problem, try to discover the cause of the water seal loss. It could be caused by superheat, sudden pressure fluctuations, or the trap being installed so that the water seal runs out by gravity.

Another solution is to try installing a check valve ahead of the trap. If the steam blow persists, check for dirt or wear on the valve and linkage. Replace the valve, the seat, and the lever. If this fails, check the bucket. If the bucket or lever is distorted, the trap failure was probably caused by water hammer. Trace the source of the problem and try to eliminate it.

If the trap will not pass condensate, check that the maximum operating pressure marked on the trap is not lower than the actual pressure to which it is subjected. If it is, the valve cannot open. In-

stall a valve and seat assembly with the correct pressure rating. Make sure that it has sufficient capacity to handle the maximum load. Check the internal parts and be sure that the air vent hole in the bucket is not obstructed. This could cause pockets of air to be trapped in the vent. This condition is called air binding.

4.3.2 Liquid Expansion Thermostatic Trap

If a liquid expansion thermostatic trap blows steam, check for dirt or wear on the valve and seat. If wear has occurred, replace all of the internal parts. Remember that this type of trap is not self-adjusting to changes in pressure. If it has been set to close at a high pressure, it may not close off at a lower pressure. Try adjusting the trap to a lower setting, making sure that it does not waterlog excessively. If it does not appear to react to this change, replace all of the internal parts. If a trap will not pass condensate, check that it has not been adjusted to a setting that is too cold.

Steam Trap Maintenance

GOING GREEN

Nonfunctioning steam traps are a significant cause of lost energy. Steam traps must be inspected and cleaned in accordance with the manufacturer's instructions at regular operating intervals.

4.3.3 Balanced Pressure Thermostatic Trap

If a balanced pressure thermostatic trap blows steam, isolate the trap and allow it to cool before inspecting it for dirt. If the seat is wire-drawn, replace all of the internal parts, including the thermostatic element. The original parts have probably been strained by continuous steam blow.

If the valve and seat seem to be in good order, check the element. You should not be able to compress it when it is cool. Any flabbiness of the element indicates failure. Flattening of the convolutions indicates water hammer damage. If the water hammer cannot be eliminated at its source, a stronger type of trap must be used.

If the trap will not pass condensate, the element may be overextended due to excessive internal pressure, making it impossible for the valve to lift off its seat. An overextended element could be caused by superheat or by someone opening the trap while it is still very hot and before the vapor inside has had time to condense.

4.3.4 Bimetal Thermostatic Trap

If a bimetal thermostatic trap blows steam, check for dirt and wear on the valve. A bimetal trap has limited power to close because of its method of operation, and the valve may be held off its seat by the accumulation of soft deposits.

This type of trap is usually supplied preset. Check that any locking device on the manual adjustment is still secure. If it is not secure, see if the trap responds to adjustment. If cleaning has no effect, replace all of the internal parts.

The valve on a bimetal trap is on the downstream side of the valve orifice, so these traps tend to fail in the open position. Failure to pass cold condensate indicates either gross maladjustment or complete blockage of the valve orifice or built-in strainer.

4.3.5 Float and Thermostatic Trap

If an F and T trap blows steam, check the trap for dirt at the main valve and the air vent valve. If the trap has a steam lock release, check to make sure that it is not opened too far. It should be open no more than one quarter of a full turn.

Make sure that the valve mechanism has not been knocked out of line either by rough handling or water hammer, preventing the valve from seating. Check that the ball float is free to fall without fouling the casing, which would cause the mechanism to hang up.

Test the air vent the same way as the element of a balanced pressure trap is tested. If the internal parts of a float trap must be replaced, install a complete set as supplied by the manufacturer.

If the trap will not pass condensate, check that the maximum operating pressure marked on the trap is not lower than the actual pressure to which the trap is subjected. If it is, the valve cannot open. Install a valve and seat assembly with the correct pressure rating. Make sure that this has sufficient capacity to handle the maximum load. If the ball float is leaking or damaged, it is probably caused by water hammer. Also, be sure that the air vent or steam lock release is working correctly.

4.3.6 Thermodynamic Disc Trap

If a thermodynamic disc trap blows steam, the trap will probably give a continuous series of abrupt discharges. Check the trap and strainer for dirt and wipe the disc and seat clean. If this does not correct the problem, it is possible that the seat faces and disc have become worn. The extent of wear is evident by the amount of shiny surface that replaces the normal crosshatching of machining.

If records show that thermodynamic traps on one particular installation suffer repeatedly from rapid wear, the trap may be oversized, associated pipework may be undersized, or there may be excessive back-pressure.

If the trap will not pass condensate, the discharge orifices may be plugged with dirt. This is most likely caused by air binding, particularly if it occurs regularly during startup. Check the air venting arrangements of the equipment. In extreme cases, it may be necessary to install an air vent parallel with the trap, or to use a float trap with a built-in air vent instead of a thermodynamic trap.

4.4.0 Troubleshooting Steam Traps

To diagnose and solve problems with steam traps requires listening to the noises they make and measuring temperature and pressure. The temperature of steam can be measured with handheld or remote-sensing pyrometers that read in degrees centigrade or Fahrenheit. Steam is produced at 212°F when at atmospheric pressure. When the steam systems are functioning, the temperature will be substantially higher, as it is confined inside the system. The pressure will be higher as well. Pressure readings are in two scales, absolute (psi) and gauge (psig). Absolute pressure is read from zero up, which is to say that normal atmospheric pressure is 14.7 psi at sea level. Gauge pressure is relative to atmospheric pressure at sea level; therefore, zero psig is equal to 14.7 psi absolute. Pressures below zero gauge are expressed in inches of mercury. Steam trap manufacturers have excellent books and web documents that detail how to troubleshoot and repair steam traps.

4.4.1 Diagnostic Methods

Three basic diagnostic methods of reading a steam system are sight, sound, and temperature. Another process, less used to date, is the use of a conductance probe, which compares the difference in electrical conductivity between condensate and live steam.

Each of the three main methods has advantages and limitations. Sight is fairly immediate; it is easy to tell whether there is condensate coming from a trap. If the condensate line is open, or if a test valve has been added on the condensate line, the condensate can be vented to the atmosphere. It will be obvious that the trap is removing condensate from the line. If live steam is going into the condensate line, the trap has failed open, and it is necessary to resolve the problem. If no condensate is coming from the line, the trap may have failed closed. The limitation of sight diagnosis is that the worker must be able to tell the difference between flash steam and live steam, and it is necessary for the worker to check by activating a test valve. It is also possible for the trap to be overloaded or loaded from the return system, producing an apparent open failure.

> **WARNING!**
>
> In a high-pressure steam system, the condensate will flash back to steam when the pressure drops as the result of exposure to the atmosphere through a leak. This exposure causes a rapid change in volume, which can result in the release of high-pressure steam.

The two most commonly used diagnostic tools are temperature sensing and sound, either at human hearing ranges or ultrasound. Note that temperature and sound together offer more information than one or the other alone.

From a diagnostic point of view, there are two general types of traps. Traps either flow or dribble continuously or intermittently (open and close). F and T traps discharge fairly continuously if they are working properly. Inverted bucket, bellows thermostatic, thermodynamic, and most thermostatic traps flow intermittently.

A few common characteristic symptoms of traps when tested might include the following:

- A thermal analysis will show the inlet as the area at the highest temperature for any of the traps. In most functioning traps, the condensate outlet, whether it is a closed return or an open return, should be cooler than the inlet. Remember that the condensate will most frequently be at or near steam temperature. However, it should still be at a lower temperature than the steam inlet. The live steam inlet should be hotter than the condensate exiting the trap. If this is not the case, the trap is probably malfunctioning.

- Normal failure for an inverted bucket trap is open failure. The trap loses its prime, and steam and condensate blow through steadily. If this happens, the continuous rushing sound of the blow-through may be accompanied by the bucket banging against the inside of the trap. The temperature of the trap and of the condensate line will rise because it is no longer holding condensate. The sound of the bucket linkage rattling inside the trap may indicate that the trap is beginning to loosen up and needs checking. With ultrasound equipment, the inverted bucket or thermodynamic traps would show a cyclic curve on the screen in normal function.

- Normal failure for a thermodynamic trap is open; that is, passing steam. Usually, the disk clicks shut audibly once for each cycle. When the disk is cycling normally, it will shut about 4 to 10 times a minute. When the trap fails, the disk no longer clicks as it rises and falls, and steam blows through. If the disc produces a continuous, rapid, rattling sound, the disc is worn, and more problems will probably develop. If a continuous hissing sound is produced, the trap is not cycling. Again, the condensate line will heat up due to escaping steam and the absence of cooler condensate.

- Bellows or bimetallic thermostatic traps operate on a difference in temperature between condensate and steam. Both types are intermittent in operation. When there is little condensate, they remain closed most of the time. When there is a lot of condensate, for example at startup, they may run continuously for a long time. In the case of bimetallic traps, misalignment may allow steam leaks. With sound testing equipment, a constant rushing sound in the closed part of the cycle would indicate a leak.

- Float and thermostatic traps tend to fail closed. A leaky float will not float on the condensate, or the float may have been crushed by water hammer. In either case, the F and T trap will remain closed and will not cycle. The trap will remain cool, and a temperature sensor will reveal that. With listening equipment, a failed float will be silent in the absence of discharging condensate.

- The thermostatic element in an F and T trap is normally quiet; a rushing sound would indicate that the element had failed open. A rattling or metallic clanging noise might mean that the mechanical linkage had sustained some damage.

The sight method uses a test valve that vents the process steam to the atmosphere for visual inspection. This test is subjective and depends on skill and experience. Most traps are dealing with condensate that is at steam temperatures; that is, somewhere above the boiling point of water. When condensate is released from the pressure of the steam system in the trap or in a condensate release, a certain amount of the condensate will immediately turn into flash steam. The difference is fairly visible once you have seen it a few times; the live steam comes out at first as a hard, blue-white straight flow, often with a clear bluish area at the outlet. Flash steam usually billows and spreads more quickly, as much of the pressure is dissipated in the initial release. The sound probes are either audible-range or ultrasonic-range probes that pick up noise from the steam system and send an audible signal to headphones. The technician performing this test must also have skill and experience at interpreting the noise as leaks, discharges, or other steam trap problems. *Figure 44* shows a portable ultrasonic tester.

Temperature-sensing tests are done with infrared thermometers (*Figure 45*), also known as pyrometers, that are handheld for external readings, or use mounted thermowells (*Figure 46*) that house devices to read the steam temperature. A thermowell is a permanently installed well or cavity in a process pipe or tank into which a glass thermometer or thermocouple can be inserted. The thermometer or thermocouple can be temporarily or permanently installed.

03306-13_F45.EPS

Figure 45 Pyrometer.

03306-13_F46.EPS

Figure 46 Thermowells.

03306-13_F44.EPS

Figure 44 Ultrasonic tester.

Additional Resources

Mechanical Products Manual, 14th Edition, available at **www.spiraxsarco.com**.

4.0.0 Section Review

1. Installation of an inverted bucket trap requires it to be _____.

    a. primed before use
    b. mounted with the outlet up
    c. mounted with the inlet horizontal
    d. mounted with the inlet vertical

2. Steam traps must be installed with a strainer downstream from the trap.

    a. True
    b. False

3. A problem that can be isolated by installing a check valve upstream of an inverted bucket trap is a trap that is _____.

    a. blowing steam
    b. not passing condensate
    c. causing water hammer
    d. not venting

4. In a malfunctioning trap passing steam, the outlet temperature is usually _____.

    a. the same as the inlet temperature
    b. lower than the inlet temperature
    c. 10°F higher than the inlet temperature
    d. at least 25°F higher than the inlet temperature

SUMMARY

Steam systems generate and distribute steam used for comfort heating and commercial and industrial processes. Steam heating systems can be separate systems used only to heat buildings, or they can provide energy in a combined system for both comfort and process use.

Boilers used in steam systems are divided into the two broad categories of low-pressure and high-pressure boilers. Low-pressure steam boilers are normally used in residential and commercial applications for both heating and process use. High-pressure boilers are typically used in larger systems for college campuses, hospitals, and industrial applications.

Steam piping systems are either one pipe or two pipe in design. The piping approach is determined by the requirements to supply steam to the terminals and by the destination of the condensate. Condensate can be returned by gravity when appropriate, but many systems will require mechanical assistance when the condensate is to be returned to the boiler for re-use.

Steam traps provide a means to separate live steam from condensate, trapping the steam inside the terminal until it has condensed, thereby giving up the vast majority of the energy it contains before being allowed to pass through the trap. Steam traps automatically open an orifice, drain away accumulated condensate, and then re-close once the condensate is removed. They can be classified based on their operating characteristics as mechanical, thermostatic, or thermodynamic. Steam traps that fail can be quite costly in terms of energy loss, so the proper maintenance of steam traps is crucial to control energy costs and ensure good system performance.

1. The amount of latent heat required to vaporize or evaporate, 5 pounds of water into steam is _____.
   a. 970 Btus
   b. 1,114 Btus
   c. 1,294 Btus
   d. 4,850 Btus

2. The latent heat of condensation is defined as _____.
   a. the heat removed during the change of state from steam to water
   b. the heat gained during the change of state from water to steam
   c. the heat lost in changing water to ice
   d. any heat added after ice changes to water

3. The majority of maintenance attention in a steam system is generally devoted to the _____.
   a. piping system
   b. pressure regulating valves
   c. steam traps
   d. heat exchangers

4. Steam boilers _____.
   a. have exactly the same type of controls as hot water boilers
   b. are entirely filled with water
   c. must operate at extreme pressures only
   d. are partially filled with water

5. A boiler with a maximum operating pressure of 75 psi would be classified as a _____.
   a. high-pressure boiler
   b. low-pressure boiler
   c. process boiler
   d. medium-pressure boiler

6. An assembly generally installed between the boiler and an external pressure gauge to protect the gauge from contacting raw steam is a _____.
   a. strainer
   b. siphon
   c. heat exchanger
   d. trap

7. Steam-to-water heat exchangers designed for potable hot water use are of _____.
   a. stainless steel construction
   b. aluminum construction
   c. hermetically sealed construction
   d. double-walled construction

8. When the condensate return line is above the boiler water line, it is called a _____.
   a. steam line
   b. Hartford loop line
   c. dry return line
   d. wet return line

9. Most two-pipe steam systems use a _____.
   a. mechanical condensate return
   b. gravity condensate return
   c. fixed-orifice trap
   d. very long startup cycle

10. In a vacuum return system, excessive vacuum pump operation can be caused by _____.
    a. too much condensate
    b. low steam pressure
    c. air leaks
    d. heat exchangers

11. To maintain and ensure a sufficient water supply is available for the boiler, the condensate receiver tank is equipped with a _____.
    a. flash reservoir
    b. makeup water valve
    c. thermodynamic trap
    d. vacuum pump

12. When more than one condensate line is connected to a flash tank, each line should be equipped with its own _____.
    a. check valve
    b. pressure reducing valve
    c. condensate pump
    d. sight glass

13. Collected flash steam can be discharged into a _____.
    a. floor drain
    b. vacuum pump
    c. boiler
    d. low-pressure main

14. Operating steam boilers at pressures significantly lower than their design operating pressure increases the likelihood of _____.

    a. excessive leaks
    b. thermal shock and hammering
    c. water droplet carryover
    d. trap damage

15. Steam system piping that is too large results in _____.

    a. excessive pressure drop and increased system noise
    b. increased trap failures
    c. high installation costs and greater heat loss
    d. condensate pump overload

16. Steam supply piping for a system designed to operate at 5 psi should generally be designed for a pressure drop no greater than _____.

    a. 0.3 psi
    b. 0.5 psi
    c. 1.0 psi
    d. 2.0 psi

17. Mechanical steam traps take advantage of the difference in density between steam and _____.

    a. air
    b. steam condensate
    c. gases
    d. debris

18. Fixed orifice steam traps are advantageous because they _____.

    a. are practical under most conditions
    b. have very precise flow rates at all times
    c. have no moving parts
    d. always pass condensate at startup

19. Bimetallic steam traps function based on the difference between condensate and steam _____.

    a. air volume
    b. temperature
    c. weight
    d. pressure

20. The three basic diagnostic methods for steam systems, especially traps, are _____.

    a. pressure, temperature, and time
    b. temperature, flow, and sound
    c. current, pressure, and sound
    d. sight, sound, and temperature

Drip leg: A drain for condensate in a steam line placed at a low point or change of direction in the line and used with a steam trap.

Dry steam: The steam that exists after all the water has been vaporized into steam at its saturation temperature (being superheated).

Flash steam: Formed when hot condensate is released to a lower pressure and re-evaporated.

Saturated steam: The pure or dry steam produced at the temperature that corresponds to the boiling temperature of water at the existing pressure (saturation).

Trim: External controls and accessories attached to the boiler itself, such as sight glasses and water feeder controls.

Turndown ratio: In steam systems, the ratio of downstream pressure to upstream pressure, usually applied to steam pressure-reducing valves. More than a 50 percent turndown ratio is generally considered a large pressure reduction.

Water hammer: A condition that occurs when hot steam comes into contact with cooled condensate, builds pressure, and pushes the water through the line at high speeds, slamming into valves and other devices. Water hammer also occurs in domestic water systems. If the cause is not corrected, water hammer can damage the system.

Additional Resources

This module presents thorough resources for task training. The following resource material is suggested for further study.

The Control of Boilers, 2nd Edition. Sam G. Dukelow. Research Triangle Park, North Carolina: International Society of Automation (ISA).

High Pressure Boilers, 4th Edition. Fredrick M. Steingress, Harold J. Frost, Daryl R. Walker. Orland Park, Illinois: American Technical Publishers.

Low Pressure Boilers, 3rd Edition. Fredrick M. Steingress, Daryl R. Walker. Orland Park, Illinois: American Technical Publishers.

Mechanical Products Manual, 14th Edition, available at **www.spiraxsarco.com**.

Piping for High Pressure Boilers: The Installation and Inspection of High-Pressure Boiler Piping for Code Compliance with the ASME and National Board Code Requirements. Steve Kalmbach. Tucson, Arizona: Wheatmark®.

Figure Credits

Topaz Publications, Inc., Module opener, Figures 26, 38

Courtesy of Fulton Boiler Works, Inc., Figures 6, SA02

Courtesy of ECR International, Inc., Figure 7

Kele, Figure 8

Xylem Applied Water Systems, Figure 13

Courtesy of **PexSupply.com**, Figure 14

Courtesy of Honeywell International, Figure 15

Anderson Greenwood, Figure 16

Leslie Controls Inc., Figures 17, 37

Courtesy of Reznor/Thomas & Betts, Figure 19

McQuay International, Figures 20, 21

Courtesy ITT, Figures 22, 24, 25, 32

Courtesy of NES Company Inc., Figure 28

Watson McDaniel, Figure 41 (photo)

Courtesy of UE Systems Inc., Figure 44

Fluke Corporation, Figure 45

Weed Instrument Co., Figure 46

Answer	Section Reference	Objective
Section One		
1.b	1.1.0	1a
2.a	1.2.0	1b
Section Two		
1.c	2.1.0	2a
2.b	2.2.1	2b
3.b	2.3.1	2c
Section Three		
1.b	3.1.2	3a
2.a	3.2.4	3b
3.c	3.3.0	3c
4.b	3.4.1	3d
Section Four		
1.a	4.1.1	4a
2.b	4.2.0	4b
3.a	4.3.1	4c
4.a	4.4.1	4d

NCCER CURRICULA — USER UPDATE

NCCER makes every effort to keep its textbooks up-to-date and free of technical errors. We appreciate your help in this process. If you find an error, a typographical mistake, or an inaccuracy in NCCER's curricula, please fill out this form (or a photocopy), or complete the online form at **www.nccer.org/olf**. Be sure to include the exact module ID number, page number, a detailed description, and your recommended correction. Your input will be brought to the attention of the Authoring Team. Thank you for your assistance.

Instructors – If you have an idea for improving this textbook, or have found that additional materials were necessary to teach this module effectively, please let us know so that we may present your suggestions to the Authoring Team.

NCCER Product Development and Revision

13614 Progress Blvd., Alachua, FL 32615

Email: curriculum@nccer.org
Online: www.nccer.org/olf

❏ Trainee Guide   ❏ Lesson Plans   ❏ Exam   ❏ PowerPoints   Other _____

Craft / Level: _____ Copyright Date: _____

Module ID Number / Title: _____

Section Number(s): _____

Description: _____

_____

_____

_____

Recommended Correction: _____

_____

_____

_____

_____

Your Name: _____

Address: _____

_____

Email: _____ Phone: _____

03304-13

Retail Refrigeration Systems

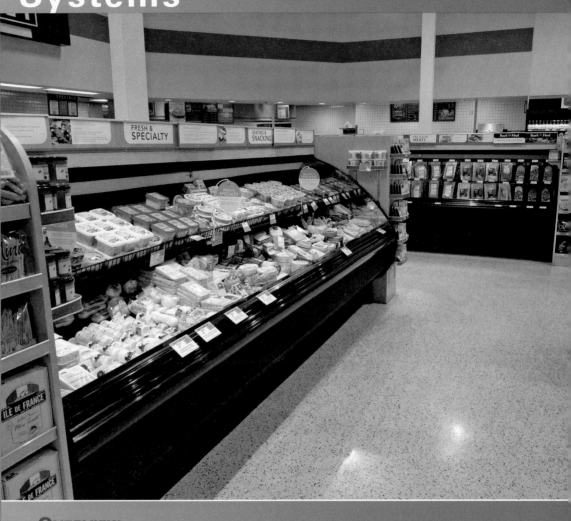

OVERVIEW

Retail refrigeration systems are everywhere. Stop by any convenience store—you will immediately see a large reach-in cooler filled with snacks and beverages. Need some ice for that cooler? Grab a bag or two from the reach-in freezer. When you go grocery shopping, you will see row upon row of refrigerated cases containing meats, dairy products, and all sorts of pre-packaged foods.

In fact, refrigerated cases and refrigerated dispensers of all kinds are widely used in retail establishments that sell or serve food and beverages. The specialized field of retail refrigeration offers many career opportunities for HVAC/R service technicians.

Module Eleven

Objectives

When you have completed this module, you will be able to do the following:

1. Describe retail refrigeration applications and the related refrigeration cycle.
   a. Describe the operation of the refrigeration cycle for medium-temperature systems.
   b. Describe the operation of the refrigeration cycle for low-temperature systems.
   c. Describe the various approaches to defrosting.
2. Identify various types of refrigeration equipment used in retail refrigeration applications.
   a. Identify reach-in coolers and freezers.
   b. Identify walk-in coolers and freezers.
   c. Identify various types of commercial ice machines.
3. Identify and describe components related to the operation of retail refrigeration systems.
   a. Identify and describe the primary components used in retail refrigeration applications.
   b. Identify and describe secondary components and accessories used in retail refrigeration applications.
   c. Identify and describe common refrigeration system controls.
4. Explain the basic principles of maintaining and troubleshooting various retail refrigeration systems.
   a. Explain how to maintain and troubleshoot a typical reach-in freezer.
   b. Explain how to maintain and troubleshoot a typical cubed-ice machine.

Performance Tasks

Under the supervision of your instructor, you should be able to do the following:

1. Clean an ice machine.
2. Isolate a fault in refrigeration equipment or an ice machine.
3. Set up a defrost timer for a freezer.

Trade Terms

Phase change
Slinger ring
Unit cooler

Industry-Recognized Credentials

If you are training through an NCCER-accredited sponsor, you may be eligible for credentials from NCCER's Registry. The ID number for this module is 03304-13. Note that this module may have been used in other NCCER curricula and may apply to other level completions. Contact NCCER's Registry at 888.622.3720 or go to **www.nccer.org** for more information.

Contents

Topics to be presented in this module include:

Figures and Tables

1.0.0 BASIC REFRIGERATION SYSTEM OPERATION

Objectives

Describe retail refrigeration applications and the related refrigeration cycle.

a. Describe the operation of the refrigeration cycle for medium-temperature systems.
b. Describe the operation of the refrigeration cycle for low-temperature systems.
c. Describe the various approaches to defrosting.

Trade Terms

Phase change: The conversion of refrigerant from one state to another, such as from liquid to vapor.

How many times each day does refrigeration impact your life? How different and difficult would life be without mechanical refrigeration? In earlier times, ways of preserving foods were limited to curing, dehydrating, or using natural resources such as snow and winter ice to avoid spoilage.

If not for bacteria, many foods could remain edible for long periods of time. By the mid-1800s, it was understood that bacteria are responsible for most food spoilage. After it was discovered that bacteria are unable to grow at lower temperatures, development of modern approaches to refrigeration began in earnest. Throughout the 1800s, advances continued in methods to protect and preserve foods. In 1880, a ship carrying meat from Australia to England was fitted with an apparatus to cool the meat for the voyage. However, the meat was accidentally frozen. That unexpected result led to the intentional freezing of meat for future shipments.

Today's retail refrigeration systems and applications are evident at every turn. Although the preservation of a variety of products remains the primary function, style and convenience built into refrigerated fixtures play a significant role in marketing as well. Quiet, visually appealing equipment attracts the consumer, increases sales, and helps to reinforce the feeling of confidence that contents are safe and properly preserved.

Many types and styles of mechanical refrigeration systems are used to cool or freeze foods and other perishables. In most cases, consumable products reach the retail level already cooled or frozen.

The retail fixture then is used to maintain those conditions while providing easy access to the products. Operation of systems used for these applications is basically the same as that of comfort cooling systems, with the addition of components and controls necessary to achieve the desired refrigerated effect in the fixture's environment. *Figure 1* shows a basic air-cooled refrigeration system flow diagram.

Primary refrigerant circuit components are the compressor, evaporator, condenser, and expansion (metering) device. However, when the circuit is applied to refrigeration, a number of other factors are different. The installed location of primary components and interconnecting lines, evaporator temperatures, condensing environment, and types of refrigerants used are all significant variables.

Operation of the basic refrigeration cycle and its components are described in detail in *HVAC Level One*. Reviewing that material may assist in understanding differences in the operating cycle when applied to lower-temperature applications.

Refrigeration applications are generally classified by required space temperature for a given application. These classifications are important to remember because some components, such as compressors and expansion devices, are classified for duty according to the following temperature ranges:

- *Above 60°F* – Air conditioning/high-temperature refrigeration
- *30°F to 60°F* – Medium-temperature refrigeration
- *-40°F to 30°F* – Low-temperature refrigeration

The basic refrigeration cycle for most applications is similar to that for air conditioning. However, many technicians are uneasy about servicing refrigeration systems because they have spent much of their career working with comfort cooling systems where a 40°F evaporator temperature is the established standard. In refrigeration, different evaporator temperatures must be reached to achieve the desired storage temperature. In reality, while simple changes in a few components yield far different results, the theory of operation remains the same.

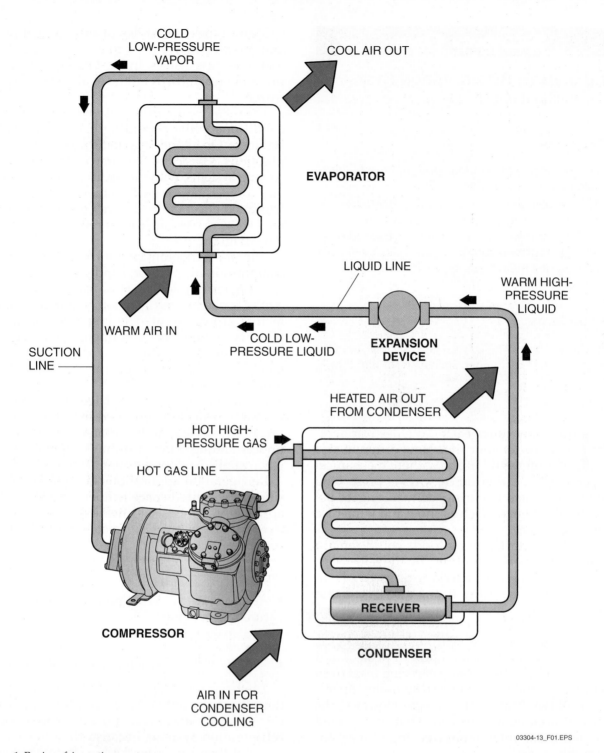

COLD
LOW-PRESSURE
VAPOR

COOL AIR OUT

EVAPORATOR

LIQUID LINE

WARM HIGH-
PRESSURE
LIQUID

WARM AIR IN

COLD LOW-
PRESSURE LIQUID

EXPANSION
DEVICE

SUCTION
LINE

HEATED AIR OUT
FROM CONDENSER

HOT HIGH-
PRESSURE GAS

HOT GAS LINE

RECEIVER

COMPRESSOR

CONDENSER

AIR IN FOR
CONDENSER
COOLING

03304-13_F01.EPS

Figure 1 Basic refrigeration system.

1.1.0 Medium-Temperature Refrigeration Cycle

Refer to *Figure 2* for an example of a typical medium-temperature refrigeration cycle using HFC-134a as the refrigerant.

Assume that the system is for a retail fixture inside a conditioned space, such as a convenience store. The air temperature in the vicinity of the fixture and unit condenser is 75°F, and the de-sired box temperature is 38°F. In actual practice, these conditions can vary dramatically throughout the course of a day.

The following numbers correspond to numbered points shown in *Figure 2*. Note that this example is theoretical and does not account for normal pressure drops through piping or components in a normally operating system.

1. When the refrigerant leaves the expansion device, it is a mixture of liquid and vapor. The vapor results from some liquid flashing at the expansion device, absorbing heat to cool the remaining liquid to the desired evaporator temperature. A 75 percent liquid to 25 percent vapor volume can be considered reasonably normal.

To achieve the desired box temperature of 38°F, an evaporator temperature lower than 38°F is needed. For this example, use an evaporator temperature of 24°F at an HFC-134a pressure of 21.4 psig. This creates a temperature difference (ΔT) of 14°F between the refrigerant-saturated suction temperature in the evaporator coil and the desired box air temperature.

2. As the refrigerant mixture flows through the evaporator, it absorbs heat from the warmer air in the box and continues to change phase from liquid to vapor. The saturation temperature of the refrigerant remains unchanged until the **phase change** has been completed. The mixture of liquid and vapor will remain at 24°F until all liquid has returned to the vapor phase.

To ensure the maximum amount of evaporator area is providing usable refrigeration, the point where the phase change is complete is ideally located in the last 3 to 7 percent of the coil. Once phase change is complete, the vapor can begin to increase in temperature and superheat can be measured.

3. Refrigerant exiting the evaporator coil and entering the suction line should now be in a single state—vapor. The completion of the phase change process can be proven by the existence of measurable superheat. Temperature and pressure measurements taken at this point should reveal that the refrigerant has increased in temperature above that shown for saturation on the pressure-temperature (P-T) chart. This provides evidence that the phase change is complete and ensures that damaging liquid refrigerant is not on its way to the compressor.

In this example, the refrigerant has gained 8°F of superheat for a suction line temperature of 32°F. It is important to note that the suction line will often form frost, which should not be considered an indication of the presence of liquid refrigerant.

4. The vapor returning to the compressor picked up only a very small amount of heat (2°F) as it traveled through the suction line. In actual practice, this value can vary dramatically, depending on the length of the suction line, quality of insulation, and the ambient temperature in the area of travel. In a small reach-in retail case, the suction line may be as short as a few inches, while other applications with remote condensing units may require much longer lines.

Figure 2 Typical medium-temperature refrigeration cycle for HFC-134a.

5. Following the compression process, highly superheated vapor now exits the compressor. In this example, the discharge vapor temperature has reached 160°F at a pressure of 143 psig as the direct result of compression. A quick check of the P-T chart reveals that the superheat is now at 52°F. All of this superheat must be removed from the vapor before condensation back to the liquid state can begin. A small amount of this superheat is often lost to the surrounding air as the vapor travels to the condenser. The hot gas line is rarely insulated,

and a large temperature difference between the line and the surrounding air is common.

6. As the refrigerant vapor enters the condenser coil, superheat is quickly removed initially. As the refrigerant temperature continues to fall toward the temperature of the condenser air, the DT between condenser air and refrigerant is reduced and the process slows somewhat. All superheat must be removed and the vapor temperature must fall to the saturation temperature for HFC-134a at the given pressure before the condensing process can begin. For this example, the appropriate condensing temperature would be 108°F.

7. When the vapor has cooled to 108°F in the condenser, condensation begins and the temperature will remain unchanged throughout the process. As refrigerant progresses through the condenser, the ratio of liquid to vapor constantly changes until only liquid remains.

8. When the phase change is complete, refrigerant liquid can begin to experience measurable reductions in temperature and be subcooled. In this example, the refrigerant temperature has fallen to 93°F at the exit of the condenser. Because the saturation temperature for HFC-134a is 108°F at the pressure of 143 psig, the subcooling is calculated to be 15°F.

   Remember that without measurable subcooling as an indicator, the condensation process has not been completed and some amount of vapor remains in the liquid. This causes a serious loss of refrigerating capacity in the system, because any vapor that travels through the expansion device and into the evaporator is virtually useless in absorbing any significant amount of heat.

9. As is the case with other refrigerant lines connecting primary components, the heat loss or gain in the liquid line can be large or small. In this example, the liquid line is very short and has had a temperature reduction of only 1°F. If the liquid line travels through an area of great temperature difference, a far larger increase or decrease in temperature may occur.

10. At this point, the liquid refrigerant has passed through the expansion device, having a pressure drop of over 121 psig in the process. Although the refrigerant was subcooled and entirely in its liquid state as it entered the expansion device, some of the refrigerant quickly flashed into the vapor state, absorbing heat and cooling the remaining liquid to

the proper saturation temperature for the new pressure condition of 21.4 psig. The cycle is complete and ready to begin again.

1.2.0 Low-Temperature Refrigeration Cycle

Although the cycle used to produce freezing temperatures is not significantly different, an example using HFC-404A refrigerant is provided in a briefer format. Remember that HFC-404A, although classified as a near-azeotropic blend, has a very small glide. This glide can be considered insignificant in normal servicing of the system.

Figure 3 represents a typical low-temperature refrigeration circuit, such as that used in a reach-in freezer. For this example, assume that the condenser entering air temperature is 75°F, since the condenser is located indoors. However, the desired box temperature will now be +10°F.

1. To achieve the desired box temperature of +10°F, a much lower saturated refrigerant temperature must be created in the evaporator. For this example, use an evaporator temperature of –5°F, at an HFC-404A pressure of 28.9 psig. This creates a temperature difference (ΔT) of 15°F between the refrigerant in the evaporator coil and the desired box air temperature.

2. The mixture of liquid and vapor in the evaporator will remain at –5°F until all liquid has returned to the vapor phase. To ensure that the maximum amount of evaporator area is providing usable refrigeration, the point where the phase change is complete is ideally located in the last 3 to 7 percent of the coil. When the phase change is complete, the vapor can begin to gain heat and superheat can be measured.

3. Refrigerant exiting the evaporator coil and entering the suction line should now be in a single state—vapor. The completion of the phase change process can be proven by the existence of measurable superheat. Temperature and pressure measurements taken at this point should reveal that the refrigerant has increased in temperature above that shown for saturation on the P-T chart, providing evidence that the phase change is complete. This ensures that damaging liquid refrigerant is not on its way to the compressor. In this example, the refrigerant vapor has increased in temperature 6°F (superheat) following the phase change.

Figure 3 Typical low-temperature refrigeration cycle for HFC-404A.

°F	R-404A	°F	R-404A	°F	R-404A	°F	R-404A
−10	24.6	24	62.0	58	121.7	92	210.7
−8	26.3	26	64.8	60	126.0	94	217.0
−6	28.0	28	67.7	62	130.5	96	223.4
−4	29.8	30	70.7	64	135.5	98	230.0
−2	31.7	32	73.8	66	139.7	100	236.8
0	33.7	34	76.9	68	144.4	102	243.6
2	35.7	36	80.2	70	149.3	104	250.6
4	37.7	38	83.5	72	154.3	106	257.8
6	39.8	40	86.9	74	159.4	108	265.1
8	42.0	42	90.4	76	164.6	110	272.5
10	44.3	44	94.0	78	169.9	112	280.1
12	46.6	46	97.6	80	175.4	114	287.9
14	49.0	48	101.4	82	181.0	116	295.8
16	51.5	50	105.3	84	186.7	118	303.8
18	54.0	52	109.2	86	192.5	120	312.1
20	56.6	54	113.3	88	198.4		
22	59.3	56	117.4	90	204.5		

PRESSURE-TEMPERATURE CHART

03304-13_F03.EPS

4. Following compression, highly superheated vapor exits the compressor. In this example, the discharge vapor temperature has reached 160°F at a pressure of 264 psig.

5. All superheat must be removed and the vapor temperature must fall to the saturation temperature for HFC-404A at the given pressure before the condensing process can begin.

For this example, the appropriate condensing temperature would be 108°F. When the vapor has cooled to 108°F in the condenser, the temperature will remain unchanged throughout the process of condensation. As refrigerant moves through the condenser, the ratio of liquid to vapor constantly changes until only liquid remains.

6. In this example, the refrigerant temperature has fallen to 93°F at the exit of the condenser. Because the saturation temperature for HFC-404A is 108°F at a pressure of 264 psig, the subcooling is 15°F.

7. At this point, the liquid refrigerant has passed through the expansion device, having a pressure drop of roughly 235 psig in the process. The cycle is complete and ready to begin again.

Refrigeration cycles are generally the same. Differing temperatures and pressures in various systems result from characteristics of the different refrigerants and metering device used, but the process remains unchanged. Armed with correct information and an understanding of the cycle, a technician should have no trouble understanding system operating characteristics and identifying when the refrigeration cycle is not operating as designed.

One important note about superheat values for refrigeration applications when compared to comfort cooling. Although typical desired superheat values for comfort cooling may be 10°F to 20°F, refrigeration systems generally operate at lower superheat values. As saturated suction temperatures are reduced, compression ratios become higher and compressor energy efficiency is significantly reduced. Energy efficiency ratios (EERs) for medium-temperature applications are typically around 7.0 to 8.0, while low-temperature systems may operate with EERs as low as 4.0.

As a result, gaining as much capacity per watt from the system as possible is essential for low-temperature applications. One way to achieve this is through reduced superheats, exposing as much evaporator internal volume as possible to boiling liquid refrigerant. All significant heat exchange takes place in this phase-change process.

Once the liquid has boiled into vapor, little heat is exchanged as it superheats. Superheat values of 6°F to 12°F, measured at the evaporator, are considered acceptable for high- and medium-temperature applications. Superheat values for low-temperature applications may range from 3°F to 8°F to maximize the refrigerating effect in the evaporator. Great care must be taken to ensure adequate compressor protection from liquid floodback exists when low superheat values are chosen. However, there is much to gain in energy efficiency.

1.3.0 Defrost Systems

As long as moisture is inside the refrigerated space, it will condense on the surface of the evaporator, as it does in comfort cooling systems. However, when the coil surface is below freezing, frost and ice will quickly form, slowing and eventually blocking airflow through the coil. The coil must be defrosted to resume proper operation.

In retail refrigeration applications, there are four primary defrost methods:

- Off-cycle defrost
- Timed defrost
- Electric defrost
- Hot-gas defrost

Regardless of the defrost approach used, several factors regarding frost accumulation on the evaporator coils are common. Fixtures that were selected and designed to operate at high air-to-refrigerant temperature differences will generally build frost faster, since their coil temperature will likely be well below the dew point for the box. This high air-to-refrigerant temperature difference (TD or ΔT) can be used when the product cannot dry out, such as canned or bottled drinks. This encourages greater moisture condensation on the coil, but is more energy efficient. Fixtures with low air-to-refrigerant temperature differences will not accumulate as much condensate nor frost as quickly. Foods that can be damaged through dehydration must be refrigerated with fixtures having low air-to-refrigerant TDs.

Another factor in frost accumulation is related to fixture usage and infiltration of outside air. Most retail refrigeration fixtures experience high customer usage. In hot, humid weather, a fixture that is opened many times each day will constantly be exposed to hot, moist air. Not only does this significantly increase the volume of airborne moisture in the enclosure, infiltration of hot

Refrigerant Phase-Out

The phase-out of refrigerants that contain chlorine means these refrigerants are becoming harder to find and far more expensive. To satisfy demand for these scarce refrigerants, people have resorted to smuggling chlorine-based refrigerants into the United States. This is a federal crime that can result in fines and/or imprisonment. Any service technician who knowingly purchases a smuggled or illegal refrigerant is also subject to prosecution.

air replacing refrigerated air greatly increases the required refrigeration operating time as well.

Conditions like this can overwhelm and interfere with fixture performance, in spite of defrost cycles. Consider the potential difference in performance of an identical reach-in refrigerated fixture installed in two different locations, one with little consumer traffic in Arizona, and the other in a busy, 24-hour convenience store in coastal Florida.

Understanding that all heat added to the refrigerated enclosure during the defrost cycle, regardless of the source, must again be removed when the process is complete is important. This fact has a large effect on the refrigeration load. In addition, the time a system is projected to spend in defrost mode must be subtracted from the available time for refrigeration. For example, if the total refrigeration load for a fixture is calculated as 48,000 Btus over a 24-hour period (2,000 Btuh), but the unit is anticipated to operate in defrost for 2 hours per day, then the refrigeration equipment must be able generate this much capacity in 22 hours instead of 24 hours. That would require 2,182 Btuh of capacity instead of 2,000.

1.3.1 Off-Cycle Defrost

Off-cycle defrost is a simple passive defrost method. In fixtures that maintain temperatures at or above 36°F, the coil is simply allowed to defrost during the normal off cycle. Fixtures in this category will usually operate at a coil/saturated suction temperature of 16°F to 31°F, depending on the desired humidity levels for the stored product and box design.

Because the fixture temperature is above freezing, the coil will build frost, but should defrost naturally when the unit has cycled off. The evaporator fan continues to run during the off cycle.

One assist to this natural defrost approach is to ensure that the fixture temperature is not set lower than needed. Maintaining the box even a few degrees colder than necessary can slow or defeat the natural defrost process, resulting in frozen coils and rising box temperatures. When

this occurs, the fixture requires shut down for a period long enough to rid the coil of ice.

1.3.2 Timed Defrost

Systems operating at slightly lower temperatures than those using simple off-cycle defrost will require a longer period to clear accumulated ice. Fixtures that operate at 32°F to 36°F often benefit from this approach. Typically, a 24-hour timer (*Figure 4*) with normally closed contacts, placed in series with the compressor controls, is set for a reasonable period of time, yet much longer than the typical off cycle. The evaporator fan continues to run during the off cycle.

Fixture temperature may rise a few degrees above normal during the extended period, but will quickly recover. Like all defrost scenarios, setting the timer for a defrost period that coincides with the fixture's lightest period of usage (late at night, for example) helps to prevent the fixture temperature from rising beyond an acceptable level. Several cycles per day may be required.

1.3.3 Electric Defrost

Electric defrost systems are common for medium- and low-temperature fixtures that operate at temperatures too cold for off-cycle or timed defrost to be effective, and for freezers in the retail refrigeration class. Electric defrost components are inexpensive, reliable, and simple to troubleshoot and repair.

Figure 5 shows a typical arrangement for resistive defrost heaters installed on both the face and back of an evaporator coil. For proper operation, the heater must be in good thermal contact with the coil fins to effectively distribute heat and minimize the required defrost period.

A timer stops refrigeration equipment and evaporator fans during the defrost, then energizes the electric heater. In many cases, the shape and size of the defrost heater are unique to a given evaporator coil and manufacturer. This requires original parts when replacement is necessary. Some coil designs require that the heater be

---

Infiltration

Badly worn door gaskets and hinges can seriously affect the performance and energy efficiency of a refrigerated fixture. The constant infiltration of warm and moist outside air through poorly fitted doors and gaskets can prevent even the best fixture from reaching its setpoint. It can also increase frost accumulation, further reducing coil heat transfer and increasing refrigeration equipment run-time and operating costs. Maintaining and replacing door gaskets and hinges with high wear and abuse is as important to a fixture's efficient function as having proper refrigerant charge.

Figure 4 24-hour defrost timers.

placed deep inside the coil assembly, making repair and replacement harder, but resulting in an effective and rapid defrost.

Some units are equipped with radiant electric heaters (*Figure 6*). The resistance heater wire is spirally wound and enclosed inside a quartz tube, with the wires passed through the ends of the sealed tube cap. Using infrared radiant heat, defrost can be accomplished without the heater being in contact with the coil. These heaters are capable of reaching a high temperature very quickly and are corrosion resistant. However, the quartz enclosure can be fragile.

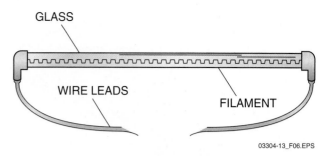

03304-13_F06.EPS

Figure 6 Infrared quartz heater assembly.

When electric defrost is required, condensate removed from the coil must also be maintained above freezing until it has completely left the cold environment. Most condensate collection pans, either as part of the evaporator coil assembly or as a separate collection point inside the fixture, must also be equipped with electric heaters such as the one shown in *Figure 7*.

Some installations with the drain line is routed through a freezing area may require the condensate drain line to be kept above freezing as well. For this, electric pipe heating cable (heat tape) is attached to the exterior of the line.

Electric defrost systems also use a 24-hour timer for defrost control, except for large and/or complex systems. Electromechanical timers have a dial assembly, as shown in *Figure 8*, allowing the timer to be configured for the time, duration, and number of defrost periods during a 24-hour period. The inner defrost dial sets the duration

03304-13_F05.EPS

Figure 5 Evaporator electric heat arrangement.

JIFFY CLIP

03304-13_F07.EPS

Figure 7 Condensate pan heater.

of the defrost period; defrost is terminated after the period has elapsed. The timer contacts then return to normal position and the refrigeration cycle resumes. The contacts can generally handle the electrical load of the defrost heater(s) directly; no additional relay or contactor is needed in retail fixtures.

Electric defrost is often terminated by a temperature sensor or line-mounted thermostat. By monitoring the temperature of the coil, or refrigerant suction line close to the coil, the thermostat indicates that the coil has reached the correct temperature. Defrost can be terminated before the timer has reached its allowed maximum time. The proper setpoint for temperature termination may differ among products, since this is based on the manufacturer's chosen location and laboratory testing. Line-mounted thermostats are generally non-adjustable, so using the correct replacement is important.

The final choice of defrost control and termination is often an option provided to the end user, either as a standard feature or at an added cost. Defrost termination can therefore be time-only, temperature-only, or a combination of the two.

1.3.4 Hot-Gas Defrost

Although not as common as the other defrost methods, hot-gas defrost is widely considered to be the fastest and most efficient. Because of the additional refrigerant circuit components required, this method generally costs more, making it less popular. It is typically used only on large systems.

Hot-gas defrost uses solenoid valves to send discharge gas directly from the compressor to the evaporator, bypassing the metering device. The hot gas quickly condenses in the cold evaporator as the frost melts. The latent heat of condensation raises the coil surface temperature above freezing. Care must be taken in these systems to avoid compressor overload and the potential for liquid refrigerant to return to the compressor.

TIME OF DAY INDICATOR

DEFROST CYCLE DURATION

ADDITIONAL PINS

TRIP PIN

03304-13_F08.EPS

Figure 8 Typical defrost timer.

Additional Resources

Refrigeration and Air Conditioning, An Introduction to HVAC/R, Latest Edition. Larry Jeffus. Upper Saddle River, NJ: Prentice Hall.

1.0.0 Section Review

1. In a medium-temperature refrigeration unit, phase change from liquid to vapor takes place in the _____.

   a. suction line
   b. evaporator
   c. liquid line
   d. condenser

2. A typical superheat range in low-temperature refrigeration applications is _____.

   a. 3°F to 8°F
   b. 6°F to 12°F
   c. 8°F to 16°F
   d. 10°F to 20°F

3. Condensate collection pans located in cold areas must be kept from _____.

   a. sweating
   b. dripping
   c. freezing
   d. corroding

2.0.0 RETAIL REFRIGERATION EQUIPMENT

Objectives

Identify various types of refrigeration equipment used in retail refrigeration applications.

   a. Identify reach-in coolers and freezers.
   b. Identify walk-in coolers and freezers
   c. Identify various types of commercial ice machines.

Trade Terms

Unit cooler: A packaged unit containing evaporator coil, metering device, and evaporator fan(s) that serves as the evaporator section for a refrigeration appliance.

Retail refrigeration equipment and systems include systems that are relatively mobile and can be installed in a variety of locations with a minimal amount of external utilities. Most retail systems, for our purposes, would be considered packaged or unitary. Although the refrigerated cases in supermarkets may be considered under the retail heading, they represent some of the most complex systems in the industry and are rarely self-contained. As a result, they are not reviewed in great detail here.

One important consideration for many packaged retail refrigeration units is their environment. Many manufacturers specify that indoor units are designed for indoor spaces maintained at 75°F and 55 percent relative humidity. Although most units are placed in areas that conform to these requirements, many are not. Technicians and end-users should not expect reliable operation from units placed in areas that are excessively hot and humid.

In this section, we will review the following types of equipment and their features:

- Reach-in coolers and freezers
- Merchandising walk-in systems
- Commercial ice makers

2.1.0 Reach-In Coolers and Freezers

Reach-in coolers and freezers are available in styles and types to suit almost any merchandis-

ing need. Despite the large variety available, refrigeration cycles and operating sequences are similar. The refrigerated enclosure is insulated with urethane or polystyrene materials, either in board-form or sprayed.

Figure 9 represents a typical two-door reach-in unit. Some typical features and options for these units include the following:

- Interior lighting
- Self-closing doors that can be locked open for stocking
- A variety of shelving options
- Shelf moldings to allow for price tags
- Digital thermometers
- A unitary condensing unit that can be removed for service or replacement
- Adjustable leveling legs

Figure 10 shows a typical under-counter unit used to store food or beverages. Because units of this type are not directly accessible by consumers, appearance is less important than a durable and maintainable design.

Typical options and features for this style include the following:

- Stainless steel cabinet construction
- A unitary condensing unit that can be removed for service or replacement

03304-13_F09.EPS

Figure 9 Two-door upright reach-in merchandiser.

Figure 10 Under-counter reach-in unit.

- Casters for easy movement, or adjustable legs
- Epoxy-coated evaporator coils to protect the aluminum from corrosion
- Exterior thermometers

Condensing units in packaged systems are mounted either above or below the refrigerated case, depending on manufacturer preference and fixture style. Medium-temperature units may incorporate either off-cycle or timed defrost, with electric defrost available as a factory-installed option. Low-temperature freezer units are equipped with electric defrost almost exclusively. Units using HFC-134a or HFC-404A as a refrigerant dominate today's market. The metering device is usually a capillary tube or thermostatic expansion valve (TEV).

2.1.1 Ice Merchandisers

Ice merchandisers for bagged retail ice have a simple design and function. Models are designed for indoor or outdoor use in sizes ranging from roughly 25 cubic feet to over 100 cubic feet of internal volume. Indoor models (*Figure 11*) are generally equipped with a glass door with internal lighting. Outdoor units (*Figure 12*) have metal doors that can be locked and are built to withstand the elements and possible physical abuse. Both styles are usually controlled by a simple thermostat at box temperatures of 20°F to 25°F.

Ice merchandisers are generally offered with two major options—cold-wall construction, which has no automatic defrost, and unitary evaporator assemblies, which have electric defrost. In cold-wall models, the cabinet walls are actually plate evaporators, with the serpentine evaporator tubing bonded to the metal sheets. No evaporator fan is used and defrost is accomplished when the merchandiser is empty. Simply shut down the unit

Figure 11 Indoor ice merchandiser.

Figure 12 Outdoor ice merchandiser.

and allow it to warm to ambient temperature. An example of this type is shown in *Figure 12*. Those that use unitary evaporator coil assemblies have a defrost timer and use electric defrost elements on the evaporator coil. *Figure 11* is an example of this type; the coil assembly is visible at the top of the box, where the evaporator fan is visible.

2.2.0 Merchandising Walk-In Systems

Many of today's convenience stores feature merchandising walk-in cooler and freezers (*Figure 13*). These units are simply walk-in coolers or freezers with one or more walls replaced by doors, allowing the consumer direct access to products.

Specialized racks placed directly behind the doors allow for rapid stocking from within the walk-in unit itself. Products are received and stored inside the same enclosure, never leaving the refrigerated area except in the consumer's hands. Well-stocked racks at the door opening help to block the consumer's view of perishables in storage. The merchandising rack system often has product replacement feeding, allowing a new product to fall or slide into place as one is removed.

Walk-in coolers generally use **unit coolers** and remote condensing units, located either outdoors or indoors on top of the refrigerated enclosure. Because the refrigeration capacity need is often larger than smaller self-contained fixtures, semi-hermetic compressors are often used. TEVs are most often used as the metering device. The doors of these units are often equipped with a means to quickly defog the glass after the door has been opened and closed. Freezers have low-wattage electric heaters around the doorframe to prevent the door from freezing closed.

03304-13_F13.EPS

Figure 13 Walk-in merchandiser.

2.3.0 Commercial Ice Makers

Packaged ice-making equipment can be divided between two primary types: cubed-ice machines and flaked-ice machines. Both types are widely used in restaurants, convenience stores, hotels, and other point-of-use locations. Typically, ice is formed on a surface at roughly 10°F. This is the target surface temperature for many ice machines.

The two types of ice machines have some common features. They are designed to operate automatically, control their own water supply, freeze water into the desired shape and size, collect the ice in a storage bin, and shut down when the bin is full. In most cases, the bin is simply an insulated enclosure with a tough plastic or vinyl liner, refrigerated only by the ice it contains. A drain removes melted ice water. Smaller ice makers may also be placed on top of a small bin. Such units are equipped with a mechanical drive to dispense small amounts of ice on demand. These small units are often coupled with beverage dispensing systems accessible to the consumer.

Most retail-sized ice machines feature air-cooled condensers, but water-cooled models are available. Self-contained units can usually be placed in ambient temperatures between 40°F and 115°F. Ambient temperature can significantly affect the speed at which ice is made.

The temperature of the supply water can also impact production speed because warmer water requires more cooling before ice formation can begin.

Care must be taken when locating air-cooled models to prevent recirculation of condenser air. Larger ice makers use remote condensing units to remove rejected heat from the conditioned space. For obvious reasons, most ice makers are not designed to operate in environments that approach or drop below freezing. If they are shut down during the winter but left in an area prone to freezing, the water supply must be turned off and water drained from all internal components.

2.3.1 Cubed-Ice Machines

Cubed-ice machines (*Figure 14*) generally range in capacity from 65 pounds to 2,000 pounds of ice per 24-hour period. Storage bins mated with ice machines commonly store between 120 pounds and 1,500 pounds of ice. Selection of cuber and bin combination is based on customer requirements. Applications in which large quantities of ice are used randomly versus those with steady consumption throughout the day would result in different cuber/bin combinations.

Figure 14 Cubed-ice machine with bin.

Cube sizes and shapes vary significantly, depending on the customer's needs. Although the machines themselves operate in the same way, the grid where ice is formed is changed to produce the desired shape and size. Although sizes vary by manufacturer, a half-dice cube may be 3/8" × 7/8" × 7/8" deep; a dice cube would be 7/8" × 7/8" × 7/8" deep; and a full cube would be 1⅛" × 1⅛" × 7/8". Other sizes and shapes are also available, such as nuggets, tubular nuggets, and even gourmet shapes such as an octagon.

Cubed-ice machines have four basic modes of operation: fill, freeze, harvest, and drain-down. In the fill mode, new water is supplied for the next batch of cubes. In the freeze mode, water applied to the evaporator freezes. When the ice has reached a predetermined size, the evaporator is defrosted to quickly loosen the ice from the evaporator, melting only a small portion of the ice in the process. The harvest mode is like a typical defrost mode in other refrigeration applications. Drain-down, or purge, rids the machine of minerals left behind when the water freezes.

Some units harvest the ice in a single solid sheet, which falls from the evaporator onto an electrical wire grid. The grid is heated and the wires melt through the ice in knife-like fashion, separating the ice sheet into cubes of various sizes, such as 1" × 1". Figure 15 provides an example of this design. After the cubes are cut by the wire grid, they fall into the storage bin.

The other popular design (Figure 16) either sprays water into cube-shaped chambers permanently attached to a horizontal plate evaporator, or water is pumped to the top of a vertically positioned evaporator of the same design and allowed to cascade down by gravity. When the harvest mode begins, hot refrigerant gas (or possibly electric heat) is directed to the plate evaporator to loosen the cubes, which fall into the bin. The hot gas condensing in the coil quickly adds heat to the back of the evaporator, loosening the cubes. This is the same approach used to defrost a heat pump. With either style, ice cube size is an available option.

An ice thickness sensor is typically used on a cubed-ice evaporator. It is set at a precise distance from the evaporator. As ice builds on the plate, the cascading water will eventually make contact with the sensor. When the sensor has been wet-

Figure 15 Cubed-ice maker with an electrical cutting grid.

High-Quality Ice

Water quality is important to the quality of ice. Aerated ice and ice with a high mineral content may appear cloudy and unappealing. The entering water supply should be filtered and potentially conditioned to help avoid these problems.

Cubed ice should also be very hard, a result of being sufficiently subcooled before harvesting. Ice that is harvested too early will be smaller and will melt quicker.

Figure 16 Cubed-ice maker with cube-shaped chambers on the evaporator plate.

ted continuously for 10 seconds, indicating the ice has reached proper thickness, the harvest mode begins. Water flow then ceases.

Cubed-ice makers often have a complex sequence of operation to ensure that a consistent ice cube is produced under a variety of operating conditions. Shutdown of the process is done by either a temperature sensor located near the top of the bin or by using a photoelectric eye that is blocked by the presence of ice. Both indicate a full bin.

2.3.2 Flaked-Ice Machines

Ice flakers (*Figure 17*) operate in a continuous ice-making mode. Although they may look similar to cube machines, they are different in their operation. When the unit is started and the evaporator assembly reaches the proper temperature for freezing, it continues to operate until the bin is full.

The freezing assembly for flaked-ice makers (*Figure 18*) consists of an insulated vertical cylinder.

The outer walls of the cylinder form the evaporator, which is filled with boiling, low temperature refrigerant. Ice forms on the cylinder's inner walls, which are kept bathed with a consistent supply of water. An auger, driven by a constantly and slow turning gear motor drive, scrapes the ice from the walls of the cylinder. The flaked ice is pushed out of the top of the cylinder by the auger and falls into the bin.

As is the case with cubed-ice machines, flaked-ice units are now generally designed around HFC-134a and HFC-404A refrigerants.

03304-13_F17.EPS

Figure 17 Flaked-ice machine with bin.

03304-13_F18.EPS

Figure 18 Flaked-ice machine operation.

Additional Resources

Refrigeration and Air Conditioning, An Introduction to HVAC/R, Latest Edition. Larry Jeffus. Upper Saddle River, NJ: Prentice Hall.

2.0.0 Section Review

1. The storage box temperature range in ice merchandisers is typically _____.

   a. −10°F to 5°F
   b. 5°F to 15°F
   c. 10°F to 20°F
   d. 20°F to 25°F

2. What type of metering device is typically used with walk-in coolers?

   a. Thermostatic expansion valve
   b. Capillary tube
   c. Metering piston
   d. Bypass piston

3. The harvest mode of a cubed-ice machine closely resembles the _____.

   a. heating mode of a heat pump
   b. purge cycle of a flaked-ice machine
   c. cooling mode of a heat pump
   d. defrost mode of a heat pump

SECTION THREE

3.0.0 RETAIL REFRIGERATION SYSTEM COMPONENTS

Objectives

Identify and describe components related to the operation of retail refrigeration systems.

   a. Identify and describe the primary components used in retail refrigeration applications.
   b. Identify and describe secondary components and accessories used in retail refrigeration applications.
   c. Identify and describe common refrigeration system controls.

Trade Terms

Slinger ring: A ring attached to the outer edge of the fan that picks up water as the fan rotates and slings it onto the condenser coil.

03304-13_F19.EPS

Figure 19 Refrigeration condensing unit.

Components found in retail refrigeration systems are similar or identical to components found in comfort cooling systems. Since retail refrigeration equipment operates at lower temperatures than comfort cooling systems, the components have to accommodate that difference. For example, compressors must be rated for refrigeration duty, which means they must be able to operate reliably at higher compression ratios.

Different refrigerants are often used. Refrigerants that are ideal for comfort cooling may not be suited for refrigeration applications. In comfort cooling, fixed-orifice metering devices such as capillary tubes can provide adequate refrigerant flow to the evaporator. In refrigeration systems, capillary tubes cannot provide good refrigerant flow control over the wider temperature ranges in which refrigeration units may have to operate. For that reason, expansion valves with their better control of refrigerant flow are more commonly used for retail fixtures.

3.1.0 Primary Refrigeration Cycle Components

Many retail refrigeration applications use air-cooled condensing units (*Figure 19*), a package assembled on a single chassis that contains the compressor, condenser and condenser fan(s), and a variety of controls and/or compressor start components.

These units can be used in self-contained refrigeration fixtures or they can be remotely installed and piped to the fixture. Used indoors, self-contained fixtures reject heat from the box, as well as heat of compression, to the space in which they are installed. This additional heat can be a significant comfort problem if not taken into consideration when loads for the comfort cooling system are calculated.

Although more costly to install, condensing units installed outdoors have the advantage of rejecting heat outside of the conditioned space. In most cases, remote installation is reserved for larger applications such as walk-in coolers or freezers, as well as larger ice machines. Installations in which a single condensing unit serves multiple evaporators or fixtures will be encountered.

3.1.1 Compressors

For retail refrigeration use, hermetic and small semi-hermetic reciprocating compressors (*Figure 20*) have long dominated the market. However, scroll compressors (*Figure 21*) are also making their way onto the scene, with design changes that improve their suitability for lower temperature applications.

Early attempts to use scroll compressors in refrigeration applications resulted in overheated and failed compressors because of high compression ratios and very low density of the returning refrigerant vapor. However, several manufacturers have now developed scroll compressors for the refrigeration market.

For low-temperature applications, many scroll compressors are fitted with a liquid or vapor in-

Figure 20 Semi-hermetic and hermetic compressors.

jection device, designed to provide for improved internal cooling and reduced refrigerant discharge temperatures. Many scroll compressors used in refrigeration are not equipped with internal pressure relief valves. Therefore, providing reliable high-pressure safety controls is a must. Control systems that use a discharge line thermostat help prevent compressor damage or failure from overheating. They are generally set to open at 250°F to 260°F.

Systems that use hermetic reciprocating compressors can be fitted with additional refrigerant flow control components for capacity control. Semi-hermetic compressors with multiple cylinders can be used with integral capacity control to relieve compression in one or more cylinders. Scroll compressors are now available with internal capacity controls that modulate compressor capacity in small increments between 10 and 100 percent. This reduces energy consumption and provides very stable temperature control.

3.1.2 Condensers

Most retail refrigeration systems use air-cooled condensers because of the ease of installation, mobility, and simplicity of the design. The condenser can be mounted on the condensing unit, within inches of the compressor, or remotely installed. Aluminum fin or copper tube coils are typical (*Figure 22*).

Proper condensing pressures and temperatures must be maintained for the refrigerant circuit to operate effectively. As condensing temperatures rise, capacity and efficiency are reduced. Some retail systems operate in a conditioned environment, but many do not. This must be considered

Figure 21 Digital scroll compressor.

in final equipment selection. Even when placed in a conditioned space such as a floral showroom or convenience store, it should not be assumed that the condensers will remain clean. Indeed, when used near cooking apparatus, air-cooled condensers can foul rapidly from airborne grease depositing on the coil fins. Regular condenser cleaning must be an integral part of retail refrigeration system maintenance.

Reduced condensing temperatures can provide increases in energy efficiency. However, low condensing pressures can result in a loss of performance because of insufficient refrigerant flow through the expansion device. When condensers or condensing units are exposed to outdoor conditions, this condition often occurs in winter.

Either air-side or refrigerant-side controls can be used to maintain minimum required condensing pressures. On the air side, the condenser fan(s) can be cycled on and off by pressure, or the condenser fan speed can be modulated for a more stable result. On the refrigerant side, controls installed in the refrigeration circuit can be used to control the active surface of the condenser by flooding a portion of its internal volume with liquid refrigerant.

FIN-AND-TUBE CONDENSER

HOT REFRIGERANT VAPOR INSIDE TUBES

COOL AIR PASSES OVER COPPER TUBES ABSORBING HEAT FROM REFRIGERANT

WARM LIQUID REFRIGERANT

03304-13_F22.EPS

Figure 22 Air-cooled fin-and-tube condenser.

3.1.3 Evaporators

Many types and styles of evaporators are necessary in retail refrigeration systems. Although most are very typical air-cooling models, built much like comfort cooling coils, evaporators used for liquid cooling and ice making can be unique in their construction and design.

Forced-draft, finned-tube evaporators are most commonly used in smaller retail refrigeration fixtures and systems. For larger applications, such as walk-in freezers or coolers, a unit like the one in *Figure 23* is used. The evaporator coil is generally a packaged assembly consisting of the coil, evaporator fan, metering device, and defrost heaters (when needed). These are often referred to as unit, or unitary, coolers. Unlike most comfort cooling applications that require blower wheels to overcome resistance to airflow in duct the system, these units generally use free-blow (no ductwork) propeller-type fans made of plastic or aluminum.

As in comfort cooling, condensation occurs on the evaporator coil surface when coil temperatures are below the dew point of the enclosure. A small amount of condensate must be collected and drained out. For portability of small refrigerated fixtures, condensate is often collected in a small heated pan installed in the fixture cabinet; an electric heating element embedded in the pan re-evaporates the condensate back to the air. For coils that operate below freezing, the condensate collection pan on the coil assembly generally uses an electric heater to prevent freezing and block-

03304-13_F23.EPS

Figure 23 Walk-in refrigeration unit cooler.

Improving Energy Efficiency

GOING GREEN

Newer models of retail refrigeration equipment use scroll compressors to improve energy efficiency. Scroll compressors have no pistons to compress gas, so they achieve 100-percent volumetric efficiency, which provides reduced energy costs in many applications. Re-expansion losses that typically occur in reciprocating models are eliminated. Similarly, losses at valves are eliminated, because suction valves are not required.

age of condensate flow. The condensate line itself, where it passes through the refrigerated space, also requires freeze protection. Heat tape, wrapped around the outside of the line, prevents freezing.

Some units may also use this condensate to advantage. When collected in a pan under the condenser coil, condensate can be picked up by the condenser fan blades fitted with a **slinger ring**. This slings water droplets onto the surface of the condenser coil to enhance the condensing process. Moisture is evaporated back to the surrounding air in the process.

Maintaining humidity in the refrigerated space is important, depending on the intended use of a refrigerated fixture. If storing fresh produce, flowers, and similar products, humidity levels must remain high to prevent dehydration of the product. In other cases, especially when the stored product is sealed, low humidity levels may be needed to prevent fog from forming. Proper selection of the evaporator is essential to maintain the desired condition.

To maintain higher humidity levels, evaporator coils must operate at an extremely narrow TD between the desired storage temperature and the refrigerant (4°F to 8°F, for example). To meet this requirement, coil assemblies generally have large surface areas with high airflow rates. Conversely, when low humidity levels are desirable, the evaporator coil is often smaller and operates with low airflow values. In these applications, the air-to-refrigerant TD may be as high as 20°F to 30°F. Unitary evaporator coils that operate at 30°F and below generally form frost and will require some method of frost removal.

Many commercial cubed-ice machines use evaporator coils similar to simple plate evaporators. A length of tubing is firmly attached to a plate on one side, while the opposite side is used to produce the desired cube size and shape. A variety of cube shapes and sizes are available.

Figure 24 shows one such evaporator assembly in which the water flows over the vertical plates to form individual cubes. In other units, the evaporator assembly is a single-sided vertical grid cell plate. Water cascades down the face of the grid, building ice cubes in the cell.

3.1.4 Expansion (Metering) Devices

Metering devices are covered in greater detail in an earlier module. Briefly, an expansion device controls the flow of refrigerant into the evaporator, which provides the pressure drop that lowers the boiling point of the liquid to the desired temperature. It also meters the flow of liquid re-

03304-13_F24.EPS

Figure 24 Commercial cubed-ice machine evaporator.

frigerant into the evaporator to match the rate of refrigerant boiling into vapor. Typically, only refrigerant vapor should leave the evaporator. At the same time, most coil internal surfaces should be in contact with boiling liquid for maximum capacity.

Although fixed metering devices such as capillary tubes (*Figure 25*) are used, they are not as widely used in retail refrigeration because they cannot effectively meter refrigerant in many load conditions and may over-feed or underfeed refrigerant. They are also very sensitive to the amount of refrigerant charge in the system. Very small changes in the refrigerant charge have a significant effect on the operating superheat. For that reason, TEVs are widely used in retail refrigeration applications.

In *Figure 26*, a TEV controls the amount of refrigerant flow using the superheat temperature

STRAINER

LIQUID ➡

CAPILLARY TUBE

TO EVAPORATOR

03304-13_F25.EPS

Figure 25 Capillary tube.

Figure 26 Thermostatic expansion valve.

sensed at the evaporator outlet. The device provides a flow rate to match the evaporator load. As the evaporator load changes, the TEV responds by opening or closing. TEVs are not nearly as charge-sensitive as fixed devices. Using TEVs ensures that a specific superheat value can be maintained effectively under a wide variety of conditions, even if the refrigerant charge is not precise.

Electronic expansion valves, or EEVs, (*Figure 27*) are gaining popularity in the industry because they are not refrigerant-specific. By installing different orifice sizes, each valve body can be used to cover many capacity ranges. They function through a wide range of loads and are remotely adjustable. EEVs offer precise operation and flexibility at extremely low superheat values, allowing maximum use of the evaporator's heat transfer surface. However, they are presently used in more critical applications than most common retail units represent.

Figure 27 Electronically controlled expansion valves (EEVs).

These valves can be very compact, offering thousands of possible positions, called steps. This term is derived from using a stepper motor to open the valve in very small steps. While older EEVs were equipped with motors offering a few hundred steps or less, over 6,000 steps are now possible. This provides very precise refrigerant metering. Valve positioning is controlled by inputs from one or more temperature sensors and microprocessor-based controls. Many valves can change positions at a rate of 200 steps per second.

EEV Superheat Control

There are two basic approaches to controlling superheat with EEVs. Superheat is all about pressure-temperature relationships. These relationships are specific to each refrigerant, so programming the valve requires a pressure transducer at the coil outlet, a temperature sensor, and the input of the proper P-T chart into the software of the microprocessor control. This is considered the most accurate method, since it is directly related to the thermodynamics involved.

Another simpler approach uses two less expensive temperature sensors. This method discounts the pressure-temperature relationship completely. One sensor is located at the evaporator inlet, and the other at the evaporator outlet. Software compares the two and assumes any rise in temperature through the coil is superheat. This method is not as precise and may require that a slightly higher superheat value be programmed in as a safety net. One problem that arises relates to the position of the evaporator inlet sensor: to remain accurate, it must be located in a position that is always exposed to refrigerant in its liquid state. However, this two-temperature method can be used with any refrigerant without reprogramming the valve for a specific compound or blend.

Systems with traditional TEVs are often equipped with a solenoid valve installed in the liquid line upstream of the TEV. During the compressor off cycle, the solenoid valve closes to prevent refrigerant migration to the evaporator. Once in the evaporator and exposed to the cold temperatures of the enclosure, the refrigerant condenses. It can then return to the compressor as floodback and cause damage. Since EEVs offer the advantage of tight closure during the off cycle, they eliminate the need for a solenoid valve. Expect to see more uses of EEVs as the industry strives to increase the energy efficiency of retail refrigeration products.

3.2.0 Other Refrigeration Circuit Devices and Components

A variety of supporting components are used in refrigeration systems. Although some of the components discussed here may be used in comfort cooling applications, they are more common in refrigeration applications.

3.2.1 Receivers

Receivers provide a place where liquid refrigerant can be stored in refrigeration units. Because variations in load conditions can cause rapid changes in refrigeration load, having an adequate supply of refrigerant to satisfy the needs of the evaporator under all load conditions is important.

Many small receivers use a back-seating shutoff valve on the outlet. Receivers are also equipped with a dip tube, drawing liquid refrigerant from the bottom to ensure that pure liquid is sent to the metering device. The liquid refrigerant level inside will change in relation to the system flow rate at any given time.

Receivers are used only on systems with a TEV or other modulating metering device. They are of little value in systems with a fixed metering device.

3.2.2 Accumulators

Because refrigeration systems tend toward lower superheat values, the chance of liquid refrigerant leaving the evaporator becomes higher. The accumulator (*Figure 28*) is basically a trap designed to hold liquid refrigerant in the system suction line.

Accumulators are also used in some comfort cooling applications and heat pumps. The accumulator uses a U-shaped dip tube to separate liquid from vapor. The compressor draws refrigerant vapor from one end of the U-tube. The other end is contained within and terminates near the top

Figure 28 Suction accumulator.

of the vessel to ensure that only vapor is drawn from the accumulator. A small orifice at the lowest point in the tube allows small amounts of trapped compressor oil and liquid refrigerant to return to the compressor without damaging the compressor.

3.2.3 Crankcase Pressure Regulators

To prevent overload of the compressor motor, crankcase pressure regulating (CPR) valves are installed in the suction line just upstream of the compressor, as shown in *Figure 29*. The valve controls the maximum pressure at the compressor suction inlet.

Refrigeration equipment must often be started up when the entire fixture is at room or ambient temperature, even though the fixture may be designed to maintain –10°F. This load is well outside of the compressor's normal operating range for the application. This same extreme load can occur following a defrost cycle, especially in systems that use hot-gas defrost. The CPR valve ensures that refrigerant vapor is fed to the compressor in a reasonable volume that prevents it from being overloaded.

The valve setpoint is adjusted by changing the internal spring pressure so that the valve moves towards its closed position on a rise of suction pressure. CPR valves are very simple to set by starting the system in a known-overload condition while checking compressor current with an

THERMOSTATIC EXPANSION VALVE (TEV)

EVAPORATOR

CONDENSER

CPR

LIQUID
LINE
DRIER

COMPRESSOR

RECEIVER

03304-13_F29.EPS

Figure 29 Crankcase pressure regulating valve.

THE PRESSURE HERE
IS MAINTAINED AT
A HIGHER LEVEL.

HI-TEMP.
EVAPORATOR

TEV

EPR

LOW-TEMP.
EVAPORATOR

TEV

CHECK
VALVE

CONDENSER

THIS SECTION
OPERATES
AT A LOWER
PRESSURE.

LIQUID
LINE
DRIER

COMPRESSOR

RECEIVER

0304-13_F30.EPS

Figure 30 Evaporator pressure regulator.

ammeter. The valve spring can then be adjusted to a lower pressure setting while watching the actual current draw of the compressor. When the current falls to, or drops very slightly below, the rated load amperage for the compressor, the valve is properly set.

3.2.4 *Evaporator Pressure Regulating Valves*

Some larger retail refrigeration systems may use a single compressor to maintain temperature in multiple compartments or fixtures that have individual temperature settings. The evaporator pressure regulating (EPR) valve allows evaporators serving higher temperature areas to maintain proper refrigerant pressures for their application and, in turn, the saturated suction temperature of the refrigerant. One example of such an installation is shown in *Figure 30*.

The pressure in the common suction line serving all of the fixtures must be low enough to properly refrigerate the coldest fixture. When the pressure rises above the set point, EPRs mounted on each of the evaporators operating at higher temperatures respond to the inlet pressure of the evaporator to open wider, allowing more refrigerant to flow through the coil. When evaporator pressure falls, the valve moves toward its closed

position. The evaporator will maintain the proper pressure that corresponds to the designated saturated suction temperature.

3.2.5 *Refrigerant-Side Head Pressure Control*

Refrigeration units installed in locations subject to low ambient temperatures generally require a method of maintaining head pressure. Cold air entering the condenser causes a reduced condensing pressure. If the pressure is allowed to fall too low, the pressure drop through the metering device may not be sufficient to ensure adequate refrigerant is fed to the evaporator.

A head pressure control valve is the most accurate and stable means of ensuring that an adequate head pressure is maintained (*Figure 31*). Although condenser fan cycling controls are often used for this function, condensing pressures rise and fall through the control's range. As a result, the constantly changing condensing pressure causes the flow rate of the metering device to be unstable. The refrigerant-side head pressure control maintains condensing pressures in a far more stable manner.

Figure 32 demonstrates the typical installation of this valve. As temperatures at the condenser air inlet fall, a corresponding decrease in the condensing pressure occurs. When this pressure falls to the valve's set point, compressor discharge vapors are routed through port B of the valve and into the liquid line to the receiver through port R. The action causes a rise in pressure at the condenser outlet, which impedes the flow of refrigerant liquid out of the condenser. Condensed liquid begins to back up, effectively reducing the available condensing surface.

Head pressure control valves are often nonadjustable. To effectively operate a head pressure control valve, the system must include a receiver, since this device requires a larger volume of refrigerant during low ambient conditions than is needed during periods of high load and high condensing pressures. Additional refrigerant, held in reserve in the receiver, is needed to partially flood the condenser coil as the valve modulates to maintain appropriate pressure.

This type of valve is usually refrigerant-specific and is factory preset at an appropriate value

03304-13_F31.EPS

Figure 31 Refrigerant-side head pressure control valve.

03304-13_F32.EPS

Figure 32 Head pressure control valve installation position.

Pilot-Operated EPRs

Normally, EPRs are installed as close to the evaporator as possible. If they must be installed away from the evaporator or downstream from a riser, a pilot-operated EPR like this one is typically used. The pilot line is connected to the evaporator to monitor the pressure and position the valve.

03304-13_SA02.EPS

for the application and refrigerant used. It has a sensing dome, as shown in *Figure 31*, which must be exposed to the ambient temperature. Other models and styles that can be adjusted in the field are available. Head pressure control valves are generally not a single valve, but two separate valves that operate independently. Since retail refrigeration systems are generally compact, the type shown here is far more popular.

3.3.0 Common Refrigeration System Controls

Some typical controls for retail refrigeration systems include thermostats, pressure controls, time delay relays, and solenoid valves. Again, while these are similar to comfort cooling controls, they are often applied in different ways in retail refrigeration applications and may not have the same appearance as commercial models.

3.3.1 Thermostats

Thermostats sense temperature at a given point and respond with an action. While many different styles are available, some of them are designed to be very equipment-specific.

Most refrigeration applications use remote-bulb thermostats for adjustment outside the refrigerated space. The basic operation of the remote-bulb thermostat is demonstrated in *Figure 33*. Even those located inside the refrigerated space may use a very short capillary tube rolled up behind the switch assembly.

Electronic temperature controls are also very popular in the retail refrigeration industry. They offer reliable operation and a wide temperature range, allowing identical controls to be installed on high-, medium-, or low-temperature systems. Installing a small temperature sensor and routing flexible wiring back to the control is also much simpler than routing a long capillary tube through complex cabinet components.

A typical thermostat features a setpoint range of –30°F to 130°F and an adjustable differential of 1°F to 30°F. The range can be modified by using different temperature sensors. An LED indicator is often used to provide status. Normally open/normally closed operation is field selectable.

Some applications for thermostats in refrigeration include the following:

- Controlling the refrigeration circuit and compressor to maintain desired fixture temperature
- Providing a low temperature safety control for the refrigeration circuit
- Delaying the restart of evaporator fans following a defrost cycle. This allows refrigeration equipment to operate long enough to re-freeze any remaining moisture on the evaporator coil before the fans start, preventing water from being blown off of the coil.
- Cycling of the condenser fan(s) based on ambient temperature
- Terminating the defrost cycle because of evaporator coil or suction line temperature
- Shutting down an ice-making process because of bin temperature

Thermostats for these applications are available with a wide variety of ranges and differentials.

Figure 33 Remote-bulb thermostat.

They can be adjustable or have a fixed setting chosen for specific applications.

Single-stage thermostats provide an action for a single temperature setting, while multi-stage thermostats respond to two or more temperature points. Those used as safety controls often must be reset manually. Manual-reset controls prevent equipment damage by keeping the equipment off until a service technician repairs the condition that caused the safety control to open.

3.3.2 Pressure Controls

Pressure controls are also applied to refrigeration systems in many different ways. As shown in *Figure 34*, internal construction of pressure switches varies widely. Many gauges and some pressure switches rely on the operating principle of the Bourdon tube, in which pressure applied to a small curved tube causes it to try to straighten. Others use different diaphragm designs to create internal movement of the switch.

Like thermostats, pressure switches perform a variety of functions and are available in different pressure ranges to suit the application. Adjustable switches with a wide operating range are generally larger than fixed devices. They can use a single high- or low-pressure input, or accept two separate pressure inputs and react to the differential between the two.

Some of the functions that pressure switches provide include the following:

- Controlling refrigeration equipment by the pressure sensed in the suction line. Since suction pressure is related to saturated suction temperature, pressure switches can provide start-stop control of refrigeration circuits (primarily those using fixed metering devices)
- Controlling one or more condenser fans by condensing pressure
- Shutting down the compressor because of excessively low pressures caused by problems such as loss of charge or evaporator fan failure
- Shutting down the compressor caused by a high-pressure condition resulting from condenser fan failure, refrigerant overcharge, or other problems
- Pump-down control for systems equipped with a liquid line solenoid valve controlled by a thermostat. When the valve closes, the compressor continues to run until most of the refrigerant has been pumped out of the evaporator and suction pressure falls to the setting of the pressure switch
- Monitoring the differential between the suction pressure and oil pump discharge pressure

BOURDON TUBE FUNCTION

C-TYPE BOURDON TUBE **SPIRAL BOURDON TUBE** **HELICAL BOURDON TUBE** **BOURDON TUBE PRESSURE GAUGE**

03304-13_F34.EPS

Figure 34 Bourdon tube function.

as proof of lubricating oil pressure on semi-hermetic compressors

3.3.3 Time-Delay Relays

Time-delay relays are used to delay a switching action or ensure that a sequence of actions takes place at the proper time. Many are very compact, such as the one shown in *Figure 35*, but still have an adjustable time-delay range.

Most time-delay relays used in refrigeration applications can be specified as follows:

- *Delay-on-make* – When energized, contacts remain in their existing position (usually closed) until the specified time period elapses, then they change position. These relays are often used to delay a particular step in a sequence of operation or to prove that a specific condition has existed for a predetermined period before allowing the system to respond.
- *Delay-on-break* – When de-energized, the relay contacts immediately change position (usually from closed to open) until the time period has elapsed. Then they return to the previous position (usually closed). Delay-on-break timers are often used as compressor short-cycle protectors by opening their contacts when de-energized, preventing the compressor from starting until the time-delay period (usually two to five minutes) has elapsed.
- *Bypass duty* – One frequent application for the bypass timer is to bypass a low-pressure switch for a short period of time after a compressor starts. Several factors (all normal) can cause a short-term period of low suction pressure when

a compressor first starts. Briefly bypassing the low-pressure switch allows suction pressure to build to normal levels and prevents a nuisance shutdown of the compressor. Bypass-duty relays are usually of the delay-on-make style, but not in all cases.

The normally-closed switch contacts of the bypass timer (*Figure 36*) are placed in parallel with the low-pressure switch, providing an electrical path around the switch until the selected time period has elapsed. Timer contacts then open to allow the low-pressure switch to regain control when the system is in stable operation.

More sophisticated timers can provide multiple timing events or a repeated sequence of timed events for special applications, such as ice makers. Timed functions can also be built into printed circuit boards and microprocessor controls.

3.3.4 Solenoid Valves

Solenoid valves (*Figure 37*) are used to deny or permit the flow of either liquids or gases in refrigeration systems. This particular valve has extended fittings that allow the valve to be brazed into the line without overheating and damaging the valve body.

Solenoid valves are selected by refrigerant flow rate and pressure drop through the valve. The resulting choice may also fit the system line size, but line size is not the primary factor.

Solenoid valves can be provided with a manual override stem, allowing the technician to temporarily open the valve manually when a coil fails or the valve sticks closed (*Figure 38*). Although most

03304-13_F35.EPS

Figure 35 Time-delay relay.

03304-13_F36.EPS

Figure 36 Bypass timer wiring.

Figure 37 Solenoid valve.

solenoid valves are normally closed and are energized to open, types that are normally open and close when energized are also available. The solenoid coil is typically replaceable and available in all commonly used voltages.

Applications for solenoid valves include the following:

- Provide flow control of liquid refrigerant to the metering device that is controlled by the fixture thermostat. This prevents refrigerant migration during the off cycle and enables a pump-down circuit on systems so equipped.
- Providing independent temperature control on systems with a single compressor serving multiple evaporators. The valves are used in the liquid line to each evaporator.
- Controlling liquid refrigerant flow to injection-cooling TEVs.
- Controlling the flow of hot gas refrigerant for defrost, capacity control, or low-ambient condenser circuiting.

Figure 38 Solenoid valve with manual override.

Solenoid Valve Service Tool

One solenoid valve manufacturer has developed a multi-function service tool for use with its products. The body of the tool is a powerful permanent magnet that simulates the coil of a solenoid valve. When placed over the plunger's enclosure tube, the magnet can open a valve even with a 300-psi pressure differential across it. The body of the tool acts as a spanner for taking apart the valve body. It also has a manual stem adjustment tool for solenoid valves that can be manually opened.

3.0.0 Section Review

1. What device is often installed with scroll compressors used in low-temperature applications to improve compressor reliability?

   a. Suction unloader device
   b. Hot-gas bypass valve
   c. Liquid or vapor injection device
   d. Cylinder unloader device

2. What is the purpose of the orifice in the U-tube of an accumulator?

   a. Oil or liquid refrigerant return
   b. Pressure equalization
   c. Refrigerant separation
   d. Pressure relief

3. Delay-on-break time delays are often used to prevent _____.

   a. nuisance cycling of the low-pressure switch
   b. nuisance cycling of the high-pressure switch
   c. compressor short-cycling
   d. evaporator fan short-cycling after shutdown

SECTION FOUR

4.0.0 TROUBLESHOOTING RETAIL REFRIGERATION SYSTEMS

Objectives

Explain the basic principles of maintaining and troubleshooting various retail refrigeration systems.

   a. Explain how to maintain and troubleshoot a typical reach-in freezer.
   b. Explain how to maintain and troubleshoot a typical cubed ice machine.

Performance Tasks 1 through 3

Clean an ice machine.

Isolate a fault in refrigeration equipment or an ice machine.

Set up a defrost timer for a freezer.

03304-13_F39.EPS

Figure 39 Reach-in freezer.

Other than a difference in work environments, all basic troubleshooting skills previously presented for troubleshooting comfort systems apply. Troubleshooting and maintenance procedures for two different retail refrigerant systems are examined: a reach-in freezer and a cubed-ice machine. Note that systems can differ widely, so specific sequences of operation and service information for a given system should always be consulted before beginning the troubleshooting process.

4.1.0 Troubleshooting and Maintenance of a Reach-In Freezer

For a typical reach-in freezer (*Figure 39*) with an electric defrost, a lack of simple maintenance is one of the primary causes of performance problems. Although the periodic maintenance required for a reach-in freezer is not involved, it is critical to proper operation. The most important of these tasks is to inspect and maintain the evaporator and condenser coils.

Depending on where the unit is located, the condenser coil can become fouled very quickly and cause high condensing pressures. In the early stages, this condition may only raise the unit's operating cost without affecting performance. But as condenser coil fouling worsens, performance will begin to drop.

Evaporator coils generally do not foul as quickly, but are subject to becoming clogged.

The aluminum fins have the potential to rapidly corrode or deteriorate due to the off-gassing of certain foods. The fins then become coated with aluminum oxide or similar residue, reducing heat transfer.

Generally, refrigeration systems of this size have a fairly small refrigerant charge. Refrigerant operating pressures should not be checked every maintenance visit unless you suspect problems. Each time pressure gauges are used, a small loss of refrigerant occurs, so constant pressure checking will eventually cause performance problems. Most motors on units of this size do not require lubrication, but they should be checked to ensure this is the case.

Before troubleshooting, review the manufacturer's data and normal operating conditions. For example, suppose the features of a typical freezer include the following:

- The unit power switch is located on the electrical box behind the lower louvered access panel. The condensing unit is also located in this area.
- The unit is equipped with an electronic temperature control that has no Off position. The electromechanical defrost timer and temperature control are located inside the same enclosure where the power switch is mounted.

- The defrost timer has been factory-preset for two defrost cycles per day at 6:00 am and 10:00 pm, with a 40-minute fail-safe defrost termination period. The defrost cycle is normally terminated by defrost thermostat located on the evaporator suction line. The three-wire thermostat will terminate defrost when a line temperature of 58°F is reached. As refrigeration resumes, it will not allow the evaporator fans to operate until the line temperature has dropped below 32°F. This prevents condensate on the coil from being blown onto the product. Should the defrost thermostat fail to terminate defrost, the timer will terminate the cycle after 40 minutes.
- The electric defrost heaters are attached to the fins of the evaporator coil using spring clamps.
- Fluorescent lamps provide interior lighting and can be turned off with a switch located inside the cabinet above the door.
- A fan switch is located at the top of the doors. The evaporator fans stop when the door is opened to prevent cold air from being blown out of the box.
- Low-wattage electric heaters are installed in the structure framework and in the perimeter of the doors to prevent freezing. The heaters are not energized until the box has reached operating temperature. The thermostat controlling their operation is also referred to as the alarm thermostat, because it can be wired to generate a visual or audible alarm when the temperature rises above 18°F. The thermostat switches to its normal position, energizing the door heaters, when the box temperature has reached 0°F.
- The refrigeration circuit is equipped with a TEV metering device and a refrigerant receiver is installed in the liquid line.
- A semi-hermetic compressor is used, and the system refrigerant is HFC-404A. To avoid overloading the compressor, a crankcase pressure regulating valve is installed and factory set at 10 psig.
- The manufacturer indicates the standard ambient temperature should be 75°F and lists the

expected head pressure as 230 to 240 psig. The expected suction pressure is 5 to 6 psig with the box temperature stable at –10°F. The design air-to-refrigerant temperature difference is 15°F, resulting in an expected evaporator air discharge temperature of –25°F.

This critical information allows technicians to recognize proper and improper operation. Before beginning any troubleshooting procedure, understand the system sequence of operation and recognize the indicators of proper and improper operation. Only then will you be able to determine when system performance is not what it should be. The wiring diagram in *Figure 40* and the troubleshooting charts in *Figure 41* are helpful troubleshooting tools.

Although not every potential situation is covered by this troubleshooting chart, it will help you quickly diagnose many of the problems you are likely to encounter. Your ability to locate and repair unusual problems will increase with hands-on experience.

> **WARNING!**
>
> It is important to remember that ice machines produce a consumable product, requiring that the unit be properly maintained and kept sanitary. Failure to properly maintain and clean these units could lead to widespread illness from contamination.

4.2.0 Troubleshooting and Maintenance of Cubed-Ice Machines

Most cubed-ice machines have a more complex sequence of operation than the reach-in freezer. Troubleshooting and properly testing a cubed-ice machine can be time consuming because it can take a long time to complete a full cycle. After any adjustments or changes are made, the unit should be put through several complete cycles to ensure that it is operating normally. An air-cooled cubed-ice machine is being used to review troubleshooting and maintenance procedures.

Silver and Biological Films

Silver is being used in products at an increasing rate to assist in the prevention of biological growth. Even clothing items can now be impregnated with silver for this purpose. Many ice machine manufacturers are adding elemental silver in the form of silver ions to plastic components such as ice bin walls to delay biological growth. One such product is known as AlphaSan®. Such additives, compounded directly into the part during manufacturing, release silver ions slowly to the surface of the plastic. Not only do they reduce or prevent biological growth, but the additives also help avoid discoloration of the part over time. They do not impact the taste and flavor of the ice, but the presence of these products should not be cause to ignore proper periodic cleaning.

Figure 40 Freezer wiring diagram.

TROUBLE	PROBABLE CAUSE	SOLUTION
Compressor runs continuously, product too warm	1. Short of refrigerant 2. Inefficient compressor 3. Dirty condenser	1. Leak check; change drier; evacuate and recharge 2. Replace 3. Clean
High head pressure	1. Cabinet location too warm 2. Restricted condenser air flow 3. Defective condenser fan motor 4. Air or non-condensable gases in system	1. Relocate cabinet 2. Clean condenser to remove air flow restriction 3. Replace 4. Leak check, change drier, evacuate, and recharge
Warm storage temperatures	1. Temperature control not set properly 2. Short of refrigerant 3. Cabinet location too warm 4. Too much refrigerant 5. Low voltage; compressor cycling on overload 6. Condenser dirty	1. Reset control 2. Leak check; change drier; evacuate and recharge 3. Relocate 4. Change drier; evacuate and recharge 5. Check power 6. Clean
Compressor runs continuously, product too cold	1. Defective control 2. Control feeder tube not in positive contract 3. Short on refrigerant	1. Replace 2. Assure proper contract 3. Leak check; change drier; evacuate and recharge
Compressor will not start, no noise	1. Blown fuse or breaker 2. Defective or broken wiring 3. Defective overload 4. Defective temperature control 5. Power disconnected	1. Replace fuse or reset breaker 2. Repair or replace 3. Replace 4. Replace 5. Check service cord or wiring connections
Compressor will not start, cuts out on overload	1. Low voltage 2. Defective compressor 3. Defective relay 4. Restriction or moisture 5. Inadequate air over condenser 6. Defective condenser fan motor 7. CRO not set properly	1. Contact electrician 2. Replace 3. Replace 4. Leak check; replace drier; evacuate and recharge 5. Clean condenser 6. Replace 7. Reset to 10 psi
Icing condition in drain pan	1. Low voltage 2. Cabinet not level 3. Defective drain tube heater 4. Defective drain pan heater	1. Check voltage at compressor 2. Check front to rear leveling; adjust legs accordingly 3. Replace 4. Replace

03304-13_F41A.EPS

Figure 41 Troubleshooting chart. (1 of 2)

TROUBLESHOOTING LIGHTING SYSTEM

Lights won't start	1. Check light switch 2. Check continuity to ballast 3. Check to see if bulbs are inserted properly into sockets 4. Check voltage
Lights flicker	1. Allow lamps to warm up 2. Check lamp sleeve for cracks 3. Check sockets for moisture and proper contact 4. Bulb replacement may be necessary 5. Check voltage 6. New bulbs tend to flicker until used
Ballast hums	1. Check voltage 2. Replace ballast

03304-13_F41B.EPS

Figure 41 Troubleshooting chart. (2 of 2)

WARNING!

Ice machine cleaning and sanitizing products can be harmful. It is a violation of federal law to use them in any manner not consistent with their stated purpose. Always wear rubber gloves and goggles/face shield when handling and using these products.

4.2.1 Cleaning and Maintenance

Properly performing a maintenance inspection and cleaning of an ice machine can be a significant benefit to the technician as well as to the equipment. The technician can become familiar with the location of components and observe normal operations without the pressure of trying to correct a problem.

Primary maintenance activities for cubed-ice machines are outlined in the list that follows. Note that the process presented contains general instructions. Cleaning or maintenance should not be performed without access to unit-specific manufacturer's instructions.

Interior cleaning and sanitizing – This maintenance should be conducted at least every six months for most applications. If problems appear before this interval occurs, other preventive actions may be needed, such as the installing improved water filtration or relocating the unit to a conditioned area (from an excessively hot area) to help impede bacterial growth.

Ice machine cleaners help remove lime scale and mineral deposits left on surfaces from the water supply. Ice machine sanitizers disinfect and remove algae and slime. Always ensure that cleaning and sanitizing agents are used according to the manufacturer's recommendations and are used in their proper concentrations.

For this process, the ice machine should be shut down following a harvest operation so that the evaporator is clear of ice. All ice should be removed from the bin. Thorough cleaning requires significant disassembly of the unit and a lot of time to complete. Schedule these activities to ensure that an adequate supply of ice will be on hand.

Ice machine cleaners should be added and circulated throughout the water distribution system using the water pump. Most manufacturers provide a cleaning control switch position for this operation. The cleaner should remain in circulation for at least 30 minutes unless directed otherwise.

Disassemble and remove all components possible that are in contact with water—water curtains, distribution tubes, water sumps, water level probe, and pump assembly. Clean these components manually using fresh solution, allowing them to soak 15 to 20 minutes for heavily concentrated mineral deposits. Use a soft-bristle brush or sponge for cleaning. With the components removed, use fresh cleaner solution to wipe down all accessible surfaces of both the internal ice machine and the bin. Then, rinse or wipe down again with clean water.

Prepare sanitizer solutions and again clean removed components and wipe down all accessible internal surfaces of the unit and bin. Reassemble the unit and add sanitizing solution to the water sump. Circulate the sanitizing solution through the water distribution system for approximately 30 minutes. Disassemble the unit a second time and wipe down removed components and accessible areas of the bin and unit. Rinsing with clear water is generally not needed.

Reassemble the unit, restore power and water, and return the unit to service. Observe the unit's operation through several ice cycles to ensure that all components have been properly reinstalled and are working as expected.

Ice machine cabinet and bin exterior – External surfaces should be cleaned for an attractive appearance. Customers will be turned off by a poorly maintained exterior.

Air-cooled condenser coils – As with all condensers, clean these using a soft brush and vacuum. A coil cleaning solution may be needed if grease or other deposits are evident. Some ice machines are equipped with washable filters to help protect the coil; these can be washed using a mild soap and water solution. Straighten bent fins with a fin comb.

Water filters and conditioning systems – Many systems are equipped with water filters to remove odor, taste, and particulates from the water supply. Filter elements should be replaced during the cleaning/sanitizing process, or when fouled.

Specialized water treatment systems are often installed to help prevent biological growth by adding controlled amounts of chlorine dioxide to the water. These products do not prevent mineral deposits, nor should they be considered a replacement for proper manual cleaning and sanitizing. However, they may reduce the amount of time required for a cleaning operation.

4.2.2 Operation and Troubleshooting

Before beginning a troubleshooting process, ensure that the condenser coil and other heat transfer surfaces are clean. Fouled surfaces and clogged coils will cause abnormal operation and high refrigerant pressures. Pressure gauges should be connected only when necessary. Since the refrigerant charge is often very small, even slight losses that occur when gauges are connected and the hoses are filled may have a significant effect. The proper refrigerant charge is vital to their proper operation.

The electrical system of the ice machine (*Figure 42*) includes circuits for the compressor, water pump, condenser fan(s), water and refrigerant solenoid valves, control transformer, control relays, and microprocessor (if used). Basic electrical troubleshooting principles apply to these items when a problem exists.

Knowing and understanding the precise timing and sequence of operation is essential. Although units may appear to be identical, the sequence of events may differ. Always ensure that proper literature is on hand for the equipment being serviced.

The following sequence of operation applies to the cubed-ice machine shown in *Figure 42*. Also refer to the energized parts chart (*Figure 43*) for details of the operating sequence of that same cubed-ice machine. A watch is helpful when timing each event.

- The toggle switch must be in the Ice position and the water curtain covering the evaporator plate must be in position before the unit will start.
- The water pump and water dump solenoid are energized for 45 seconds before the compressor starts to purge the unit of old water. The harvest valve is also energized during the water

purge and remains energized for an additional 5 seconds after the purge process stops.

- The compressor starts after the water purge process and remains running throughout the ice-making and harvest cycles. The water fill valve is also energized. The condenser fan motor circuit receives power, but the fan cycling control may cause it to cycle on and off.
- The compressor runs for 30 seconds before the water pump starts to chill the evaporator plate. The water fill valve remains on until the probe indicates that a sufficient water level exists in the sump.
- The water pump starts after the 30-second period. Water is evenly distributed across the top of the plate and trickles down into the individual cube cells. Unfrozen water collects in the evaporator trough. Some water remains behind as ice in the cells, continuing to build in thickness as the cycle continues. The water fill valve may cycle open one more time to replenish the water level.
- As ice builds, water flow is pushed out slightly away from the evaporator plate. Eventually, the flowing water makes contact with the ice thickness probe. After roughly 10 seconds of constant contact proving that ice has formed, the harvest cycle can begin. The unit allows at least 6 minutes to elapse in the ice-making cycle before initiating a harvest cycle.
- The harvest valve is opened, allowing hot compressor discharge gas to circulate through the evaporator. At the same time, the water pump continues to run as it did during the ice-making cycle, but the water dump valve is energized for 45 seconds and the water is purged from the sump. The water fill valve is energized during the last 15 seconds of this period to help flush out the existing water. After the 45-second period, the water pump, fill valve, and dump valve are all de-energized.
- The harvest valve remains open, warming the evaporator plate and breaking the bond between the ice and the plate. Cubes are formed as a single sheet with a thin bridge between the individual cubes. The sheet slides out and breaks apart into individual cubes as it falls into the storage bin. The water curtain swings out as the sheet falls, actuating the bin switch in the process. If the curtain and the bin switch remain open for more than 30 seconds, this is a signal that the ice has not completely fallen out and the unit stops, assuming that the storage bin is now full. When the bin is not full, the actuation of the bin switch is momentary and is a signal to the unit to return to the ice-making cycle.

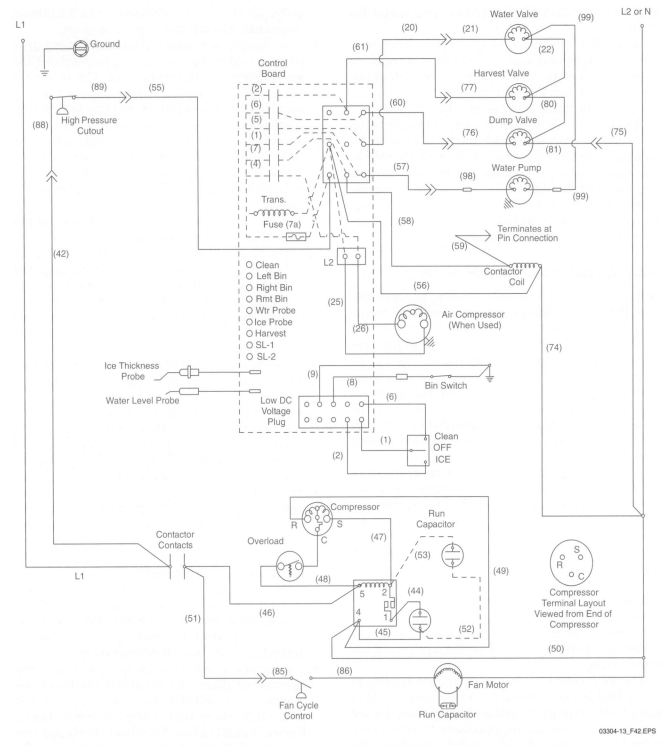

Figure 42 Cubed-ice machine wiring diagram.

- If the bin switch is open for more than 30 seconds and the unit shuts down, it must remain off for at least 3 minutes before it will restart. When enough ice is removed from the bin to allow the curtain to swing closed, ice making can again begin.
- The maximum amount of time allowed to complete the ice-making cycle is 60 minutes, after which the unit will automatically initiate a harvest cycle.
- The maximum amount of time allowed for a harvest is 3½ minutes after which the unit will return to the ice-making cycle.

Over time, water vapor will collect and freeze on the back of the evaporator plate. To rid the unit of this accumulation, after 200 freeze/harvest

Ice Making Sequence of Operation	Control Board Relays							Contactor	Length of Time
	1 Water Pump	2 Harvest Valve (Left) HRP Valve	3 Harvest Valve (Right) (When Used)	4 Air Comp. (When Used)	5 Water Inlet Valve	6 Water Dump Valve	7 Contactor Coil Liquid Line Solenoid	7A Compressor and 7B Compressor Fan Motor	
Initial Start-Up 1. Water Purge	On	On	On	35 sec. off 10 sec. on	Off	On	Off	Off	45 seconds
2. Refrigeration System Start-up	Off	On	On	On	On	Off	On	On	5 seconds
Freeze Sequence 3. Prechill	Off	Off	Off	Off	May cycle on/off during prechill	Off	On	On	Initial start-up is 60 seconds 30 seconds thereafter
4. Freeze	On	Off	Off	Off	Cycles off then on one more time	Off	On	On	Until 10 seconds water contact w/ice thickness probe
Harvest Sequence 5. Water Purge	On	On	On	On after 35 sec.	30 sec. off 15 sec. on	On	On	On	Factory set at 45 seconds
6. Harvest	Off	On	On	On	Off	Off	On	On	Bin switch activation
7. Automatic Shut-off	Off	Off	Off	Off	Off	Off	Off	Off	Until bin switch recloses and 3 minute delay

03304-13_F43.EPS

Figure 43 Energized parts chart.

cycles have been completed, the microprocessor will initiate a warm-water rinse in the following sequence:

- Both the Clean and Harvest LEDs come on, indicating that the unit has entered a warm-water rinse cycle.
- The compressor and harvest valve are energized.
- The water pump is energized.
- The water inlet valve is energized until it makes contact with the water level probe.
- The water is warmed by this process for 5 minutes, then the compressor and harvest valve are de-energized.
- The water pump remains on for an additional 5 minutes, then is de-energized.
- The process can be terminated by moving the control switch to Off, then back to Ice.

Most ice machines are test operated at the factory and adjusted as necessary. The following tests and adjustments should be conducted, after cleaning and sanitizing at initial startup, and after a prolonged shutdown:

- The water level sensor is designed to maintain proper water level in the sump and is not adjustable. If the water level appears incorrect, check the probe for physical damage and clean it if necessary.
- The ice thickness probe is factory set to maintain an ice bridge thickness of ⅛" (*Figure 44*). This amount is sufficient to keep the ice cubes in a sheet until the mass falls from the evaporator, yet thin enough to break apart when the sheet falls to the bin. If the ice bridge appears to be too thick or too thin, set the ice thickness probe distance using the adjusting screw. Begin with a ¼" gap between the thickness probe and the evaporator as a starting point.

This unit is controlled by the microprocessor, which precisely handles the large number of timed events. Microprocessor controls are very reliable, so if one fails, technicians are not expected to troubleshoot the control to the component level. However, because this type of control can be costly, be sure the control has failed before replacing it.

Before condemning the microprocessor, make sure the control is receiving correct input signals and delivering the correct output signals. If it is doing both, and the equipment is not responding, the problem lies somewhere other than the microprocessor control.

Here are some other troubleshooting hints:

- Always listen carefully to all of the customer's complaints. While they may not understand details of the operating sequence, valuable information can be gained from their observations. In addition, technicians who develop good listening skills are often the first to advance.
- Pay close attention to the unit when it is operating properly. If you are familiar with normal unit operation, use sight, sound, and touch (carefully) to detect abnormalities before using tools and instruments.
- Develop a logical approach to troubleshooting all systems. Patience and logical thinking are important troubleshooting tools at your disposal, especially when working with ice makers and other complex systems. Each step depends on the proper execution of previous steps. Without patience and the use of available data to confirm suspicions, technicians can be easily fooled into an incorrect diagnosis.

One of the best ways to evaluate a problem is by using the proper sequence of operation and a watch to time each step in the cycle. Look for events that do not occur when expected. Use the LEDs on the microprocessor to help determine what it is attempting to do, then see if controls external to the board respond properly. When they do not, test the affected components for proper operation.

Figure 44 Ice thickness check.

To prevent loss of refrigerant, service gauges should only be applied when absolutely necessary. When refrigerant must be added, or when refrigerant cycle components must be replaced, use proper refrigeration service practices. Always weigh in the refrigerant when possible to ensure accuracy. Also, always remember to charge zeotropic refrigerants in their liquid state. Remember that charging by subcooling is used for TEVs, while superheat is used when charging units with fixed restrictors. Superheat and subcooling should be checked even when the charge has been weighed in to ensure proper system operation. Indeed, subcooling and superheat values should be checked any time gauges have been connected to the system.

Additional Resources

Refrigeration and Air Conditioning, An Introduction to HVAC/R, Latest Edition. Larry Jeffus. Upper Saddle River, NJ: Prentice Hall.

4.0.0 Section Review

1. A very important step in maintaining a reach-in freezer is _____.

   a. changing compressor oil
   b. lubricating fan motors
   c. checking refrigerant charge
   d. keeping coils clean

2. Circulate ice machine cleaner through the water distribution system for at least _____.

   a. 30 seconds
   b. 10 minutes
   c. one-half hour
   d. one hour

SUMMARY

Retail refrigeration equipment provides an essential service—the preservation of consumable and perishable goods. This segment of the HVAC/R trade includes a variety of equipment and systems designed for specific needs of the products to be maintained. Although there are similarities in the refrigeration circuit when compared to comfort cooling systems, a much wider variety of refrigerants and operating parameters are found in the retail refrigeration classification. Retail refrigeration equipment is often placed in direct view of the consumer, providing direct access to food products. As a result, the equipment should look good as well as operate reliably.

Systems that operate under the classification of high-temperature refrigeration are often very simple in design and require limited troubleshooting skills. However, creating storage temperatures below freezing requires equipment capable of eliminating frost buildup through a variety of approaches. Components that enable equipment operation in severe environments also increase the overall complexity for the service technician.

Commercial ice-making systems generally represent the most complex equipment to be serviced, with many individual and well-timed steps required for the equipment to operate properly and produce the required amount of ice. Troubleshooting these systems requires equipment knowledge and a logical troubleshooting process. Reviewing product-specific literature is critical so that the technician can provide high-quality service and repairs.

1. Which temperature application operates at an evaporator temperature below –30°F?

   a. High-temperature
   b. Medium-temperature
   c. Moderate-temperature
   d. Low-temperature

2. What occurs with the refrigerant liquid/vapor mixture in the condenser until all vapor has condensed to a liquid?

   a. Begins to vaporize
   b. Remains at same temperature
   c. The mixture freezes
   d. Decreases in pressure

3. A technician can determine if all refrigerant in a condenser has changed to its liquid state by _____.

   a. the presence of some superheat
   b. a sudden increase in pressure
   c. a sudden decrease in pressure
   d. the presence of some subcooling

4. What process happens immediately after all superheat is removed from a refrigerant?

   a. Subcooling
   b. Vaporization
   c. Condensation
   d. Oil separation

5. One way to help compensate for the increased energy consumption in retail refrigeration systems, as compared to comfort cooling systems, is to _____.

   a. operate at low superheat
   b. reduce the level of subcooling
   c. reduce the refrigerant charge
   d. increase operating voltage

6. Defrost cycles should be programmed _____.

   a. for periods of light usage
   b. at least 6 times per day
   c. for periods of heavy usage
   d. for long times in dry climates

7. Which defrost method does not require physical contact with the evaporator to melt the frost?

   a. Resistive electric heaters
   b. Hot gas defrost coils
   c. Infrared electric heaters
   d. Nichrome electric heaters

8. Ice merchandisers using cold-wall construction are defrosted _____.

   a. with hot gas
   b. manually
   c. electrically
   d. twice a day

9. In commercial ice makers, ice is formed on a surface with a temperature near 0°F.

   a. True
   b. False

10. Ice thickness sensors on cubed-ice machines function _____.

    a. by instant contact with the ice
    b. based on water temperature
    c. based on refrigerant pressures
    d. by contact with water for a specified time

11. Condensate from a refrigerated fixture can be _____.

    a. thrown on the condenser coil for re-evaporation
    b. recycled to form new ice cubes
    c. placed back into the refrigerated area
    d. used to cool the compressor

12. Which device has the refrigerant feed rate determined by inputs from temperature sensors?

    a. Electronic expansion valve
    b. Capillary tube
    c. Thermostatic expansion valve
    d. Refrigerant injector

13. The primary function of the suction accumulator is to _____.
    a. store refrigerant for system operation at heavy load
    b. remove oil from refrigerant in the suction line
    c. store liquid refrigerant to prevent compressor damage
    d. prevent refrigerant vapor formation during defrost

14. Allowing the head pressure to get too low as a result of low ambient conditions can cause _____.
    a. low flow through the metering device
    b. the compressor to overheat
    c. damage to the air-cooled condenser
    d. the metering device to freeze

15. Electronic temperature controls generally _____.
    a. have a very narrow operating range
    b. are applied to low-temperature systems only
    c. use capillary tubes filled with refrigerant
    d. are easier to install than remote bulb thermostats

16. What component is typically bypassed with a bypass timer when the compressor starts?
    a. Fixture thermostat
    b. Compressor contactor
    c. Low-pressure switch
    d. Defrost thermostat

Figure 1

03304-13_RQ01.EPS

17. Refer to *Review Question Figure 1*. When the fan switch is closed, the other device that controls power to the evaporator fans is the _____.

    a. defrost thermostat
    b. defrost timer motor
    c. alarm thermostat
    d. defrost heaters

18. Any brand of household sanitizer can be used to clean and sanitize a cubed-ice machine.

    a. True
    b. False

19. Refer to *Review Question Figure 2*. The ice machine compressor is de-energized during which operating event(s)?

    a. The harvest water purge cycle.
    b. The initial startup water purge cycle and the automatic shut-off cycle.
    c. The freeze prechill cycle.
    d. The initial startup water purge cycle and the harvest water purge cycle.

20. When charging a refrigeration unit with a zeotropic refrigerant, you should _____.

    a. replace the compressor oil
    b. charge in the vapor state
    c. charge in the liquid state
    d. use any blended refrigerant

Ice Making Sequence of Operation	Control Board Relays							Contactor	Length of Time
	1 Water Pump	2 Harvest Valve (Left) HRP Valve	3 Harvest Valve (Right) (When Used)	4 Air Comp. (When Used)	5 Water Inlet Valve	6 Water Dump Valve	7 Contactor Coil Liquid Line Solenoid	7A Compressor and 7B Compressor Fan Motor	
Initial Start-Up 1. Water Purge	On	On	On	35 sec. off 10 sec. on	Off	On	Off	Off	45 seconds
2. Refrigeration System Start-up	Off	On	On	On	On	Off	On	On	5 seconds
Freeze Sequence 3. Prechill	Off	Off	Off	Off	May cycle on/off during prechill	Off	On	On	Initial start-up is 60 seconds 30 seconds thereafter
4. Freeze	On	Off	Off	Off	Cycles off then on one more time	Off	On	On	Until 10 seconds water contact w/ice thickness probe
Harvest Sequence 5. Water Purge	On	On	On	On after 35 sec.	30 sec. off 15 sec. on	On	On	On	Factory set at 45 seconds
6. Harvest	Off	On	On	On	Off	Off	On	On	Bin switch activation
7. Automatic Shut-off	Off	Off	Off	Off	Off	Off	Off	Off	Until bin switch recloses and 3 minute delay

Figure 2

03304-13_RQ02.EPS

Trade Terms Introduced in This Module

Phase change: The conversion of refrigerant from one state to another, such as from liquid to vapor.

Slinger ring: A ring attached to the outer edge of the fan that picks up water as the fan rotates and slings it onto the condenser coil.

Unit cooler: A packaged unit containing evaporator coil, metering device, and evaporator fan(s) that serves as the evaporator section for a refrigeration appliance.

Additional Resources

This module presents thorough resources for task training. The following resource material is suggested for further study.

Refrigeration and Air Conditioning, An Introduction to HVAC/R, Latest Edition. Larry Jeffus. Upper Saddle River, NJ: Prentice Hall.

Figure Credits

Answer	Section Reference	Objective
Section One		
1.b	1.1.0	1a
2.a	1.2.0	1b
3.c	1.3.3	1c
Section Two		
1.d	2.1.1	2a
2.a	2.2.0	2b
3.d	2.3.1	2c
Section Three		
1.c	3.1.1	3a
2.a	3.2.2	3b
3.c	3.3.3	3c
Section Four		
1.d	4.1.0	4a
2.c	4.2.1	4b

NCCER CURRICULA — USER UPDATE

NCCER makes every effort to keep its textbooks up-to-date and free of technical errors. We appreciate your help in this process. If you find an error, a typographical mistake, or an inaccuracy in NCCER's curricula, please fill out this form (or a photocopy), or complete the online form at **www.nccer.org/olf**. Be sure to include the exact module ID number, page number, a detailed description, and your recommended correction. Your input will be brought to the attention of the Authoring Team. Thank you for your assistance.

Instructors – If you have an idea for improving this textbook, or have found that additional materials were necessary to teach this module effectively, please let us know so that we may present your suggestions to the Authoring Team.

NCCER Product Development and Revision
13614 Progress Blvd., Alachua, FL 32615

Email: curriculum@nccer.org
Online: www.nccer.org/olf

❑ Trainee Guide   ❑ Lesson Plans   ❑ Exam   ❑ PowerPoints   Other _____

Craft / Level: _____   Copyright Date: _____

Module ID Number / Title: _____

Section Number(s): _____

Description: _____

_____

_____

_____

_____

Recommended Correction: _____

_____

_____

_____

_____

Your Name: _____

Address: _____

_____

Email: _____   Phone: _____

03316-13

Customer Relations

OVERVIEW

To a customer, the HVAC/R service technician is the face of a company. The customer's attitude toward the company, and the likelihood of that customer doing future business with the company, are determined largely by the customer's impressions of, and interactions with, that technician. For those reasons, a service technician must learn how to deal effectively with customers.

Module Twelve

Trainees with successful module completions may be eligible for credentialing through NCCER's National Registry. To learn more, go to **www.nccer.org** or contact us at **1.888.622.3720**. Our website has information on the latest product releases and training, as well as online versions of our *Cornerstone* magazine and Pearson's product catalog.

Your feedback is welcome. You may email your comments to **curriculum@nccer.org**, send general comments and inquiries to **info@nccer.org**, or fill in the User Update form at the back of this module.

This information is general in nature and intended for training purposes only. Actual performance of activities described in this manual requires compliance with all applicable operating, service, maintenance, and safety procedures under the direction of qualified personnel. References in this manual to patented or proprietary devices do not constitute a recommendation of their use.

Objectives

When you have completed this module, you will be able to do the following:

1. Explain the service technician's role in customer relations.
   a. Explain how personal habits, behaviors, and attitudes affect customer relations.
   b. Explain how to properly communicate with customers.
2. Describe basic conduct required for a service call.
   a. Describe how to conduct the three phases of a service call.
   b. Describe ways to handle challenging customer situations.

Performance Task

Under the supervision of your instructor, you should be able to do the following:

1. Participate in at least three different role-playing scenarios related to challenging customer service situations.

Trade Terms

Empathy
Rapport

Industry-Recognized Credentials

If you are training through an NCCER-accredited sponsor, you may be eligible for credentials from NCCER's Registry. The ID number for this module is 03316-13. Note that this module may have been used in other NCCER curricula and may apply to other level completions. Contact NCCER's Registry at 888.622.3720 or go to **www.nccer.org** for more information.

Contents

Topics to be presented in this module include:

Figures and Tables

1.0.0 THE TECHNICIAN'S ROLE IN CUSTOMER RELATIONS

Objectives

Explain the technician's role in customer relations.

 a. Explain how personal habits, behaviors, and attitude affect customer relations.
 b. Explain how to properly communicate with customers.

Performance Task 1

Participate in at least three different role-playing scenarios related to challenging customer service situations.

Trade Terms

Empathy: The ability to identify with another person's situation or feelings.

Rapport: Establishing a relationship of mutual trust.

Good customer relations are essential to the success of most businesses, including those in the HVAC industry, where each service technician represents the company and influences what customers think about the business. Keeping these impressions positive requires that all employees work toward generating customer good will. Providing good service is a part of this, but so are good personal habits, good work practices, and customer-pleasing attitudes. Customer goodwill often results in more business through referrals and repeat servicing of existing customers.

This module provides guidelines for dealing with customers during service calls. Although the examples used in the module lean toward working with residential customers, the basic principles apply to all types of service calls.

1.1.0 How Personal Habits, Behaviors, and Attitudes Affect Customer Relations

Positive habits, behaviors, and attitudes help to ensure that a customer's first impression of you and your company is a good one; that the customer's needs are clearly understood; and that appropriate customer service is provided to meet these needs. When you provide this level of effective service to your customers, you can increase your company's share of business.

Your ability to relate well to your customers is critical. As a service technician, you are likely to be the only person from your company that the customer ever sees. In the customer's eyes, you are the company, so appearances do count. First impressions are made only once, and it generally happens in the first 60 seconds. There is an old saying: "You don't get a second chance to make a first impression." Mastering the art of making a good impression will not only make your job easier, it will also make it more likely that a customer will call your company the next time HVAC service is needed. Positive responses and feedback from customers that reach your place of business also reinforce your value to the employer and provide support for pay increases.

Because customers lack technical training and test equipment, they have to rely on a service company for help with heating and cooling problems. Although they may not be able to judge the service technician's professional competence, they can, and do, form an impression based on the technician's appearance and attitudes. The customer's first glimpse of a technician can determine that customer's opinion of the worker and

It's All About Perception

Customers and potential customers base their buying decisions on their personal attitudes and perceptions. If someone sees you speeding or driving aggressively in your company vehicle, it can affect that person's willingness to do business with your company. If your truck is dirty on the outside, or cluttered and disorganized on the inside, many customers will see that as a reflection of you. The same applies to your personal appearance. If you need a shave or a haircut, if your clothes are dirty or wrinkled, or even if it is as simple as not having your shirt tucked in, some customers will be put off by it. Many customers, especially older people, could be put off by heavy tattooing. It is usually best to cover them up, rather than displaying them.

the company the technician represents. Examine your first impression. Do you:

- Practice good personal hygiene?
- Get enough sleep and look alert?
- Wear a neat, clean uniform?
- Wear shoe covers or remove shoes when you will be walking on carpeting or finished floors?
- Carry a pencil and pad to take notes?
- Park on the street, rather than the customer's driveway? While this is not always feasible, it is a good idea. Vehicles often have oil or fluid leaks that will stain a driveway.

Once the customer has formed a first impression of you and your company, your on-site work habits will confirm or change that impression. Here are some questions to ask yourself on a regular basis, rather than just once:

- Are you on time? Keep in mind that customers must often take time away from work or from other activities in order to be available for the service call. Make every possible effort to be on time for the appointment. Respect the time of your clients. If you are going to be unavoidably late, have the courtesy to call and let them know. When you arrive, apologize for your tardiness.
- Is your tool set complete and neatly packed? Check your tool kit and your truck every day before setting out on your service calls to make sure you have everything you need.
- Do you show concern, courtesy, and respect to the customer by listening carefully while the customer identifies the problem? Listening attentively may provide you with useful information about the problem. It also helps you establish rapport with the customer.
- Do you tackle the problem promptly and quietly?
- Do you remember to protect the work area? Do you carry rags for cleaning up and drop cloths to protect floors and carpets?
- Do you respect the job site by not tracking in dirt and by cleaning up after yourself?
- Do you take the time to explain carefully to the customer what's wrong and how the unit will perform satisfactorily after you've repaired it?
- Do you refrain from smoking on the premises, avoid smelling of smoke, or spitting?

- Do you respect company equipment, tools, and vehicles? Trucks and vehicles displaying the company name are rolling advertisements (*Figure 1*). Is your vehicle clean and in good repair? Are your driving habits courteous?
- Do you avoid using alcohol and drugs while on the job or around vehicles?
- Do you avoid profanity and horseplay? Customers, especially those with children on the premises, are likely to be offended by inappropriate language.
- Do you avoid wearing political or sports-related articles that might cause a confrontation?

If you are working at a commercial construction site, you must be aware of and respect the site safety requirements. This includes wearing the proper personal protection equipment (PPE) (*Figure 2*), which can include:

- Hard hat
- Ear protection
- Safety glasses
- Reflective vest
- Work gloves
- Steel-toed boots

Site managers are perfectly within their rights to refuse entry to someone who is not properly attired. The right thing to do is to learn in advance what PPE is required. Don't assume anything, because requirements vary from site to site. You will make a bad first impression if the site manager has to refuse you entry to the site. If and when that does happen, your attitude and response to

03316-13_F01.EPS

Figure 1 Keep your service van clean.

Customer Technical Knowledge and Skills

When talking with customers, never assume that they do not have technical knowledge or skills. Although many do not, some customers have a broad technical background and can readily understand technical concepts.

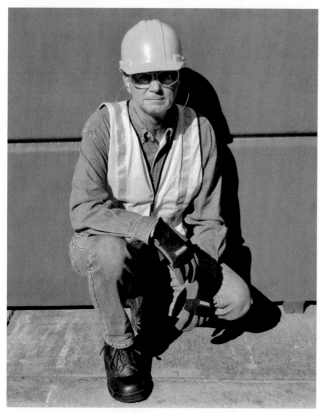

Figure 2 Construction site PPE.

it will be remembered by others long term. Every hour of every working day, you are an advertisement for your company. Your appearance and behavior must be consistent with the positive image your company wants you to reflect.

Off the job, you can increase word-of-mouth advertising (free advertising created by satisfied customers who recommend your company to friends) by having a positive attitude about your employers, co-workers, and customers. Positive statements about your job are free advertising for your company.

The internet and social media have changed the face of customer relations. A person who is dissatisfied with a service company can go to a website and lodge a complaint that hundreds of other potential customers can read. Many people these days will not even do business with a company without looking for opinions from others on websites. People living in a residential subdivision are likely to be part of an internet social group; they can instantly share a positive or negative opinion of a contractor to hundreds of their neighbors. Both good and bad news travel at lightning speed through social media (*Figure 3*). Unfortunately, negative information typically spreads farther and faster than positive information. One reason for this is the power of anger

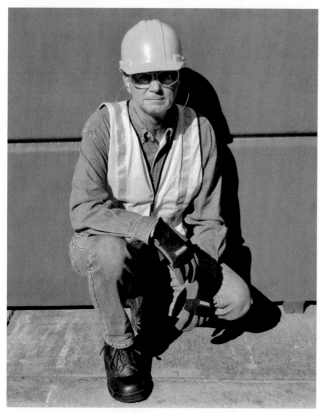

Figure 3 Social media has increased an individual's network of friends and acquaintances to levels never before possible.

and/or resentment as motivators. Both are powerful motivators for unhappy customers to spread their dissatisfaction. Happy, pleased clients are not as likely to access social media to share this information, but will share it when it is the context of a conversation.

1.2.0 Customer Communication

It is important to know the technical requirements of your job. Knowing the steps in completing a typical call is important, too. Confidence is an important trait to convey to a customer. However, you can gain a little extra leverage by learning how to interact well with customers. This may make all the difference for your company and may determine whether you or your competitor is called the next time. When you practice good customer relations throughout the service call, you give the customer confidence in your service and your company and stimulate that customer to call you again. One of the most important things to remember about communication is that it is a two-way street. Be sure to let customers communicate their feelings and ideas. Part of this process is allowing them to vent their frustrations about the equipment or about the inconvenience or discomfort that the equipment failure caused. For example, a customer who has who has been without heat on a cold day is likely to be very uncomfort-

---

Think About It
Personal Habits, Behaviors, and Attitudes

If you were asked to hire a new service technician for your company, what qualities would you look for?

---

able (*Figure 4*). The customer will want to let you know about it, and listening with sincerity and a concern for the situation will be required.

It is sometimes necessary to contact the customer by phone. When that happens, it's important to ensure that the communication is clear and precise. Here are some tips to follow when making a phone call to a customer:

- Before you make the call, think about what you want to say and the questions you want to ask. It's always a good idea to write them down.
- When the person answers, identify yourself and your company. Speak clearly and (if practical) take notes to be sure you are getting the correct information. Repeat it back to make sure.
- Prepare in advance in case you have to leave a message on an answering machine. Be ready to give your name and company, the purpose of your call, and a way to contact you. Speak slowly and clearly, especially when you are giving a phone number or e-mail address. Speak the numbers slowly, and then repeat the number.

03316-13_F04.EPS

Figure 4 It can be difficult to communicate with unhappy customers.

1.2.1 Keeping Communications Positive

On every service call, you should work to keep the communications positive. This means paying close attention to verbal and nonverbal elements such as:

- The technical elements of the job
- What the customer needs
- Your problem-solving expertise
- What you've done to make the customer happy

Here are some simple but effective tips:

- Do the best technical job you can, since actions speak louder than words.
- Always treat the customer with courtesy, concern, and respect. Treating others as you'd like to be treated yourself is particularly important, especially in unpleasant situations. Remember, it is always easy to be courteous when things are going well. A truly mature person is courteous even when things aren't going well.
- Treat each service call as if it were an emergency. To the owner, it is. Usually, customers do not call a service technician unless they are without heat in cold weather or without cooling in hot weather. When comfort is at stake, repair becomes a top priority in the customer's mind.
- Show **empathy** for the customer. This means putting yourself in the customer's shoes. Think about how you and your family would feel on a cold day if your furnace had not been working for many hours. Be sincere in your empathy; the customer is likely to detect insincerity and resent it.
- Leave your personal life at home. Don't let personal frustrations affect your attitude and your relations with customers. If you are having personal problems, avoid using customers as sounding boards.
- If your schedule is running late, report this to your supervisor or the customer. If you're running late, the customer must often make other arrangements to fulfill later commitments. The main thing is: don't leave the customer wondering about what is going on. Make sure they know you are running late and see if they need to reschedule.

1.2.2 The Positive Approach

The following are some important tips on keeping things positive:

- *Smile* – Be genuinely interested in the customer. You may think you were called to repair a furnace. You were really called to make the customer's home comfortable again.

- *Use positive statements with the customer* – Compliment the equipment if you know it is good. Compliment the cleanliness of the furnace area if that contributes to the operation of the equipment. Speak of your employer in positive terms, because that builds customer confidence in your company.
- *Do not criticize the client or the system* – Whenever possible, avoid negative statements. Be careful of what you say about the brand of equipment, the design of the system, or the quality of the workmanship, because you may offend the customer. Homeowners do not want to hear that you think they own inferior equipment. Even if it is inferior, fix what is possible or explain respectfully to the customer what must be replaced. Also, this is not the time to remind the customer that the problem might not have happened if they had taken your advice about signing a maintenance agreement. It's better to bring that up later, in a positive way, after you have corrected their problem.
- *Do not criticize your own organization* – In some cases, you may be called upon to service a system recently repaired by one of your co-workers. Rather than criticizing his or her work, apologize for the inconvenience and assure the customer you will get to the bottom of the problem and report your findings to them directly. Then follow through, and take responsibility for any errors on behalf of your employer.

1.2.3 Showing Concern for Customers

Another part of good, positive, on-the-job communications is showing concern for your customers. Do this by:

- *Being a good listener* – Listen carefully when the customer explains what went wrong, what was observed, etc. If you yawn, slouch, or let your eyes wander, it is a sign of inattention and is likely to aggravate the customer. Ask questions as appropriate in order to gather the information you need to solve the problem. Remember that although the customer may not have the technical vocabulary you do, the customer's own words can supply you with important clues. After you've been in the business for a while, you will probably have heard most of the equipment problems and customer complaints. The client's description may be long and rambling. Yet, it's important to stay engaged and treat each situation like it is new to you. While it may be old hat to you, it is unique and often stressful for the customer.

- *Talking in terms of the customer's interests* – Always try to look at the problem from the customer's point of view. For example, Mrs. Jones may be upset that you are late. Maybe the emergency you fixed for your last customer caused your tardiness. However, to Mrs. Jones, the important thing may be that she's running late in picking up her child from school. If this understandable concern for her child is added to concern over a furnace that isn't heating, Mrs. Jones might be short-tempered. If she speaks to you abruptly, try turning the conversation around by saying, "I know you must be concerned about my being late. I apologize. Please show me your furnace so we can get you back on schedule quickly."
- *Keeping your personal problems to yourself* – Remember, you are on the job to solve the customer's problem. Don't let your problems spill over into your dealings with customers. If you are in a bad mood because of personal problems, don't take it out on the customer.
- *Answering questions honestly, but positively* – If Mr. Brown asks if his equipment is worn out, you may reply, "It needs to be replaced," but you should avoid negative comments such as, "This pile of junk needs to be replaced."
- *Respecting your customer's opinions* – Suggest alternatives that fall short of telling a customer, "You're wrong!" Show that the customer's opinion may have value. For example, if Mr. Smith announces that the thermostat is no good and you suspect there is a wiring problem, don't say, "No, you're wrong. It's in the wiring." Instead, say, "I think it's the wiring, but I'll check out the thermostat too."
- *Not socializing while on the call* – Generally, service technicians are paid by the customer by the hour. If you explain pleasantly that you appreciate the offer of a cup of coffee, but that you're sure they'll understand if you decline, the customer will not be offended by your trying to keep the charges to a minimum. Unless it is absolutely essential, leave the mobile phone in your pocket. Better yet, leave it in the service vehicle. Customers hearing a technician engaging in any conversation that is unrelated to the job at hand will often become unhappy very quickly.

Further, you continue to show concern for your customers when you make it easy for them to contact you for repeat business. When possible, do the following:

- Leave a business card.

- Put a sticker with your company's phone number on the equipment itself or leave one with the customer for placement near the phone.
- Show the customer some simple maintenance techniques that will prolong equipment life, such as changing filters frequently (*Figure 5*). Tell the customer why this maintenance is helpful.
- Teach the owner steps for more efficient use of the equipment, if applicable. For example, show the owner how to program the thermostat (*Figure 6*) so less energy is used during times when the structure is not occupied.
- Discuss billing accurately and honestly.
- If the opportunity arises, discuss additional equipment to improve comfort and efficiency that the customer can purchase from your company. Ask for the customer's permission to notify the sales staff. Follow up by providing this lead to the sales staff.
- Perform customer follow-up, if possible.

1.2.4 *Selling Yourself and Your Company*

You may not consider yourself a salesperson, but selling is a key element of your job even though it may not always be a conscious effort. In reality, you are selling whenever you go on a service call. First, you are selling yourself because your appearance, professionalism, courtesy, and self-confidence will sell the customer on you.

Second, you are selling your company. The impression you make, along with your successful completion of the job, will determine whether the customer is likely to call your company again. Third, every service call you make is a selling opportunity. If you handle a trouble call for a customer who has no annual service contract, there is an opportunity to convince the customer that periodic maintenance will reduce or eliminate emergency calls and the discomfort that goes along with them.

You also have the opportunity to point out the need to replace aging equipment. You might be surprised how practical your customers can be. Many of them can readily understand the long-term benefit of replacing aging equipment with

03316-13_F05.EPS

Figure 5 Changing filters can increase equipment life and efficiency.

03316-13_F06.EPS

Figure 6 Show the customer how to avoid frustration and adjust the thermostat for greater comfort and efficiency.

new energy-efficient equipment and will see beyond the initial cost. It never hurts to ask. Besides, how would you feel if your company got a $100 service call and a competitor got a $5,000 contract to replace a worn out system for the same customer a year later? One way to avoid such embarrassment is to note the age of the customer's heating and cooling equipment and pass that information along to your company's sales department.

Think About It
Showing Concern for Customers

What would you do in these situations?

1. Mr. Gray owns a brand of furnace that is no longer being manufactured. You don't think parts are available to fix the burner. What would you say?
2. Mrs. Brown called for a technician to fix her furnace. As you enter the job site, she says, "It's cold in here." What questions would you ask her?
3. Mr. Green says his blower and burner are not operating in sequence and may need to be replaced. You think the problem is a faulty limit switch. How do you build his confidence in your skills without hurting his feelings?
4. Can you suggest some ways of advertising your company that are simple and free?

Additional Resources

Tools for Success: Critical Skills for the Construction Industry. Upper Saddle River, NJ: Prentice Hall.

Applied Communication Skills for the Construction Trades. Upper Saddle River, NJ: Prentice Hall.

1.0.0 Section Review

1. First impressions are usually made in the first 60 seconds of contact.

   a. True
   b. False

2. If you are going to be late for an appointment, the best thing to do is _____.

   a. just get there as quickly as you can
   b. call the dispatcher and have them reschedule for another day and time
   c. report it to your supervisor or let the customer know you will be late
   d. leave the first job unfinished in order to be on time for the next appointment

SECTION TWO

2.0.0 HANDLING SERVICE CALLS

Objectives

Describe basic conduct required for a service call.

   a. Describe how to conduct the three phases of a service call.

   b. Describe ways to handle challenging customer situations.

Performance Task 1

Participate in at least three different role-playing situations related to challenging customer situations.

There is more to a repair call than simply showing up ready to work. In a typical case, the technician starts by interviewing the customer to find out information about the problem and, in many cases, soothing ruffled feathers. Once the repair is complete, there are wrap-up tasks that must be completed.

2.1.0 Conducting the Service Call

During each service call, you can do several things to enhance the customer's image of you and your company.

The typical service call has the following three parts:

- *The opening* – Meeting the customer and identifying the problem.
- *Performing the service* – Solving the problem.
- *The closing* – Leaving the customer with a positive impression.

Remember that your objective in servicing your customer's equipment goes beyond technical competence. Your objective is to win the customer's repeat business.

2.1.1 The Opening

On most service calls, the HVAC/R technician is going to face an unhappy and uncomfortable customer. The customer may have lost heating on a cold day or cooling on a hot day. In most cases, the problem has been going on for hours, possibly for more than a day. In many cases, the customer is looking for someone to blame. If it is equipment your company installed, the customer will be happy to blame you, even if you weren't personally involved. Always remember that your real responsibility is to solve a client's problem, and they write the checks that pay your wage. An air conditioning or heating unit has never paid a bill; only satisfied customers do that.

The first order of business is to defuse the situation. Listen attentively and sincerely to the customer's complaints and try to get him or her to describe how the problem manifested itself. Engage the customer by asking questions: Is the system operating but not providing enough heating or cooling? If it is not working at all, was there any noticeable problem before it stopped working? Were there unusual noises or odors? If this is a new customer, ask about previous problems. Engaging the customer in this way not only provides you with information about the problem, it helps to make a customer feel like an ally in the battle to restore normalcy. However, be careful about using industry jargon. Using technical terms may make you seem clever, but is more likely to further annoy the customer.

As you open the service call, you are already making your first impression. Your personal appearance is good. Your vehicle looks good. You show respect for the customer by appearing promptly and politely at the door. Now what? You should do the following:

- Smile and display confidence and polite respect for the customer (*Figure 7*). Make eye contact with the customer.
- Promptly identify yourself. Many customers waiting for a service worker are naturally cautious about admitting a stranger. Identify yourself clearly and show an appropriate ID,

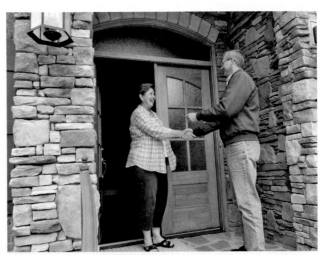

03316-13_F07.EPS

Figure 7 A pleasant greeting is a good way to start a service call.

---

if you have one. Give the customer a company business card with your name on it. If your company does not provide business cards with the names of individual service technicians, encourage the company to at least have company business cards available so that you can give one to the customer in case they need to contact the company for follow up questions or concerns.

- Be respectful. If a customer introduces himself including his first name, don't assume you can call him by his first name. Some customers resent such familiarity. Above all, concentrate on remembering the customer's name in case you have to get their attention later on.
- Understand your customer's needs. Listen to your customer to learn what the problem is. Ask questions about what the customer has seen, heard, smelled, and felt. Ask when things occurred and how often they happened. Often, the customer can help speed your diagnosis by providing clues to the equipment problem. When you listen attentively, everyone wins.

  It's important to ask open-ended questions, the kind that result in specific, information-gathering answers. You don't want to be led to the equipment without any information from the customer. Ask as many questions as you need to get the exact symptoms. Don't say, "I'm here to fix your furnace. Where is it?" Instead, say, "I came to repair your furnace. What seems to be the problem?"
- In a residential situation, do not enter the home until you are invited to do so, either verbally or by gesture. Before entering, don some disposable shoe covers (*Figure 8*). They should not be worn until you are at the entrance. Otherwise, they will get soiled as you walk from the truck to the door.

2.1.2 *Performing the Service*

While servicing the equipment, you should be adding to your customer's positive impression of you and your company. The customer has an equipment problem. You are there to solve it. Once you have listened to the customer explain the problem, you can build customer confidence by getting right to work to determine the probable cause of the malfunction. Practice good work habits. Avoid general conversation while working. Remember, many companies charge customers by the hour for service. The customer must feel that you are filling the time with productive work, not idle chatter.

03316-13_F08.EPS

Figure 8 Disposable shoe covers.

If you make a mistake that is going to lengthen the duration of the service call, do not try to cover it up by telling the customer that the problem is worse than you thought and bill them for the extra time. Be honest. Own up to your mistakes and learn from them. Although constant errors will not be tolerated by employers or clients, a great deal of respect and admiration can be gained by stepping up when an error is made. Assure your client that any additional repair costs that resulted from your negligence or error will not be included on the invoice.

It is important to arrive prepared, with the proper tools and, to the extent feasible, replacement parts appropriate to the situation. The company dispatcher should have obtained information about the nature of the customer's problem and the type of equipment. If this is an existing customer of your company, you should have access to the make and model of the equipment. This will make it easier to be prepared and avoid wasting time. If you have to run back to the shop for what the customer might think is a standard part, they might think they are getting billed for unnecessary time.

During this part of the service call, you will need your professional skills to:

- Analyze the symptoms.
- Perform necessary testing to isolate the problem.

Think About It
Customer Relations

What questions can you ask Mrs. Jones when responding to an air conditioning service call?

- Determine the solution.
- Explain the solution to the customer, including your best estimate of what it will cost.
- If the customer approves, remedy the situation.

While the work is ongoing, maintain a clean, organized work area. Not only does this look far more impressive to a customer, it also helps to ensure your work activities will be completed safely. When the work is complete, make sure that all trash is picked up and inventory your tools. Take the extra step to ensure that the area is cleaner than it was when you arrived by picking up any other trash that is nearby. Successful technicians carry simple cleaning materials to clean the cabinets of furnaces and condensing units. Make sure that the equipment serviced looks its best before departing. Many companies also provide labels with company contact information to their technicians that are applied to the equipment before departing.

2.1.3 The Closing

This very important step helps set the tone for your customer's image of you and your company. When you finish the work:

- Neatly pack your tools.
- Return the premises to its original condition. (Replace covers, wipe off dirty fingerprints, clean up drop cloths, etc.)
- Explain to the customer that the problem has been solved. Explain what parts you replaced and offer to show the customer the defective parts.
- Demonstrate that the equipment works.
- Wrap up the call by relating your service to the customer, not the equipment. Don't say: "I fixed the thermostat; your furnace will turn on now." Rather, say: "I repaired your thermostat so you can be warm and comfortable again." A positive closing like this builds confidence in you and your company, and shows that your mind is in the right place.

Following up after the repair is important, but it is often overlooked. If time permits, call the customer a day or two later to ask if the repairs were satisfactory. If you made adjustments, stop by to see if they were acceptable. Often, a simple phone call from you can be the touch that creates a repeat customer. In your company, if the service technicians don't have the time to follow up on service visits, perhaps someone from the office staff can handle these calls.

2.2.0 Handling Challenging Customers

When you deal with customers, remember: the customer wants safety, comfort, and convenience. If your habits, behaviors, and attitudes inspire that customer to believe your company will provide for their safety and comfort in a convenient manner, the customer will return to your company rather than seeking out a competitor next time.

Sometimes, the service technician encounters problems that go beyond failed machinery. For example:

- A customer may be unduly worried about safety.
- A customer may be upset about something totally unrelated to the problem, but is taking out that irritation on the service technician.
- A customer may be upset over equipment performance and your company's bill.

In cases like these, the service technician's actions and attitude will often determine whether the service visit is a successful one or not.

The following sections cover several situations that require more advanced customer relations skills. Decide how you would handle these customers and compare your ideas with those of your fellow trainees.

Remember, there may be no single right way out of an angry confrontation, but your attitudes and reactions can go a long way to help resolve even the most difficult situations.

2.2.1 Fearful Customers

Sometimes customer fears are based on experience, rumors, or misunderstandings. As the technician, you have the technical information that can calm your fearful customer. There is always a

Going the Extra Mile

People often neglect to change or clean the air filters for their equipment. Sometimes they are not physically able to do it. If you are on a service call for a new customer that does not have an annual service contract, offer to check the condition of their filters. This takes very little time and could increase the value of the service call. Make a note of the filter sizes so you can put them on the truck the next time the customer calls. If the filters have been neglected, it is a good opportunity to mention the value of a service contract.

reason why your customer has a fear. Find it and work from the facts to ease the customer's mind. If you listen with concern and take customers' fears seriously, they will feel reassured and be more willing to listen to your explanation.

Exercise Question:

Mr. Fearful thinks his furnace is leaking gas because he smelled some when the pilot light blew out. You have re-ignited the pilot and are closing the call. What can you say to calm his fear that the house will explode from a leak at the pilot light?

Exercise Solution:

Fear is real. Often you can discover the reason for it by asking the following questions:

• When does this happen? When did it start?
• When it happens, what does it look like?
• What does it sound like?
• What does it smell like?
• Before it happens, are you doing something different from what you used to do?

After you have found the answers, don't just say, "Yes, when the pilot light goes out there is a smell. It is fixed." This has not addressed Mr. Fearful's concerns for the next time it happens. Instead, explain: "When a pilot goes out, there is a smell of gas. A minute quantity escapes but will not ignite. It is also mixed with a definite odor that alerts you to re-light the burner. The small amount of gas will not cause a fire. The vent pipe will exhaust it. Your family is not in any danger." Possibly suggest that a gas sniffer alarm and carbon monoxide sensor be installed to give the customer peace of mind.

Since this explanation not only addresses the technical problem, but also the fear associated with it, the customer will be more satisfied.

2.2.2 Opinionated Customers

Opinions are beliefs based not entirely on facts, but also on what seems probable in a person's mind. Everyone has opinions. Everyone also likes to get approval for opinions held. If your opinions are ridiculed, you feel hurt and offended.

It is important that customers do not feel their opinions have been challenged or belittled. With a customer, it's best to turn the conversation to the technical problem at hand (even when you agree with the opinion).

Exercise Question:

Mrs. Opinionated has had a bad week, including the failure of her air conditioner. She's trying to draw you into her earlier problems and her opinions about them. You're trying to finish quickly because you have two other service calls before lunch. What do you say?

Exercise Solution:

Taking what seems to be a negative approach might offend her. So don't say: "Yeah, we all have problems. My problem is fixing air conditioners. Where is it?" Instead, say: "I've had times like that myself. I can help you with one of your problems, though. Please show me your air conditioner so I can help you solve that problem." This offers sympathy while at the same time diverting the customer to the task at hand.

2.2.3 Argumentative Customers

Some customers are just the argumentative type. They want you to challenge their opinions, just to get an argument going. You're sensitive to the customer, and you don't want to seem aloof. You don't want to express negative opinions either. You also know that some topics are really argument-prone, such as politics, religion, children, and in-laws. Here, too, the safest approach is to avoid controversy by diverting the conversation to the task at hand.

Exercise Question:

Mr. Argumentative has no airflow from his furnace. You've discovered a broken fan belt. He wants to pull you into a discussion of local politics. He thinks the mayor is a crook, but he senses you disagree and wants to argue. You do not want

---

Think About It

Closing

After you've replaced the refrigerant charge in Mrs. Jones' air conditioner, what closing statement can you make?

Verify That the System Works

One of the most important things you can do at the end of a service call is to demonstrate to the customer that the system works. Don't leave them wondering about it; show them.

---

to be drawn into this discussion. How would you handle this?

Exercise Solution:

The best way to handle this is to state a general fact that will not offend, and to avoid prolonging or expanding the political discussion while you gently turn the conversation to the problem you're there to solve. (Even if you disagreed with the customer, the technique would still be same: shifting attention to the fan belt problem.) Say, "Some people feel that way. But my feeling is that you'll be a lot more comfortable after I replace this fan belt."

2.2.4 Sloppy Customers

As you make service calls, you'll see all kinds of housekeeping: some spotless; some cluttered; some downright dirty. Always, it's best to avoid offending your customer. When admitted to the job site, avoid being judgmental in your words and gestures. Since you're a visitor, the owner will often offer an excuse or reason if things are messy. In some homes, the owner's sloppy habits can adversely affect the operation of the equipment.

You can handle such sensitive situations in the following ways:

- *Focus on the task at hand* – Don't look like you're judging the job site. Avoid comments like "What a mess!" and raised eyebrows or rolled eyes that say the same thing. Assume that the owner is the unfortunate victim of several small domestic catastrophes that will be remedied soon.
- *Put the owner at ease* – If you must reply, avoid degrading and comparing them unfavorably. Say: "This happens at our home, too."

- *Be diplomatic* – If it's cluttered in front of the furnace, move only what's necessary and try to restore things when you're finished. Don't be obvious in moving things. As you clear space, avoid grunts, groans, and comments. If obstructions are blocking the flow of combustion air, let the customer know that this could be a safety hazard. Also, if you see exposed detergents, bleach, or chemicals near a furnace, point out the hazards.
- *Don't sound like an accuser* – Avoid "you" statements. Talk about what's best for the health of the furnace or make up a similar situation that you solved, using your own family as an example.

2.2.5 Angry Customers

In the last section, you were advised to avoid "you" statements to a customer when it might suggest a negative judgment; for example, accusing them of bad housekeeping. There is, however, a good use of a "you" statement, and that's when the customer has a right to be angry.

When equipment breaks down twice in three days, or when the customer has tried to meet the service technician two days in a row without success, the customer may feel justified in venting anger at the first available target: the technician who appears at the door. What should you do?

First, let the angry customer vent his or her emotions. Next, acknowledge the customer's feelings. Sometimes, this means saying something sympathetic like, "I'd feel that way, too." Finally, assure the customer that you'll do your best to solve the problem.

Think About It
Mr. Sloppy

Mr. Sloppy takes you into his home to fix the air conditioner. The musty air smells of cat litter; dusty newspapers fill all the chairs; the kitchen looks like a scene from a disaster movie. Mr. Sloppy says, "Isn't this place a mess?" What would you say?

a

03316-13_SA01.EPS

Exercise Question:

Mrs. Smith's furnace was repaired just last week. It has broken down again, leaving the family without heat for two days. The children have colds. The Smiths have already paid a sizable sum for the repair that did not last. As you walk in the door, Mrs. Smith snarls at you. What do you say?

Exercise Solution:

Here is an appropriate answer: "Hello, Mrs. Smith. I hear your furnace isn't working again. I'm sure you're very concerned about your children's comfort. I know just how you feel. My air conditioner quit twice on me last summer during a heat wave. I'll get to the problem as quickly as I can. Let's see what the matter is." You have shown concern for her feelings and worries; you have acknowledged that she is justified in her concern; and you have assured her that you will address the problem quickly and to her satisfaction.

2.2.6 Critical Customers

Sometimes you'll encounter a customer who can't be satisfied, no matter what you say or do. Often this happens because the customer is under a lot of stress or has had a bad experience that hasn't been dealt with properly. When this customer explodes, you may be the nearest target. How can you handle these situations with sensitivity and tact?

You've heard of cases where a parent, angry about a bad day at work, comes home and yells at the kids. This displacement of anger can also happen when a person has an unresolved conflict that is upsetting.

- A man who refused to pay extra for a weekend call now has a mess on his hands because the pipes in his poorly insulated laundry room are frozen.

- A woman having trouble with her sales accounts at work doesn't really want to be home waiting for a service technician when she should be dealing with her office problems.

The first step in handling these situations is to remember that everyone has bad days. Approach the customer with the attitude that someday you will feel out of sorts. On these days, you hope those around you will make allowances and be tolerant. Give your customer the same understanding you'd like to have at these times.

If the customer explodes, don't take it personally. After all, you may have just been a convenient target. You may not be responsible for the problem. Review what you have done. If your work was correct, shrug off the comments silently.

If your customer is justifiably upset, let the customer vent some anger. Empathize with the feelings, explaining that you would probably feel the same under the same circumstances. Reassure the customer that you will fix the HVAC technical problem as quickly as you can, so at least part of the day will be a positive experience.

Get the job done as quickly as you can. Get out of the situation as soon as possible without offending the customer. Stick to the facts. Focus on the mechanical problem as you solve it.

2.2.7 Customers with Unresolvable Problems

Once in a while, you will have a customer whose problem you cannot resolve. For example, the wiring may be too outdated to support the air conditioner the customer wants installed, or the existing furnace may be inadequate to heat a newly expanded space. What should you do?

Here is a poor response: "Sorry, sir. Your electrical service is not adequate to handle this air conditioner. I can't do electrical work."

Think About It
Mr. Negative

Mr. Negative is unhappy. The builders who just completed his home left several jobs unfinished. He begins a tirade about builders and service workers in general, including you, in his speech, even though you've only just arrived. The tirade begins anew after you have already started work on his furnace. How do you handle this difficult customer?

03316-13_SA02.EPS

A better response might be: "I found that your electrical service was too small to handle the new air conditioner. We can't do this kind of work, but here is the name of an electrical company we often deal with. Or, you can check the phone book for the names of other electricians. If you prefer, we can subcontract a licensed electrical contractor to do the job, then after the electrical service is fixed, we'll be glad to come back to install the air conditioner."

A typical homeowner might prefer to contact a company that is recommended, rather than searching out several competing companies and taking a chance on the quality of their work. Your positive, helpful attitude can leave a favorable impression, even if the customer already has an electrician. You have identified the problem and proposed a good solution, then offered to complete the job when the customer is ready.

2.2.8 Customers Who Request Help With Odd Jobs

Sometimes a customer asks a service technician to perform a task that is outside company regulations. In these situations, explain why you can't help. Then suggest another way to solve the problem.

Exercise Question:

While working on her furnace, Mrs. Oddjob asks you to move her freezer. What do you say?

Exercise Solution:

A good answer is: "I'm sorry, Mrs. Oddjob, but company regulations won't let me handle appliances we don't service. I know you're anxious to move that freezer. Perhaps a neighbor can help you. It wouldn't be fair to my later customers if I spent too much time at any one place. As you know, you're billed by the hour for my time. I know you want to keep your costs down." This response recognizes that the customer has a problem, but explains why you can't help her solve it.

Think About It

Mr. Knowitall

Mr. Knowitall wants to use his existing furnace to heat the home addition he just proudly completed himself. He has asked you to check out the ducts and connect the furnace, since he has already made up his mind it can handle the additional load. He's an expert at everything; a legend in his own mind. The addition is 950 square feet on the north side, with two outside walls. You determine that the existing furnace is inadequate to service the additional load. How do you approach this client with the news?

03316-13_SA03.EPS

Additional Resources

Tools for Success: Critical Skills for the Construction Industry. Upper Saddle River, NJ: Prentice Hall.

2.0.0 Section Review

1. It is a good idea to establish rapport with a customer by chatting about general topics while you are working on their equipment.

   a. True
   b. False

2. If an angry customer starts berating you because equipment your company installed has stopped working, you should _____.

   a. immediately turn around and leave
   b. shift the blame to others in your company
   c. empathize with the customer
   d. tell the customer you don't want to hear it

SUMMARY

Knowing how to diagnose and repair failures in HVAC/R equipment is only half the battle. The HVAC/R technician, as the company's representative, must know how to properly interact with customers. When customers call-in for repair service, it usually means that they are suddenly without heat, air conditioning, or refrigeration, so to them it is a crisis. They are likely to be uncomfortable and may be angry as well. In order to make a good first impression, the technician needs to be careful to be on time, well-groomed, and have a positive, friendly attitude. It is very important to communicate with the customers in positive ways that reflect concern for their problems. Listening patiently and attentively to customers is a good way to earn their confidence.

Learning how to deal with difficult customers is an important aspect of a technician's job. Technicians are likely to encounter customers who are angry, critical, demanding, or opinionated. It is important to learn how to deal with each type.

Review Questions

1. A customer begins forming a first impression about you and your company _____.

   a. after you have fixed their equipment
   b. after you have submitted your bill
   c. as soon as you appear at their premises
   d. as soon as you speak

2. You arrive at a service call wearing a NY Yankees ball cap. Unfortunately, the customer is a Boston Red Sox fan who hates the Yankees, so he immediately starts insulting the Yankees. Looking back on this incident, what do you think you would do in the future?

   a. Turn around and walk away because you won't do work for a Red Sox fan.
   b. Leave the ball cap in the truck to avoid the possibility of confrontation.
   c. Argue the relative merits of the two teams.
   d. Brush past the customer and find the equipment that needs to be repaired.

3. When speaking with a customer, always _____.

   a. start by telling them about your training and experience
   b. break the ice by chatting about local politics
   c. tell the customer if you think their equipment is low-quality
   d. show courtesy, concern, and respect

4. One way to make a good first impression is to use a lot of technical terms when talking to the customer.

   a. True
   b. False

5. A customer is giving a long, boring description of a problem. Because it is obvious that the customer has no technical knowledge, you should _____.

   a. tune the customer out and start thinking about how you will approach the problem
   b. nod appreciatively and pretend you are interested even if you are thinking about something else
   c. listen carefully because it is part of establishing rapport with the customer
   d. talk about similar problems you have encountered and how you solved them

6. During the introduction at the door, your customer says: "Hi, I'm Tom Collins." That means you can call him by his first name.

   a. True
   b. False

7. After telling the customer how long it will take you to complete a repair, you accidently tear a length of flex duct in the system. Replacing it will lengthen the service call considerably and the customer is paying by the hour. What should you do?

   a. Tell the customer about the mistake and explain that the repair costs will not be included in their bill.
   b. Tell the customer that you discovered an additional problem with the system that will lengthen the service call and increase their cost.
   c. Tell the customer the problem was caused by another technician who previously serviced the system.
   d. Ignore the damage because it will eventually result in another service call and increase your company's billing.

8. Mrs. Smith is worried because she has read about families dying of carbon monoxide poisoning from a defective furnace. How should you respond?

    a. tell her to relax and stop worrying

    b. explain that it hardly ever happens

    c. calm her and use factual information to ease her mind

    d. ask your supervisor to contact her

9. In order to access a furnace, a lot of clutter must be moved from in front of it. The way to deal with this problem is _____.

    a. refuse to continue working until the customer moves it

    b. move it out of the way and leave it there when you are finished work

    c. call your supervisor and explain the problem

    d. move it out of the way and return it when the work is finished

10. A customer was supposed to have had an electrical circuit installed to serve a new humidifier you have come to install. The electrical work is not done and the customer asks you to do it, even though you are not an electrician. What should you do?

    a. Go ahead and do the wiring because it's an easy job.

    b. Advise the customer to have the work done by a licensed electrician and provide a reference.

    c. Temporarily run an extension cord from the nearest electrical outlet so you can do the humidifier installation.

    d. Contact your supervisor and see if it is okay to do the electrical work.

Trade Terms Introduced in This Module

Empathy: The ability to identify with another person's situation or feelings.

Rapport: Establishing a relationship of mutual trust.

Additional Resources

This module presents thorough resources for task training. The following resource material is suggested for further study.

Tools for Success: Critical Skills for the Construction Industry. Upper Saddle River, NJ: Prentice Hall.
Applied Communication Skills for the Construction Trades. Upper Saddle River, NJ: Prentice Hall.

Figure Credits

Section Review Answer Key

Answer	Section Reference	Objective
Section One		
1. a	1.1.0	1a
2. c	1.2.1	1b
Section Two		
1. b	2.1.2	2a
2. c	2.2.6	2b

NCCER – *HVAC Level Three* 03316-13

NCCER CURRICULA — USER UPDATE

NCCER makes every effort to keep its textbooks up-to-date and free of technical errors. We appreciate your help in this process. If you find an error, a typographical mistake, or an inaccuracy in NCCER's curricula, please fill out this form (or a photocopy), or complete the online form at **www.nccer.org/olf**. Be sure to include the exact module ID number, page number, a detailed description, and your recommended correction. Your input will be brought to the attention of the Authoring Team. Thank you for your assistance.

Instructors – If you have an idea for improving this textbook, or have found that additional materials were necessary to teach this module effectively, please let us know so that we may present your suggestions to the Authoring Team.

NCCER Product Development and Revision
13614 Progress Blvd., Alachua, FL 32615

Email: curriculum@nccer.org
Online: www.nccer.org/olf

❏ Trainee Guide   ❏ Lesson Plans   ❏ Exam   ❏ PowerPoints   Other _____

Craft / Level: _____   Copyright Date: _____

Module ID Number / Title: _____

Section Number(s): _____

Description: _____

_____

_____

_____

_____

Recommended Correction: _____

_____

_____

_____

_____

Your Name: _____

Address: _____

_____

Email: _____   Phone: _____

Glossary

Actuator: The portion of a regulating valve that converts one type of energy, such as pneumatic pressure, into mechanical energy (for example, opening or closing a valve).

AL-CU: An abbreviation for aluminum and copper that is commonly marked on electrical connectors to indicate the device is suitable for use with conductors made with either metal.

Annual Fuel Utilization Efficiency (AFUE): An industry standard used to define furnace efficiency.

Atomize: Reduce to a fine spray.

Base speed: The rating given on a motor's nameplate where the motor will develop rated horsepower at rated load and voltage.

Bleed control: A valve with a tiny opening that permits a small amount of fluid to pass.

Boiler horsepower (Bohp): One Bohp is equal to 9.803 kilowatts. Also, it is the heat necessary to vaporize 34.5 pounds of water per hour at 212°F. This is equal to a heat output of 33,475 Btus per hour or about 140 square feet of steam radiation.

Break-away torque: The torque required to loosen a fastener. This is usually lower than the torque to which the fastener has been tightened.

Burnout: The condition in which the breakdown of the motor winding insulation causes the motor to short out or ground electrically.

Cad cell: A light-sensitive device (photocell) in which the resistance reacts to changes in the amount of light. In an oil furnace, it acts as a flame detector.

Capacitance boost: A procedure used to start a stuck PSC compressor. It involves momentarily connecting a start (boost) capacitor across the run capacitor of the stuck compressor in an attempt to start the stuck compressor by increasing the starting torque.

Connector: A device used to physically and electrically connect two or more conductors.

Cooling compensator: A fixed resistor installed in a thermostat to act as a cooling anticipator.

Coping: The upper masonry course of a chimney designed to keep water out.

Deadband: A temperature band, usually 3°F, that separates heating and cooling in an automatic changeover thermostat.

Demand defrost: A defrost control strategy that typically does not depend on time to determine when a defrost cycle is needed. Instead, the system responds directly to current conditions, such as temperature or pressure drop through the outdoor coil, to determine when frost is present.

Differential: The difference between the cut-in and cut-out points of a thermostat.

Direct-acting valve: A valve in which the solenoid operates the valve port.

Downdraft: A situation in which outdoor air is forced down a chimney, preventing proper draft. Downdraft can be caused by not extending the top of the chimney above the roof line.

Drip leg: A drain for condensate in a steam line placed at a low point or change of direction in the line and used with a steam trap.

Droop: A mechanical condition caused by heat that affects the accuracy of a bimetal thermostat.

Dry steam: The steam that exists after all the water has been vaporized into steam at its saturation temperature.

Dump zone: Area used as a point to bypass unneeded supply air from a zoning system. The dump zone is typically in the conditioned space, but not an area where maintaining the space temperature at a particular setpoint is required. A dump zone generally has no supply air grilles, but a return grille is often installed to relieve room pressure and return mixed air back to the system.

Electric-pneumatic (E-P) relay: A three-way pneumatic solenoid valve.

Electronically commutated motors: A variable-speed electric motor based primarily on a DC-driven permanent magnet design that responds to microprocessor inputs to achieve the desired speed.

Empathy: The ability to identify with another person's situation or feelings.

Fault isolation diagram: A troubleshooting aid usually contained in the manufacturer's installation, startup, and service instructions for a particular product. Fault isolation diagrams are also called troubleshooting trees. Normally, fault isolation diagrams begin with a failure symptom then guide the technician through a logical decision-action process to isolate the cause of the failure.

Field wiring: Wiring that is installed during installation of the system; for example, such as a thermostat hookup to unit.

Firetube boiler: A boiler in which the flue gases are contained inside the tubes, with the heated water on the outside.

Flame rectification: A process in which exposure of a sensing rod to a flame produces a tiny current that can be sensed and used to control the gas valve.

Flame-retention burner: A high-efficiency oil burner.

Flash economizer: An energy-conservation process used in chillers that accomplishes intercooling of the refrigerant between the condenser and evaporator using a flash chamber or similar device. Flash gas of coolant produced in the flash chamber flows directly to the second stage of the chiller compressor, bypassing the evaporator and the first compressor stage.

Flash steam: Formed when hot condensate is released to a lower pressure and re-evaporated.

Free cooling: A mode of economizer operation. During cooling operation, the system compressor is turned off and the indoor fan is used to bring outside air into the building to provide cooling.

Insulated gate bipolar transistor (IGBT): A type of transistor that has low losses and low gate drive requirements. This allows it to be operated at higher switching frequencies.

Invar®: An alloy of steel containing 36 percent nickel. It is one of the two metals in a bimetal device.

Joist: Framing member used to support floors and ceilings.

Label diagram: A troubleshooting aid usually placed in a convenient location inside the equipment. It normally depicts a wiring diagram, a component arrangement diagram, a legend, and notes pertaining to the equipment.

Ladder diagram: A troubleshooting aid that depicts a simplified schematic diagram of the equipment. The load lines are arranged like the rungs of a ladder between vertical lines representing the voltage source. Normally, all the wire color and physical connection information is eliminated from the diagram to make it easier to use by focusing on the functional, not the physical, aspects of the equipment.

Lithium bromide: A salt solution commonly used in absorption chillers.

Lockout: A safety feature that prevents a furnace from automatically cycling if the flame fails.

Lug: A wiring terminal used on wire sizes No. 8 and larger. Also a term used to describe a screw-type terminal that simply applies pressure to a stripped wire, without the need for a crimp-on terminal.

Mechanical cooling: The cooling provided in the conventional manner by the compressor and mechanical refrigeration system.

Modulating gas valve: A gas valve that precisely controls burner capacity by adjusting the flow of gas to burners in small increments.

Net stack temperature: A value needed to perform a combustion efficiency analysis, calculated by subtracting room temperature from the temperature measured in a furnace's vent or vent connector.

Nominal size: A means of expressing the size of a bolt or screw. It is the approximate diameter of a bolt or screw.

Over-firing: A situation caused by an oversized nozzle in which excess oil is pumped into the combustion chamber.

Phase change: The conversion of refrigerant from one state to another, such as from liquid to vapor.

Pilot-operated valve: A valve in which the solenoid operates a plunger, which in turn creates or releases pressure that operates the valve.

Pneumatic-electric (P-E) relay: A pressure switch in which a pneumatic signal causes an electrical change.

Polling: A network communications arrangement where a VVT system seeks information and needs from individual zones, determines what demands have priority, and then responds accordingly. Essentially, members of a polling system are voters.

Pressure-dependent operation: A VVT-system approach where dampers are positioned based on demand only, or operate in two positions. The result is that the design volume of air may or may not be flowing through the damper at any time. The volume of air flow through each damper is dependent on the duct pressure; it fluctuates as the pressure fluctuates. A minimum damper position is set, but air volume is not a controlling factor, and so the minimum damper position is a fixed point.

Pressure-independent operation: A VVT-system approach where dampers are positioned by demand and by the actual volume of air flowing through the damper. To achieve a specific amount of airflow, the damper positions itself as necessary. A maximum and minimum airflow setpoint is effective, and the damper will modulate as necessary to remain within these boundaries.

Primary control: A combination switching and safety control that turns the oil burner on and off in response to signals from the thermostat and safety controls.

Rapport: Establishing a relationship of mutual trust.

Refnet joints: Engineered joints used as connection points in a VRF-system refrigerant network. Refnet joints connect the main refrigerant lines to branch runs or to indoor units.

Refractory: Materials such as ceramics designed to withstand extremely high temperatures.

Reverberator: A wire screen used to cover the ceramic grids in a gas-fired heater that operates by submerged flame.

Run-down resistance: The torque required to overcome the resistance of associated hardware such as locknuts and lock washers, when tightening a fastener.

Saturated steam: The pure or dry steam produced at the temperature that corresponds to the boiling temperature of water at the existing pressure.

Secondary coolant: Any liquid, such as water, that is cooled by a system refrigerant and then used to absorb heat from the conditioned space without a change in state. Also known as indirect coolant.

Seismic activity: Movement of the earth such as during an earthquake.

Seismic restraint: A device or method used to prevent or restrict the movement of equipment during a seismic event.

Set or seizure: In the last stages of rotation in reaching a final torque, the fastener may lock up; this is known as setting or seizing. This is usually accompanied by a noticeable popping effect.

Slinger ring: A ring attached to the outer edge of the fan that picks up water as the fan rotates and slings it onto the condenser coil.

Smoke spot test: A test to determine the level of smoke in the flue gases.

Soot: Fine carbon particles resulting from incomplete combustion. An accumulation of soot on heat exchangers will reduce combustion efficiency and result in increased carbon monoxide production.

Submerged flame: A method of gas-fired heater operation that uses burning gas within two-layer porous ceramic grids.

Superheated steam: Steam that has been heated above its saturation temperature.

Temperature differential: The difference in temperature between the flue gases and the gases above the combustion chamber.

Tensile strength: The maximum stress a material can endure before being pulled apart.

Terminal: A device used for connecting wires to an object.

Termination: The connection of a wire to its destination.

Thermal economizer: An energy-conservation process used in chillers in which warm condensed refrigerant is brought into contact with the coldest (inlet) water tubes in a condenser or heat exchanger. This causes the condensed refrigerant to subcool to a temperature below the condensing temperature.

Thread classes: Threads are distinguished by three classifications according to the amount of tolerance the threads provide between the bolt and nut, and are designated by a number and letter combination that indicates the degree of fit for a thread.

Thread series: The combinations of diameter and pitch in threads that are applied to specific thread diameters.

Thread standards: An established set of standards for machining threads.

Torque: Applying a specific amount of twisting force to a threaded fastener.

Trim: External controls and accessories attached to the boiler itself, such as sight glasses and water feeder controls.

Turndown ratio: In steam systems, the ratio of downstream pressure to upstream pressure, usually applied to steam pressure-reducing valves. More than a 50 percent turndown ratio is generally considered a large pressure reduction.

Unit cooler: A packaged unit containing evaporator coil, metering device, and evaporator fan(s) that serves as the evaporator section for a refrigeration appliance.

Venturi tube: A short tube with flaring ends and a constricted throat. It is used for measuring flow velocity by measurement of the throat pressure, which decreases as the velocity increases.

Vibration isolator: A device that prevents the transmission of vibrations from operating machinery to surrounding objects.

Water hammer: A condition that occurs when hot steam comes into contact with cooled condensate, builds pressure, and pushes the water through the line at high speeds, slamming into valves and other devices. Water hammer also occurs in domestic water systems. If the cause is not corrected, water hammer can damage the system; a banging noise in water pipes that occurs when the water supply solenoid valve on a humidifier abruptly shuts off water to the humidifier.

Watertube boiler: A boiler where the heated water is contained inside the tubes, with the flue gases on the outside.

Weatherized furnace: A furnace that has been well insulated and has a weather-resistant case as specified by government standards.

Wiring diagram: A troubleshooting aid, sometimes called a schematic, that provides a picture of what the unit does electrically and shows the actual external and internal wiring of the unit.